TECHNOLOGICAL APPLICATIONS OF DISPERSIONS

SURFACTANT SCIENCE SERIES

1. Nonionic Surfactants, *edited by Martin J. Schick* (see also Volumes 19 and 23)
2. Solvent Properties of Surfactant Solutions, *edited by Kozo Shinoda* (out of print)
3. Surfactant Biodegradation, *by R. D. Swisher* (see Volume 18)
4. Cationic Surfactants, *edited by Eric Jungermann* (see also Volumes 34 and 37)
5. Detergency: Theory and Test Methods (in three parts), *edited by W. G. Cutler and R. C. Davis* (see also Volume 20)
6. Emulsions and Emulsion Technology (in three parts), *edited by Kenneth J. Lissant*
7. Anionic Surfactants (in two parts), *edited by Warner M. Linfield* (out of print)
8. Anionic Surfactants: Chemical Analysis, *edited by John Cross* (out of print)
9. Stabilization of Colloidal Dispersions by Polymer Adsorption, *by Tatsuo Sato and Richard Ruch*
10. Anionic Surfactants: Biochemistry, Toxicology, Dermatology, *edited by Christian Gloxhuber* (see Volume 43)
11. Anionic Surfactants: Physical Chemistry of Surfactant Action, *edited by E. H. Lucassen-Reynders* (out of print)
12. Amphoteric Surfactants, *edited by B. R. Bluestein and Clifford L. Hilton* (out of print)
13. Demulsification: Industrial Applications, *by Kenneth J. Lissant*
14. Surfactants in Textile Processing, *by Arved Datyner*
15. Electrical Phenomena at Interfaces: Fundamentals, Measurements, and Applications, *edited by Ayao Kitahara and Akira Watanabe*
16. Surfactants in Cosmetics, *edited by Martin M. Rieger*
17. Interfacial Phenomena: Equilibrium and Dynamic Effects, *by Clarence A. Miller and P. Neogi*
18. Surfactant Biodegradation, Second Edition, Revised and Expanded, *by R. D. Swisher*

19. Nonionic Surfactants: Chemical Analysis, *edited by John Cross*
20. Detergency: Theory and Technology, *edited by W. Gale Cutler and Erik Kissa*
21. Interfacial Phenomena in Apolar Media, *edited by Hans-Friedrich Eicke and Geoffrey D. Parfitt*
22. Surfactant Solutions: New Methods of Investigation, *edited by Raoul Zana*
23. Nonionic Surfactants: Physical Chemistry, *edited by Martin J. Schick*
24. Microemulsion Systems, *edited by Henri L. Rosano and Marc Clausse*
25. Biosurfactants and Biotechnology, *edited by Naim Kosaric, W. L. Cairns, and Neil C. C. Gray*
26. Surfactants in Emerging Technologies, *edited by Milton J. Rosen*
27. Reagents in Mineral Technology, *edited by P. Somasundaran and Brij M. Moudgil*
28. Surfactants in Chemical/Process Engineering, *edited by Darsh T. Wasan, Martin E. Ginn, and Dinesh O. Shah*
29. Thin Liquid Films, *edited by I. B. Ivanov*
30. Microemulsions and Related Systems: Formulation, Solvency, and Physical Properties, *edited by Maurice Bourrel and Robert S. Schecter*
31. Crystallization and Polymorphism of Fats and Fatty Acids, *edited by Nissim Garti and Kiyotaka Sato*
32. Interfacial Phenomena in Coal Technology, *edited by Gregory D. Botsaris and Yuli M. Glazman*
33. Surfactant-Based Separation Processes, *edited by John F. Scamehorn and Jeffrey H. Harwell*
34. Cationic Surfactants: Organic Chemistry, *edited by James M. Richmond*
35. Alkylene Oxides and Their Polymers, *by F. E. Bailey, Jr., and Joseph V. Koleske*
36. Interfacial Phenomena in Petroleum Recovery, *edited by Norman R. Morrow*
37. Cationic Surfactants: Physical Chemistry, *edited by Donn N. Rubingh and Paul M. Holland*
38. Kinetics and Catalysis in Microheterogeneous Systems, *edited by M. Grätzel and K. Kalyanasundaram*
39. Interfacial Phenomena in Biological Systems, *edited by Max Bender*
40. Analysis of Surfactants, *by Thomas M. Schmitt*
41. Light Scattering by Liquid Surfaces and Complementary Techniques, *edited by Dominique Langevin*
42. Polymeric Surfactants, *by Irja Piirma*
43. Anionic Surfactants: Biochemistry, Toxicology, Dermatology. Second Edition, Revised and Expanded, *edited by Christian Gloxhuber and Klaus Künstler*
44. Organized Solutions: Surfactants in Science and Technology, *edited by Stig E. Friberg and Björn Lindman*
45. Defoaming: Theory and Industrial Applications, *edited by P. R. Garrett*

46. Mixed Surfactant Systems, *edited by Keizo Ogino and Masahiko Abe*
47. Coagulation and Flocculation: Theory and Applications, *edited by Bohuslav Dobiáš*
48. Biosurfactants: Production · Properties · Applications, *edited by Naim Kosaric*
49. Wettability, *edited by John C. Berg*
50. Fluorinated Surfactants: Synthesis · Properties · Applications, *Erik Kissa*
51. Surface and Colloid Chemistry in Advanced Ceramics Processing, *edited by Robert J. Pugh and Lennart Bergström*
52. Technological Applications of Dispersions, *edited by Robert B. McKay*
53. Cationic Surfactants: Analytical and Biological Evaluation, *edited by John Cross and Edward J. Singer*

ADDITIONAL VOLUMES IN PREPARATION

TECHNOLOGICAL APPLICATIONS OF DISPERSIONS

edited by
Robert B. McKay

Ciba Pigments
Paisley, Scotland

Marcel Dekker, Inc.   New York • Basel • Hong Kong

Library of Congress Cataloging-in-Publication Data

Technological applications of dispersions / edited by Robert B. McKay.
p. cm. -- (Surfactant science series; v. 52)
Includes bibliographical references (p. –) and index.
ISBN 0-8247-9180-0 (acid-free)
1. Emulsions. 2. Suspensions (Chemistry) I. McKay, Robert B. II. Series.
TP156.E6I53 1994
660'.2945--dc20 93-50099
CIP

This book is printed on acid-free paper.

Marcel Dekker, Inc.
270 Madison Avenue, New York, New York 10016

Current printing (last digit):
10 9 8 7 6 5 4 3 2 1

PRINTED IN THE UNITED STATES OF AMERICA

Preface

Dispersion science is of fundamental importance to a wide variety of technologies. Indeed there are few chemically based technologies in which it does not feature in some role. This book aims to substantiate these statements by giving insight into a range of well established technologies of diverse character and in which dispersed particles play a key role. It is intended to provide a technological background for later books in the Surfactant Science Series on specific aspects of dispersion science.

The term *dispersion* in its broadest sense describes any system in which at least one phase is distributed in finely divided form throughout a continuous second phase. However, the technologies discussed in this book place emphasis on dispersions of solid particles in liquid or solid media. Emulsions have been discussed in previous volumes of this series.

The chapters are overviews of specific technologies written by technical experts. They illustrate the effects of dispersions in application and the means necessary for optimal achievement of these effects. The book should be of interest to formulators and users of dispersions, suppliers of components, industrial and academic scientists, and students of colloid science.

The major application areas exemplified are surface coating, optical effects, uses of polymer dispersions, strength enhancement of inorganic materials, recording and display of information, and biological systems. Some of the major components of dispersions, especially pigments, are also discussed. Two major applications, namely ceramics (hence aspects of clays, and related materials) and agricultural chemicals, are the subjects of separate books in the series (in preparation), and so are mentioned only briefly here. The chapter on polymer dispersions for special purposes widens the scope of the book to include specific aspects of various other important applications, some of which could have become chapters in their own right. Clays have been discussed briefly in the chapter on paper coatings. The extensive literature quoted by the authors provides a further source of information.

Some of the technologies included are interrelated. For example, the technologies of paints, printing inks, plastics, and man-made fibers are in part dependent upon pigment technology, and vice versa, and through this have much in common. There have been few texts that deal with such interrelated technologies collectively, and one aim of the book is to fill such gaps.

The *disperse phase* has a specific function in all the applications, through factors such as chemical constitution, solid state and liquid-crystalline properties, physical structure, particle size, surface character. In some cases it does not survive as such in end use and in others it is transformed. The *continuous phase* may be solid or liquid, and in the latter case may be transformed by cooling, loss of solvent, or polymerization into a solid matrix or binder for the disperse phase. *Interfacial properties* are of course critically important to the effectiveness of dispersions, through their influences on dispersibility, dispersion stability, rheology, particle interactions, and interactions with substrates.

It was difficult to decide the order of chapters. Paints were chosen as the subject of the first chapter because they are probably the earliest man-made dispersions and the most ubiquitous today. Then follow chapters on other technologies that entail or relate to dispersion of powders in liquids. The latter part of the book deals with more complex dispersions—first, those where changes are induced within the disperse phase by external influences during application and, second, those for probably the most complex application of all, biological systems.

Some points of terminology require comment. The terms *aggregate* and *agglomerate* commonly used in dispersion technology present difficulty. IUPAC recommends that *aggregate* be used as a general term to describe any structure formed by the cohesion of colloidal particles [1]. *Agglomerate* is often used in colloid science and technology to describe small, strongly coherent aggregates. In pigment and pigment user industries, for example, the terms are used in an almost converse way—*aggregate* describing a strongly coherent cluster of colloidal particles and *agglomerate* a more loosely coherent assembly of particles, usually in a powder. In this book *aggregate* has been used in the general sense recommended by IUPAC, with some indication of strength of cohesion of constituent particles where appropriate. The term *agglomerate* has been avoided.

Traditional terminology of some technologies is at odds with more generally accepted terminology in dispersion science. For example, photographic emulsions and emulsion paints are really dispersions of solid particles, as indeed are many so-called polymer emulsions. In paper coating the terms *pigment* and *color* are applied to clay and calcium carbonate, even though, strictly speaking, such solids are colorless and have little direct opacification effect.

I should like to thank Ciba Pigments for their encouragement and support.

Robert B. McKay

REFERENCE

D. H. Everett (for IUPAC), Pure and Applied Chemistry, *31*:579 (1972).

Contents

Preface *iii*
Contributors *ix*

1. Paints 1
Andrew Doroszkowski

2. Titanium Dioxide Pigments 69
John G. Balfour

3. Colored Inorganic Pigments 105
Hans Sander

4. Organic Pigments 143
Robert B. McKay, Abul Iqbal, and Bernhard Medinger

5. Dispersion of Pigments into Printing Inks 177
Graham Battersby

6. Dispersions in Plastics and Synthetic Fibers 205
Gérard Häuptli and Roger C. Sykes

7. Paper Coatings 245
Derek J. Priest

8. Dispersions of Polymers for Special Purposes 271
John G. Langley

9. Strength Enhancement of Inorganic Materials Through Dispersion 335
Kevin Kendall

10. Liquid Crystal Dispersions 349
John L. West

11. Use of Monosize Dispersions in Silver Halide Photographic Emulsions 373
Trevor Maternaghan

12. Magnetic Recording Materials 443
David H. Davies

13. Dispersions as Pharmaceutical Dosage Forms 457
Duncan Q. M. Craig

14. Milk as a Food Dispersion 499
David S. Horne

Index 539

Contributors

John G. Balfour Technical Service Department, Tioxide Europe Limited, Billingham, Cleveland, England

Graham Battersby Coates Lorilleux Research, Coates Lorilleux International, Orpington, Kent, England

Duncan Q. M. Craig Centre for Materials Science, The School of Pharmacy, University of London, London, England

David H. Davies Ampex Recording Media Corporation, Redwood City, California

Andrew Doroszkowski The Paint Research Association, Teddington, Middlesex, England

Gérard Häuptli Consultant, Binningen, Switzerland

David S. Horne Hannah Research Institute, Ayr, Scotland

Abul Iqbal Research Pigments Division, Ciba-Geigy Ltd., Marly, Switzerland

Kevin Kendall Department of Chemistry, University of Keele, Keele, Staffordshire, England

John G. Langley Allied Colloids Ltd., Bradford, West Yorkshire, England

Trevor Maternaghan Research and Development, ILFORD Limited, Mobberley, Cheshire, England

Robert B. McKay Ciba Pigments, Paisley, Scotland

Bernhard Medinger Research Pigments Division, Ciba-Geigy Ltd., Marly, Switzerland

Derek J. Priest Department of Paper Science, UMIST, Manchester, England

Hans Sander Pigments and Ceramic Department, Bayer AG, Krefeld, Germany

Roger C. Sykes Pigments Division, Ciba-Geigy Ltd., Basel, Switzerland

John L. West Liquid Crystal Institute, Kent State University, Kent, Ohio

TECHNOLOGICAL APPLICATIONS OF DISPERSIONS

1

Paints

ANDREW DOROSZKOWSKI The Paint Research Association, Teddington, Middlesex, England

I.	Introduction	2
	A. Paint: a short history	3
	B. The paint industry: the economic framework	5
II.	General Nature of Paints	8
	A. Paint types	8
	B. Paint systems	11
III.	Components of Paint	12
	A. The disperse phase	14
	B. The continuous phase	16
	C. Polymers used in the paint industry	17
IV.	The Paint-Making Process	21
	A. Dispersion machinery and grinding conditions	22
	B. Millbase formulation and let-down	39
	C. Assessment of fineness of grind	43
	D. Mixing	44
V.	Properties of Paints	45
	A. Color and the state of pigment dispersion	45
	B. Pigment flocculation and other properties	50
	C. Assessment of pigment flocculation	52
	D. Particle packing	53
	E. Flow properties	55

A fine reference to terms used is found in *Paint/Coatings Dictionary*, published by Federation of Societies for Coatings Technology, Philadelphia (1978).

VI. Paint Application 58
VII. Future Trends 59
References 64

I. INTRODUCTION

The ubiquitous nature of paint and our continual use of this unique material mask its complexity and evolution. It is often treated with condescension and lack of wonderment at its properties. Frequently it is referred to in a derogatory way, for example "as exciting as watching paint dry." And most house decorators consider themselves experts on paint because they can use it to decorate their homes to a high standard. This in itself should provoke thought. Here is a product that can be used by any layman, with no preliminary training. It can be applied over various complex surfaces under a variety of climatic conditions ranging from cold and dry to hot and humid atmospheres, with a variety of implements such as a brush, roller, or spray gun, onto horizontal, vertical, and even upside-down surfaces, and still produce a satisfactory finish. It can withstand everyday wear and tear for a prolonged period and still protect the substrate from corrosion as well as producing a pleasing aesthetic finish resistant to the bleaching and degrading properties of sunlight. Is this not indeed a remarkable product!

In reality, "paint" is not just a simple substance that can be easily defined; it is any of a multitude of diverse materials, hence it is frequently called a "surface coating" in an attempt to encompass its varied nature. However, by trying to generalize the nomenclature for different materials that are readily identified as paint, lacquer, finish, coating, etc., impinges on the term "coatings," which is frequently used to describe metal veneers such as galvanized or electroplated metals and metallized plastics, for which the metal is deposited from its vapor or sputtered onto the surface. One would never dream of calling such materials "metal-painted"; they are (metal) coatings. Moreover, the term "paint" is readily differentiated in most people's minds from ink: but in the Oxford dictionary ink is defined as a substance used for marking paper (which, as will be seen from Chapter 5, is not strictly correct); hence one might suggest that the difference lies in the thickness of the coating—a paint being a heavy coating. How should one categorize artists' watercolors? The difference between an ink and a watercolor paint is not the pigments used, in their state of dispersion, or in the binder of the pigment, if any, or indeed in the thickness of the coating. Similarly, one might say a dye is used for coloring textiles, wood or plaster, and not an ink. Are not frescoes stained fresh plaster? Does one dye or paint them? Colored chalks are also known as pastels (compressed pigment and chalk), yet whereas one *writes* with chalk, one *paints* with pastels. A paint might be defined as a pigmented adhesive; indeed all polymers used as adhesives are also used in paints, but a colored adhesive is not a paint.

To summarize, it is suggested that there is no clear-cut, simple definition for what is commonly called a paint. There are a number of similar materials that are known as paints, varnishes, coatings, adhesives, dyes, and inks, and the nomenclature is traditionally based on why and where they are used. Therefore the terms "paint," "surface coating," "finish," etc. will be used freely and interchangeably throughout this chapter. The reader is asked to be flexible and not be constrained by the name but see it as describing a colloidal dispersion of a coloring matter in a continuous phase that may contain dissolved, or dispersed, polymers, even though such a definition will also apply to an ink. While dyes are more in the nature of "colored molecules" in solution than colloids.

A. Paint: A Short History

Paint is one of the oldest artifacts of man and has been made since prehistoric times. It is thought that it was first made by grinding earth pigments (iron and manganese oxides) in naturally occurring hollow stones, with a bone acting as a pestle. Animal fats or egg white and water might have been added to act as a binder, thus allowing the pigment to adhere to the cave walls to produce paintings such as those found at many sites in the Perigord, France, such as Grotte de Font-de-Gaume (Lascaux being referred to the most frequently, but closed to the public since 1963), and in Spain (Altamira) and Australia (Obiri rocks in Arnhem Land [1]).

One can only speculate as to why early humans decorated caves. The early painter could be considered a chemist, technologist, and artist combined, since he had to develop the pigments and binders for his creations. The earliest synthetic pigment is thought to be Egyptian blue, a copper silicate, found in the tomb of Perneh (2650 B.C.), made by heating malachite with sand and lime in a controlled manner. The Greek word *pharmakeia* when referring to pigments or substances used in paintings also illustrates this interrelationship. Another ancient pigment, Naples yellow, found in Assyria around 800 B.C., was formulated by heating lead carbonate or lead oxide with antimony oxide. These pigments were dispersed in hot wax and left to dry, although the exact details are not known. This is thought to be the oldest known form of paint; it was used by the Egyptians (and Greeks) and is called "encaustic" [2,3]. Ink for writing on papyrus or vellum was made from carbon black or red ocher in water, while the media for paints—gum tragacanth, honey, and egg white and yolk—were used in aqueous form; however, little is known about paint-making techniques in antiquity. A first century B.C. treatise by Vitruvius gives some insight into fresco paintings [2], where colors dispersed in egg albumen were applied to fresh plaster, which on drying fixed the pigments into the surface and produced great permanency. Dioscorides' *The Greek Herbal*, written in the first century, mentions oils, gums, and resins. Drying oil is said to have been used in ancient China as a binder for painting on wood, but not in Egypt [3]. In the later civilizations of Greece and Rome a wider variety of pigments were known. An analysis of the pigments used in Roman Britain (from a London excavation site) was made by Latner [4], who lists 20 different colors, although lapis lazuli (ultramarine) and white lead, known to have been used by the Romans, were not found.

Paint technology for the next millennium did not appear to progress beyond what was known in Greco-Roman times. A monk, Theophilus Presbyter, somewhere around the eleventh century is usually credited for having made the earliest records of paint formulations that had been in use for some considerable time. Theophilus's treatise describes the grinding of pigments in oils and the preparation of a varnish, for which one part gum (amber?) is melted and two parts of a hot oil (linseed) are slowly added. The mixture is cooked further, with samples being withdrawn from time to time and tested for clarity [5]. White lead was used in quantity; for example, 100 lb was used for painting the vaulting of the Treasurer's house at Windsor in 1356, and whitewash (whiting bound with casein) was used for the Keep at the Tower of London, the "White Tower," in Norman times [5]. Jacobus de Tholeto's manuscript (1440) is the first instance of lead driers being cooked into a varnish [5]. The Cennino Cennini fifteenth century manuscript *Il Libro Dell Arte* describes many of the techniques and pigments used at the time, but apart from some recipes for pigments and dyes does not add much to what was known according to Armitage. In Medieval to early Renaissance times the powerful craft guilds made their paints according to their own formulas. These compositions were guarded secrets that were handed down from father to son (cf. the speculation as to the formulation of the varnish used by the Stradivari for their violins). Painters and stainers were engaged in decorating ship superstructure and sails, painting sculptures, and decorating the walls of castles, churches, and manors. The distinction between "paynter" and "stayner" was that a painter painted on metal, timber, plaster, and glass whereas a stainer painted on canvas, linen, and cloth [5]. Toward the end of the medieval period, painters began to be recognized either as artists, who painted pictures, or as "seilers" or house painters, whose efforts were confined to varnishing woodwork or painting plaster walls. "Vulgar" painting was carried out by plasterers, and although they were not prevented from painting, their operations were restricted to the use of whiting, red lead, yellow ocher, and black ground in size only, by a Bill to the Court of Common Council presented by the London Paynter-Stayners in 1599 [6].

From the Tudor to the end of the Stuart period, great changes took place in Britain as the rapid expansion of trade and industry led to the loosening of the craft guilds' restrictive practices and the creation of independent manufacture. The industrial expansion, which continued in the eighteenth century, brought about the practice of disclosing a description of the process in a patent, thereby enabling us to learn about the paints of the time. In 1790 a trade directory for London included as tradesmen or manufacturers 25 oil merchants, 10 manufacturers of varnishes or japans, 120 oil men, 90 oil and color men, 75 painters or japaners, and six makers of oiled silk or oiled cloth [7].

The materials used in paint and varnish making, as published by Jean Watin in 1773, were relatively sophisticated. These were based on various combinations of fossil gums and resins in oil (oleoresinous materials) and continued in use up to the twentieth century [8]. The mid 1800s saw the introduction of ready-mixed paints, whereas prior to this craft expertise was required to correctly dilute the pigment pastes. Although Alexander Parkes took out a patent to cover the use of solutions of nitrocellulose as protective coatings in 1855, it was not until 1882, with the discovery of amyl acetate as solvent, that nitrocellulose could be used as an effective film former [9]. In 1912, the New Explosives Company (now ICI Paints) was selling a range of nitrocellulose-based products including clear and pigmented coatings for metal finishing and leather and paper finishes [10]. During World War I (1914–1918), activity was confined mainly to "dopes" for aircraft of that period. After the war, the development of the mass production car industry was an ideal outlet for the surplus of guncotton, being produced. The 1920s saw the introduction of glycerol-modified rosin and oil-soluble phenolics [11] and the development of alkyd-based coatings, which superseded oleoresinous materials. Then came urea formaldehyde-alkyd blends in the 1930s followed by polyurethanes and melamine–formaldehyde thermosetting paints [11]. After World War II came styrene-butadiene latex paints, followed by acrylic latexes in the 1950s.

B. The Paint Industry: The Economic Framework

The economic framework of the paint industry in terms of the relative size of the various paint markets, world production, and paint output by country is presented here in tabular and graph form for easier perception.

Today, the manufacture of paint is considered a mature industry, with a gradual annual increase in production reflecting economic growth as shown in Fig. 1 (with projections to the turn of the century). Figure 2 gives a breakdown of regional paint production by country. Table 1 shows the status of the paint industry by sector (in monetary terms) in comparison with other industries (using the United Kingdom as representative). Table 2 lists the world's top paint producers. It must be noted that this is a very volatile situation, since acquisitions, sales, and mergers are continually changing the order. Even large paint companies such as De Soto, which was among the top 10 U.S. producers in the late 1980s, can not only drop out of the league table but more or less disappear by 1991 through sales of their component markets to various competitors, while others may suddenly appear.

Paint companies are gradually becoming larger and more global through acquisitions and licensing agreements, thereby reducing the total number of paint manufacturers (mainly middle-sized companies). In 1992 the top 10 companies accounted for about 30% of world production, and they are expected to account for 50% at the end of the decade, although there will probably always be a place for small entrepreneurial "niche" paintmakers.

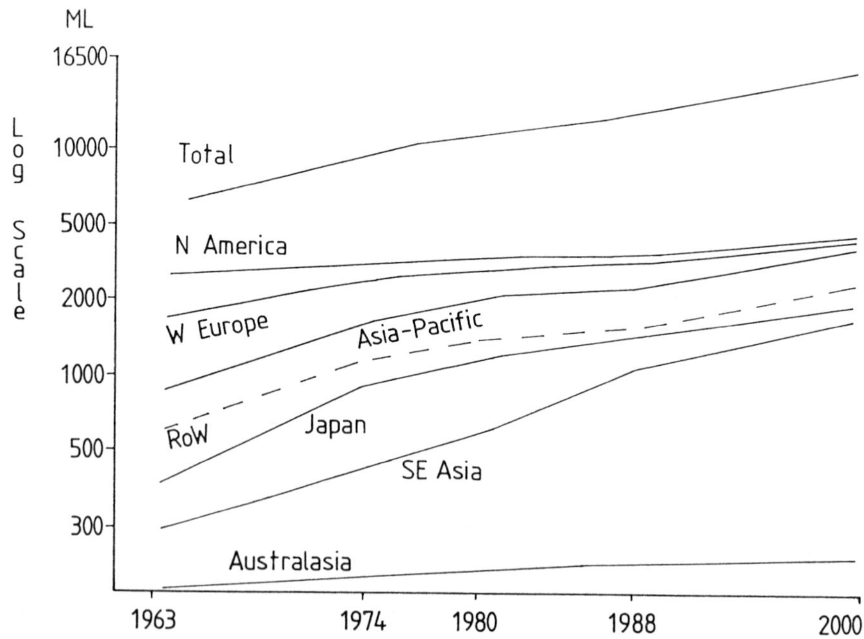

FIG. 1 World paint production over the years. (Courtesy ICI.)

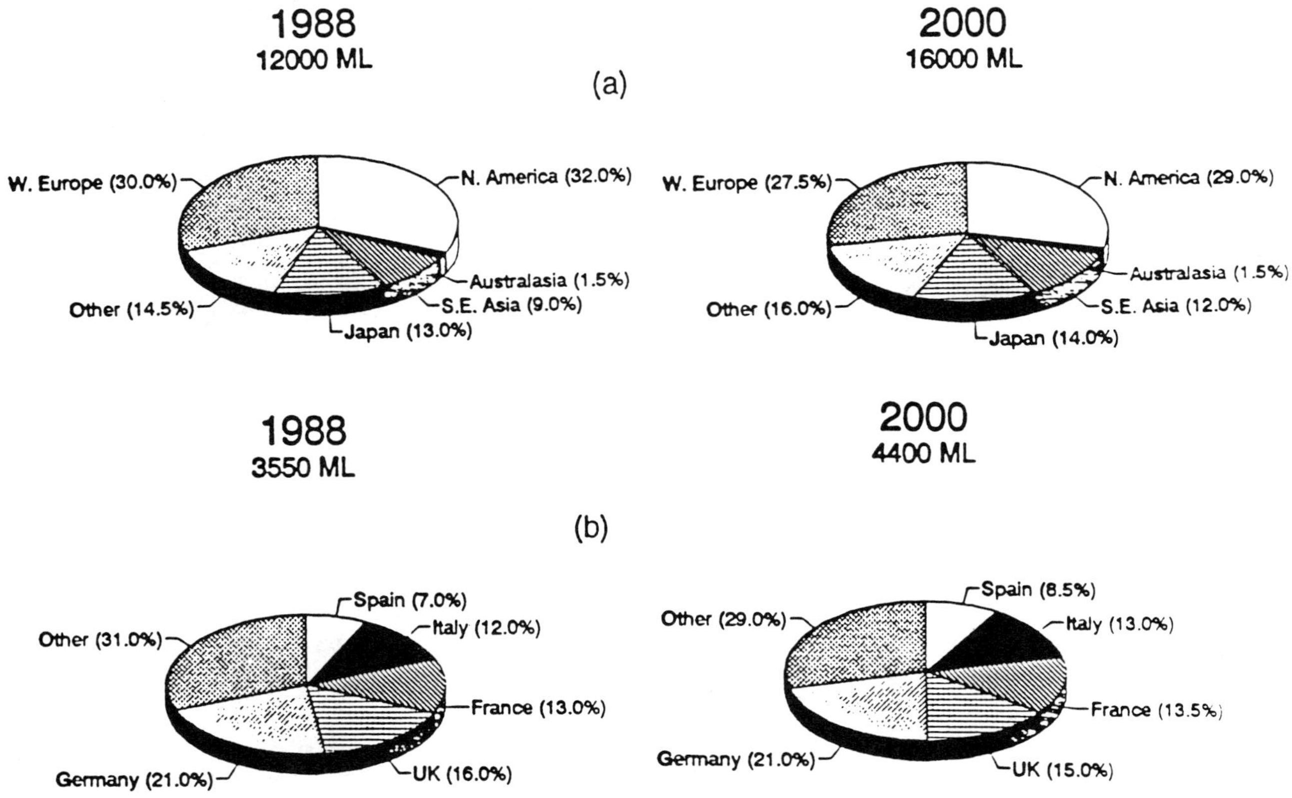

FIG. 2 (a) World paint markets (in millions of liters) by region: "SE Asia" includes China, Indian subcontinent, and southeastern Asia. "Other" includes Africa, Latin America, and Middle East but excludes eastern Europe and former USSR at 15% of world total. (b) West European markets by region. "Other" includes Austria, Benelux, Greece, Portugal, Scandinavia, and Switzerland; with projections to the end of the decade. (Courtesy ICI.)

TABLE 1 Status of the U.K. Paint Industry: A Sectorial View 1987–1988

	Sales of principal products (£M)	% of Total chemical sales
Paints and varnishes	1,512	6.25
Cosmetics and toiletries	1,464	6.05
Soap and synthetic detergents	1,063	4.39
Dyestuffs and pigments	938	3.88
Formulated adhesives	427	1.76
Printing inks	277	1.14
Pharmaceutical products	4,681	19.34
Organic chemicals	3,627	14.98
Synthetic resins and plastics	2,906	12.01
Inorganic chemicals	1,652	6.82
	18,547	76.62
Total chemical industry	22,206	100.00

Source: Ref. 12.

II. GENERAL NATURE OF PAINTS

A. Paint Types

There are many different types of paint, and these can be grouped in various ways, such as waterborne paints, solvent-based paints, and high solids paints. However, they are usually classified as to where they are used, that is, according to their market. For example, *decorative* (sometimes called architectural) paints are used to decorate, and preserve, houses or buildings.

Automotive paints and *marine* paints are self-descriptive as to their types.

Industrial paints are those used by industry in factory finishes, but the term usually excludes automotive, marine, and heavy duty coatings. They include a wide variety of products such as those used to coat consumer durables, wire enamels and coil coatings used to paint sheet metal (coil) such as steel or aluminum for caravan panels. Also included in this group are " can coatings," paints subject to stringent food safety (FDA) regulations and used to coat the inside (and outside) of food or beverage cans.

Refinish paints are coatings used in the repair of cars after road accidents or accidental damage during the car manufacturing process. For example, if final factory inspection reveals faulty paintwork, it cannot be repaired using the original (OEM) paint; it is not possible to stove the car in the oven (unlike the car shell) because the normal stoving temperature will damage the trim—the wiring, car seats, plastic components, etc. Hence a

TABLE 2 The World's Top Paint Producers, 1991, by Output

Company	ML
1. ICI (UK)	805
2. Sherwin-Williams (USA)	533
3. PPG (USA)	515
4. BASF (Germany)	500
5. Akzo (Netherlands)	490
6. Nippon Paint (Japan)	350
7. Casco-Nobel (Sweden)	350
8. Courtalds (UK)	300
9. Kansai (Japan)	280
10. DuPont (USA)	265
11. Valspar (USA)	200
12. Sigma (Belgium)	180

Source: Ref. 12.

refinish paint is used that can be dried (cured) in the presence of trim at a lower temperature.

An indication of the relative size of the markets for these paints is shown in Fig. 3.

Within a paint group, decorative paints for example, there are further divisions according to kind, such as gloss, semimatte or matte finishes, which may be waterborne or solvent-based, as well as (decorative) primers and undercoats that are different from automotive primers and undercoats.

In general, the full name of a paint will reveal where and how it is used, as in "cathodic automotive electroprimer" or "industrial cathodic clearcoat." The former is used to apply a primer coat by a cathodic (as opposed to anodic) electrodeposition process onto car bodies, while the latter is used to apply a clearcoat finish onto a metal object (such as a silver-plated trinket) in a factory process.

The term "varnish" (or "lacquer") is usually reserved for materials that produce clear films and can be thought of as an unpigmented paint, because although they may be colored they are generally transparent (cf nail-varnish), unlike paint, which is opaque. There are many types of varnish, and these are grouped according to their use—marine varnish, wood varnish, etc. The clear films used on vehicles are not usually called varnishes (though, until recently, "coach varnish" was used to brush onto bus paintwork to give a high gloss finish) but are known as clearcoats.

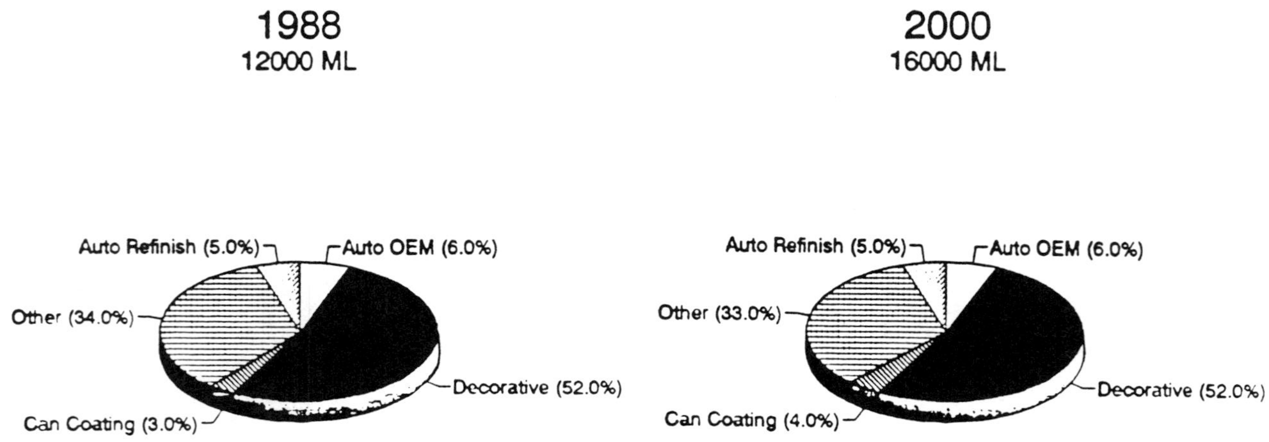

FIG. 3 World paints by market size, with projections to the end of the decade. (Courtesy ICI.)

B. Paint Systems

The diversity of paint arises from the properties required of a coating, which cannot be met in a single product because they are constrained by the nature of the substrate on which the paint is used. There is no such thing as an "universal paint," and there probably never will be. For example, a paint that will have good durability on a wooden substrate will be unsuitable as a car finish and vice versa. Seasoned timber expands and contracts, anisotropically, by several percent depending on the wood and the relative humidity of the atmosphere. Therefore it requires a paint that can withstand this amount of movement; hence the paint must be very elastic. However, such an elastic material is inherently soft. A car finish, on the other hand, has to dry quickly, withstand stone-chipping, dirt impaction, and retain its high gloss; hence it must be hard. Thus the two sets of requirements are mutually incompatible. Furthermore, the desired properties of a car finish frequently cannot be met within one product. This problem is overcome by using several coats of different paints to build up a laminate or paint system as illustrated in Fig. 4 to achieve the required objective.

FIG. 4 Cross-sectional micrograph of multicoat automotive paints showing the various layers (bar = 25 μm). (a) Substrate; (b) zinc-coated mild steel; (c) phosphate pretreatment; (d) electroprimer; (e) primer surfacer; (f) primer sealer; (g) basecoat with aligned aluminum flakes; (h) clear topcoat. (Courtesy Longman Group UK Limited [13].)

The first coat shown is a pretreatment or conversion coating. It comprises an inorganic layer of complex zinc and iron phosphates [14], producing a crystalline layer about 1 μm in thickness, obtained by the reaction of phosphoric acid with the iron substrate in the presence of soluble zinc phosphate salts. Its purpose is to produce corrosion resistance and to act as a key for the paint onto the metal surface. An electroprimer (epoxy resin-based) is then, usually cathodically, deposited when the car shell is immersed in a dip tank and a dc current is applied to deposit a uniform paint coating, on both horizontal and vertical surfaces, of about 20 μm thickness. The purpose of the electroprimer is to adhere strongly onto, and protect, the substrate against corrosion. A primer-surfacer (polyester-based) is then spray-applied to a thickness of about 40 μm. Its purpose is to in-fill small metal imperfections such as file or disk marks arising from the metal pressing of the components that may be "telegraphed" through to the air interface by the electroprimer. Hence it must have good filling properties (high solids) and good sandability (as well as mechanical properties) to produce a smooth, even surface before the spray application of a topcoat (acrylic-based). The topcoat's main task is to produce an aesthetic appearance along with durability. It frequently consists of two separate coats, (1) a basecoat, which is a pigmented layer (a "solid" color) or may contain leafing aluminum flakes to produce a metallic paint which has flop [15], that is, the paint changes color depending on the angle of view, and (2) a clearcoat, which is a clear (transparent) film that contains UV-absorbing material and produces a very high gloss.

Gloss is measured instrumentally (with a glossmeter), as an intensity of image, which is dependent on more than just smoothness of surface. It approximates to what is perceived by the eye—more a distinctness of (reflected) image [16].

In describing the various stages of painting a car, it is hoped that the reader acquires some feel for how different paints arise in the various markets. Putting these different paints together produces a composite, or strata (or multicomponent system), of paints which have the required physical and chemical properties that are difficult to come by in a single product. However, at present, much research effort is spent in trying to develop coatings that self-stratify to produce the equivalent of these different paints (hence properties) from a single product [17].

III. COMPONENTS OF PAINT

Paint is a colloidal system and therefore consists of a disperse phase (the pigment) and a continuous phase (the polymer solution, known as the binder). However, the polymer is not always in solution; it can be in the form of small particles (i.e., also a disperse phase) in what is commonly called an emulsion paint where the continuous phase is the nonsolvent, water. Strictly speaking, the term "emulsion paint" is misleading (in this instance) and the paint is more correctly called a latex paint, since an emulsion is defined as one liquid dispersed in another. True emulsion paints do, however, exist.

The components that make up a paint are schematically shown in Fig. 5 along with representative concentrations (which will vary according to specific requirements) to give some idea of composition. The scheme covers many different types of paint; for example, a latex paint would have no driers, solvent, or antiskinning agent, and the polymer (resin) would be in dispersion, as opposed to solution, form, whereas an oil-based gloss paint

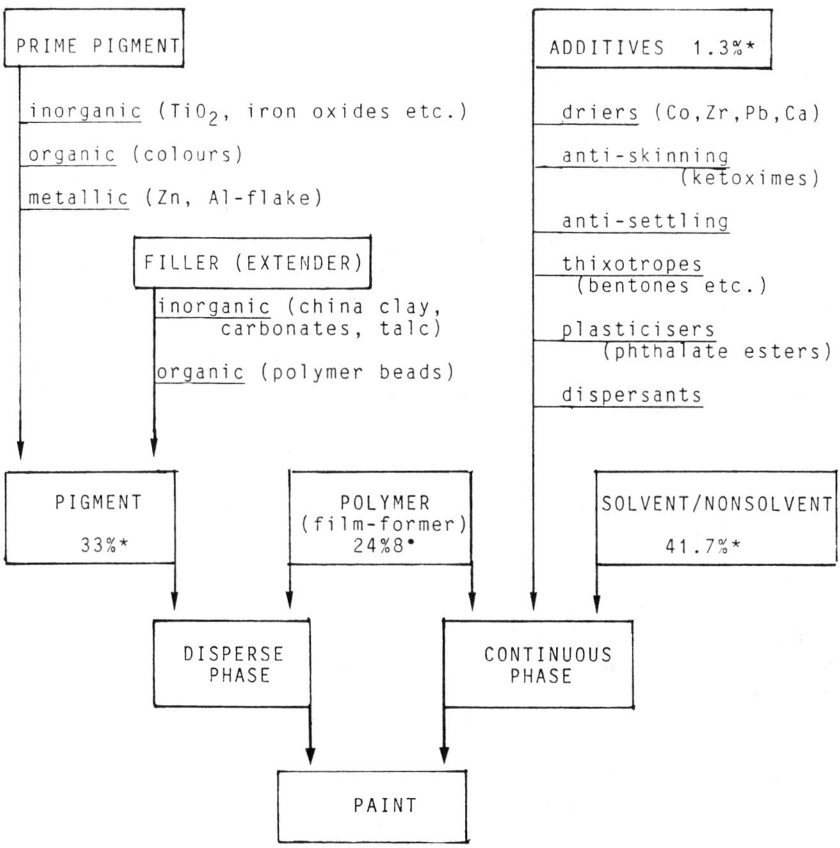

FIG. 5 The components of paint. *By weight.

will have its polymer in solution form and have antiskinning agents, driers, and antisettling agents, but no fillers, and its concentration of pigment will be lower. Table 3 gives raw material use by the U.K. paint industry in 1988, as a typical example of material usage.

A. The Disperse Phase

In all paint systems the pigment is always the disperse phase; however, it may not be the sole disperse phase. Dyes are not normally used in paint because they are generally in molecular subdivision and as such are not sufficiently lightfast. They also have a tendency to "bleed" (diffuse) into fresh paint if a soluble dye-containing film is coated with a solvent-containing paint. Although dyes such as lakes are used, they are particulate because of their mordant.

The most important requirement of the pigment disperse phase is the introduction of color to the polymer matrix, so that when the paint is applied it will mask (opacify) the substrate as well as producing a pleasing appearance. The pigment is also required to enhance durability; for example, there are known anticorrosive pigments that will inhibit the onset of corrosion of a metal substrate. Pigments such as (treated) TiO_2 can also enhance the paint's durability by absorbing the harmful UV radiation that would degrade the polymer matrix.

Titanium dioxide (rutile form, used to make paint white) is the most important pigment used in the paint industry, and its annual consumption in the production of paint is 75% greater than that of all other pigments put together [18], hence the separate chapter on TiO_2 pigments in this book. The yellow and red pigments come next; these consist of inorganic and

TABLE 3 Raw Material Use by U.K. Paint Industry, 1988 as a Typical Example of Material Usage

Material	Tons	% Vol	% Value
Solvents	273,000	31.7	15.2
Resins	200,000	23.4	33.7
White pigments	140,000	16.3	34.6
Extenders	115,250	13.4	1.4
Water	86,000	10.0	—
Other pigments	33,750	3.9	13.8
Additives	12,000	1.4	1.0
Total	860,000	100.0	99.7

Source: Ref. 12.

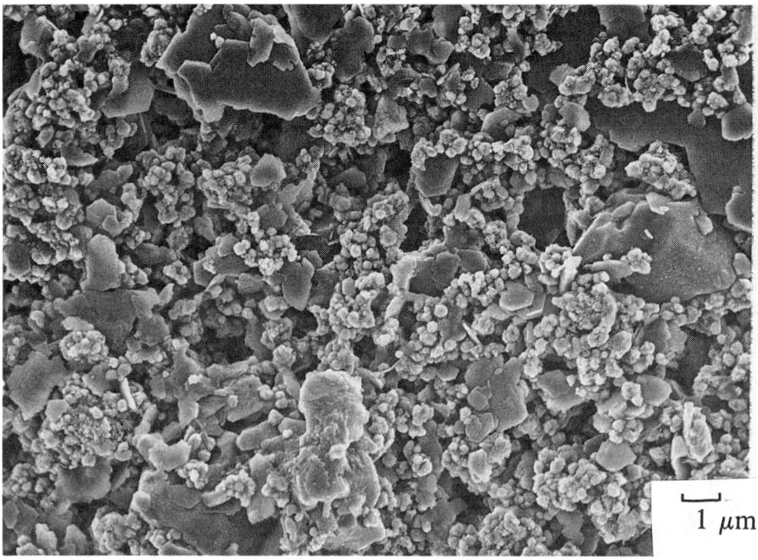

FIG. 6 Micrograph of a matte emulsion paint after oxygen plasma etch showing TiO_2 (27%), china clay (16%), and calcium carbonate extenders (8%). (Courtesy of SCM Chemicals Europe.)

organic pigments, the inorganics being greater in consumption (tonnage) than the organics, though not in unit price.

Extenders (extending pigments or fillers), when used, are also part of the disperse phase. Their purpose is to extend the pigment content, that is, to fill the polymer matrix—hence their name. These materials, for example, talcs, carbonates, and china clays, are colorless, and their refractive indices are close to that of the organic medium so they do not opacify by scattering light to any useful extent. Extenders do not have pigmentary properties per se; they are used to modify the physical properties of a paint film in order to impart sandability in an undercoat as well as film-build or to produce a matte finish as in an emulsion paint (see Fig. 6). They also modify the rheology of the system by contributing to the disperse phase volume, since the overall viscosity (η) is the product of the continuous phase viscosity (η_0) and a function of disperse phase volume (x), $\eta = \eta_0[f(x)]$.

Not all extenders are inorganic; organic extenders are sometimes used when certain properties are required (see Fig. 7). For example, polystyrene particles are claimed to enable paint to be made at critical packing without the deterioration of film properties normally associated at this disperse phase concentration [19].

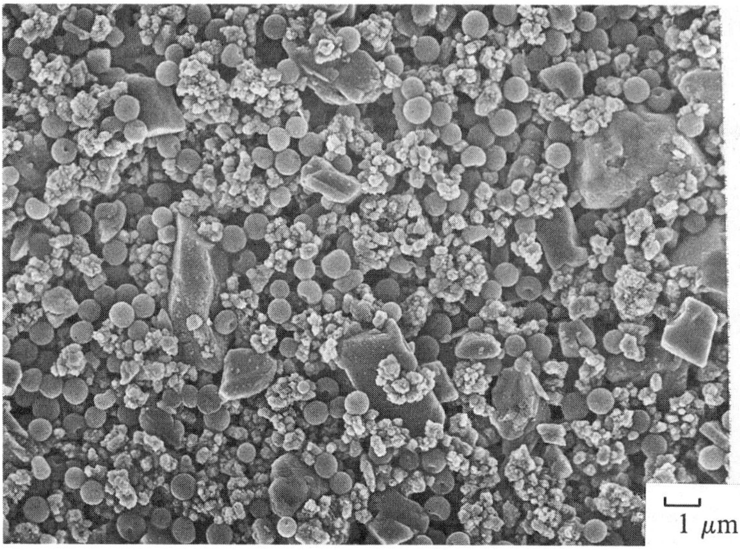

FIG. 7 Micrograph of a matte emulsion paint after oxygen plasma etching showing TiO_2 pigment with organic (polymer beads) and inorganic extenders. (Courtesy of SCM Chemicals Europe.)

Polymer particles such as aqueous latexes or nonaqueous dispersion polymers (NADs), which comprise the film-forming material, may also be part of the disperse phase of liquid paint.

The common denominator in all disperse phases is that the small particles are stabilized in suspension (in the continuous phase) by the presence, or adsorption, of specific entities on the particles' surfaces to give colloid stability. This may be achieved in one of two general ways, by electrostatic stabilization or steric (or polymer) stabilization (or both). A third possibility in the form of "solvation forces" is sometimes claimed, but Overbeek [20] has expressed doubts as to the capability of these solvation forces to be of practical significance in preparing stable dispersions due to their relatively short range character.

B. The Continuous Phase

The continuous phase (or serum) consists of the polymer film-forming material dissolved in a range of solvents and nonsolvents as well as additives, which comprise the minor constituents as shown in Fig. 5. The choice of polymer determines the paint film properties and as such is the most important factor in the formulation of a paint.

The continuous phase, unlike the disperse phase, is the phase in which most of the changes occur as a paint dries. These changes can be either physical, such as the evaporation of solvent as in lacquer drying, or chemical, as in autoxidation. Both of these processes bring about a decrease in volume due to the liquid–solid transition. When the area is fixed by an immobile substrate, then the volume decrease must come about mainly through a decrease in film thickness, which translates into anisotropic stresses that develop within the film. This volume contraction on drying has been utilized in producing decorative effects such as crackle finishes and in the tensioning of the canvas on gliders by the use of "dope."

C. Polymers Used in the Paint Industry

The terms "resin" and "polymer" are used interchangeably in this chapter, whereas one could argue that all polymers are resins, not all resins are polymers, since some resins are oligomers. The difference in name is really of historical nature. Resins have been known as such from old, to describe gumlike materials such as shellac, copal, kauri, and other fossil gums used in paint making. Now "resin" is used to describe synthetic materials such as alkyds, phenol formaldehydes, and poly(methyl methacrylates). It was not until the 1920s and Staudinger's work that the term "polymer" was accepted to describe the high molecular weight material made up of covalently bonded repeat units of monomer(s), although Berzelius may have coined the word "polymer" earlier. Figure 8 is a schematic representation of the main polymers used in the paint industry according to importance and type of finish.

(a) Alkyds. Alkyds are one of the oldest types of synthetic resins and are still probably the single largest group of polymers used in coatings. They are best described as highly branched, randomized, oil-modified polyesters. Although alkyds are polyesters, the term "alkyd" is reserved for oil-modified materials. They are made by condensing multifunctional aromatic acids with polyols and fatty acids. [the fatty acid is frequently used as the triglyceride (the oil)]; hence an alkyd may comprise phthalic anhydride, pentaerythritol, and soybean oil; the molar ratios are balanced to give a residual balance of COOH to OH groups to prevent gelation [21,22]. They are classified according to their "oil length," the amount (and type) of oil incorporated, as well as residual hydroxyl and carboxyl values. Long-oil alkyds (>60% oil w/w) are made using unsaturated oils, such as linseed or soybean oil, to make soft, flexible resins that air-dry by autoxidation of the double bonds in the oil and are used in undercoats and topcoats for decora-

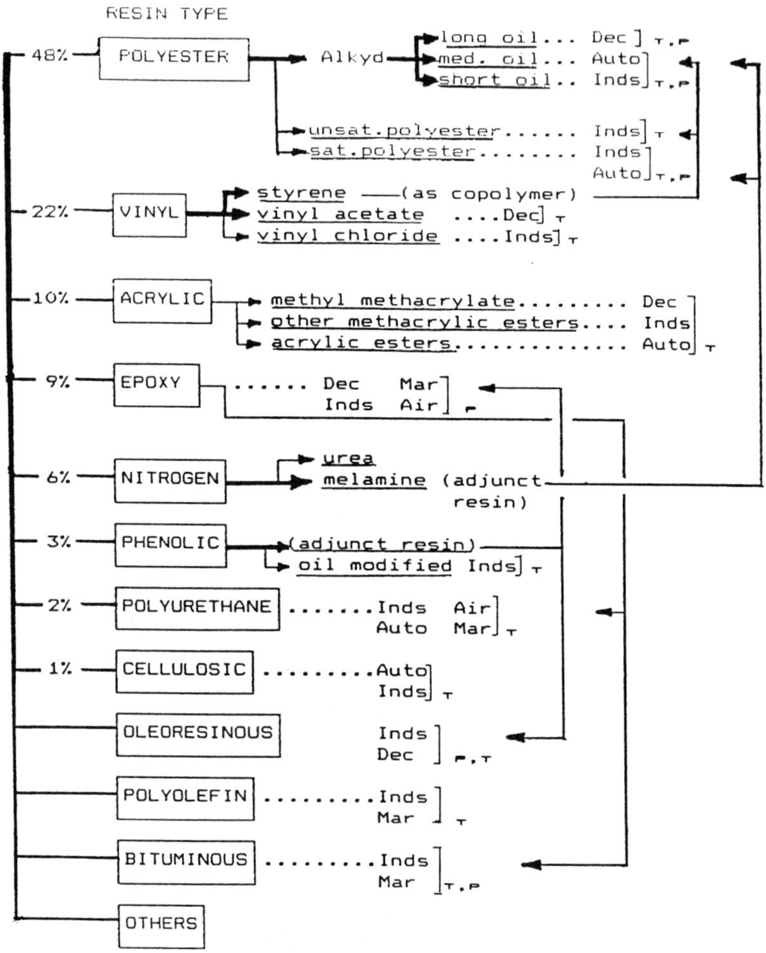

FIG. 8 Polymers used in the paint industry by chemical type and market. Dec = decorative; Inds = industrial; Auto = automotive; Mar = marine; Air = aircraft; P = primer; T = topcoat. Bold arrows indicate major use; lighter arrows, minor use. (Courtesy Longman Group UK Limited [13].)

tive paints. The drying process, catalyzed by “metal driers” [23] gives rise to the formation of scission products such as higher alcohols, aldehydes, and acids that impart the unpleasant odor associated with freshly painted surfaces.

Medium-oil (40–60% oil w/w) and short-oil (<40% oil w/w) alkyds rely on cross-linking reactions with another resin via their hydroxyl groups, with maybe some autoxidation (short-oil alkyds might incorporate saturated oils, e.g. rapeseed oil, hence no autoxidation). These and the short-oil alkyds are less flexible and more durable. They are used in automotive and industrial paints in conjunction with some other adjunct resin (e.g., amino resin) as stoving finishes.

(b) Polyester Resins. The term "polyester resins" is reserved for oil-free resins. They can be either saturated or unsaturated and are made from di- or trihydric alcohols with di- or tribasic acids; they may also contain monofunctional acids (e.g., benzoic) to modify their properties.

Unsaturated polyesters contain vinyl unsaturated bonds and are typically maleic/fumaric- containing materials and have reactive solvents such as acrylic or styrene monomers. They are used in refinish two-pack systems or glossy wood finishes. Cure is usually initiated using a redox couple containing a peroxy free radical generator.

Saturated polyesters can be both hydroxy and carboxy functional and are frequently made using isophthalic acid (cf. orthophthalic in alkyds). They are coreacted with other dibasic acids (e.g., adipic) to soften the polymer; alternatively, longer chain polyols (e.g., neopentyl glycol) can be used. These polyesters are solventborne and are reacted through their hydroxyl groups with other polymers such as melamine-formaldehyde resin to produce automotive or industrial paints.

(c) Vinyl Resins. This category should include the esters of acrylic and methacrylic acids, but since they comprise such an extensive range of materials they are generally classified on their own. Vinyls such as styrene are used as reactive solvents (in polyesters) or as comonomers to produce polymers such as styrene-maleic anhydride copolymer. Vinyls also include poly(vinyl chloride) and poly(vinylidene chloride or fluoride), which can be spray-dried to produce powders and used as solvent-swollen materials called organosols (or plastisols) that are used to make thick coatings for the industrial market. The powders (e.g., PVF_2) can also be used as dispersions in solution polymers to give highly durable finishes on sheet metal. Poly(vinyl acetate) and poly(vinyl versatate) and their copolymers are generally used as dispersion polymers in the form of latexes for emulsion paints in the decorative market.

(d) Acrylic Resins. These polymers cover a wide range of resins from mainly poly(methyl methacrylate), used as a solution polymer in highly durable "lacquer-dry" automotive topcoats (thermoplastic), to approximately 50% butyl acrylate containing dispersion polymers (aqueous latexes). Thermosetting polymers are made by incorporating various functional groups, which can be introduced via the ester group of the (meth)acrylic

acid and which will react with other appropriate functional group–containing resins. The physical properties in terms of hardness/softness of the copolymer can be manipulated through comonomer addition to give materials of the required glass transition temperature (T_g). The T_g of the amorphous copolymer can be estimated [24] from the individual T_g values of the constituents in a manner analogous to determining the total resistance of a series of parallel resistors in an electric circuit. While these polymers are mainly used in dispersion form as aqueous latexes in emulsion paints, they are also made as nonaqueous dispersions (NADs) and used as such.

(e) Epoxide Resins. The epoxy (oxirane) group is incorporated into a number of different polymers to act either as a reactive functional group during the setting of the paint film or as a functional group for the attachment of other material in order to modify the properties of the polymers. Thus epoxy resins are one of the most adaptable and versatile types of polymers used in coatings. They range from two-pack systems, which are mixed just prior to use, utilizing the high reactivity of primary (and secondary) amines with the oxirane group for room temperature curing finishes, to thermosetting, one-pack, finishes that cure at elevated temperatures. One of the most common epoxy resins is based on the condensation of diphenylol propane with epichlorohydrin to produce polyethers with terminal oxirane groups. These materials are frequently extended, or modified, with substances such as fatty acids (to give epoxy esters) or with acrylics to form graft copolymers. They can be blended with other resins (utilizing their residual OH groups) such as M/F or phenol-formaldehyde resins to make tough anticorrosion primers. Epoxies are used with reactive coal tar pitches to produce flexible, chemically resistant heavy duty coatings for pipelines and marine coatings.

(f) Amino (or Nitrogen) Resins. These are terms used to describe materials containing two or more amino groups, especially urea and melamine. The amino groups are methylolated using formaldehyde, and in turn the methylol groups are etherified to varying degrees with low molecular weight alcohols such as methanol or butanol to increase their hydrocarbon solubility. The greater the etherification and the higher the alcohol used, the greater the hydrocarbon solubility. These urea-formaldehyde (U/F) and melamine-formaldehyde (M/F) resins are used as adjunct resins to cross-link with hydroxyl-containing polymers to produce thermosetting systems on stoving.

(g) Phenolic Resins. These are mainly low molecular weight materials made by reacting a phenol (or substituted phenol) with formaldehyde to produce resins containing many methylol groups. They are usually adjunct resins for cross-linking other polymers, mainly by the reaction of the meth-

ylol with free OH groups on other polymers, although the methylol groups will also self-condense. They are frequently used with modified epoxy and oleoresinous polymers.

(h) Polyurethanes. These are classified according to the isocyanate used in their preparation, which can be either aromatic or aliphatic. Aromatic isocyanates are more reactive and constitute the bulk of polyurethane resins. They are used as adjunct resins (with free isocyanate groups) to cross-link hydroxyl-containing polymers at room temperature, as for example in two-pack aircraft or automotive refinish paints. Alternatively, they may contain "blocked isocyanate" (no free isocyanate) groups in one-pack systems; on stoving, the urethane (blocked isocyanate) group unblocks to react with free hydroxyl or amino groups to cross-link the polymer. Polyurethanes are also used in oil-modified form such as urethane alkyds to increase the hardness of decorative paints and varnishes.

(i) Cellulosics. Various derivatives of cellulose are used in coatings. They range from nonaqueous plasticizers/modifiers, such as cellulose acetate butyrate (CAB) or nitrocellulose used as a lacquer-dry topcoat in automotive refinish paints and woodfinish lacquers, to water-soluble additives such as thickeners (carboxy- or hydroxy ethyl cellulose) in emulsion paints.

For a more comprehensive description of polymers used in the paint industry and their chemistry, the reader is referred elsewhere [25,26].

IV. THE PAINT-MAKING PROCESS

The paint-making process can be subdivided into a number of stages which consist of millbase manufacture (pigment dispersion), millbase let-down, mixing, tinting, and adjustment to specification as shown schematically in Fig. 9.

The millbase, or pigment dispersion concentrate, can be prepared as a batch process using ball mills or as a continuous process using bead mills (a premix stage is required to produce a pumpable slurry for the bead mill); and is known as the grind stage. This is the essential pigment dispersion stage where the pigment particles undergo wetting, separation into their individual particle sizes, and colloidal stabilization. Generally there is negligible particle size comminution, the grinding process being one of applying shear forces to break up pigment aggregates into individual units, as predetermined by the pigment manufacturer. The quantities of pigment, grind resin, and solvent required to establish millbase working (though not necessarily optimum) conditions are determined by empiricism rather than theory—the objective being to produce a stable, fine dispersion of pigment(s) in the minimum of solvent and resin within a given time. Once the required

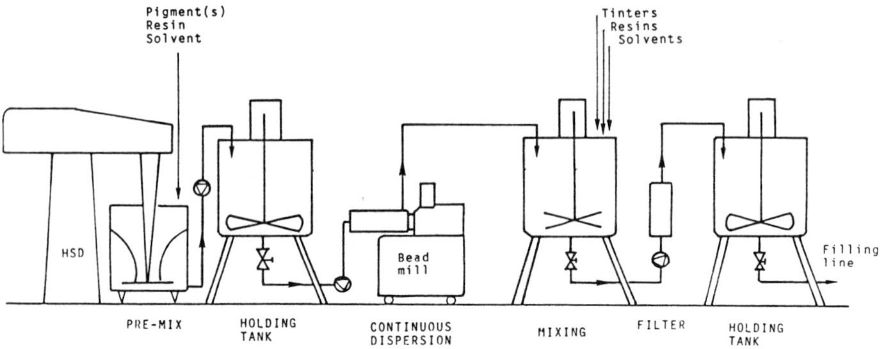

FIG. 9 Schematic diagram of the paint-making process.

fineness of grind has been achieved, further resin is gradually added to minimize colloidal shock (see later), preferably within the dispersion equipment. The mixture is then transferred to a mixing vessel, where further resins, additives, solvents, and tinters are added to make up the paint to its specification. It is then filtered, placed into holding vats, and checked prior to delivery to the filling line. The cleanness and strength of color developed by the pigments in a paint is dependent not only on the resins but also on the dispersion conditions and the dispersion equipment used.

A. Dispersion Machinery and Grinding Conditions

Some idea of the variety of dispersion machinery available for pigment dispersion can be obtained by looking through the current *Polymer Paint Colour Year Book*, which is produced annually. There is a large array of dispersion machinery dependent on the quality and scale of millbase production, rheological properties of the millbase, suitability for continuous or batch manufacture, control of temperature (temperature can have an adverse effect on the color of some pigments [27]), and solvent loss, not to mention ease of cleaning of the mill. Hence it is proposed to describe here only some of the equipment used in paint making and why it is used.

Improvements in the dispersibility of pigments and in health and safety regulations, not to mention automation, have brought about changes in the type of equipment used as well as improvements in equipment design. Old workhorses such as the ball mill are still used, even though they may be slow in dispersing pigments compared to media mills. They remain in use because they are established and reliable and do not require close supervision, whereas roll mills are out of favor because of solvent emission, throughput, and the need for intensive skilled labor in their operation.

Dispersion equipment can be divided into two extremes of action represented by impact and shearing mechanisms, which require different millbase viscosities. Most machinery, however, operates as a hybrid of the two, with a bias toward the shearing mode as summarized in Table 4.

1. Ball-Mills

The ball mill is probably the oldest equipment for batch manufacture of millbases in the paint industry. It is still used by some of the largest paint producers to make high quality dispersions; however, it probably would not be installed in a modern factory. One of its advantages is that it is a closed system and hence there is no solvent loss during milling, but there is a temperature build-up in large-scale ball mills that does not occur in small mills. To overcome this problem, production ball mills are cooled, and laboratory mills are placed on heated racks to simulate production conditions, but this also precludes the use of low boiling point solvents. Ball mills can range in size from 500 mL to several metric ton (see Fig. 10). This enables the paint chemist to determine the optimum conditions for dispersion using small-scale laboratory experiments, with small porcelain

TABLE 4 Mode of Action in Milling Equipment

Impact	Hybrid	Shear	Millbase viscosity
		Pug mixer (sigma blade)	high
		Triple-roll mill	
		Colloid mill	
		Single-roll mill	
	HSD		
	Attritor		
	Bead mill		
	Ball mill		
High speed impingement mill (Kady mill)			
Ultrasonics[a]			low

[a]Cavitation being classified as a negative form of impaction.

FIG. 10 Production-scale ballmill. (Courtesy OBS Machines Ltd.)

jars and appropriately sized steatite balls, and make adjustments with gradual scale-up in size to production volume.

The ball mill consists of a cylindrical container with its internal diameter approximately equal to its length. It is charged with appropriately sized balls to 60% of its volume (assuming that voids do not exist). Thus when the mill is charged with millbase so that it fills the voids between the balls, the total volume occupied is 60% of the mill's capacity; it is then said to have a charge-to-voids ratio of 1 : 1. The mill is rotated about its horizontal axis on rollers that have their rotational speed adjusted for the size of the mill. The mechanical shear to break down pigment aggregates is produced by the cascade of balls down one side of a "hill" produced by the rotation of the mill, as shown schematically in Fig. 11. The mill's rotational speed for optimum cascade conditions is given by the relationship

$$\mathrm{rpm}_{\mathrm{opt}} = (37 - 3.3R)/R^{1/2} \tag{1}$$

where R is the ball mill radius measured in feet [28, p. 413]. The formulator has to choose the grinding medium (steatite or steel) and the charge-to-voids ratio, not to mention "sighting" the millbase formulation.

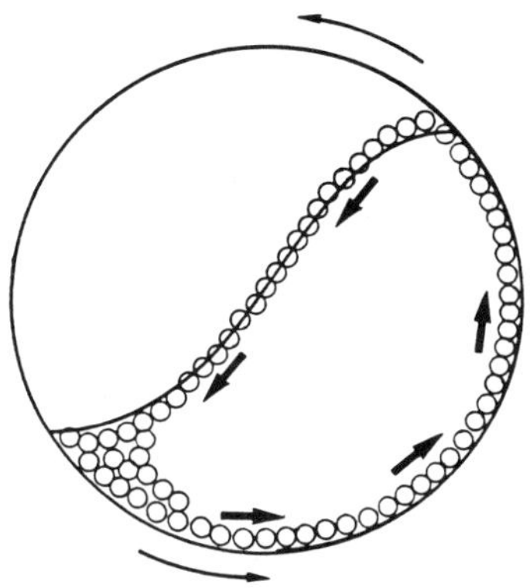

FIG. 11 Cascade of balls in ball mill.

(a) Steatite or Steel. The choice is simple if a white (TiO_2) millbase is to be made. Steel balls cannot be used because the fine metal powder produced by wear of the grinding medium will mar the whiteness of the millbase. However, other grinding materials such as glass, porcelain, pebbles, (steatite), alumina, and zirconia (in order of increasing density) can be used. Generally speaking, the greater the density difference between the millbase and the grinding medium, the faster the dispersion process. The viscosity of the millbase, which is probably non-Newtonian (and undergoes change with dispersion progress), must also be taken into account since it will affect the cascade of balls; the thicker the millbase consistency, the denser the medium required. However, if the millbase is too thin and/or there is too large a difference in density, then overheating, loss of efficiency, and excessive wear of the grinding material will occur. With experience it is possible to tell how well the mill is operating by the sound of the cascading balls.

Rahter [29] has published the ball sizes required (dependent on ball density and millbase viscosity) to produce efficient ball mill dispersion as shown in Table 5. This interdependency can be expressed by Patton's empirical equation [28, p. 424]

$$\eta = 10\rho + 72d \tag{2}$$

where η is the viscosity in Kreb units (KU; units of Stormer viscometer [30]), ρ is the ball density (g/cm^3), and d is the ball diameter (in.).

(b) Charge-to-Voids Ratio. Ideally the charge-to-voids ratio should be 1 : 1, but if greater throughput is required, then the ratio can be increased

TABLE 5 Required Ball Size for Desired Millbase Viscosity

Ball diameter		Porcelain, $\rho = 2.38$		High-density porcelain, $\rho = 3.3$		Steel, $\rho = 7.85$	
in.	mm	KU	Pa.s	KU	Pa.s	KU	Pa.s
1/2	13	55–63	0.2–0.35	63–72	0.35–0.6	101–111	1.75–2.3
5/8	16	63–77	0.35–0.7	70–80	0.55–0.85	114–124	2.4–3.1
3/4	19	78–87	0.8–1.2	94–104	1.35–1.85	125–134	3.2–4.0
1	25	96–105	1.5–1.9	111–121	2.3–2.95	141–151	4.7–
1¼	32	109–118	2.15–2.7	125–135	3.2–4.1		
1½	38	120–129	2.85–3.6	135–146	4.1–5.2		
2	50	137–146	4.3–5.2				

[a]KU values are approximately related to pascal-seconds.
Source: Ref. 29.

to 1.2 : 1. If comminution of the pigment particles is required, then the charge-to-voids ratio is reduced to 0.9 : 1 and grind time is prolonged.

The advantage of ball milling is that it requires little supervision and can handle "difficult" pigments that are not so readily handled by some of the other equipment. Moreover, a premix stage is not necessary; frequently the resin solution and pigment can be added simultaneously, though it might be preferable to make one or two incremental pigment additions to the mill. The disadvantages are that it is a slow process, taking 10–20 h, and sometimes longer, to achieve dispersion. The mills are also difficult to clean, so generally special mills (and grinding media) have to be reserved for white millbases only.

(c) Planetary Ball Mills. These are small ball mills which are placed vertically into special holders at the ends of a horizontal rotor within a safety cage. When the rotor arm turns, the ball mills spin at the end of the turning arm like planets around the sun, producing a grinding action similar to that of a normal ball mill cascade due to the centrifugal forces created by the circular motion (which magnifies the effect of the density differential between the millbase and the grinding material). The advantages of this type of mill are that it can handle very difficult thixotropic bases and it is fast; 20 min grind time can be comparable to an overnight grind in a normal ball mill. They are used for dispersing small batches of very expensive pigments that are difficult to process, or in the laboratory where speed is essential. However, the disadvantages (apart from cleaning, which is the same as for a normal ball mill) are that at least two ball mills must be used in symmetrical alignment. If a six-slot piece of equipment is being used, the ball mills must be placed in balance around the axis of rotation. The mill base weight and the *grinding material weight* must be the same in each ball mill; the apparatus is complex and requires watchful supervision as well as maintenance. With the advent of more readily dispersible pigments, this equipment has become displaced by attritors.

2. Attritors

Originally the attritor was developed for other than paint industry application, where particle size reduction was an important requirement. Its fast dispersing capabilities led to its introduction into the coatings industry, and it is slowly displacing the much older ball mill (as are the bead mill and high speed disperser). There are different forms of attritors—batch, continuous, or circulating in action. They are fast, safe, simple to operate and have low power consumption and good temperature control.

An attritor can handle a wide range of viscosities up to the highest that will flow under shear. Millbase density does not impose the same limitations as it does with conventional ball mills since the attritor can process millbases that are denser than the grinding medium, which usually ranges in size from 3 mm to 13 mm diameter balls of the same composition as those used in ball mills. However, as in ball milling, selection of the grinding material to be used is based on similar considerations; (1) generally, high density media give better results—preferably the grind medium should be denser than the millbase, and the more viscous bases should be used with the higher density media; (2) the harder the grind medium, the better the grinding efficiency; (3) small grinding media cannot readily break up large particles; (4) whites should not be used with steel ball grinding media.

The early attritors (e.g., the Szegvari attritor) were batch-type machines and could usually be charged directly into the grinding vessel or by combined premix and grind. Continuous models were a later development, followed by circulating models; a separate premix unit is required with both of these machines.

The *batch attritor* consists of a vertically mounted steel (or porcelain-lined) container with a central rotating shaft that operates at speeds intermediate between those of ball mills and bead mills, that is, at moderate speeds between 60 and 300 rpm depending on machine size—the smaller the unit, the higher the rpm. The shaft is equipped with several horizontal "fingers," which are about 85% of the vessel's internal radius and exert a stirring action of the grinding medium, usually balls, which force the grinding medium to tumble randomly through the container. The grind medium charge occupies about two-thirds of the vessel and covers the top sweep arm [31]; sufficient millbase is used to just cover the grind medium. The greatest grinding action occurs about two-thirds of the radius from the center.

The grinding time [32] is related to the ball diameter and agitator speed by

$$t = kd/n^{1/2} \tag{3}$$

where t is the grinding time to reach a certain median particle size; k is a constant dependent on slurry, grind medium, and attritor; d is the ball diameter, with a lower limit due to momentum considerations; and n is the rpm of the rotor.

The *continuous attritor* consists of a vertical grinding vessel whose length is several times its diameter. It is jacketed for temperature control and is tapered, with the lower portion being slightly smaller in diameter than the top. Premixed millbase made by paddle stirring (or by a high speed disperser) is fed into the bottom of the vessel by a variable pump to enable control of the dwell time in the grinding chamber, thereby determining the fineness of grind. Discharge is from the top through a medium-retaining mesh. High or low speed machines are used where the shaft rpm is adjusted relative to the tank diameter to provide a constant tip speed. For example, with a 10-gal grind tank, low speed (200 rpm) continuous attritors with bar agitators are commonly used for processing medium to high viscosity (over 25,000 mPa.s) material using 6 mm balls. High speed models (800 rpm) with disk impellers and using media of about 3 mm can give large throughputs of material of less than 5000 mPa.s viscosity [33].

The *circulating attritor*: the disadvantage of a batch attritor is that the batch size is determined by the mill volume, whereas with the continuous attritor this constraint does not apply. In both machines the dwell time can be adjusted to process the most difficult particles, but once set for a given batch it cannot be adjusted during operation. With either machine there is a possibility that some softer (or smaller) particles (say in a mixed pigment co-grind) will be overground, resulting in a wide particle size distribution. Combining the features of the batch and continuous models to provide a narrower finished particle size distribution led to the development of the circulating attritor in the 1970s. Its design is similar to that of the batch attritor; except for a small freeboard (about 25–75 mm deep, depending on mill size), the vessel is entirely filled with grind medium. The grind vessel is jacketed for temperature control, and rod-type agitator arms rotate at about twice the speed of batch models. The millbase is pumped (positive displacement pump) upward through the grind medium to a horizontally fitted discharge arm at the top of the tank, from where it is returned by gravity to the hold tank, which is about 10 times the size of the grind unit and also serves as the premix tank. The rate of throughput can be varied, and the optimum rate for each product is determined by trial.

The circulation rate of this attritor is many times the flow rate through a continuous attritor (the contents of the holding tank are circulated every 6 min), a high circulating rate being essential for efficient operation. Better uniformity in particle size is obtained because larger particles are ground to a greater extent. The larger the pigment particle, the more frequent its deflections and the longer its route (and time) in passing through the grinding medium; hence the larger the particle, the more the grinding action per pass. The degree of grinding on successive passes is reduced in proportion to the decrease in particle size, thus enabling batches of initially differing particle sizes to become products with a narrow particle size distribution. Furthermore, ingredients requiring different degrees of processing can be added sequentially, and if the millbase thickens, more dispersant can be added.

The various types of attritors are shown in Fig. 12. Continuous attritors can be included in the broad "media mill" classification, which covers a wide variety of designations including bead mills, shot mills, agitated media

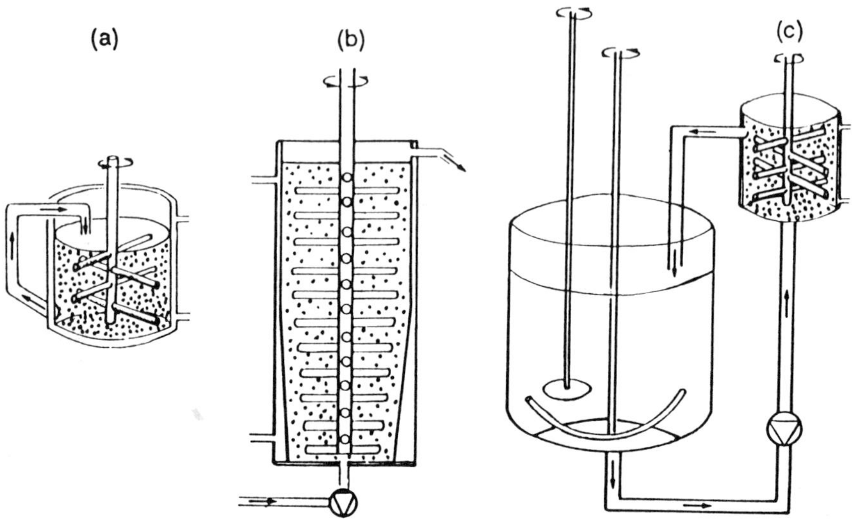

FIG. 12 Three types of attritors: (a) Batch; (b) continuous; (c) circulating.

mills (see Fig. 13), attrition mills, and others. They are also known by a number of proprietary names such as Perl mills, Molinex, Rotomill, and Dynomill.

3. Bead Mills

The disadvantages of the ball mill, such as difficulty in emptying the millbase (especially when it is pseudoplastic), cleaning, slowness of grind, and mill size led to the introduction of bead mills about 40 years ago. Hochberg's original patent, to DuPont [34], specified 20–40 mesh Ottawa sand (a round, impurity-free quartz sand) as the grinding medium, and the mills were known as sand mills. With development, improvements were obtained using glass (Ballotini) and other synthetic beads, and hence the change in name to bead mills (or media mills).

Bead milling is a two-stage process, as premixing is essential. This can be carried out with a paddle mixer or more preferably with a high speed disperser. The greater the degree of dispersion achieved at the premix stage, the shorter the dwell time required in the bead mill. The essential features of a bead mill are shown in Fig. 14. The grind vessel length is at least twice the chamber diameter. A motor drives the central shaft at a high speed. The shaft is fitted with a series of disks (which are about 70% of chamber diameter in size) and are made of an abrasive-resistant material (hard steel) to agitate the beads. It revolves at a speed dependent on the dimensions of

FIG. 13 Interior structure of an LME25 media mill showing agitator arm. The mill is capable of a throughput of 50–500 L/h. (Courtesy of Netzsch Mastermix Ltd.)

the unit, that is adjusted to give disk tip speeds of about 2000 ft/min (10 m/s). A feed pump transfers the millbase slurry to the bottom of the grind chamber, which is jacketed for temperature control, and the pumping rate is adjusted to achieve the desired degree of dispersion via the dwell time. The millbase is forced upwards through the grind medium at a constant

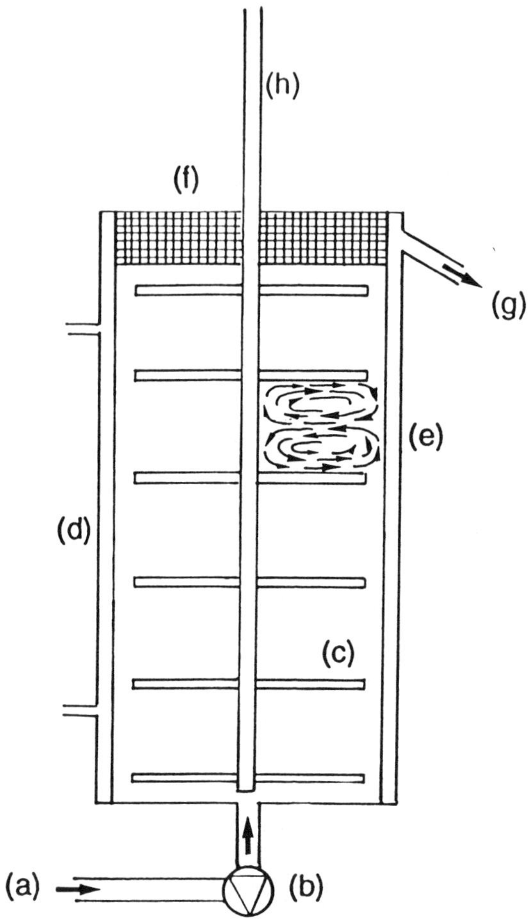

FIG. 14 Schematic drawing of a continuous beadmill, showing essential features. (a) Entry of premixed millbase; (b) pump; (c) impeller disks; (d) water-jacketed containing vessel; (e) typical flow pattern for millbase between two impeller disks; (f) screen for retaining grinding media but allowing passage of ground millbase; (g) exit port for ground millbase; (h) drive shaft to motor.

flow rate and emerges at the top, while the grind medium is retained by a screen. The velocity of the millbase and medium is higher in the areas between the rims of the disks and vessel wall than between the vessel wall and shaft, thereby producing a series of velocity gradients. These velocity gradients form a "double-doughnut" effect between the disks, which will also develop shear forces to break down the aggregates as well as the normal shear effect produced by the rotating media.

As in other forms of media milling, the charge-to-voids ratio and the bead size are important. Bead sizes usually range from 1 to 3 mm, the lower size limit being set by the fineness of the screen required to effectively retain the grind medium. Similar considerations about the bead density in relation to millbase apply as described previously, the general-purpose workhorse being the Ballotini bead.

There are many variations in bead mill design, such as disk shape and number of disks. The main differences are whether they are vertical open-top, closed-top, or horizontal mills. In open-top mills, discharge is through a medium retaining screen where it can be readily seen and checked; however, solvent evaporation can occur, which leads to deposits that may partially blind the screen, and flow problems can arise with thixotropic (especially pseudoplastic) materials. The closed-top vertical mill is designed to overcome solvent loss and screen blinding in that the screen is submerged in the discharge and operates under a slight applied pressure, overcoming some of the viscosity problems. However, a packing gland has to be fitted where the shaft passes through the cover, and this increases the complexity (and cost) of the mill. Horizontal mills (Fig. 15) are a further development; they have the advantage that settlement of the grind medium does not create start-up problems, the grind medium remains uniformly distributed throughout the chamber, allowing higher density media to be more readily used as well as a higher medium-to-charge ratio.

In the laboratory, when formulating with new materials or studying dispersion characteristics it is convenient to use a batch grind. This is readily achieved using a 5 : 3 ratio of 1 mm diameter Ballotini beads (apparent volume) to millbase volume in a simple 1 liter (or more) open metal container (with a water jacket) and a single disk whose diameter is 70% of the container diameter. The disk is inserted into the chuck of a fixed commercial drill equipped with varying diameter pulleys and driven by a belt to obtain a disk tip speed of 10 m/s. However, with the increased availability of commercial small-scale laboratory equipment fitted with process sensors, simple devices as described above are becoming less frequently used.

4. High Speed Dispersers

These units (see Fig. 16) consist essentially of a flat disk that rotates (horizontally, in a container) at a very high tip speed, 20 m/s and higher. The disk can have various sawblade-like perimeters (or protrusions as in a Diaf disk). There are many minor differences in the arrangement of the teeth (in terms of their shape and alignment) at the disk's periphery (see Fig. 17), and the various manufacturers claim that these help in the dispersion process.

FIG. 15 Fifty-liter Supermill horizontal beadmill (Courtesy OBS Machines Ltd.)

The essential feature in obtaining dispersion using a high speed disperser (HSD) is the creation of the rolling vortex (or doughnut), without surging or splashing, by the correct positioning of the dispersion disk and correct choice of size in relation to the container [35]. This is usually in the ratio of 1 : 3*D*, where *D* is the disk diameter (see Fig. 18). If the millbase flow is poor, a larger ratio disk is preferable (1 : 2.5*D*). The disk is immersed so that it is approximately 0.5–1*D* from the bottom of the container, which is filled to a depth of 2*D*; if the viscosity drops during dispersion and upsets the vortex, some compensation can be obtained by lowering the disk. Ideally the rotation speed of the shaft should be variable to provide mixing at low speeds during the initial loading stage when the continuous phase is first charged. The pigment is added in incremental stages, and the disk speed is increased. With proper circulation, the millbase spends most of its time passing through the high energy zone around the disk to produce a dispersion in about half an hour without the need of grinding material. However, the HSD, while suitable for preparing dispersions for emulsion paints and primers, has its limitations, and for high quality finishes requires refinement by further dispersion in a bead mill or attritor. The main feature of an HSD is that pigment dispersion is obtained mainly through the shear-

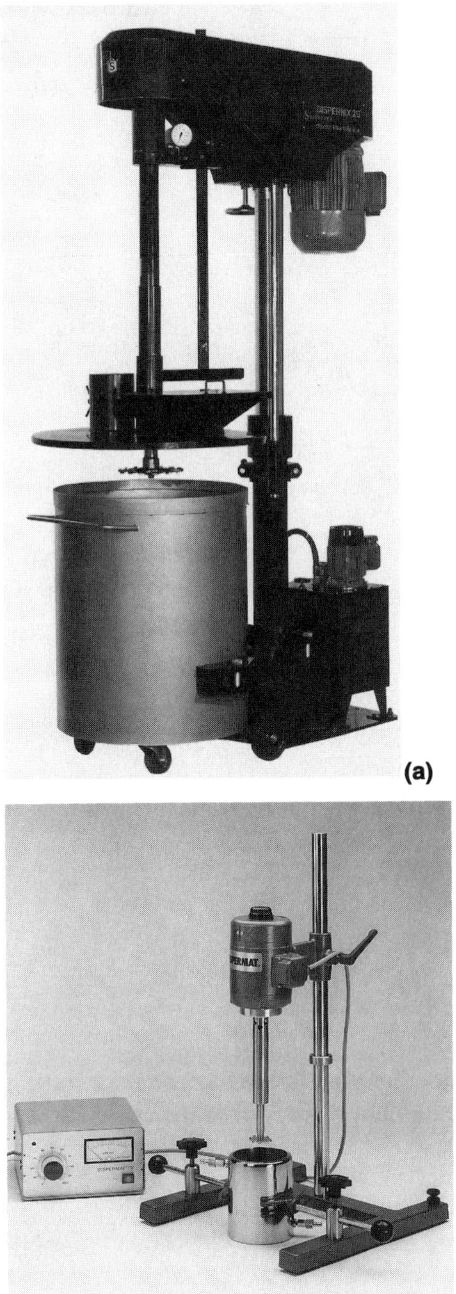

(a)

(b)

FIG. 16 (a) Production-scale 25 hp Dispermix high speed disperser. (Courtesy OBS Machines Ltd.) (b) Laboratory-scale Dispermat high speed disperser. (Courtesy VMA-Getzmann GmbH.)

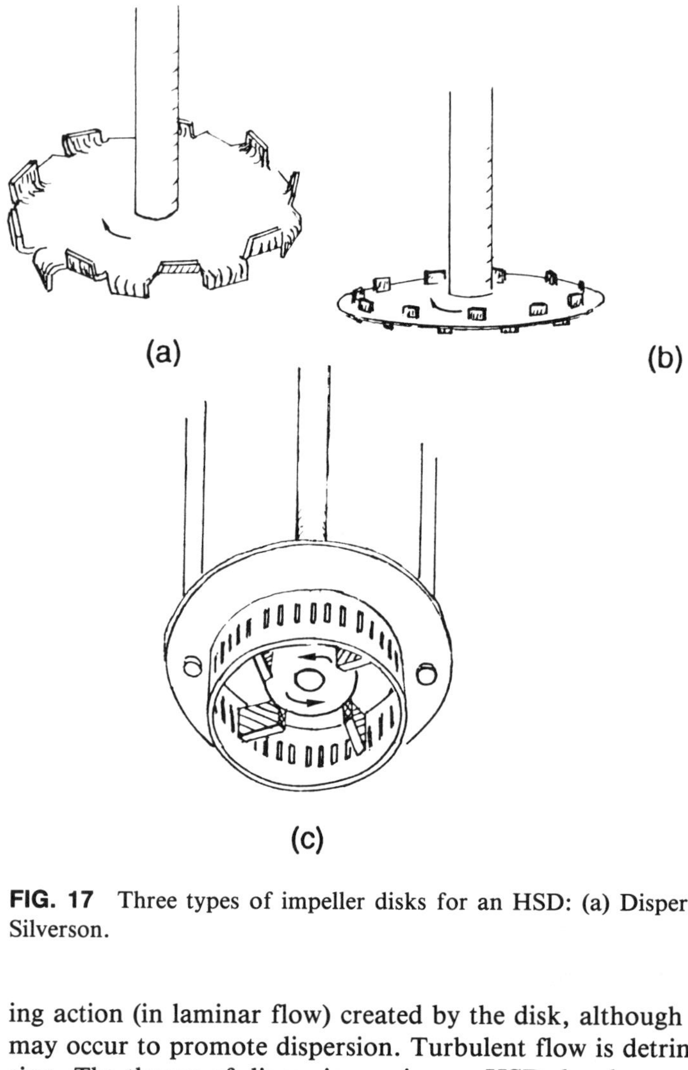

FIG. 17 Three types of impeller disks for an HSD: (a) Dispermat; (b) Diaf; (c) Silverson.

ing action (in laminar flow) created by the disk, although some cavitation may occur to promote dispersion. Turbulent flow is detrimental to dispersion. The theory of dispersion, using an HSD, has been well presented by Patton [36], who stresses the difference between mixing and dispersing. In the former, turbulence is beneficial but not in the latter.

5. Pug Mixer

There are various forms of pug mixers employing, for example, sigma blades [37, p. 997], which rotate at different speeds from each other, or ellipsoidal blades, whose loci of revolution ensure self-cleaning. These dispersers are rarely used in the paint industry now. They operate by a kneading and shearing action and are used to produce stiff pastes such as putties and woodfillers. The quality of pigment dispersion, which may be satisfactory for some purposes, is not as good as that produced by bead milling, for example. Their relatively low throughput and difficulty in cleaning have led to them being used for special purposes only.

6. Roller Mills

There are various forms of roll mills such as one-roll, two-roll, and three-roll mills. These operate by dispersing pigments through their shearing action as the millbase passes through the nip between the rollers (or roller and bar in one-roll mills) and are described by Patton [28, p. 388], Payne [37, p. 997] and Berg [38]. A high viscosity material is required for effectiveness as well as to produce a tack. Roller mills are not used in a modern paint factory because of their low volume throughput, need for constant skilled operator attention, lack of solvent containment, and high maintenance requirement. However, that is not to say that a three-roll mill cannot produce a high quality dispersion where color strength and cleanness of color are of primary importance.

7. Colloid Mills

These mills are old established machines, introduced in the 1920s and used mainly for emulsification (as their name implies). They also can be used for pigment dispersion because of their high shearing action when a flowable (premixed) millbase passes through a narrow gap between a truncated cone (rotor) revolving at high speed in a closely fitting stator. They are not generally used in large volume paint manufacture nowadays, although they are still commercially available and are used in the food and pharmaceutical industries. Figure 19 shows schematically how the mill operates in a vertical configuration; horizontal colloid mills are also available.

8. High Speed Impingement Mills

High speed impingement mills (or kinetic dispersion mills, e.g., Kady mill) are not used (generally) in the paint industry. They require a very low viscosity millbase to operate by an impaction mechanism, which in turn requires a low pigment concentration with a high solvent content, thereby severely limiting paint formulation. A comparison of their dispersion properties in relation to other machinery as well as the optimum operating conditions is given by Cremer [39]. Although the configuration of the mill disk described by Kew [40] (see also Patton [28], p. 498) is reminiscent of a Silverson disperser, it operates at approximately twice the tip speed of Silverson rotor blades (shown in Fig. 17).

9. Ultrasonics

Although studies have been carried out to explore the possibilities of introducing ultrasonics into paint manufacture [41], the author is unaware of any use of ultrasonics per se to produce pigment dispersions in paint manu-

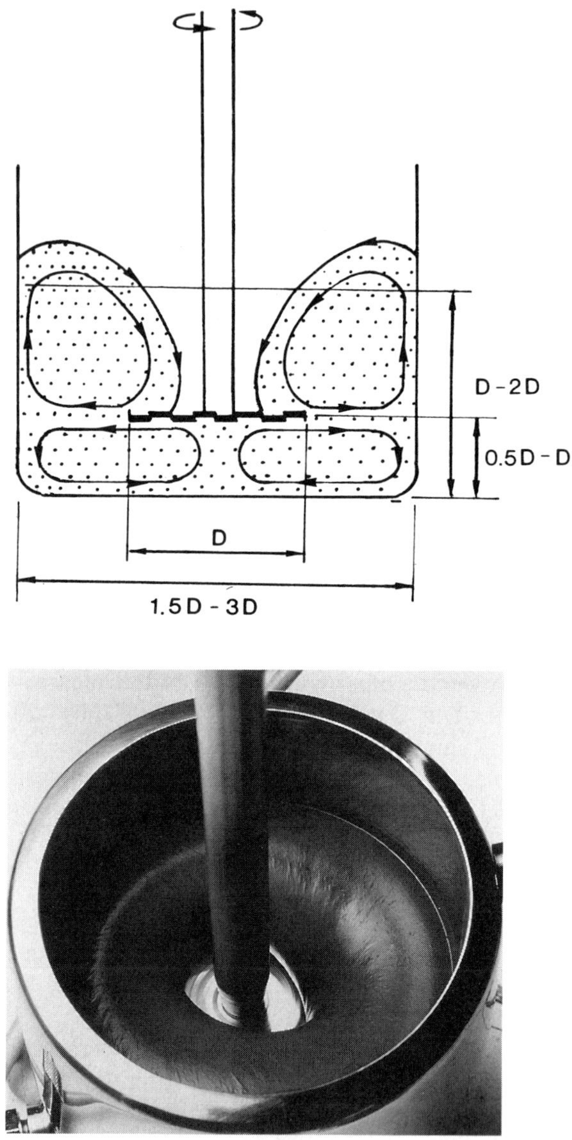

(a)

(b)

FIG. 18 (a) High speed disperser showing correct positioning of impeller for optimum dispersion conditions; dimensions given in terms of impeller disk diameter *D*. (b) An impeller disk in action at optimum dispersion conditions. Note the "rolling doughnut" flow. (Photo courtesy of VMA-Getzmann (GmbH.)

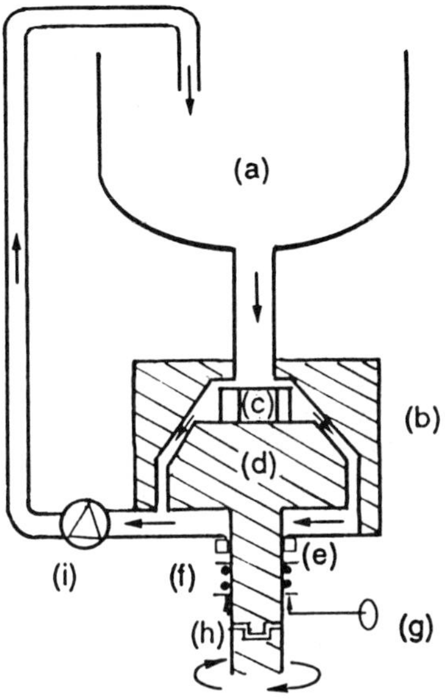

FIG. 19 Schematic drawing of a colloid mill. (a) Container; (b) water-jacketed stator; (c) turbine blades (channel material downward); (d) rotor; (e) seal; (f) thrust bearing, maintains constant gap; (g) worm gear for gap adjustment; (h) coupling to motor shaft; (i) pump.

facture. This is due to the low continuous-phase viscosity (and beneficial high surface tension) requirements in the use of ultrasonics, which are suitable for aqueous systems. Paint making, however, requires versatility, which will include the use of viscous nonaqueous polymer solutions with low surface tensions (not to mention vapor pressure); these will hinder the cavitation effect necessary to produce pigment dispersion by ultrasonics. However, ultrasonics through the use of either an ultrasonic bath or probe is invaluable in producing pigment dispersion in very small scale laboratory experiments and low viscosity conditions. High speed rotor/impeller dispersers with auxiliary ceramic collars to produce ultrasound are claimed by their manufacturers to be effective pigment dispersers. In the author's experience they have been found unsatisfactory as pigment dispersers, although they are excellent liquid emulsifiers.

B. Millbase Formulation and Let-Down

Millbase formulation is based on determining the dispersion conditions required to process the maximum amount of pigment in a mill to produce a "good" dispersion—the smallest particle size obtainable with the given pigment in the most stable colloidal dispersion possible. This involves the selection of the grind resin (if more than one resin is used in making the paint), its concentration (vehicle solids content), the choice of solvent, and the incorporation of an additional dispersant if required.

To achieve pigment dispersion, a number of different physical chemistry and chemical engineering processes have to be taken into account. The wetting of the pigment surface (and its nature) by the continuous phase, which will entail the displacement of air from the pigment surface; the penetration of pigment aggregates by the liquid and their break-up by mechanical forces; the adsorption of the wetting agent, perhaps with subsequent displacement of the wetting agent by the colloid stabilizer and re-equilibration. It is not germane to discuss these topics in depth in this chapter, and the reader is directed elsewhere [42]. It is important to note, however, that a good wetting agent is not necessarily a good colloid stabilizer (and vice versa) and the polymer adsorption from solution, unlike gas adsorption, can also be athermal and even endothermic. Therefore calorimetry is a dangerous guide in selecting a polymer for dispersion properties. The choice of resin for pigment dispersion purposes is empirical, guided by an understanding of colloid stabilization principles.

The selection of the dispersing resin is frequently based on the Daniel flow point method [43]. This consists of titrating a known amount of pigment with a resin solution, using a burette for incremental addition while mulling the pigment vigorously with a palette knife or spatula. The minimum volume of resin solution to produce a soft coherent mass defines the "wet point," and the further (minimum) addition of resin solution to produce flow (or the falling off) from a vertical spatula blade determines the "flow point." The process is repeated with various resin solution concentrations, usually between 15% and 45% vehicle solids content, to select the vehicle solids content (VS) which maximizes the pigment concentration. On examining the choice of resins (and VS) for dispersion, selection is based on the combination that incorporates the maximum amount of pigment used. An indication of the dispersion properties of the resin is also given by the separation of the wet point from the flow point. The greater the difference, the poorer the dispersion properties of the resin solution. The pigment and resin solution composition selected is then proportionally adjusted to comply with scale considerations of the grinding mill to be used. For example, Tatman [44] proposed the following relationship, using Daniel's wet point (DWP), to formulate a millbase composition:

$$W_v/W_p = (0.9 + 0.0069S + 2.5\eta)\mathrm{DWP}/100 \tag{4}$$

where W_v/W_p is the pigment-to-binder ratio (by weight), S is the vehicle solids (by weight), and η is the vehicle viscosity (mPa.s).

Tatman's empirical relationship is more pertinent than Guggenheim's equation (on which it is based) since the wet point is that of the resin and solvent being used, while Guggenheim's relationship is based on an arbitrary oil adsorption measurement, which may have no bearing on the materials in question.

Although the Daniel flow point is a simple titration process, it is subjective (i.e., operator-dependent). Hence there are various instrumental adaptations, such as the one described by Farkas [45] using the Brabender plastograph (essentially a torque-measuring pug mill), in which the maximum torque recorded corresponds to the wet point and the flow point is set at an arbitrarily subsequent low torque value, thus placing millbase formulation on a firmer footing.

The flow point (or wet point) method is only a guide to the choice of resin concentration (and type) required for dispersing a pigment. In practice, small-scale milling experiments must be carried out. It is worth noting that sometimes when a dispersion is being ground the millbase takes on a "floppy" viscous consistency (a highly flocculated dispersion). The natural tendency is to add more solvent to reduce the apparent viscosity, because calculations based on the quantities of polymer and pigment used (dependent on the pigment surface area of around 10 m^2/g, varying with particle size) and polymer molecule cross-sectional area (about 0.5–1 m^2/mg [46]) would indicate that there is a manyfold excess of resin molecules to satisfy adsorption considerations. Usually, it is better to add resin and not solvent, for while the calculations may be correct, they presuppose that the resin comprises a homogeneous species of molecules. This may not be the case, for the preferred dispersing species within the resin composition may be there only in a very small concentration—as for example the phthalic glyceride fatty half-ester present in alkyds [47]. An alternative possibility is that the pigment has a very low surface energy and does not adsorb polymer chains onto its surface to produce adequate "steric stabilization" (i.e., polymeric stabilization) but functions by a "depletion stabilization" mechanism [48] instead. Both of these possibilities favor polymer addition instead of solvent addition to enhance deflocculation.

When formulating millbase compositions it is important to remember that if the resin concentration is very low, pigment flocculation on let-down (addition of more resin or solvent) may occur; this is known as colloidal shock. While there are general recommended procedures to minimize colloidal shock [49], a continuous phase with a higher solids resin content is less prone to shock than one with a low concentration; hence there is a general preference for compositions with higher vehicle solids. In 1961, before steric stabilization was understood, Daniel published a series of recommendations [50] on how to avoid colloidal shock (pigment flocculation). They were the following.

1. Avoid adding solvent or resin directly; premix them before adding, and preferably use the same VS concentration as in the millbase for the first aliquot addition.
2. Never dump the bulk of the let-down vehicle into the millbase; instead, dilute the millbase slowly and under agitation all the time. If this is not possible (as in a ball mill), add the resin in small increments with mixing in between.
3. Avoid the addition of very cold and/or highly viscous vehicles to warm millbases.
4. In paints with more than one solvent, consider which solvent should go where. The weaker solvents should be used in the millbase that will have a large amount of vehicle added.
5. Avoid the addition of gel-like substances (thixotropes) to thin bases. Reduce their gel strength prior to their incorporation into the base by diluting them with vehicle.
6. Minimize slow laminar flow; use high shear postmixing as much as possible.

To this might be added: When mixing, always dilute the millbase (under agitation) with the vehicle, and not vice versa.

When one considers the criteria for good steric stabilization in terms of polymer solvation, adsorption, and anchoring to the pigment surface, the above recommendations make sense. They are designed to avoid collapsing the stabilizing barriers around the pigment particles by osmosis of the solvent into the more concentrated resin being added. The main criterion in steric (polymer) stabilization is the free energy of dilution, which produces the osmotic pressure generated in the steric barriers when the barriers start to overlap. This in turn is dependent on the polymer surface concentration and configuration, and not just steric barrier size [51], as well as on the polymer–solvent interaction (Flory's chi parameter).

Attempts to specify the size of the steric barrier surrounding particles to produce dispersion stability can be misleading [52]. For example, if the best available value for the effective Hamakar constant for TiO_2 in aliphatic hydrocarbon, $A_{12} = 5 \times 10^{-20}$ J [53] were to be used as presented in Ref. 52, then one would obtain a value of 51 nm for the steric barrier thickness; that is, when two adsorbed polymer sheaths overlap, the free energy of dilution would keep the particles separated by at least twice 51 nm. This in turn must mean that the adsorbed polymer must extend more than 51 nm from the surface, and whether in chains (or loops) its size in terms of polymer molecular weight would suggest at least many (tens of) thousands of daltons. Simple practice is not in accord with this; small oligomeric chains have been found to be effective stabilizers [47,51,54]. Making the

Hamakar constant value one order of magnitude lower, the steric barrier size requirement is reduced by a tenth, bringing it more in line with observed dispersion stability (at the expense of the "best" Hamakar value). This can then be justified by introducing London van der Waals "retardation" effects [55] or other modifying effects such as the "Vold" effect as suggested by Ruckstein [54] (which in itself is flawed and corrected by Osmond et al. [56]). To summarize, it is unwise to use the Hamakar values literally, because there is uncertainty in the values for the Hamakar constant due to interpolation sensitivity. For a fuller discussion of the Hamakar constant on colloidal stability and on the failure of charge stabilization in organic media of low polarity, the reader is referred elsewhere [57].

C. Assessment of Fineness of Grind

There are a number of ways to assess the fineness of grind of a pigment dispersion. The simplest quick test for assessing this is with a Hegman gauge [58] or Grindometer. This consists of a stainless steel block (about $170 \times 65 \times 15$ mm) that has a wedge-shaped channel about 12.5 mm wide and 127 mm long cut into the flat face so that the maximum depth is 100 μm, reducing to zero at the other end, with depth marks evenly spaced beside the channel. A sample of millbase is placed at the deep end, and by means of a steel scraper the dispersion is quickly spread down the channel. Assessing the depth at which a "pepperyness" first becomes visible when the liquid millbase is viewed obliquely gives an indication of the residual aggregate size. The gauge leaves much to be desired. It can only be used with dispersions that produce glossy wet films and that do not dry during the measurement, thus making aqueous dispersions and other volatile component systems very difficult to assess. Needless to say it only gives an indication of the residual large particle sizes, when frequently the interest may be in the fines. However, with experience it is a very useful tool as a guide to dispersion. Millbase specifications are frequently defined in Hegman gauge terms; for example, the dispersion must be "off-scale" for an oil-based gloss paint.

The advantage of the Hegman gauge is that it takes only a matter of seconds to make an assessment, which is invaluable during pigment grinding. A better assessment can be made by direct observation of the fineness of grind, by viewing the dispersion as a thin film compressed between a cover slip and a glass slide, using an optical microscope. Disk centrifugation of diluted paint samples [59] is another method that has been applied to determine the particle size distribution after grinding of color pigments and TiO_2.

Tinting strength is an alternative way of assessing pigment dispersion, albeit indirectly [60]. The basis is that the tinctorial strength of a pigment dispersion increases as the dispersion becomes finer. Hence an assessment can be made by comparing the color obtained when a fixed small amount of a colored tinter is added to a fixed amount of a white dispersion (being ground). If colored pigment is being ground, then a white base can be used as the tinter. The tinting strength, as a measure of dispersion, can be evaluated instrumentally, but this takes time. The evaluation can be simplified and speeded up on a qualitative (relative) basis as, for example, when two samples of the millbase are being compared during a grind, by adding identical portions of the tinting solution to the two samples of the millbase. The two tinted samples of paint produced may appear similar in color in their separate containers. However, placing the two paints on a glass panel (adjacent to each other) and spreading them down the panel side by side, touching but not mixing, enables the eye to compare them discerningly when viewed upside-down through the glass. Differences in tinting strength are then readily apparent, but not if the two colors are separated even by a space of 1 mm, since the human eye is much better at making relative color judgments than it is at making absolute ones.

D. Mixing

Once the pigment dispersion has been prepared by grinding and let-down, it is then tinted to the required color. At this stage driers and other additives may be added, such as methyl ethyl ketoxime and dipentine (antiskinning agents in oil-based paints) or antisettling agents and silicones (flow additives).

It is not unusual to have about 10 "color bases" that can be blended in varying amounts to give a nearly infinite array of colors. Sometimes a number of pigments are ground together in a resin solution to give the color and are then fine-tinted to the required hue. Generally it is not advisable to try to obtain the exact color with one pigment, since there may be batch-to-batch variations. A colored paint usually includes both white and black pigments so that its tone can be altered by varying these ratios, as well as two colored pigments (preferably in adjacent "color space") so that its hue can be adjusted. The exact color can be represented within a visually uniform color space by the coordinates of a solid ellipsoid, using the "Adams coordinate" system. In this system the long axis represents the extremes of black and white at the poles, and the short axes the complementary colors, blue-green and red-yellow, thus enabling calculation of the required amount of tinter to add to produce the desired color. This is achieved by using the semiempirical theory based on light scattering and absorption properties known as the Kebulka-Munk theory, and its extension by Duncan to the additivity of Kebulka-Munk absorption and scattering coefficients and their relationship to pigment concentration. A readable, simplified outline of this concept is given by Bullett [61], and a more detailed presentation is given by Judd and Wyszecki [62].

Whereas a colored paint is usually a blend of different pigment dispersions, a pure "white" paint only contains a light-scattering pigment as there must be no light absorption from the visible spectrum. The light-scattering properties of a pigment are dependent on the refractive index difference between the pigment and its environment (i.e., the binder) as well as its particle size. Rutile TiO_2 is the pigment normally used in white paints because of its high refractive index (2.71) and because it does not absorb light in the visible spectrum (off-whites have traces of light-absorbing pigments added). It is produced to have an average particle size of about one-half the wavelength of green light (middle of the visible spectrum) to maximize light scatter. If the average particle size is smaller, then the white takes on a blue tone; if larger, then a yellowish-brown tone; therefore an increase in effective particle size by flocculation of the dispersion will affect the color of the paint.

Other resins may also be added to the mix at this stage to give thixotropy, such as alkyd-amide resins or bentones (organically treated montomorillonite clays). Resins such as oil-modified polyurethanes might also be added to oil-based paints to give faster drying times or harder films when dry, or nitrogen resins to thermosetting compositions. This mixing of pigment dispersions and resins may lead to color problems related to changes in pigment dispersion stability as re-equilibration of adsorption by the pigment takes place.

V. PROPERTIES OF PAINTS

A. Color and the State of Pigment Dispersion

The color of a paint depends on the scattering of light as well as light absorption. Both of these properties are dependent in different ways on particle size. The scattering of light by a pigment is dependent, inter alia, on pigment particle size, a, where there is an optimum size for maximum scattering properties. On either side of this value there is a decrease in scattering properties. Light absorption, on the other hand, is inversely proportional to particle size. This is because the cross section of a particle is proportional to a^2 and the number of particles per unit volume is proportional to a^{-3}, hence the total cross section offered to light is a^{-1}. The state of pigment dispersion affects the particle size, for while the pigment may be ground to a specific size, any flocculation will give a larger "apparent" size, which will change the optical properties.

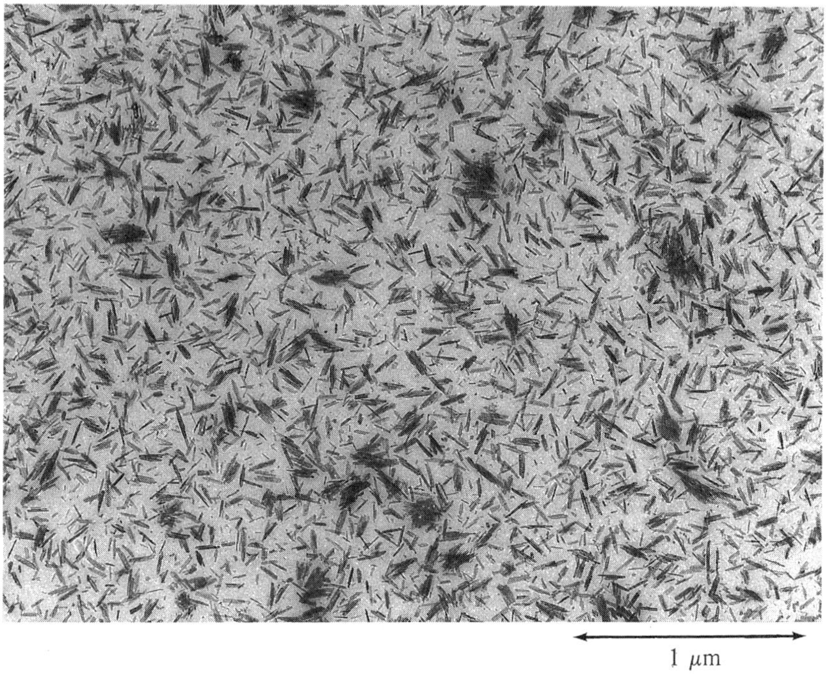

FIG. 20 Micrograph of a good dispersion of "transparent" red iron oxide; bar indicates 1 μm. (Courtesy of ICI Paints.)

For example, if a "transparent" red iron oxide pigment is not fully dispersed and deflocculated in a dry film [e.g., as used in automotive finishes based on poly(methyl methacrylate) polymer] it will produce a red film that is hazy and partially transparent instead of clear, bright, and transparent; that is, it has a light-scattering component instead of being solely light-absorptive. Although the particles are too small to scatter light per se (see Fig. 20), even though their refractive index (2.9) is comparable to that of rutile TiO_2, their flocculation forms larger particles that are capable of producing some light scatter, hence the haze and loss of clarity.

The simplest way to gain an understanding of the effect of pigment flocculation on optical properties is to examine an idealized arrangement of purely light scattering centers. This optimum can be represented by a regular array of uniform particles whose diameter is one-half the wavelength of light (0.25 μm for a material such as TiO_2) and spaced by a particle diameter—effectively giving a volume concentration of about 9%. If all the particles flocculated to form the same array, but of doublets as shown in Fig. 21, then there would be a halving of the number of light-scattering centers, hence a 50% reduction in light-scattering properties (i.e., in opacity) for a paint whose opacity is derived solely by the uniform scattering of light. Furthermore, the light-scattering centers would now be larger than the optimum size and there would be a further reduction due to this effect. However, this is a very simplistic view, and the real loss in scattering efficiency due to pigment flocculation is variable. While it is accepted that it is substantial in highly flocculated systems, the extent of loss is open to debate. In a white paint the TiO_2 particle is not of a single uniform size, nor does it flocculate uniformly into doublets; the pigment volume concentration is necessarily higher (>9%) to increase total light scatter, as opposed to scattering efficiency; the refractive index between the scattering centers and the continuum is also a function of the TiO_2 concentration, making the theoretical quantification of loss of pigment effectiveness difficult.

The changes in optical properties due to changes in the state of pigment dispersion brought about by flocculation can also be seen in practice as a color drift problem upon storage of a paint. This frequently occurs when

(a)

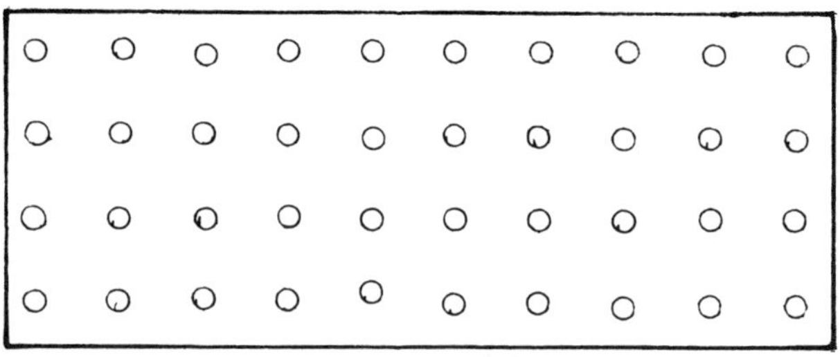

(b)

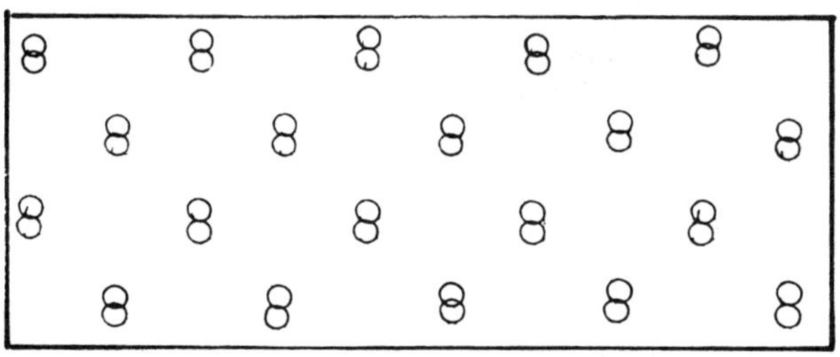

FIG. 21 Schematic representation of the effect of flocculation on particle distribution in a film. (a) Deflocculated; (b) flocculation into doublets.

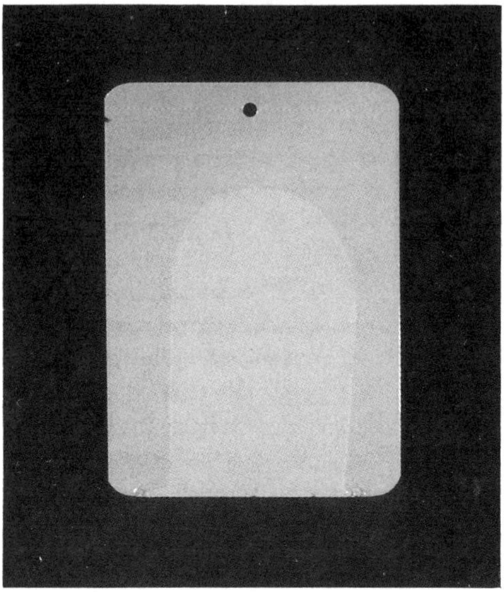

FIG. 22 Spray and pour applications of the same paint showing difference in color indicating pigment flocculation.

different tinters are mixed to form a "color," which "drifts" with time so that the color is lighter (or darker) on sample cards prepared at different times with the same paint. An indication of what is happening can be obtained by spreading the paint on a panel, then rubbing a portion of the panel until the paint is dry, and comparing the color of the rubbed portion with that of the undisturbed portion. The onset of very high viscosity conditions when the paint is nearly dry effectively sets the pigment in its deflocculated state produced by the mechanical deflocculating effect of rubbing. This color change produced by rubbing is known as flooding, the color of the rubbed portion indicating which pigment is flocculating. A variation of this test is to compare the colors developed by sprayed and poured samples of paint on the same panel, as shown in Fig. 22. The high shearing and solvent loss in spraying deflocculate and set, respectively. The color drift problem can frequently be overcome simply by passing the finished paint through a bead mill, where the high shear generated effectively accelerates [63] the re-equilibration of the adsorbing species on the pigment surface and stabilizes the changing state of dispersion and hence color.

It is interesting to note that color change due to flocculation does not appear to exist in artists' oil paints. These are made so that the pigment loading is at the critical volume concentration, where all the pigment particles are already in contact with each other. Hence it is immaterial if the particles touch because of attraction forces or they are forced into contact because of a shortage of continuous phase.

Another problem associated with colloidal properties is known as "floating," which occurs when two pigments behave independently of each other (have different mobilities and hence are usually of very different particle sizes). The finer pigment may rise to the surface (float), in convection currents created by solvent evaporation, disproportionately to its concentration, producing a surface color different from that of the bulk. Alternatively, this defect may impart a mottled appearance by the formation of hexagonal colored cells, which can vary in size and are known as Benard cells (see Fig. 23). These vortex cell patterns were first quantitatively studied by Benard [64] although they were first commented on in the seventeenth

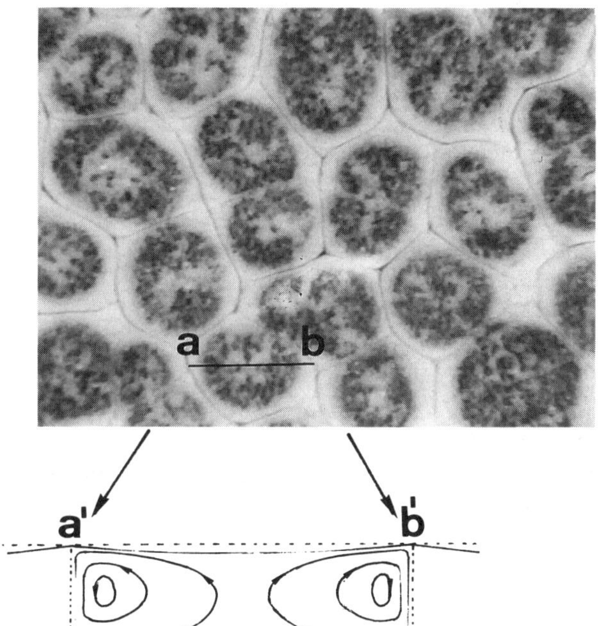

FIG. 23 Schematic cross-sectional view ($a'b'$) of section ab (=0.5 mm) in micrograph, showing convection flow lines that led to pigment separation in formation of Benard cell pattern.

century [65]. Bell [66] probably coined the term to describe the "uneven flotation" created by vortex patterns set up by convection currents. Their formation is shown schematically in Fig. 23. This dispersion (color) problem can be overcome by coflocculating the two pigments by means of a suitable additive, thereby suppressing the differential mobility.

B. Pigment Flocculation and Other Properties

The effect of pigment flocculation on the optical properties of pigment dispersions is generally undesirable except in the case of floating. Flocculation, however, can have some advantages in relation to other properties, for example, resuspension of sediments and antisag properties. Pigment particles settle out of suspension at a rate dependent on their size and density and according to Stokes' law. If well deflocculated they form a hard, compact sediment that is difficult to resuspend. (Deflocculated particles give observable sedimentation only if they are sufficiently large and dense to settle at a rate greater than about 10^{-5} mm/s, to counteract diffusion and convection effects.) If flocculation is present, on the other hand pigment sedimentation is rapid, but the deposited material is soft, loosely packed, and readily resuspended by stirring. Furthermore, the viscosity of a dispersion of noninteracting (i.e., deflocculated) particles is Newtonian and dependent upon only the disperse phase volume and the continuous-phase viscosity, up to moderate dispersion concentrations.

If flocculation is present, the interaction of the particles induces a very high viscosity at low shear rates. This confers desirable antisag properties when paint is applied onto a vertical surface. Yet the high shear rates generated during brushing break down the interactions to give a low viscosity, which enables ease of application.

To appreciate the effects of flocculation it is necessary to have some idea of the rate of flocculation and perhaps its strength, at least whether it is strong or weak and whether strength should refer to the individual "pair-potential" attraction or to the system as a whole.

Smoluchowski [67] showed that the rate of decrease in the number of particles undergoing irreversible flocculation, in the absence of interparticle interactions, could be expressed as

$$1/N = 1/N_0 + Kt \tag{5}$$

where N_0 is the initial number of particles, N the number at time t, and K a rate constant. Therefore $2N = N_0$ at time $t_{1/2}$ defines a half-life to be

$$t_{1/2} = 1/KN_0 \tag{6}$$

If the rate constant is diffusion-controlled, then $K = K_D = 4k\mathrm{T}/\eta$ and

$$t_{1/2} = \eta/4kTN \tag{7}$$

where k is the Boltzmann constant; T is temperature (K), and η is the viscosity of the medium.

Table 6 shows the effect of particle size, concentration, and continuous-phase viscosity on the rate of pigment flocculation. Despite the likelihood that the exact rate of flocculation may be somewhat different when hydrodynamic interactions are taken into account and that all collisions may not result in capture, Smoluchowski's theory nevertheless gives a good indication of what is happening and has been well confirmed by experiment [68,69].

The above calculations show how fast flocculation of a pigment, such as TiO_2, can take place in a typical paint. Furthermore, when flocculation is due to London van der Waals forces (secondary minimum flocculation), the attractive energy (V_A) is inversely proportional to size:

$$V_A = -Aa/12d \qquad \text{if } d < a \tag{8}$$

where A is the effective Hamakar constant, a the radius of the sphere, and d the interparticle distance. However, since the number of particles in a given volume is proportional to a^{-3}, the total attraction energy is proportional to a^{-2}. Hence a fumed silica (particle size about 0.014 μm) is used as an additive to give "false body" in preference to a larger particle size material (even though the larger particles may have a stronger pair-potential attraction). Only a small amount is required (a few percent) to produce the desired effect, as in clear, thick polyester wood finishes (or in black piano finishes).

While the use of flocculation by fine particle size extenders to produce high viscosity at low rates of shear produces good antisag properties, the high speed of setup by the same token prevents the flow-out of brushmarks. Therefore it is preferable to introduce thixotropy with a slower time re-

TABLE 6 Half-Life of Particles Undergoing Rapid Flocculation as a Function of Size and Concentration in a Medium of 0.1 Pa.s Viscosity at 25°C

Half-life (s)	% Disperse phase vol. (0.2 μm diam.)	Half-life ($\times 10^{-6}$ s)	% Disperse phase vol. (0.014 μm diam.)
7.6	1	2400	1
1.5	5	480	5
0.76	10	240	10
0.5	15	160	15
0.38	20	120	20

sponse, in brushing finishes, by means of a different mechanism such as hydrogen bonding in thixotropic resins (amide-modified alkyds).

Pigment flocculation in a wet film produces a coral-like structure of particles within the fluid paint. As the paint shrinks on drying (due to solvent loss) its surface collapses onto the skeletal structure of particles, producing an unevenness in the surface of the dry film. This leads to a loss in gloss [70] as well as giving an increased "coin marring" property because more of the pigment protrudes through the surface of the paint [71], making it more abrasive. The flocculation of the pigment also leads to a loss in the durability of the paint film [72].

When the overall effect of pigment flocculation in a paint is assessed, it is seen to be undesirable, since most of its beneficial properties can be obtained by other means, leaving only its detrimental effects.

C. Assessment of Pigment Flocculation

Several methods of assessing flocculation have already been described, all of them qualitative. Direct quantitative measurement such as obtaining the number of particles making up a floc (cf. micelle number) is rare but has been accomplished using proton straggling [73]. Visual counting methods using electron micrographs are too subjective, although qualitative visual assessment using an optical microscope and observing the presence or absence of Brownian motion is effective (and of course allows observation of cluster formation). Image analysis techniques have been claimed to overcome subjectivity and used to quantify pigment dispersion [74]. Usually, however, flocculation is measured indirectly and related to some property such as light scattering by TiO_2, for example, in the IR region (to emphasize scattering by clusters), as in the "flocculation gradient" method [75]. The larger the gradient in a plot of measured backscatter versus film thickness, the greater the degree of flocculation of *similar* films as shown in Fig. 24.

The deviation from Newtonian viscosity is another effective approach in the assessment of flocculation in the fluid state. For if the continuous phase is Newtonian, then any deviation from this must be due to flocculation. Hence comparison of the viscosity at infinite shear rate (where flocculation effects have been eliminated) to that at finite rates of shear enables determination of a shear rate gradient as a measure of flocculation [76].

There are also a number of simple methods for assessing the state of dispersion in liquid paints, including the blot test shown in Fig. 25 in which the paint is first diluted with solvent (as in a sedimentation test) and then spotted on filter paper [77]. The larger the "halo," the poorer the dispersion, which can be recorded in an analogous way to R_f values in chromatography. An alternative to this method, described by Smith and Pemberton

[78], is claimed to give better results. Another simple test for TiO_2 dispersions is to view the element of a clear light bulb, for example, a torch bulb, through a thin film of the dispersion. If the incandescent element has a red tinge, then the dispersion is deflocculated; if blue, then it is flocculated (cf. Zigmondy's red/blue assessment of the "gold number" [79]).

However, whereas a measure of flocculation/deflocculation may be obtained in the liquid state, this does not necessarily mean that the same measure pertains to the dry state. For while it may seem obvious that a flocculated dispersion in the liquid state will dry in that condition, a deflocculated dispersion, on the other hand, can flocculate on drying. Therefore extrapolation from the liquid to the dry state requires caution. Relative sag ratings of a paint [47] overcome this problem because they can give a comparative assessment of flocculation as a paint dries.

In summary, the state of pigment dispersion/flocculation, while readily recognized and qualitatively assessed [80], is still very difficult to quantify. It involves particle packing, which depends on particle mobility and concentration, and has important consequences on multiple light-scattering and/or absorption effects that are not readily quantified in real systems. This begs the question as to whether, apart from rheological considerations, flocculation is important at critical packing.

D. Particle Packing

The packing of contiguous solid particles is important in determining the properties of a paint film. For noninteracting identical spheres the maximum random packing density is 64% by volume, obtained by vibrating a bed of particles [81]. Loose random packing is about 52%, while hexagonal close packing yields 74%. If large and small particles are mixed, then it is possible to obtain even higher volume fractions because the small particles fill the interstices between the larger particles. When the voids between the closely packed particles are just filled by resin with no excess, the condition is known as the critical packing volume concentration (CPVC); its value is variable and dependent on the materials used. For example, a resin in dispersion form (a latex) will not be able to penetrate the voids in a hypothetical CPVC array of particles as effectively as a solution polymer; hence latex paints tend to have a lower CPVC. Even if a solution polymer is considered, a certain amount of polymer is immobilized on the surfaces of the particles; although particle packing volume is size-independent, the effect of the adsorbed layer of polymer will exert an effect on the CPVC. The adsorption of polymer, especially by the smaller particles, will extend their effective radius, thus increasing the effective disperse phase volume concentration and forcing a lower CPVC. Furthermore, if the particles are

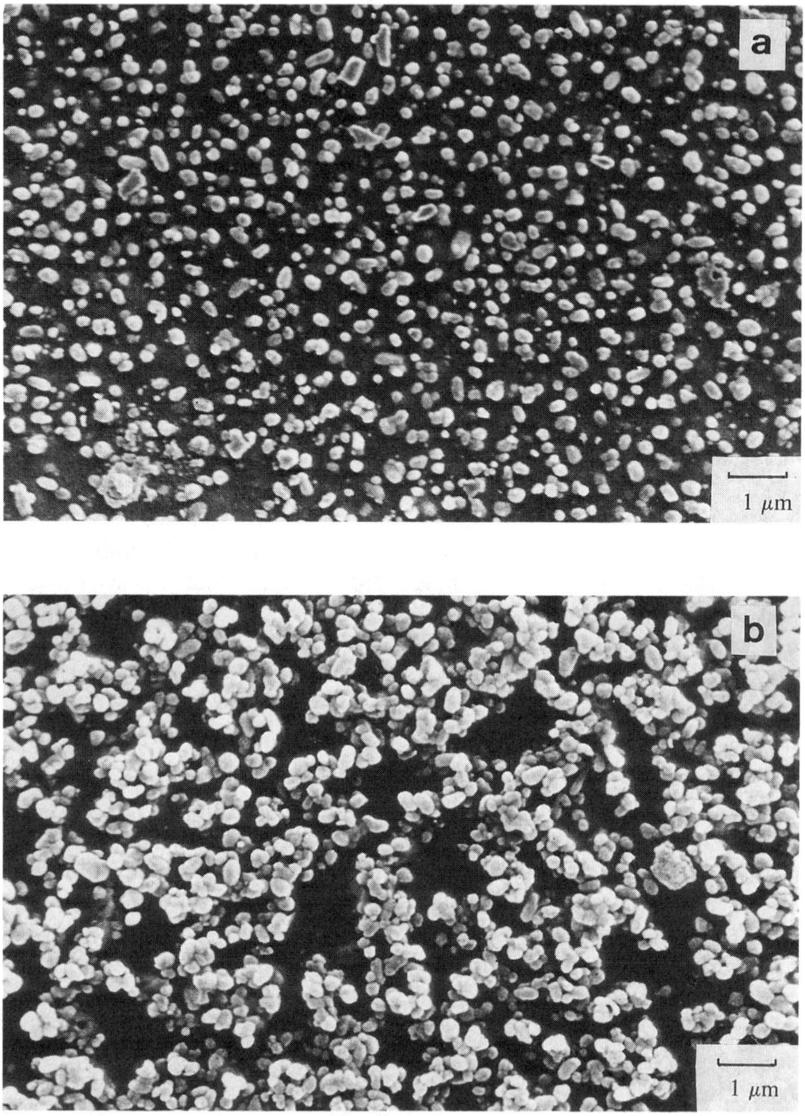

FIG. 24 Electron micrographs of a gloss paint after oxygen plasma etching showing TiO_2 pigment with different degrees of pigment flocculation. (a) dispersed, floc. grad. = 0.25; (b) flocculated, floc. grad. = 0.76. (Courtesy of Tioxide Ltd.)

aggregated, the internal volume of the aggregate will take up binder without increasing particle separation, and hence the CPVC is further reduced.

The properties of a dried paint film hinge on the CPVC. If the system is above the CPVC, then voids occur in the film, and these will act as light-scattering centers, thereby increasing the total scattering properties of the paint film. However, the porosity is also increased, an effect that will lead to increased water transmission, and this can decrease the corrosion-preventive properties of the paint. The increased permeability will also reduce the blister potential of the paint on a wooden substrate by allowing the wood to "breathe" more freely. It will, however, make the film less resistant to staining by allowing the ingress of dirt as well as producing a loss in the adhesion to the substrate and a decrease in the mechanical properties through film embrittlement. Hence the importance of the CPVC, which is a function of the pigments used, the state of pigment dispersion, the binder, and other factors, and why so much is published on its determination [82]. A useful concept pioneered by Bierwagen and Hay [83] is the reduced PVC (Λ), defined as Λ = PVC/CPVC. They showed that Λ is a parameter that controls many film properties and that coatings for particular applications are most usefully formulated to have characteristic values of Λ.

E. Flow Properties

The packing of spheres as discussed in terms of CPVC is not only important in dry films, but also extends to liquid conditions in viscosity considerations as discussed by Lee [84] in the context of the crowding factor in the Mooney equation relating dispersion viscosity with disperse phase volume [85]. There are over 400 published equations relating dispersion viscosity to disperse-phase volume and continuous-phase viscosity in one form or another. This illustrates the difficulty of the task of defining the flow properties of dispersions of particles such as pigments, which are not necessarily spherical or noninteracting. Furthermore, moderately concentrated pigment dispersions (such as solvent-based paints) may not only be viscous, but can also have an elastic component due to pigment flocculation [86,87] and/or polymer chain interactions. Rheological measurements are, however, very useful in determining the effectiveness of dispersants and dispersing machinery [51,88,89].

While it is possible to observe and perhaps even measure these effects, there is no theoretical approach available to predict reliably from first principles the rheological behavior of dispersions such as those pertaining to a practical paint. It is essential to know the rheological requirements of a paint to produce a useful product and understand what affects its rheology.

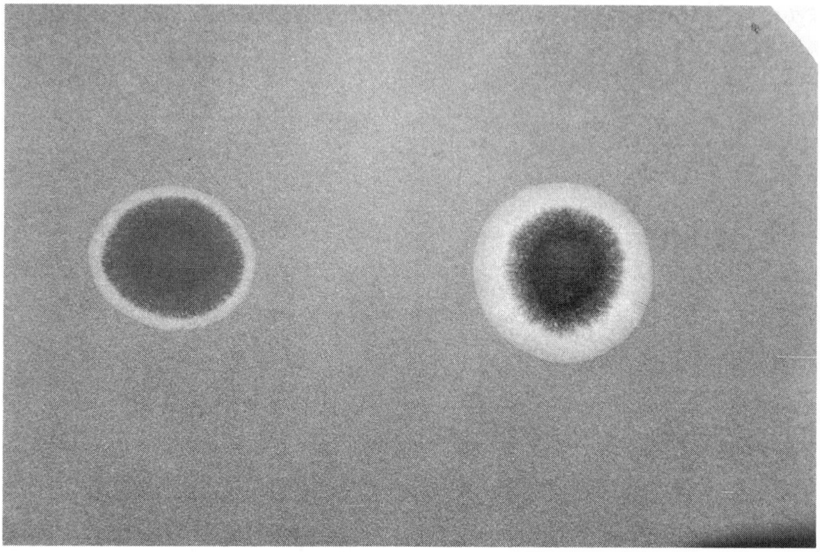

FIG. 25 Blot test of pigment dispersion on filter paper. The larger "halo" indicates the greater degree of flocculation.

Consider a simple paint that is going to be applied by a brush, for example. The movement of a hand-held brush over a flat surface is at about 10 cm/s. If the thickness of the paint under the brush is 10 μm, the velocity gradient (shear rate) is 10,000 s^{-1}. Then if the paint has a viscosity, defined as the ratio of shear stress to shear rate, of 0.1 Pa.s (1 poise) it will produce a shear stress of 10^3 Pa. If a "4-in." brush were used to apply the paint and the area under the brush was 40 cm^2, it would require a force of 4 N (equivalent to a weight of 400 g or nearly 1 lb!) to spread the paint—an arm-aching task when painting all day. Hence the need for a low viscosity paint (the lower the better) when high shear rates are employed in its application onto a surface. There is a limit to how low the viscosity can be made because it is a function of the product of the continuous-phase viscosity and disperse-phase volume. The continuous-phase viscosity in turn is governed by the soluble polymer concentration, its molecular weight, and solvent "quality." The disperse phase is dependent on the hydrodynamic volume (a function of particle size, shape, and adsorbed polymer [90]) even when flocculation effects are ignored, and it has to be moderately concentrated for opacifying reasons. However, if the paint has a low viscosity, only very thin films can be applied to a vertical surface, as the film will tend to sag. The brush will be difficult to load, and the paint will drip from the brush, making painting a messy undertaking.

The necessity of a low continuous-phase viscosity immediately brings about a constraint on the polymer molecular weights of the resins used, which have to be low in comparison to those used in the plastics industry. Since the viscosity of a polymer solution is dependent on the weight-average molecular weight, there should be no high molecular tail in the distribution, as this would have a disproportionate effect on solution viscosity. Furthermore, the presence of dissolved polymer of very high molecular weight can produce viscoelastic properties that will prevent atomization if spray-applied and cause the formation of "cobwebs" [91] or lead to a spatter instead of a fine mist as discussed by Waite et al. [92]. The low viscosity consideration also has an important bearing on solvent choice. Polymers at low concentration have their chains expanded to the maximum in a good solvent and increase the viscosity of the solution to a greater extent than in a poor solvent. However, at high polymer concentrations the solution viscosity is greatest in a poor solvent and lowest in a good solvent. Hence when paints are highly diluted, as in spray application, a nonsolvent mixed with a good solvent can be used to increase the viscosity after application (to prevent sagging) through differential evaporation rates. High molecular weight polymer, however, is desirable for durability properties and can be introduced in dispersion form without contributing greatly to the rheological properties, as in latex paints or NADs.

Very high viscosities, on the other hand, are desirable to stop paint sagging on vertical surfaces, to prevent paint dripping, to improve transfer of material onto a brush during dipping, and to prevent pigments from settling in containers during storage—conditions that are associated with very low rates of shear ($<1\ s^{-1}$). These disparate requirements can be met by a non-Newtonian system that has a low viscosity at high rates of shear and a high viscosity at low shear rates, that is, a shear thinning fluid. Shear thinning properties exist in pseudoplastic fluids, but the instantaneous viscosity recovery after high shear conditions will prevent low stress recovery, as required in the flow-out of brushmarks and the removal of other application irregularities produced in paint films. Thus it is important to have a time element in viscosity recovery after high shear, that is, thixotropy, to have sufficient time at low viscosity for brushmark flow-out but insufficient for sagging to occur. Pigment flocculation usually induces very rapid thixotropic recovery (toward pseudoplastic properties), yet although it is a step in the right direction it is not the best way of introducing desirable non-Newtonian flow properties to paint. The use of thixotropic resins and other proprietary materials with slower recovery times is better.

Liquid paint films (as opposed to bulk paint) are not static systems.

During and after application, there is a loss of solvent that markedly changes polymer solution behavior (as well as disperse-phase volume and interparticle interactions) and hence rheology. Even when a "touch dry" coated object is placed in an oven to be force dried at 120°C or undergo chemical reaction at 170°C, there are two major opposing effects: (1) the onset of flow due to the large temperature rise and (2) the setting of the film by the loss of residual solvents or by condensation of the polymers due to chemical reaction. There are many good analytical rheometers that can be used to study the various rheological (including elastic) properties of constant-composition paint. These instruments have well-defined shear rates and oscillatory frequencies with cone-and-plate or concentric cylinder geometry (e.g., the Weissenberg rheogoniometer or the Bohlin VOR rheometer) or controlled stress (e.g., the Carrimed CS rheometer). There are no equivalent instruments available to study the most important rheological changes taking place as a film dries either at room temperature or during stoving. There are a few instruments such as the TNO film viscometer [93] and the interference microscope [94] that relies on Orchard's [95] analysis of brushmark flow-out and can be used to study the changing viscosity in drying paint films. Neither of these methods is able to resolve the components of rheology and they present viscosity as a simple equivalent Newtonian viscosity.

The control of viscosity on the shop floor (by necessity) relies on simple instrumentation such as standard efflux cups used to control paint dilution with solvent prior to spray application. The structures of gel paint and solid emulsion paint are likewise determined with simple structure-measuring devices.

VI. PAINT APPLICATION

Prior to application, paint is usually diluted to the required viscosity for the intended mode of application as specified by the manufacturer. There are many different ways of applying a paint to a surface, ranging from simple spreading with a brush or roller to paint electrodeposition in dip tanks to control film thickness and minimize paint wastage.

Paint electrodeposition (sometimes misleadingly called electrophoretic painting) is unusual in terms of utilizing colloid dispersion properties as it relies on controlled flocculation of the dispersion onto an object, which is made an electrode in the process. Electrodeposition paints are usually pigmented emulsions of solution polymer that can be colloidally stabilized by anionic or cationic and even nonionic groups [96]. The coating process is dependent on the electrolysis of water, which creates an acidic boundary layer around the anode and an alkaline layer around the cathode. The emulsions used are designed to be readily dispersed in ion-free water and usually require the minimum of agitation to form stable dispersions, which are generally made by the phase-inversion process. If the emulsions are unstable to acidic conditions they produce an "anodic electrocoat," and if unstable to alkali conditions, a "cathodic electrocoat." When a dc electric current is passed, the emulsion particles in the boundary layer (pH gradient formed around the electrode on the electrolysis of water) flocculate or "electrocoagulate" and are deposited onto the surface of the electrode, thereby gradually insulating it. The insulation of the electrode by the coagulum ensures throw properties orders of magnitude greater than metal electrodeposition as well as a controlled thickness of the deposited layer. The deposited polymeric layer, having been coagulated, is no longer self-dispersible and can be rinsed with water sprayed at a pressure of 10–15 psi to remove and recover any uncoagulated material for reuse. A good description of the electrodeposition process is given by Beck [97].

Table 7 lists the various methods of applying a paint, with associated processes, and presents a brief summary of their origins.

VII. FUTURE TRENDS

Perusal of Table 7 shows that there is a gradual and continual change taking place in methods of applying paints. This evolutionary process is taking place not only in application processes but also in the very nature of paint and its properties. The changes that have occurred in the past are not necessarily readily detected, since some of them, such as the introduction of the electrodeposition process of primers, were only of benefit to the industrial consumer. Other changes, such as the introduction of acrylic finishes to replace nitrocellulose automotive paints in the late 1950s to early 1960s, were readily apparent to the end user in the form of glossier cars, and the introduction of "basecoat/clear" automotive finishes is a much more recent example. Similar changes have taken place in the decorative market. If one were to mention "distemper," the younger adult would probably think of some form of canine or feline disease and not the antecedent of today's latex paints—yet it was not until after World War II, that latex paints became available. Some paints are no longer made because of the toxicity of their components, even though they may have had some superior quality such as the metal-corrosion protection of red lead/linseed oil paints.

Health and safety considerations are much more important today. Toxicity, health and safety, and other environmental considerations are defining current trends. Lead driers in autoxidative systems have been replaced by zirconia driers. Solvents based on ethylene glycol derivatives (e.g., methyl cellosolve) are being replaced by similar solvents based on propylene glycol;

TABLE 7 Methods of Applying a Paint and Their Date of Introduction

Process	Origin	Reference
Brushing	Oldest known tool; Egyptians used reed brushes, Greeks used rabbit's foot or tail of small animal. 1437—mention of brush made from 1 lb of bristles bound to stick to whitewash castle. 1840—first record of flat "stucco" brush in England.	Soderberg [98]
Dipping	Origin unknown, mentioned as widely in use in 1905.	Sabin [99, p. 281]
Tumbling or barrel painting	Exact date not known. Method adapted from rotary churn of dairy industry. Widely used in 1920s. First mention of barrel plating in 1911.	Wampler [100]
Hand roller	Patented in Great Britain in 1882, but came into general use only after World War II via United States.	Chatfield [101]
Spray gun (air)	Spray equipment mentioned in 1905 as being operational for 10 years in United States, principally for painting freight cars (two men could spray paint a freight car in 3–4 min). 1907—spray gun used in furniture industry. 1922—Adopted for automotive industry on Oakland model, superseding dipping as chief method of application.	Sabin [99, p. 253]; von Fischer [102, p. 446]
Silk screen process	Originally used 2000 years ago in Japan. Introduced to France (Lyon) in 1850, but press for carbon tissue silk screen printing introduced 1921.	MacKenzie [103]

TABLE 7 Continued

Process	Origin	Reference
Flow coating	One of the oldest production application methods. Original Model T Ford and Dodge were coated by low pressure method without automatic controls of today. Eclipsed by spray gun and introduction of rapid-drying nitrocellulose lacquers.	Tatton and Drew [104]
Infrared oven	ca. 1938—introduced by Ford Motors at Baton Rouge plant, reducing stoving time from 1 hr to 10 min.	von Fischer [102, p. 444]
Coil coating	1935—Used in USA for prepainting of aluminum coil for Venetian blinds, but only narrow strips. 1950s—coil lines with speeds of up to 60 ft/min on strips up to 15 in. wide. 1970—speeds of 600 ft/min on 66-in. widths.	Lye [105]
Aerosol paint packs	1948—introduced in USA	Armitage [106, p. 117]
Electrostatic spray application	1948—introduced in USA by Ransburg Co. (mentioned in 1946 book as experimental development); achieved over 90% deposition compared to 75% wastage previously. Ransburg No. 1, used air pressure; Ransburg No. 2, airless.	Anon. [107]; Armitage [106, p. 136]
Curtain coating	Swiss bicycle engineer sold invention to U. Steineman; used first in chocolate industry in Europe; introduced to Britain in 1957.	Letsky [108]
Airless spray gun	1958—introduced in USA (hand gun)	Tilney [109]

TABLE 7 Continued

Process	Origin	Reference
Powder coating	Available from mid 1950s, applied by fluidized bed process.	Hester et al. [110]
	1962—introduction of electrostatic powder spray gun by S.A.M.E.S.	Harris [111]
Electrodeposition, anodic	1919—patented by Davey.	Machu [112]
	1930—patented by Cross & Blackwell.	
	1963—First commercial application in Belgium by Ford to coat wheels and car bodies and by ICI with Pressed Steel Co. in U.K. to coat petrol tanks.	
Electrodeposition, cathodic	First used in industrial E/C in early 1970s; introduced to automotive industry by PPG in 1976.	
Fluidized bed process	First attempts by E. Gemmer in Germany 1952; first patents began to appear early 1960s.	Ranney [113]
UV curing	Originated in ancient Egypt in mummification process; "oil of lavender" containing high mol.wt. bituminous material ("Syrian asphalt") became cross-linked on exposure to sunlight.	Pryce [114], Lott [115], O'Hara [116]
	1945—patent to du Pont de Nemours using benzoin photoinitiator for vinyl monomer.	
	Late 1960s—use in wood finish high gloss polyester/styrene finishes.	
	1970—UV-cured inks used for printing packaging materials in USA; in Europe a year later.	

TABLE 7 Continued

Process	Origin	Reference
Electron beam curing	Research with Van de Graaff accelerator 1958–1960 at Tube Investments Research, Saffron Walden. British patent applied for 1960, granted 1964. Ford Motors patent published 1966. 1967—first commercial application named Electrocure by Boise-Cascade, Idaho for wood-based materials. 1970—first fully commercial finishing line on plastic car parts, Ford plant, Saline, Michigan. 1972—first appearance in Europe (Holland) for painted hardboard.	Hoffman et al. [117], Davison [118]

Source: Courtesy of Dr. Helma Jotischky, Paint Research Association.

isoparaffins are being recommended as replacements for white spirit. Producers of marine antifouling additives based on organotin compounds are under pressure to replace them with less environmentally harmful materials, as are those who use fluorocarbon propellants in aerosol paints. The use of pigments that contain leachable toxic metal ions (even if the toxic metal is highly insoluble such as cadmium yellow) is frowned upon.

There is a strong trend to redesign coatings toward high solids and waterborne systems for industrial use to decrease solvent emission and obviate the need to install expensive equipment to control emissions to the atmosphere, which is being increasingly controlled by legislation. For this reason, powder coatings (paint in powder form [110]) are regarded as the fastest growth area in industrial and automotive paints, as no solvent is required in their application. Although their rate of growth is about 5% per annum, they account for only 2% of the total paint market [119].

Future DIY paints will probably be waterborne. The largest paint company has stated [120] that by the end of the decade consumer needs will be satisfied by waterborne paints, and white spirit-based paints will no longer be part of their DIY product range. They also state that existing water-based gloss technology does not yet match the performance of white spirit gloss paints, thus indicating where future effort must be placed.

ACKNOWLEDGMENT

I take pleasure in thanking the PRA librarian, Miss C. Haworth, for help in finding references; Dr. Helma Jotischky for permission to use some of her compilations, and my wife Nina for patiently reading and commenting on my scribbles.

REFERENCES

1. F. Margival, *Peint., Pigm. Vernis 35*(10): 521 (1959); *36*(10): 614 (1960).
2. H.G. Friedstein, *J. Chem. Educ. 53*(4): 291 (1981).
3. F. Armitage, *Paint Manuf. 23*: 190 (June 1953).
4. J. Latner, *Paint Technol. 27*(6): 29 (1963).
5. F. Armitage, *Paint Manuf. 23*: 319 (Oct. 1953).
6. F. Armitage, *Paint Manuf. 24*: 57 (Feb. 1954).
7. F. Armitage, *Paint Manuf. 24*: 153 (May 1954).
8. J. Boxall, *Paint Manuf. 48*: 23 (June 1978).
9. J. Boxall, *Paint Manuf. 48*: 25 (July/Aug. 1978).
10. Anon., *Polym. Paint Colour J.*, May (1979), p. 530.
11. G.E.F. Brewer, *Plast. Surf. Finish. 71*(6): 57 (1984).
12. H. Jotischky, *The Paint Industry: The Economic Framework*, Paint Research Assoc., October 1991.
13. A. Doroszkowski and E. West, in *Multicomponent Polymer Systems* (I.S. Miles and S. Rostami, Eds.), Longman, London, 1992, p. 375.
14. D.B. Freeman, *Phosphating and Metal Pre-treatment*, Woodhead-Faulkner, 1986.
15. A. Backhouse, *J. Coatings Technol. 54*(693): 83 (1982).
16. J.H. Braun, *J. Coatings Technol. 63*(799): 10 (1991).
17. V.V. Verkholantsev, *Prog. Org. Coatings 18*(1): 43 (1990).
18. Information Research Ltd., *A Profile of the European Paint Industry* 7th ed., 1986, p. 13.
19. A. Ramig and P.F. Ramig, *J. Oil Colour Chem. Assoc. 64*: 439 (1981).
20. J.Th.G. Overbeek, in *Colloid Dispersions* (J.W. Goodwin, Ed.), Royal Chemical Society, London, 1982.
21. L.A. Tysall, *Calculation Techniques in the Formulation of Alkyd and Related Resins*, Paint Research Association, 1982.
22. T.C. Patton, *Alkyd Resin Technology: Formulating Techniques and Allied Calculations*, Interscience, New York, 1962, p. 111.
23. R.A. Hancock, N.J. Leeves, and P.F. Nicks, *Prog. Org. Coatings 17*(3): 321 (1989).

24. T.G. Fox, *Bull. Am. Phys. Soc. 1*: 123 (1956); see also R.C.C. Arridge, *Mechanics of Polymers*, Clarendon Press, Oxford, 1975, p. 84.
25. D.H. Solomon, *The Chemistry of Organic Flim Formers*, 2nd ed., 1977.
26. J. Bentley, in *Paint and Surface Coatings: Theory and Practice* (R. Lambourne, Ed.) Ellis Horwood, London, 1987, pp. 41–110.
27. W. Herbst, *J. Paint Technol. 45*(579): 39 (1973).
28. T.C. Patton, *Paint Flow and Pigment Dispersion*, 2nd ed., Wiley, New York,1979.
29. J.M. Rahter, *Paint Varnish Prod. 63*(5): 37 (1973).
30. ASTM, Stormer viscometer, D562-81, Vol 06.01, 1990.
31. B. Burgess, *Coatings Mag.* Nov./Dec. 1981, p. 66.
32. J.Johnson, A. Szegvari, and M. Li, *Polym. Paint Colour J. 172*(4075): 459 (1982).
33. B. Burgess, *Coatings Mag.* Mar./Apr. 1982, p. 58.
34. S. Hochberg (to DuPont), U.S. Patent 2,581,414 (1952).
35. G.R. Lester, *J. Oil Colour Chem. Assoc. 47*: 719 (1964).
36. T.C. Patton, *J. Paint Technol. 42*(564): 626 (1970).
37. H.F. Payne, *Organic Coating Technology*, Vol. 2, *Pigments and Pigment Coatings*, Wiley, New York, 1961.
38. I. Berg, *Polym. Paint Colour Suppl.* Feb. 1983.
39. M. Cremer, *Paint Manuf. 38*: 29 (Dec. 1968).
40. C.E. Kew, *Off. Dig. 31*(408): 70 (1959).
41. J. Stoffer, J. Gordon, M. Fahim, and L. Wallace, *Proc. 13th Waterborne and High Solids Coatings Symp., New Orleans*, 1986, p. 410.
42. A. Doroszkowski, in *Paint and Surface Coatings: Theory and Practice* (R. Lambourne, Ed.), Ellis Horwood, London, 1987, Chap. 6.
43. F.K. Daniel and Goldman, *Ind. Eng. Chem. (Anal. Ed.) 18*: 26 (1946).
44. C.C. Tatman, *J. Coat. Technol. 53*(681): 57 (1981).
45. F.K. Farkas, in *Paint and Surface Coatings: Theory and Practice* (R. Lambourne, Ed.), Ellis Horwood, London, 1987, p. 311.
46. A. Doroszkowski, R. Lambourne and J. Walton, *Colloid Polym. Sci. 255*: 896 (1977).
47. D.J. Walbridge, E.I. Scot, and C.H. Young, *Fifth Conf. Org. Coatings Sci. Technol.*, Athens, 1979, p. 526.
48. D.H. Napper, in *Polymeric Stabilisation of Colloidal Dispersions*, Academic, London, 1983, p. 14.
49. B. Burgess, *Coatings Mag.* July/August 1986, p. 24.
50. F.K. Daniel, *Off. Dig. 33*(438): 840 (1961).
51. A. Doroszkowski and R. Lambourne, *Disc. Faraday Soc. 65*: 253 (1978).
52. D.T. Wu, *Paint Ink Int. 5*(2): 36 (1992).
53. J. Visser, *J Adv. Colloid Interface Sci. 37*: 331 (1973).
54. E. Ruckstein, *Colloids Surf. 69*: 271 (1993).
55. H.B.G. Casimir and D. Poldar, *Phys. Rev. 73*: 360 (1948).
56. D.W.J. Osmond, B. Vincent, and F.A. Waite, *J. Colloid Interface Sci. 42*: 262 (1973).

57. D.W.J. Osmond and F.A. Waite, in *Dispersion Polymerisation in Organic Media* (K.E.J. Barrett, Ed.), Wiley-Interscience, London, 1975, p. 14.
58. Hegman gauge, ASTM D1210-79 (reapproved 1988).
59. M.R. Hornby and R.D. Murley, *Prog. Org. Coatings, 3*: 261, (1975).
60. J. Oyarzun, *J. Coating Technol. 55*(704): 77 (1983)
61. T.R. Bullett, in *Paint and Surface Coating: Theory and Practice* (R. Lambourne, Ed.), Ellis Horwood, London, 1987, Chap. 17.
62. D.B. Judd and G. Wyszecki, *Colour in Business, Science and Industry*, 2nd ed., Wiley, New York, 1963.
63. F. von Patat and C. Schiebeuer, *Macromol. Chem. 44*(6): 643 (1961).
64. H. Benard, Doctoral Thesis, Presented to La Faculté des Sciences de Paris, Issue No. 387, No. D'Ordre 1057, 1901.
65. L. Tysall and R.A. Paint, Prog. Rep. No. 11 (1986) referring to Huyghens' *Ouevres*, reported by Korteweg, *Nature 93*: 584 (1914).
66. S.H. Bell, *J. Oil Colour Chem. Assoc. 35*: 373 (1952).
67. M. Smoluchowski, *Z. Phys. 17*: 557 (1916); *Z. Phys. Chem. 92*: 129 (1917).
68. J.W. Lichtenfeld, Th. Panthmamanoharan, and P.H. Wiersema, *J. Colloid Interface Sci. 49*: 281 (1974).
69. A. Lips and E. Wills, *Trans. Faraday Soc. 69*: 1226 (1973).
70. H. Sander, *Farbe Lack 93*(4): 257 (1987).
71. L.A. Simpson, *J. Coatings Technol. 54*(685): 27 (1982).
72. J.G. Balfour, *J. Oil Colour Chem. Assoc. 60*(9): 365 (1977).
73. A Doroszkowski and B.H. Armitage, *XVII FATIPEC Congr. 2*: 311 (1984).
74. A. Brisson, G. L'Esperance, and M. Caron, *J. Coatings Technol. 63*(801): 111 (1991).
75. D.J. Rutherford and L.A. Simpson, *J. Coatings Technol. 57*(724): 75 (1985).
76. A. Doroszkowski and R. Lambourne, *J. Colloid Interface Sci. 26*: 214 (1968).
77. A. Doroszkowski, in *Paint and Surface Coatings: Theory and Practice* (R. Lambourne, Ed.), Ellis Horwood, London, 1987, p. 263.
78. A.E. Smith and E.W. Pemberton, *Pig. Resin Technol. 9*(11): 8 (1990).
79. R.Z. Zigmondy, *Anal. Chem. 40*: 697 (1901).
80. J. Schroder, *Prog. Org. Coatings 15*: 337 (1988).
81. J.L. Finney, *Proc. Roy. Soc. (Lond.) A319*: 479 (1970).
82. *Journal of Coatings Technology 64*(806) March 1992; issue devoted to CPVC, with papers by W.K. Asbeck; G.P. Bierwagen; F.L. Floyd and R.M. Holworth; J.C. Weaver; B.S. Skerry, C.-T. Chen, and C.J. Ray; and K.L. Hoy and R.H. Peterson.
83. G.P. Bierwagen and T.K. Hay, *Prog. Org. Coatings 3*: 281 (1975).
84. D.I. Lee, *J. Paint Technol. 42*(550): 579 (1970).
85. M.J. Mooney, *J. Colloid Sci. 6*: 162 (1951).
86. J. Schroder, *Rheology 2*(1): 40 (1992).
87. J. Mewis, *Proc. XII Int. Conf. Org. Coatings Sci. Technol. (Athens)*, 1986, p. 225.
88. R.B. McKay, *XIII FATIPEC Congr.*, Cannes, 1976, p. 428.
89. K.M. Oesterle, *XIV FATIPEC Congr.*, Budapest, 1978, p. 329.
90. R.B. McKay, *J. Oil Colour Chem. Assoc. 72*: 89 (1989).

91. D.H. Solomon, *J. Oil Colour Chem. Assoc. 38*: 284 (1965).
92. M.J. Ballard, R. Buscall, and F.A. Waite, *Polymer 29*: 1287 (1988).
93. M. Hoeflaak, *XVIII FATIPEC Congr.*, Venice, *1A*: 243 (1986).
94. M. Klarskov, A. Saarnak, and J. Jakobsen, *XX FATIPEC Congr.*, Nice, 1990, p. 271.
95. N.D.P. Smith, S.E. Orchard, and A.J. Rhind-Tutt, *J. Oil Colour Chem. Assoc. 44*: 618 (1961); *J. Appl. Sci. Res. A11*: 451 (1962).
96. A. Doroszkowski, *Colloids Surf. 17*: 13 (1986).
97. F. Beck, *Prog. Org. Coatings 4*: 1 (1976).
98. G. Soderberg, *Finishing Materials and Methods*, McKnight & McKnight, Chicago, Ill. 1952, p. 147.
99. A.H. Sabin, *Industrial & Artistic Technology*, Wiley, New York, 1905.
100. R. Wampler, *Modern Organic Finishes: Applications to Industrial Products*, Chemical Publ. Co., New York, 1946, p. 104.
101. H.W. Chatfield, *Science of Surface Technology*, Ernest Benn, London, 1962, p. 509.
102. W. von Fischer, *Paint and Varnish Technology*, Reinhold, New York, 1948.
103. F.W. MacKenzie (Ed.), *Screen Process Printing 1951*, Skinner & Wilkinson, 1951, p. 62: see also *The Application of Surface Coatings, Paint Techniques Manual*, Part 4, Chapman & Hall, London, 1965, p. 277.
104. W.H. Tatton and E.W. Drew, *Industrial Paint Application*, George Newnes, London, 1964, p. 25.
105. G.R.K. Lye, *Focus: Coated Coil 2 Conference*, Chicago, March 1977, p. 10.
106. F. Armitage, *British Paint Industry*, Pergamon, London, 1967.
107. *Australian OCCA, Surface Coatings*, NSW Univ. Press, 1974, p. 136.
108. B.N. Letsky, *Practical Manual of Industrial Finishes*, Chapman and Hall, London, 1960, p. 176.
109. R. Tilney, *Trans. Inst. Metal Finishing 43*: 113 (1965).
110. C.I. Hester, R.L. Nicholson, and M.A. Cassidy, *Powder Coating Tech.*, Noyes Data Corp., 1990, p. 3.
111. S. Harris, *The Technology of Powder Coatings*, Portcullis Press, 1976, p. 2.
112. W. Machu, *Handbook of Electropainting* (P. Neufeld, transl.) Electrochemical Publ., 1978, p. 3.
113. M.W. Ranney, *Powder Coatings and Fluidised Bed Techniques*, Noyes Data Corp., 1971, pp. 1–3.
114. A. Pryce, *J. Oil Colour Chem. Assoc. 59*: 166 (1976).
115. A.D. Lott, *J. Oil Colour Chem. Assoc. 59*: 141 (1976).
116. K. O'Hara, SME Tech. Paper No. FC75-321, 1975.
117. C.R. Hoffman, K.H. Morganstern, and S.V. Nablo, *Radiation Curing 1*(2): 4 (May 1974).
118. W.H.T. Davison, *J. Oil Colour Chem. Assoc. 52*(10): 947 (1969); *Trans Metal Finishing Org. Fin. Suppl.* 1973, p. 199.
119. J. Dansie, *Perform. Chem. 8*(1): 30 (1993).
120. Anon., *Eur. Chem. News 56*(1454): 28 (1991).

2
Titanium Dioxide Pigments

JOHN G. BALFOUR Technical Service Department, Tioxide Europe Limited, Billingham, Cleveland, England

I.	Introduction	69
II.	The Manufacture of Titanium Dioxide Pigments	70
	A. Raw materials for titanium dioxide production	70
	B. The sulfate process	71
	C. The chloride process	73
	D. Pigment coating	75
III.	Properties of Titanium Dioxide Pigment Dispersions in Application Media	75
	A. Opacity	76
	B. Color	88
	C. Gloss	90
	D. Durability	91
	E. Other properties	103
	References	104

I. INTRODUCTION

Titanium dioxide is the preeminent white pigment currently available. It is used to impart opacity and whiteness to an enormous variety of products, including paints, plastics, inks, paper, cosmetics, foodstuffs and pharmaceuticals as well as many others [1]. However, despite its established position, its commercial exploitation is relatively recent in the field of inorganic pigments, with the sale of rutile pigments only becoming established in the 1940s.

The effectiveness of titanium dioxide pigments depends on crystallographic properties, chemical purity, surface character, and the state of dispersion in the application medium. The first three of these attributes are controlled by the pigment manufacturer alone, but the state of dispersion is significantly influenced by the pigment user.

Titanium dioxide pigments are normally produced by one of two established techniques, the chloride and sulfate processes. World production is now shared almost equally between these two production routes.

II. THE MANUFACTURE OF TITANIUM DIOXIDE PIGMENTS

A. Raw Materials for Titanium Dioxide Production

Titanium is widely distributed over the surface of the earth and is, in fact, the ninth most abundant element in the earth's crust. The most important of the naturally occurring titanium ores are ilmenite and mineral rutile. Ilmenite is a black sand or rock and has the theoretical formula $FeTiO_3$. Normally, the iron is partly oxidized to the trivalent state, and the ore also contains quantities of impurities in the form of silicon compounds. The TiO_2 concentration in the ore varies from about 45% to 60% depending on its origin.

Mineral rutile is naturally occurring titanium dioxide. Its color varies from brown to reddish black, and its major impurities are iron compounds. The TiO_2 content is normally of the order of 95%.

In addition to the naturally occurring ores there are two important types of enriched ores. The first is titanium slag, which is the residue left after the extraction of iron from ilmenite or mixed ilmenite-hematite/magnetite deposits. The second is beneficiated ilmenite in which the natural ore has been treated to increase its TiO_2 concentration.

Predominantly, ilmenite and titanium slag are used in the sulfate process. Mineral rutile and beneficiated ilmenite (as well as some titanium slag and ilmenite) are used in the chloride process.

Anatase- or rutile-structured pigments can be produced. Both types may be surface treated to impart properties of importance to specific end uses. Both anatase and rutile pigments can be manufactured by the sulfate route. However, while it is not impossible to produce anatase by the chloride method, it is uneconomical, and this method is used almost exclusively for the production of rutile pigments.

B. The Sulfate Process

Although the manufacturing operation is complicated, the basic chemistry of the sulfate process depends on a series of simple chemical reactions. For example, the following apply to ilmenite:

$$FeTiO_3 + 2\,H_2SO_4 \longrightarrow TiSO_4 + FeSO_4 + 2\,H_2O$$

$$TiSO_4 \xrightarrow{OH^-} TiO_2 \cdot n\,H_2O + H_2SO_4$$

$$TiO_2 \cdot n\,H_2O \xrightarrow{\text{heat}} TiO_2 + n\,H_2O$$

A schematic diagram of the sulfate process is shown in Fig. 1. The ore is first dried and ground and then treated with sulfuric acid, normally at a strength of 85–92%; the exact strength is dependent on the composition of the ore. This digestion process can be carried out as either a batch or a continuous operation. The mixture of ground ore and sulfuric acid is agitated by compressed air, and superheated steam is blown in. At a temperature of about 160°C a vigorous exothermic reaction starts and the mixture is converted to a porous cake, which may contain ferrous, ferric, and tita-

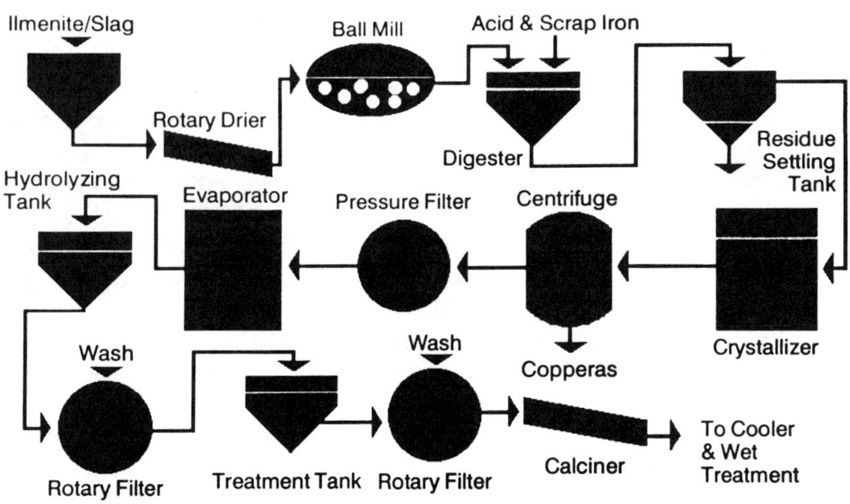

FIG. 1 Outline of the sulfate process.

nium sulfates. This cake is extracted with water or dilute acid, and any ferric iron in the solution is reduced to ferrous through the addition of scrap iron. As the iron must be in the ferrous form for subsequent handling, reducing conditions are maintained so that some trivalent titanium is formed. The liquor is then clarified by sedimentation to remove insoluble residues such as silica, zircon, and residual unreacted ore.

To ensure good clarification, an agent is added to flocculate colloidal materials. Much of the iron is then removed by cooling to about 10°C and crystallizing out ferrous sulfate. This step can be omitted if the original ore has a low iron content as in the case of titanium slag. Traces of sludge are then removed by filtration, after which the solution is concentrated.

The next stage, which involves hydrolysis to form a precipitate of hydrous titanium dioxide, is particularly critical. No ferric iron should be present, and conditions must be controlled so that the precipitate can easily be filtered and washed. In addition, it must produce crystals of the right type and size during the subsequent calcination process. Precipitation is achieved through prolonged boiling.

Nuclei may be added at this stage or later during calcination. These are prepared from pure titanium tetrachloride and can be produced in the anatase or rutile form. Although precipitates from sulfate liquors are always in the anatase form, the type of nuclei added determines whether anatase or rutile is produced during calcination. Optimum precipitation depends on the condition of the initial solution, the quality of the nuclei, and the duration of hydrolysis. The yield of hydrous titanium dioxide can be increased if the solution is diluted when precipitation is nearly complete, but care has to be taken as excessive dilution can impair the quality of the product.

The precipitate is filtered and washed with water, and the acid filtrate is recovered and recycled. The precipitate is leached under reducing conditions to remove any traces of iron. After a final wash, minor additions of materials that help to control crystal growth are made and the pulp is then calcined.

Calcination is carried out in internally fired, inclined rotary kilns through which the pulp moves slowly under gravity. As it progresses through the kiln, the wet pulp is first dried and then strongly adsorbed water and sulfur trioxide are driven off. The acidic gaseous effluent is washed to remove SO_3 to form sulfuric acid, which is recycled for use earlier in the process. The growth of crystals and, where relevant, conversion to rutile occur in the last few feet of the kiln. The temperature is carefully controlled according to the type of pigment being made and is normally of the order of 1000°C.

After calcination, the pigment is fed into a cooler. It must be cooled slowly to prevent the formation of trivalent titanium, which would have adverse effects on the final color. The unrefined pigment can then be dry-milled to break down aggregates and packed for sale as untreated pigment. However, most pigment is wet-milled and further treated to coat the particles with selected inorganic oxides before a final milling and packing. As the surface treatment or wet treatment is common to both the sulfate and chloride processes, it is described later.

C. The Chloride Process

The chloride process is described by the following chemical reactions.

$$TiO_2(\text{impure}) + 2\,Cl_2 + C \rightarrow TiCl_4 + CO_2$$

$$TiCl_4 + O_2 \rightarrow TiO_2(\text{pure}) + 2\,Cl_2$$

A schematic diagram of the chloride process is shown in Fig. 2.

The dry ore is fed into a chlorinator, in which it forms a bed, which is fluidized by an air stream. Heat is applied to raise the temperature to about 650°C. Crushed coke is fed in on top of the ore, where it ignites, increasing the temperature and enabling the initial heat source to be removed. When the required temperature is achieved, the air is switched off and replaced by chlorine, and titanium tetrachloride vapor is formed. As the reaction proceeds, the bed is continually charged with ore and coke to keep its height constant. The reaction is very exothermic, and in order to maintain a constant temperature it is necessary to cool the reacting mass. One way of doing this is to add liquid titanium tetrachloride to the chlorinator.

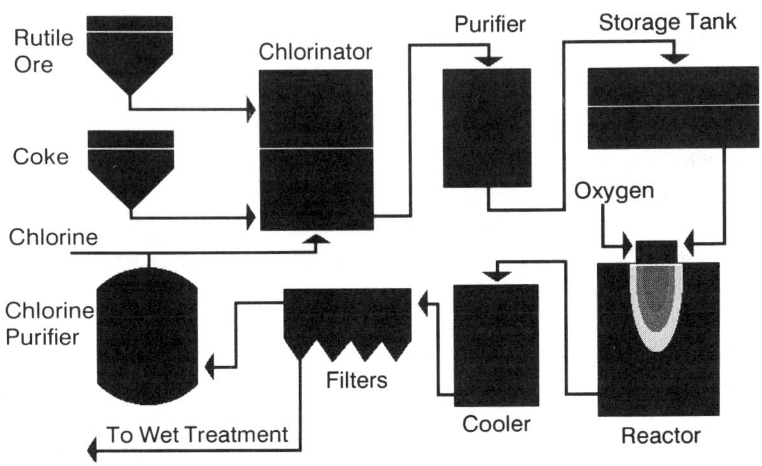

FIG. 2 Outline of the chloride process.

The gas stream is cooled, and solid matter is formed. After further cooling and cleaning, the crude titanium tetrachloride is distilled to produce a pure product suitable for oxidation to titanium dioxide.

For successful oxidation it is necessary to bring the reagent gases together at a suitable temperature and provide suitable nuclei on which the pigment particles can form. The reaction products are chilled rapidly to avoid excessive growth of pigment particles and to prevent the formation of deposits of titanium dioxide on the reactor walls. The material is collected, for example, by cyclone filters and may be heat-treated to remove traces of chlorine and control crystal size.

The heat evolved when $TiCl_4$ is burned in oxygen is not sufficient to raise the temperature of the reactants to reaction temperature and maintain that temperature. Thus, additional heat must be supplied and that can be achieved by burning an additional gas such as carbon monoxide in a preliminary stage. The flame from this gas can be used to provide a mixture of hot carbon dioxide and $AlCl_3$, which is subsequently brought into contact with the oxygen. Burners can be designed so that $TiCl_4$, oxygen, and auxiliary gases are fed through three separate ports. The burner design is complicated, and there are many variants. The nozzles must be kept cool enough to prevent the growth of TiO_2 on them, which would impair their efficiency. Oxygen may also be heated by an electric discharge and introduced into the reactor system by a plasma torch. In a further group of methods the reacting gases are passed through a hot fluidized bed of inert material such as silica and the TiO_2 is carried out with the reacted gases.

Nucleation in the reactor is necessary to promote the formation of pigmentary particles. This can be achieved by introducing a small concentration of water into the oxygen stream or produced by the combustion of hydrocarbons. Another method is to add a small amount of anhydrous $AlCl_3$ to the $TiCl_4$ feed; particles of Al_2O_3 produced by the oxidation form centers for the growth of TiO_2.

Regardless of the reactor design, the gases carrying the pigment must be cooled rapidly to below reaction temperatures so that further growth does not occur. This may be done by recycling a mixture of spent gases from a later stage or by using liquid chlorine. Separation of the pigment then follows, and finally chlorine is compressed and condensed to liquid and returned for further use.

Chloride process pigment is likely to contain adsorbed chlorine, which has to be removed, and the pigment is then neutralized by washing, heating, or treatment by steam. This may be followed by treatment with air containing steam. The reactor product usually requires milling to break down aggregates and oversize particles. Although the pigment can be used in this state, it is normal, as with sulfate process pigments, for it to be coated.

D. Pigment Coating

The properties of titanium dioxide pigment can be substantially improved by coating the pigment particles. Originally, pigments were coated to improve durability and to reduce the yellowing that occurred in certain types of paint. Subsequently, it was found that surface treatments could be used to improve the dispersiblity of pigments in various media.

Coating involves the deposition of other matter onto the surface of the pigment particles. The coating agents are white hydrated oxides such as silica, alumina, zirconia, and, in some cases, titania.

The process involves dispersion of the untreated pigment from either the sulfate or chloride process in water using, for example, a ball mill or a sand mill. The dispersion may be allowed to settle in order to remove oversize particles. It is then constantly stirred while suitable reagents, for example aluminum sulfate, are added. Adjustment of the pH precipitates the hydrated oxide onto the surface of the particles. The nature of this precipitate may vary according to the conditions of precipitation. Coatings of two or more hydrated oxides may be applied to a pigment, either separately by successive operations or simultaneously in one operation. After the coating is formed, the pigment is first washed and dried and then ground in a micronizer or other fluid energy mill.

Many titanium dioxide pigments are also given an organic surface treatment to improve their dispersiblity. This is usually carried out during the final milling operation. A variety of organic compounds are used, including polyols, amines, and silicon derivatives.

There are many variables in the coating process, and detailed procedures are well documented in the patent literature. The most important feature is that the oxides form a coating around the particles and are not present only as a simple admixture.

Variations in crystal structure (anatase and rutile), crystal size, inorganic coating, and organic treatment provide a wide range of titanium dioxide grades with differing properties.

III. PROPERTIES OF TITANIUM DIOXIDE PIGMENT DISPERSIONS IN APPLICATION MEDIA

Pure titanium dioxide is a colorless crystalline solid. As with other dioxides of *d*-block elements in its group in the periodic table, it is stable, virtually insoluble, and nonvolatile and is rendered refractory by ignition. It is amphoteric but has more acidic than basic characteristics. TiO_2 is polymorphous, existing in three fundamental crystal forms. All three forms—anatase, rutile, and brookite—exist naturally although brookite is relatively rare. Brookite does not exhibit suitable pigmentary properties so it will not be considered further. Although anatase and rutile are chemically identical and both crystallize tetragonally, they can be identified by means of X-ray diffraction.

Rutile is the more stable form and is thus most abundant. It has a more compact structure than anatase, and this gives rise to important differences in properties between the two forms. The melting point of rutile is 1825 °C. Anatase does not have a specific melting point as it is converted to rutile before its melting point is reached.

A. Opacity

The most important attribute of TiO_2 is its ability to scatter light, which gives the potential for imparting opacity to a far greater extent than can be achieved with any other common white pigment.

Pigments impart opacity [2] through absorbing or reflecting light or a combination of the two. Titanium dioxide pigments absorb very little light, as shown by the reflectance curves of anatase and rutile in Fig. 3, and consequently give opacity through reflection. In practice, opacity is achieved through multiple reflections and refractions caused by the pigment particles.

Some of the light is reflected at the surface, and some is refracted into

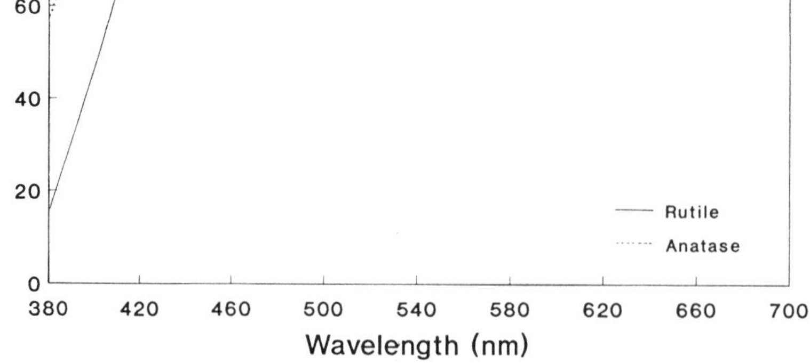

FIG. 3 Reflectance of titanium dioxide.

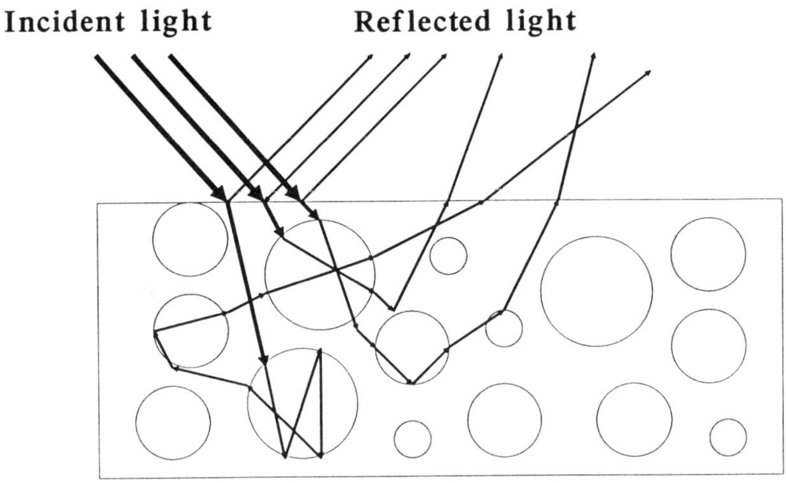

FIG. 4 Light scattering from a white paint film.

the pigment and then may be internally reflected or refracted out again. Effectively the light gets scattered in many directions. The situation becomes more complicated when there is a concentration of pigment particles and the light passes from one particle to another and eventually emerges from the film (see Fig. 4). The complexity is demonstrated by the vast number of particles. Typically, a square centimeter of a white gloss paint film may contain about 30×10^9 particles.

The reason for the high light-scattering efficiency of titanium dioxide is its higher refractive index compared with potential competitors. The refractive indices of TiO_2 and other pigments are shown in Table 1. The greater the refractive index, the greater is the ability to scatter light through reflection and refraction. A crude indication of opacifying power can be obtained using Fresnel's equation:

$$F = \frac{n_1 - n_2}{n_1 + n_2}$$

where F is reflectivity, n_1 the refractive index of the pigment, and n_2 the refractive index of the surrounding medium. Values for Fresnel reflectivity are also shown in Table 1 assuming a typical medium refractive index of 1.5. Extenders that have a low refractive index similar to that of the surrounding medium, such as china clay, give virtually no opacity.

In order to optimize the scattering properties of the pigment, it is necessary to control the size very carefully. The terms "crystal size" and "particle

TABLE 1 Refractive Indices of Some Common White Pigments

Pigment	Refractive index	Reflectivity (%)
Rutile TiO_2	2.70	8.2
Anatase TiO_2	2.55	6.7
Zinc sulfide	2.37	5.1
Antimony oxide	2.30	4.4
Lithopone 30%	1.84	1.0
Zinc oxide	2.02	2.2
White lead	2.00	2.0
China clay	1.57	0.1

size" are often incorrectly used interchangeably. *Crystal size* refers to the fundamental size of the crystal and is controlled by the TiO_2 manufacturer. *Particle size* refers to the overall size of the particles in a given system. These particles may comprise either single crystals or particles that contain a number of crystals bound together. The particle size is therefore controlled by the pigment user and is related to how well the pigment has been dispersed. In an ideal world all the particles would be single crystals, but this is never the case.

The optimum size for scattering light is about half the wavelength of light (Fig. 5), in practice between about 0.2 and 0.25 μm, although the absolute value is related to the concentration used. This explains why,

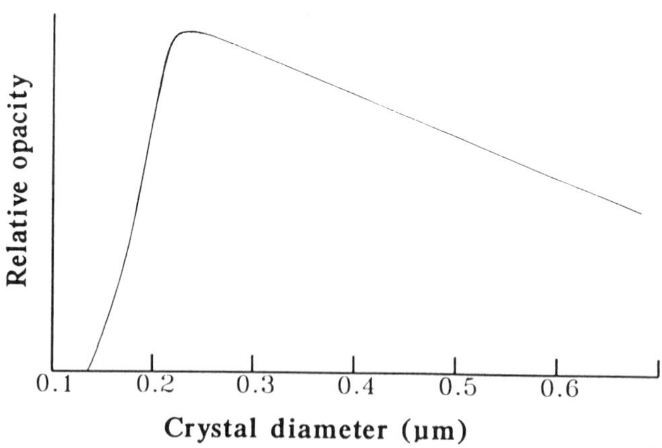

FIG. 5 Effect of crystal size on opacity.

in some applications such as plastics, a smaller size is used because the concentration is lower. Moving away from the optimum size in either direction reduces the opacity considerably. Recently, transparent grades have appeared on the market; these have a size approximately one-tenth that of pigmentary grades [3].

Opacity is obviously related to the concentration of pigment present. The relationship between opacity and pigment volume concentration (pvc) is shown in Fig. 6. Up to about 10% pigment volume concentration in the dry film, there is an almost linear increase in opacity. Above this level the individual particles do not act as efficiently and the increase in opacity becomes less than proportional. Eventually, the opacity reaches a maximum, and further increases in concentration above about 30% can lead to a decrease in opacity.

The loss of opacity is a result of what is called the crowding effect. Light can be considered to be in the form of waves, and when the pigment particles are small the light interacts with them and makes them act as if they were bigger than they actually are. Thus the particles can be considered to have a sphere of influence around them. As the concentration becomes higher, these spheres of influence overlap (Fig. 7), and each of the particles scatters light less efficiently.

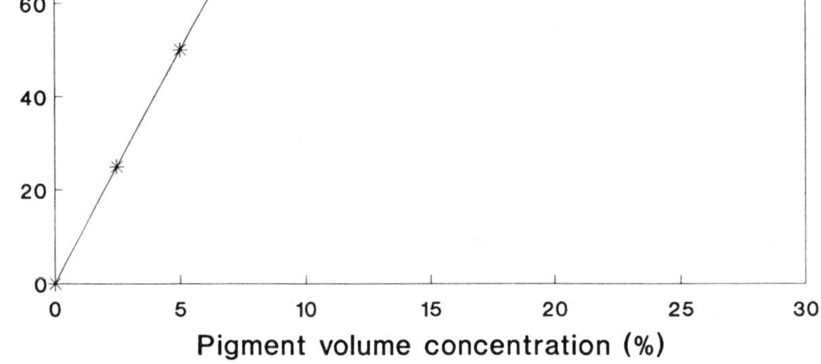

FIG. 6 Effect of pigment volume concentration on opacity.

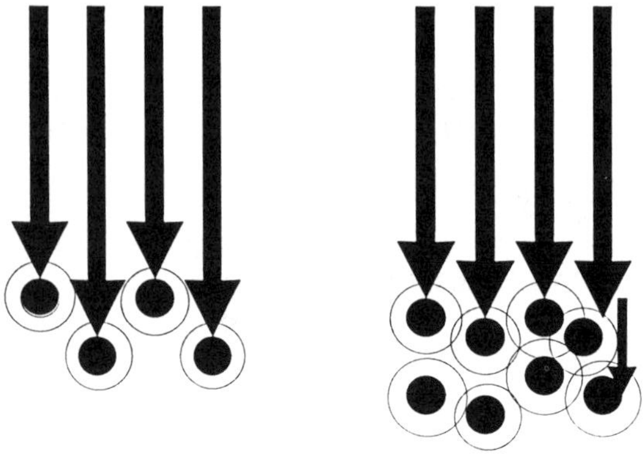

FIG. 7 Overlapping spheres of influence at high pvc's reduce scattering efficiency.

Scattering efficiency is also affected by what is described as the population effect. This is illustrated in the following hypothetical example.

A particle of size 0.23 μm has a volume of about $6.4 \times 10^{-3} \mu m^{-3}$. Reducing the size of the particle to 0.17 μm decreases the volume to about $2.6 \times 10^{-3}\ \mu m^3$. Thus, for a given concentration of TiO_2, there will be two and a half times as many particles of 0.17 μm as there are particles of 0.23 μm. At low concentrations, where the crowding effect does not operate, this reduction in size can cause an increase in opacity. This is because although the individual particles scatter less efficiently, there are sufficiently more of them to more than compensate. However, as the concentration increases, the crowding begins to predominate, and the advantage from the greater number of particles is lost. Thus, in situations where a low concentration is used, such as in many plastics applications, the use of a smaller crystal size pigment can be beneficial.

The state of pigment dispersion in the application medium is of fundamental importance. Ideally, the pigment crystals should be individually dispersed. In practice, the state of dispersion can be considered to be the extent to which particles of the pigment powder are broken down by milling and remain separated afterwards. This is affected by the degree of milling and the tendency toward pigment flocculation.

Using modern easily dispersed pigments, the effect of milling on opacity is relatively slight. Figure 8 shows the opacity development with duration of milling for a general-purpose grade of pigment in a long-oil alkyd resin and demonstrates that maximum opacity is achieved relatively quickly. This is because the majority of the loosely coherent aggregates are broken down very quickly and most of the milling time is spent breaking down a small residue of strongly coherent aggregates. This number is so small that it has little effect on opacity. However, it is still necessary to carry on milling, as this small number will be readily visible in a paint film in the form of nibs, affecting gloss and appearance. Their presence is shown in the diagram in the form of a grind gauge reading that only becomes satisfactory toward the right-hand side. No time units are included as time scales depend on the milling conditions. The relative time to achieve good dispersion and optimum opacity is virtually unaffected.

Flocculation is an effect whereby pigment particles come together after milling to form loosely bound clusters or flocculates. The predominant effect is to lower the opacity.

A flocculent paint film is shown schematically in Fig. 9. Within a flocculate the particles are very close together, their spheres of influence overlap, the scattering efficiency is reduced, and opacity is lost. The remainder of the film suffers from areas of low pvc because pigment has migrated to the flocculates and these areas have low opacity because of the lack of pigment.

The detection and measurement of pigment flocculation is not possible using a traditional grind gauge because the flocculates break down too easily. Thus, other techniques are required [4], such as the following. If thin films are prepared and the color of a tungsten light filament is viewed

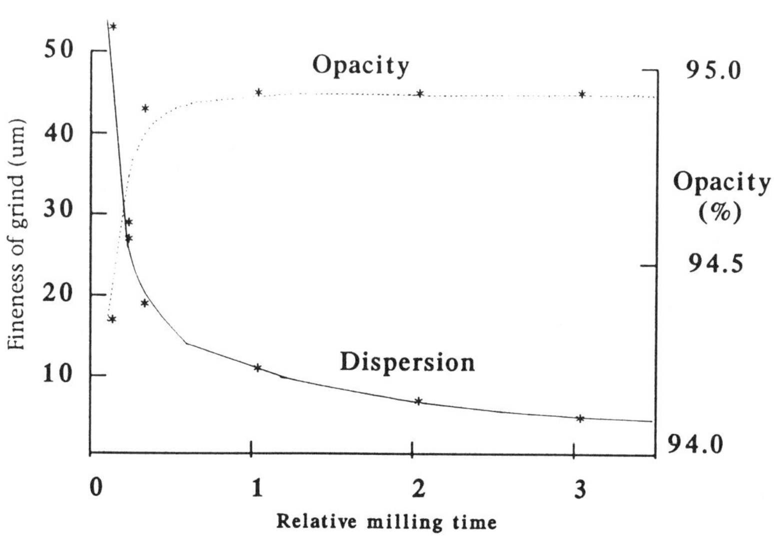

FIG. 8 Effect of milling time on opacity and fineness of grind.

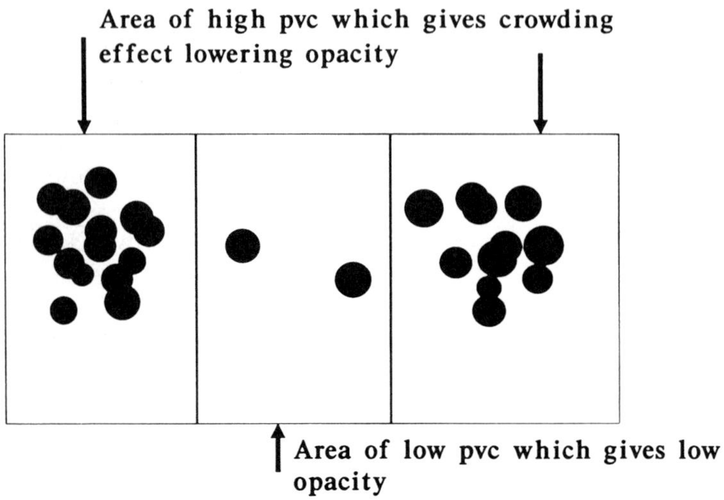

FIG. 9 Schematic diagram of a flocculent film.

through the film, the color observed gives an indication of the state of dispersion. The color is red for dispersed films and changes through lilac and blue to white as flocculation increases. The flocculates can be considered to be particles in their own right, since they scatter different wavelengths of light depending on their size. The larger the size, the greater is the wavelength preferentially scattered. The effect is summarized in Table 2.

This same effect can be seen if reflectance curves of paints with different degrees of flocculation are examined. Figure 10 shows two such curves for paints; in one the pigment is well dispersed, and in the other it is flocculated.

TABLE 2 Appearance of Filament Colors

	Well dispersed	⟶		Severely flocculated
Preferentially scattered light	Blue, green	Green, some blue, some red	Green, red	Infrared
Transmitted light	Red	Some blue, some red	Blue	White
Apparent color of filament	Red	Lilac	Blue	White

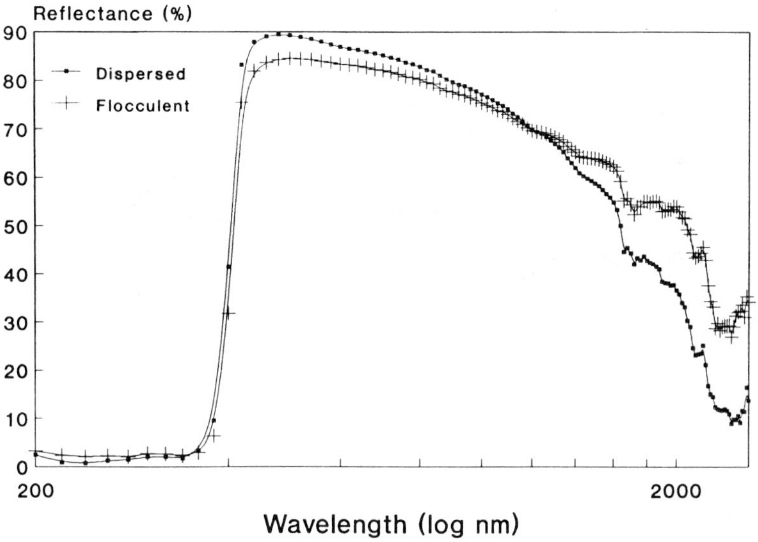

FIG. 10 Reflectance of dispersed and flocculent films.

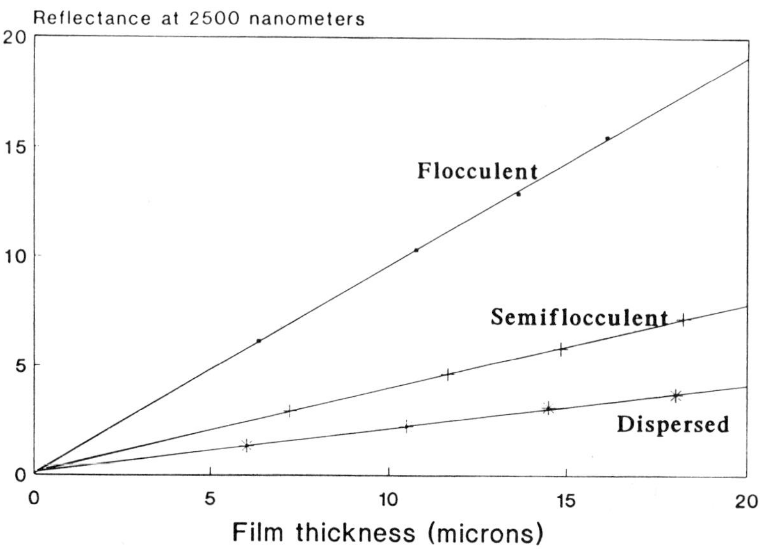

FIG. 11 Derivation of flocculation gradient.

FIG. 12a Scanning electron micrograph of etched paint film flocculation gradient = 0.21.

Well-dispersed paints have higher reflectances in the visible region, demonstrating better optical properties, but lower reflectances in the infrared. Thus, a quantitative test method for measuring flocculation has been devised. In practice, films of paint of different thicknesses are applied to clear polyester foil and the reflectance at 2500 nm is measured and plotted against thickness. Straight lines are achieved up to film thicknesses of approximately 30 μm. The gradient of the line is thus a measure of flocculation; the higher the gradient, the greater the flocculation (see Fig. 11).

The effect is also illustrated by the scanning electron micrographs of etched paint films for paints with differing degrees of dispersion (Fig. 12). The values of flocculation gradients for the same paints are shown.

Using the flocculation gradient technique, the effect of flocculation on paint properties including opacity [5] has been quantified. The effect can be seen in Fig. 13; it can be the opacity drops dramatically as flocculation increases. The values in Fig. 13 are for an alkyd gloss paint that was deliberately flocculated to different extents. It has been found that some commercially available gloss paints have flocculation gradients in excess of 1.

Using 1992 raw material costs in the United Kingdom it has been ascer-

FIG. 12b Scanning electron micrograph of etched paint film flocculation gradient = 0.52.

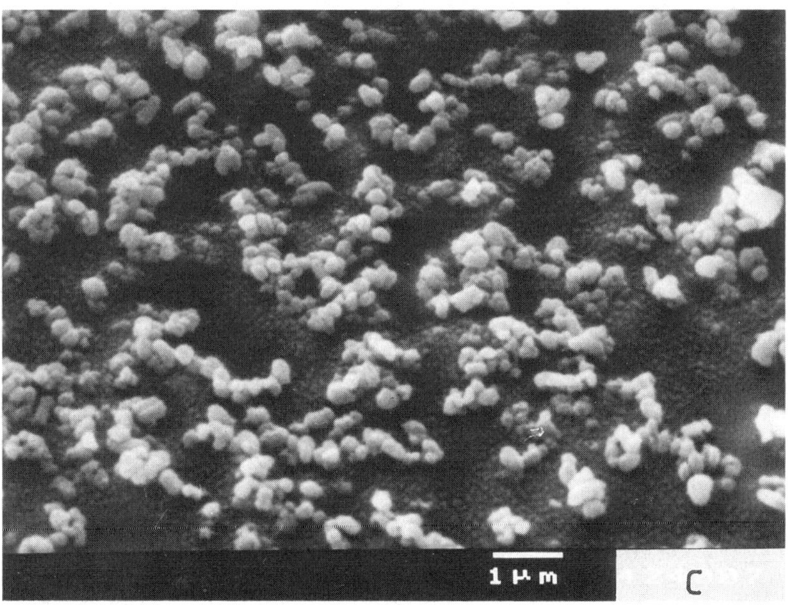

FIG. 12c Scanning electron micrograph of etched paint film flocculation gradient = 0.92.

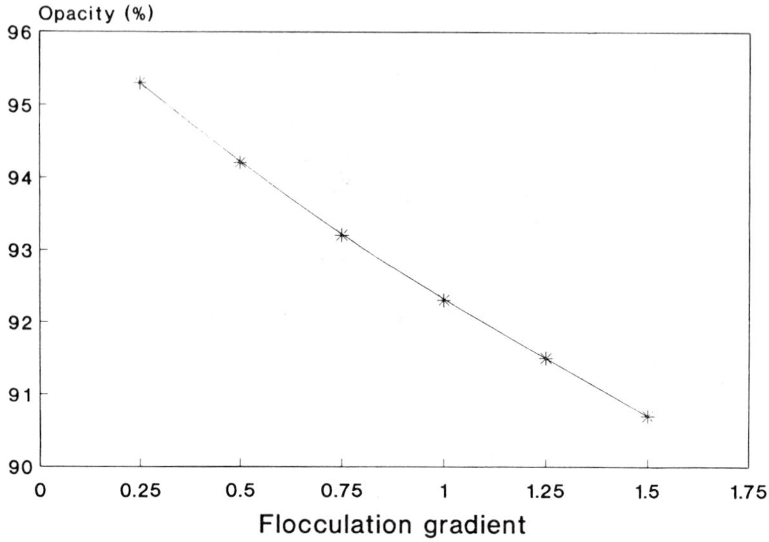

FIG. 13 Effect of flocculation on opacity.

tained that the cost penalties of making flocculent paints can be very significant as shown in Table 3. A series of paints at a pvc of 17.5% were flocculated to different extents and their flocculation gradients and opacities measured. The raw material cost was given an arbitrary value of 100 units. A series of well-dispersed paints were tested at different pigment concentrations, and the pvc required for well-dispersed paints to match the opacity of the flocculent paints was determined. For example, the 17.5% pvc paint with a flocculation gradient of 0.75 has an opacity of 93.2. This opacity could be achieved with a well-dispersed paint with a pvc of 12.8%, and the raw material cost would be 86.1. In other words, a paint of the same opacity could be achieved at a lower pvc and a cost savings of 13.9%. As was mentioned earlier, paints with flocculation gradients of the order of 1 are commercially available, representing a waste of 17% of the raw material cost.

The scattering power of a pigment depends on the difference in refractive index between the pigment particles and the medium surrounding them. If this difference is increased through lowering the refractive index of the surrounding medium, for example through the inclusion of air, opacity can be increased. Figure 14, a continuation of Fig. 6, illustrates this.

The opacity increase at about 50% pvc occurs because there is insufficient medium to surround the pigment completely, and consequently pock-

TABLE 3 Effect of Flocculation on Raw Material Cost

Paints at pvc = 17.5%			Well-dispersed paints		
Flocculation gradient	Opacity (%)	Raw material cost	pvc (%)	Opacity (%)	Raw material cost
0.25	95.3	100	17.5	95.3	100
0.50	94.2	100	14.7	94.2	91.4
0.75	93.2	100	12.8	93.2	86.1
1.00	92.3	100	11.8	92.3	83.0
1.25	91.5	100	11.0	91.5	80.6
1.50	90.7	100	10.5	90.7	79.1

ets of air are included in the film, increasing the scattering power of the pigment. This phenomenon is often exploited with emulsion paints, but in this situation most of the particles present are extenders. Extenders have a refractive index similar to those of most paint media and do not contribute to opacity in a completely bound paint. When air is included, however, the

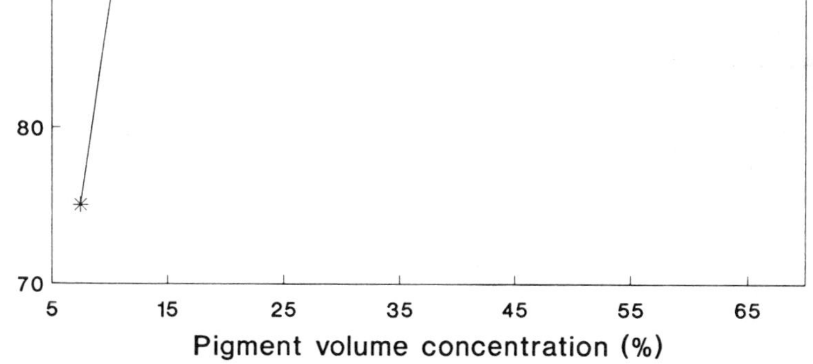

FIG. 14 Effect of pigment volume concentration on opacity. cpvc = critical pigment volume concentration.

difference between the refractive indices of extender and air is sufficient to give significant opacity. The concentration at which air becomes included in the film, known as the critical pigment volume concentration (cpvc), is a point at which a number of other properties change. For example, film integrity normally deteriorates. Paints formulated above the cpvc also show a reduced wet opacity. When the paint is wet, the pigment and extender are surrounded by medium and water, and it is only when the water evaporates that the refractive index difference becomes significant through the inclusion of air. Thus, such a paint is often characterized by a large increase in opacity on drying.

Some products are available on the market that contain encapsulated air and operate on the same principle.

B. Color

As can be seen from Fig. 3, titanium dioxide effectively reflects light across most of the visible spectrum and therefore is able to impart whiteness as well as opacity. Anatase tends to reflect more of the shorter wavelength radiation and consequently gives a bluer tone than rutile. To achieve a high degree of whiteness, it is necessary to manufacture titanium dioxide pigments to a high degree of purity, as contaminants such as colored oxides can have a significant adverse effect at very low concentrations. In principle, the chloride process offers better opportunities for purity as it is a distillation method. However, the various washing and leaching procedures of the sulfate process yield very low concentrations of unwanted contaminants.

The color is also affected by the crystal size of the pigments because as crystal size increases, the wavelengths of light that are predominantly scattered also increase. In practice, however, differences in color are more perceptible in colored systems where the titanium dioxide is mixed with colored pigments. The term "undertone" describes this effect. Undertone can be determined through milling titanium dioxide with a black pigment and assessing the resultant hue. The undertone is chiefly dependent on crystal size, and the effect is shown in Fig. 15, which is a plot of crystal size against undertone expressed as a rating on an empirical standard scale. The smaller the crystal, the bluer the undertone.

Flocculation can also have the effect of imparting yellowness as the flocculates tend to act like large particles and increase the quantity of longer wavelength scattering. The effect is shown in Fig. 16, which is a graph of *b* value (CIELab [6]) against flocculation gradient for a series of alkyd gloss paints.

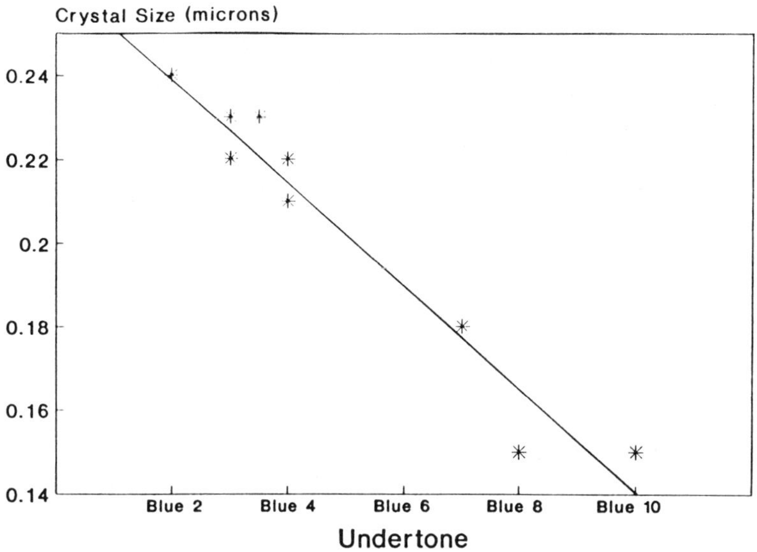

FIG. 15 Effect of crystal size on undertone.

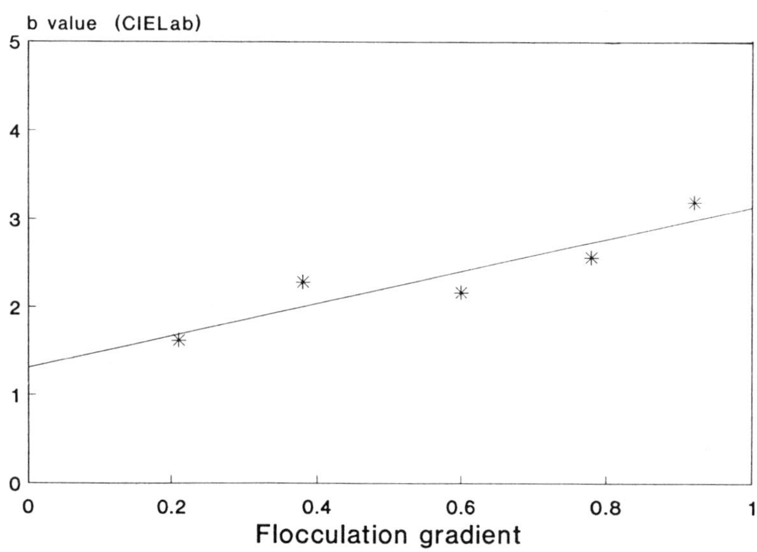

FIG. 16 Effect of flocculation on color.

C. Gloss

The gloss of paint films is a function of the refractive index of the paint, which is not normally a variable, and the roughness of the painted surface. The surface roughness is related to the flow of the film and the effect of pigmentation. The addition of titanium dioxide pigments generally leads to an increase in microroughness because of the presence of particles at the surface leading to a loss of gloss. This effect is shown in Fig. 17, which illustrates that increasing the pigment volume concentration in a vinyl acetate emulsion paint has a significant adverse effect on the 60° gloss.

In some systems, particularly with solvent-borne paints, the effect is much less dramatic, and this is thought to be related to the presence of a clear layer of unpigmented resin at the surface giving a high gloss [7]. Increasing the pigment concentration in such systems does reduce the gloss but at a much reduced rate.

Pigment flocculation also has an adverse effect on gloss because the flocculates act like large particles and protrude through the surface, increasing surface roughness and decreasing gloss.

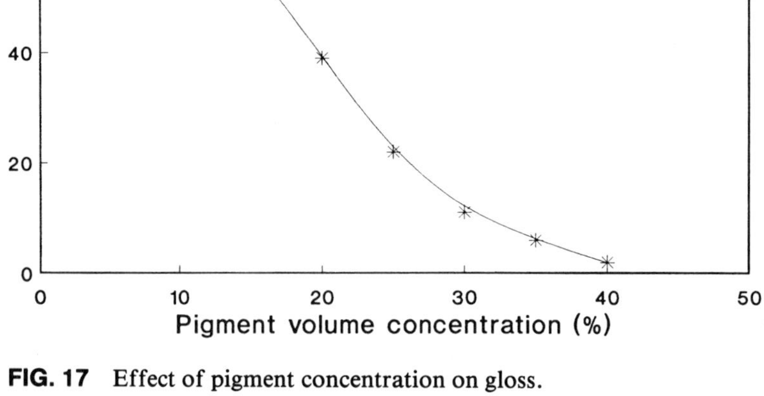

FIG. 17 Effect of pigment concentration on gloss.

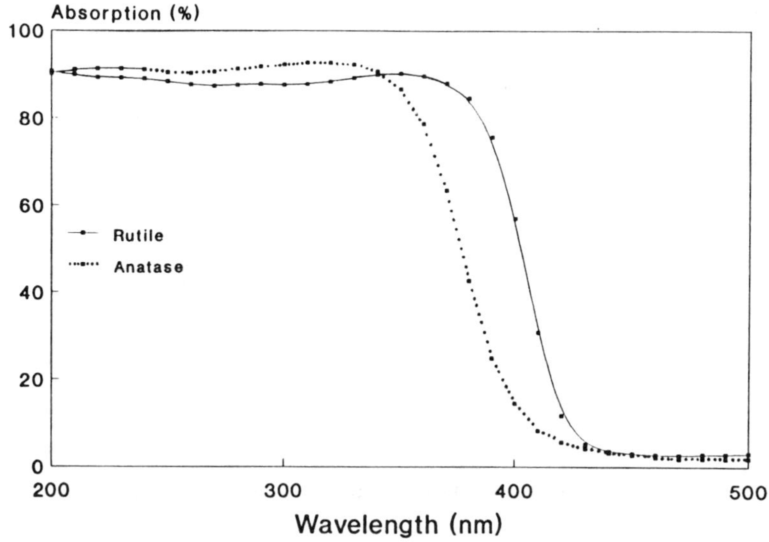

FIG. 18 Ultraviolet absorption by titanium dioxide.

D. Durability

The incorporation of titanium dioxide pigments into polymer systems can have profound effects on the durability of these systems [8]. This is primarily related to the ultraviolet absorption characteristics of the titanium dioxide. Figure 18 is an illustration of the UV absorption behavior of pressed powder samples of anatase and rutile pigments. Rutile begins to show an increase in absorption at about 420 nm whereas anatase does so at about 400 nm, which is why it tends to give a slightly bluer tone than rutile.

When titanium dioxide pigment is incorporated into a system, this absorption considerably reduces the transmission of ultraviolet, as illustrated in Fig. 19. The ultraviolet transmission of an alkyd paint pigmented with rutile can be compared with the unpigmented system shown in Fig. 20.

If a polymer system absorbs ultraviolet light and is susceptible to direct photochemical degradation, the incorporation of titanium dioxide pigments can have a beneficial effect on the durability of the system because of its ability to absorb the radiation and consequently protect the polymer. This effect is shown in Fig. 21, where the degradation for three paints, expressed as mass loss, is plotted against exposure time in a carbon arc weathering machine. They are based on a water-soluble alkyd resin and contain pig-

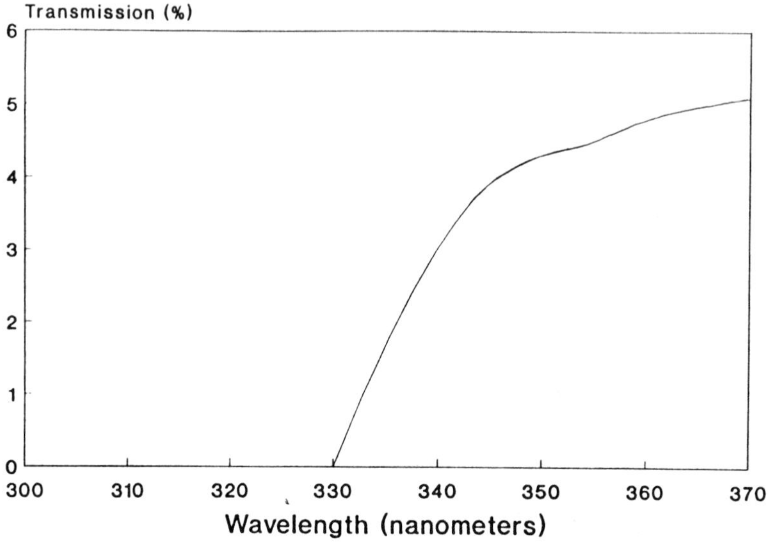

FIG. 19 Ultraviolet transmission of pigmented alkyd paint film.

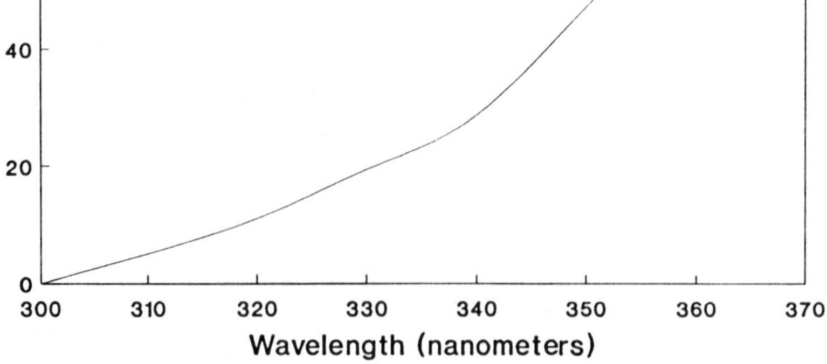

FIG. 20 Ultraviolet transmission of unpigmented alkyd paint film.

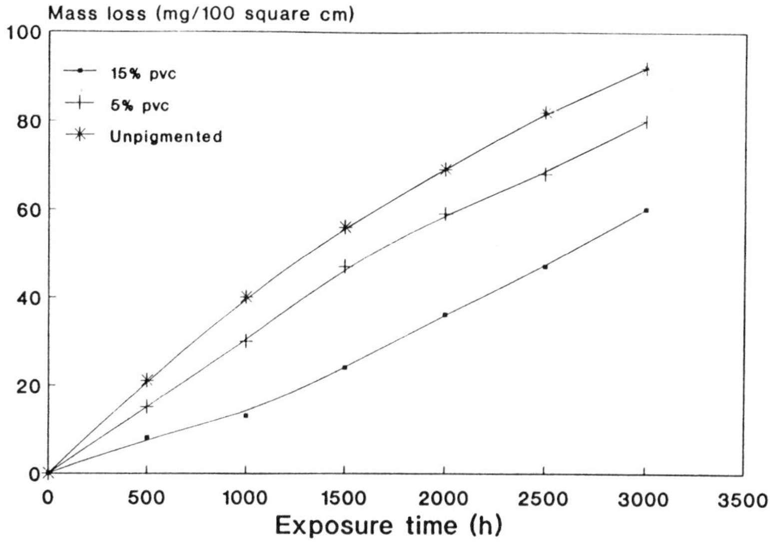

FIG. 21 Effect of pigment concentration on mass loss.

ment incorporated at pigment volume concentrations of 0, 5, and 15%. This clearly shows that in this situation the pigment is protecting the paint film. The rate of mass loss is less at 5% pvc and is further decreased when the concentration is increased to 15% pvc, which is a more realistic level of pigmentation.

Unfortunately, the ultraviolet absorption by titanium dioxide pigments can also give rise to deleterious effects. When rutile titanium dioxide is irradiated with radiation of wavelengths less than 405 nm (a quantum energy equal to or greater than 294.3 kJ/mol), electrons are excited from the valence band to the conduction band, leaving positive holes in the valence band. The holes and excited electrons are free to move within the crystal lattice of the titanium dioxide crystal where they are able to recombine. However, some electrons and holes reach the surface of the pigment particles and can the initiate chemical reactions in the surrounding polymer, thereby leading to degradation.

When the positive holes reach the crystal surface, they react with surface hydroxyl groups to form adsorbed hydroxyl radicals. Electrons reaching the surface react with adsorbed oxygen to form the O_2^- radical, which then is able to react with water to form the $HO_2\cdot$ radical.

$$TiO_2 + \text{UV light} \longrightarrow \text{excited electron} + \text{positive hole}$$

$$OH^- + \text{positive hole} \longrightarrow OH\cdot$$

$$e^- + O_2 \longrightarrow O_2^-$$

$$O_2^- \times H_2O \longrightarrow OH^- + HO_2\cdot$$

The end result of these reactions is a species with an unpaired electron, and these radicals can react with binder molecules to cause chain-breaking reactions, cross-linking leading to embrittlement, and other reactions that increase the rate of degradation.

The formation of hydroxyl and hydroperoxy radicals has been confirmed by means of electron spin resonance (ESR) [9], which depends upon the conversion of electrons from one spin state to another in a magnetic field by a radiofrequency field. The precise energy to effect this transition is sensitive to the immediate environment of the radical, thus allowing different radicals to be distinguished. This technique has proved that water is essential for the formation of radicals [10]. When pure, dry TiO_2 was irradiated under vacuum, the only signals apparent were due to Ti_3^+ and crystal defects. However, when it was irradiated in the presence of water vapor, a new signal was found that was attributed to $HO_2^{\cdot}$ radicals. This confirms the requirement for water to be present as is suggested by the above equation sequence.

To reduce the amount of photocatalytic degradation, manufacturers of titanium dioxide pigments dope the crystal with metals such as aluminum that act as hole and electron recombination sites. Also, pigment crystals are coated with hydrous oxides of aluminum, silicon, and zirconium, which destroy the hydroxy radical by providing an area for recombination.

$$4\,OH^{\cdot} \rightarrow 2\,H_2O + O_2$$

It has been shown [11] that the photoactivity of titanium dioxide is independent of wavelength at wavelengths below 405 nm. Thus, for a constant intensity, provided that the radiation is of a lower wavelength than 405 nm, photocatalytic degradation is independent of wavelength.

Thus, there are two basic degradation mechanisms in pigmented polymer systems—photochemical (dependent on UV absorption by the polymer) and photocatalytic (dependent on UV absorption by the pigment). A comparison of the effects of the two mechanisms is shown in Fig. 22. With the photochemical mechanism, the pigment particles prevent the passage of ultraviolet light so the area immediately below the pigment particles is

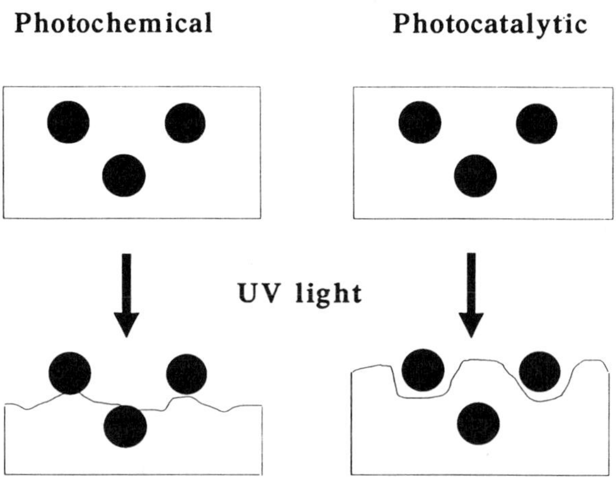

FIG. 22 Comparison of degradation mechanisms.

screened and protected. Consequently, during the early stages of weathering, the particles remain adhered to the film surface and the onset of chalking is delayed. After weathering, the surface is degraded, and the pigment particles are left standing proud on the surface. These protrusions lead to an irregular surface, which has the visual effect of lowering the gloss.

With photocatalytic degradation, which occurs at the pigment–polymer interface, the area around the particles is affected most. The film surface remains fairly flat; the effect on gloss is not so pronounced and remains higher than with the other mechanism. However, because the photocatalytic activity occurs all the way around the particles, they become detached from the surface and consequently show a degree of chalking earlier in the degradation process than with the photochemical mechanism.

Obviously, in most real systems, both mechanisms will operate. Which of them predominates, if either, will depend on the choice of pigment and resin.

Some work carried out by Day [12] illustrates the effects of different grades of titanium dioxide at various concentrations on the durability of plastics applications. The pigments used are described in Table 4.

Compression-molded plaques of rigid polyvinyl chloride (PVC) were pigmented with concentrations of 3–17% by weight and exposed for 3 years in Florida. The average chalk and gloss results are illustrated in Figs. 23 and 24. The plaques containing anatase (A) and lightly coated rutile (C) gave a high degree of chalking that increased with pigment concentration, indicating a high rate of photocatalytic degradation. In this situation the

TABLE 4 Designation of Pigments[a]

Designation	Description
A	Uncoated anatase
B	Coated anatase
C	Lightly coated (1–2%), fine crystal (0.18 μm) rutile
D	Lightly coated rutile (0.23 μm)
E	Medium coated (5–7%), highly durable, fine crystal rutile
F	Medium coated, highly durable rutile
G	Heavily coated, superdurable rutile

[a]See text and Figs. 23–26.

influence of the pigment is destructive, so the more pigment present, the greater the rate of degradation.

The durable grade (F) and the superdurable grade (G) gave much less chalking, especially at loadings above about 8%. The gloss ratings shown in Fig. 24 indicated a similar trend. This shows the pigment concentration dependence of protection from ultraviolet degradation of the polymer, indicating the sensitivity of the resin to UV radiation.

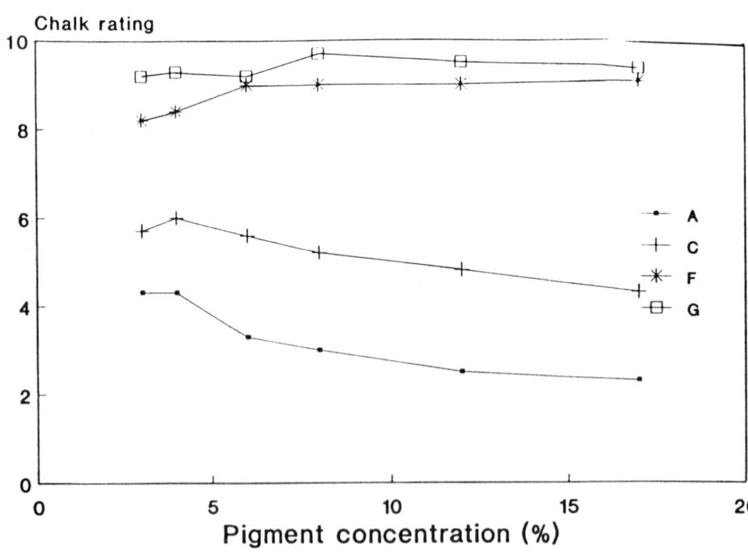

FIG. 23 Average chalking of uPVC after 3 years exposure in Florida. 10 = no chalking; 0 = severe chalking.

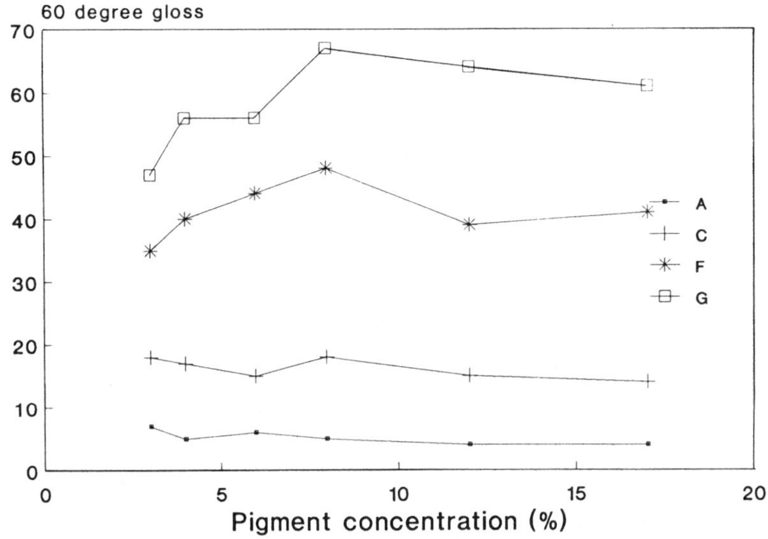

FIG. 24 60 degree gloss of uPVC after 3 years exposure in Florida.

One important characteristic of weathering is a tendency for embrittlement to occur. This has been investigated by Day using the same pigment grades (Table 4) incorporated into polyethylene. Strips of polyethylene (100 μm thick) were exposed in a carbon arc weathering machine, and the percent elongation to break was determined after various durations of exposure. Before exposure, there was little difference between grades over the concentration range of 2–8%. Figure 25 shows examples of plots of elongation at break versus exposure time. Table 5 shows the embrittlement times. It can be seen that there is a wide range of behavior, with the differences between lightly coated and superdurable rutile being as great as between anatase and lightly coated rutile. At each loading the order of improved durability matches that found in rigid PVC.

Using infrared spectroscopy, the same exposed films showed the formation of acidic and ketonic carbonyl groups and the spectra indicated that the nature of the photooxidation process was the same for all the grades, the only difference being in the rate of reaction. The rate of formation of carbonyl groups was measured by monitoring the carbonyl indices [13] at the peak absorption (1720 cm^{-1}), and plots of the type shown in Fig. 26 were prepared. It is noticeable that there is an induction time for the more durable pigments but not with the lightly coated rutile grades and anatase.

Another important aspect related to durability is the state of pigment

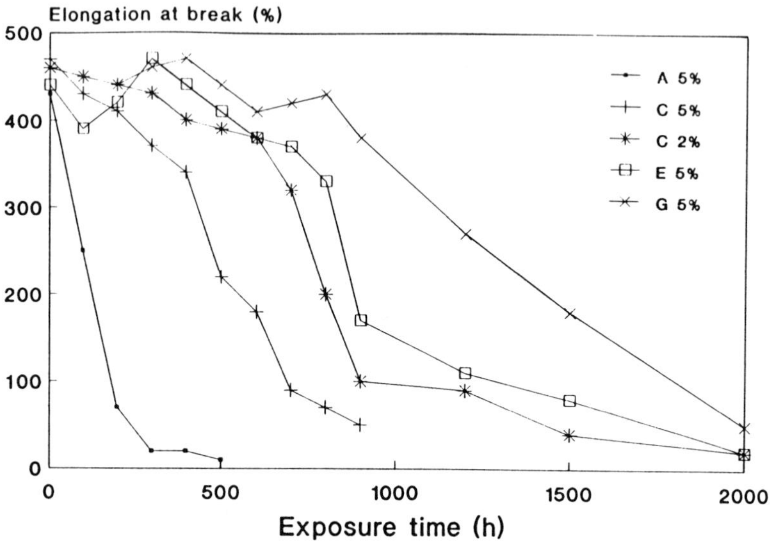

FIG. 25 Change in elongation at break of polyethylene film.

TABLE 5 Embrittlement Times for Carbon Arc Exposed Polyethylene Film[a]

	Concentration		
Pigment[b]	2%	5%	8%
A	225	125	125
B	—	250	—
C	800	475	450
D	—	525	—
E	1100	950	850
F	1250	875	850
G	1700	1300	950

[a]Times in hours to 50% reduction in percentage elongation at break.
[b]See Table 4 for details.

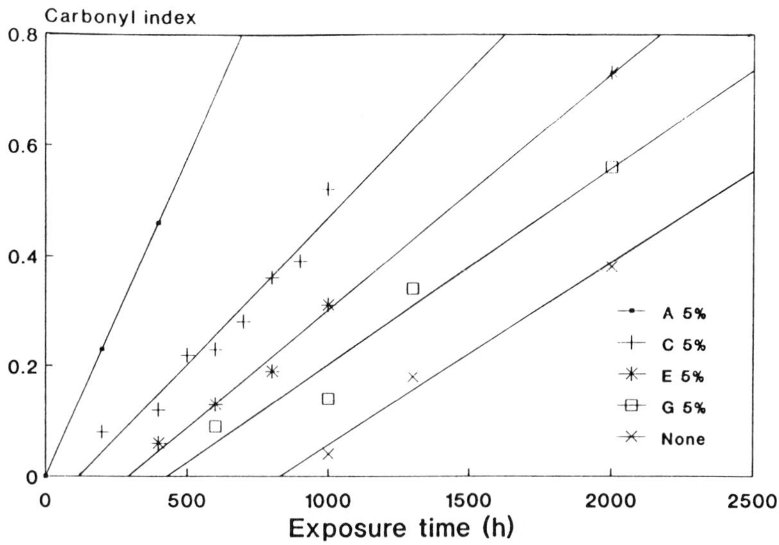

FIG. 26 Change in carbonyl index of polyethylene film.

dispersion. When pigment is initially incorporated into an application medium, the pigment particles are subjected to forces designed to break them down into individual crystals during the milling operation. This ideal situation is never achieved. There are always some residual strongly coherent aggregates. However, if the dispersion is poor and the number of such aggregates is large, they will protrude through the surface of the film, giving poor gloss. When the surface is eroded during weathering, this effect becomes more pronounced and the rate of loss of gloss becomes much greater than with a paint in which the pigment particles have been efficiently dispersed. Also, if large aggregates are present that contain clumps of pigment, with the individual particles incompletely wetted out, these can form the centers for mechanical breakdown because of poor film integrity. Normally, this kind of poor dispersion is a result of inefficient milling conditions or insufficient milling time and can be simply checked by examining the paint on a fineness of grind gauge.

Durability is also affected by pigment flocculation. Figure 27 shows the effect of flocculation on gloss retention. To put the numbers in perspective, the range of flocculation gradient values found in commercially available white gloss paints lies between about 0.2 and 0.9. The figure shows the effect on three alkyd gloss paints with differing degrees of pigment flocculation in the mildest of exposure conditions. The panels were left in a laboratory under controlled temperature (20°C) and humidity (50%) and the gloss

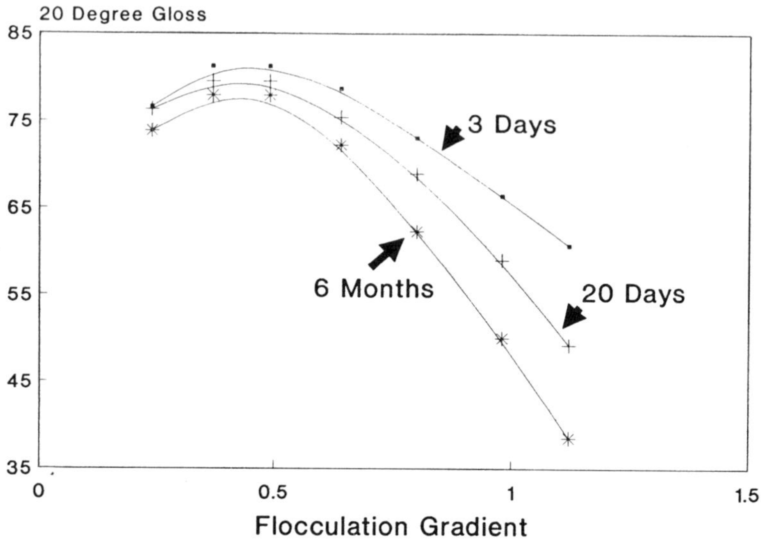

FIG. 27 Effect of flocculation on gloss retention.

measured after 3 days, 20 days, and 6 months. Even under these mild conditions the loss of gloss with increasing flocculation was very significant.

As is predictable, the effect under more severe conditions is more dramatic. Figure 28 shows what happens to paints with different levels of flocculation during exposure in an accelerated weathering machine for 1000 h. In this case the paints have flocculation gradients between 0.29 and 0.50. There is a sharp decline in gloss over the first 300 h, with the most flocculent of the paints losing nearly 50 units and the best losing about 25. The curves flatten out after longer exposure times because, especially with more flocculent paints, there is little gloss left to lose. This loss of gloss occurs when the degradation begins and the particles become exposed and protrude from the surface. With the poorly dispersed paint, the flocculates are large, and after degradation begins they protrude to a greater extent and the rate of loss of gloss is much greater.

Film thickness can have an effect. Figure 29 shows that the effect of decreasing the film thickness is a greater reduction of gloss with flocculent paints than with dispersed paints. This particular graph refers to initial gloss before any weathering has taken place. The results are to be expected because flocculates will protrude to a greater extent from a thin film than from a thicker one. However, when degradation takes place this difference

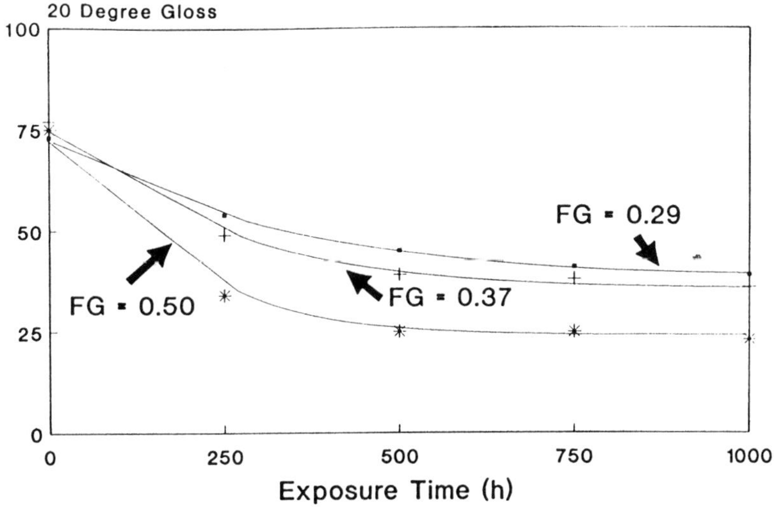

FIG. 28 Effect of flocculation on durability.

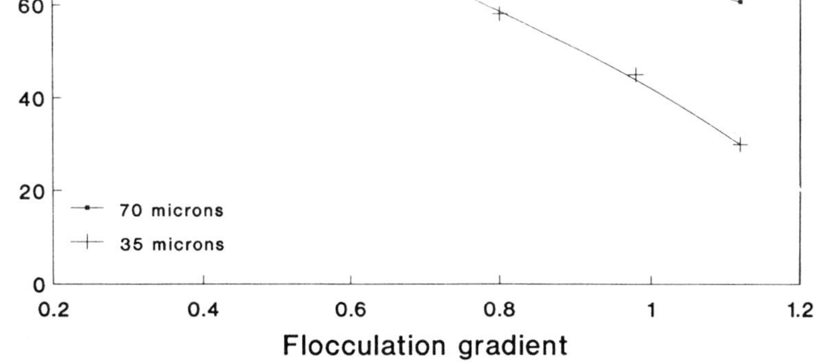

FIG. 29 Effect of film thickness and flocculation on gloss.

becomes even more exaggerated, and a combination of a flocculated pigment and a thin film will lead to a rapid rate of loss of gloss.

The overall effect of flocculation will depend on which of the two degradation mechanisms is predominant. In Figure 30, the paint on the left side is well dispersed and the particles well separated. As a result, they are able to absorb the incident ultraviolet radiation efficiently. In terms of protecting the surrounding resin this is a healthy situation; but if the photocatalytic mechanism is operating, then there are more sites available at which degradation can begin. With the flocculent paint, the individual pigment particles are clustered together and absorb the radiation less effectively. Thus, they are not as efficient at protecting but offer fewer sites for the onset of photocatalytic degradation. The overall effect is also dependent on the susceptibility of the resin to the two forms of attack.

This is, however, a theoretical situation. In practice flocculation should be minimized, as it is wasteful not to do so. When high durability is required, a well-dispersed durable pigment should be used.

In an age when some paints can be expected to last for decades, the luxury of testing for durability under natural conditions is not normally an acceptable option. Increasingly, manufacturers need to use accelerated testing as a means of comparing the durability of their products. The correlation between accelerated and natural weathering is a subject that has been studied for many years and that has generated many publications (e.g. [14]). Various aspects of weathering have been studied with a view to improving the reliability of accelerated data. In general, correlation is obtained in isolated cases, but overall absolute agreement is poor.

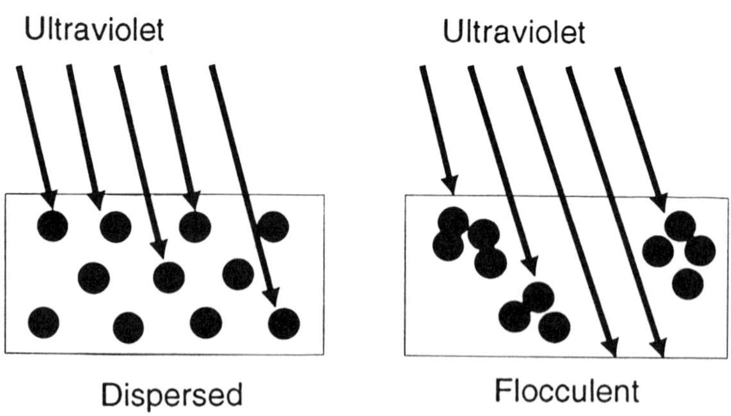

FIG. 30 Effect of flocculation on UV absorption.

To achieve results quickly, a range of equipment is available that attempts to mimic and intensify the natural conditions that cause degradation, particularly through the use of various ultraviolet light sources. These are normally based on xenon, mercury, and carbon arcs or fluorescent tubes. There are, however, a number of associated difficulties, especially with materials pigmented with titanium dioxide.

Because titanium dioxide absorbs all wavelengths below 410 nm, all the available ultraviolet can cause photocatalytic degradation. Binders, however, absorb to a lesser extent and over smaller wavelength ranges. The photocatalytic effect occurs at all wavelengths in the ultraviolet, whereas the photochemical effect can occur only where the resin absorbs. Thus the selection of wavelengths used in accelerated testing is crucial as the balance between photochemical and photocatalytic degradation may not be representative of exposure under natural conditions. The search for correlation between natural and accelerated test results is made more difficult when it is recognized that there are significant differences in results obtained when similar paints are exposed at different natural exposure sites. For example, it has been found that the relationship between exposures in Florida and those in Northern Australia is good but that that between the United Kingdom and Northern France is poor. Differences in latitude, humidity, and prevailing weather all cause differences in the nature of the irradiation and therefore in the response of pigments to weathering. Ideally, the products should be tested in the intended place of application.

E. Other Properties

Because titanium dioxide is a relatively inert material, the effect of differences between various grades is mainly limited to the properties already mentioned. Its inertness leads to a lack of toxicity, and this is typified by its use in the food, cosmetics, and pharmaceutical industries, where pure grades of anatase are generally preferred.

In some applications, particularly in textile dyes and printing inks, the hardness of titanium dioxide can cause some problems because of abrasion of manufacturing machinery, and in these situations anatase is preferred because it is not as hard as rutile and consequently exhibits lower abrasivity.

The rheology of liquid is influenced by the presence of particulate matter, and in some specific instances the choice of pigment grade can have an effect. This may occur either through specific interactions resulting from the coating on the pigment surface or through different degrees of adsorption of materials such as polymers onto the surface. The latter differences arise from variations in surface area, which are dependent on the pigment coating or crystal size. For the same reasons, pigment choice may influence the mechanical properties of solid materials, but, normally, other formulation parameters have a much greater effect than the difference between grades.

REFERENCES

1. R. R. Blakey and J. E. Hall, in *Pigment Handbook* (P. A. Lewis, Ed.), Wiley, New York, 1988, Vol. 1, p. 1.
2. J. G. Balfour, *J. Oil Colour Chem. Assoc. 73*: 225 (1990).
3. J. G. Balfour and M. J. Hird, *J. Oil Colour Chem. Assoc. 58*: 331 (1975).
4. J. G. Balfour, *Surf. Coatings Int. 75*: 21 (1992).
5. J. G. Balfour, *J. Oil Colour Chem. Assoc. 60*: 365 (1977).
6. CIE Publication No. 15, Suppl. No. 2, 1976.
7. R. D. Murley and H. Smith, *J. Oil Colour Chem. Assoc. 53*: 292 (1970).
8. J. G. Balfour, *Polym. Paint Colour J. 182*: 283 (1992).
9. B. Randy and J. F. Rabek, *Photodegradation, Photooxidation and Photostabilisation of Polymers*, Wiley, New York, 1975.
10. A. Serwicka, *Colloids Surf. 13*: 287 (1985).
11. T. A. Egerton and C. J. King, *J. Oil Colour Chem. Assoc. 62*: 386 (1979).
12. R. E. Day, *Polym. Degrad. Stab. 29*: 73 (1990).
13. N. S. Allen, A. Chirinos-Padron, and J. H. Appleyard, *Polym. Degrad. Stab. 9*: 15 (1984).
14. L. Cutrone and D. V. Moulton, *J. Oil Colour Chem. Assoc. 70*: 219 (1987).

3

Colored Inorganic Pigments

HANS SANDER Pigments and Ceramic Department, Bayer AG,
Krefeld, Germany

I.	Introduction	106
II.	Chemical and Physical Properties	107
	A. General composition	107
	B. Crystallography	107
	C. Color and crystal chemistry	110
	D. General chemical properties	112
	E. Particle characteristics	112
	F. General physical properties	114
	G. Optical properties	115
III.	Iron Oxide Pigments	117
	A. General	117
	B. Manufacture	119
	C. Properties	123
IV.	Oxidic Mixed-Phase Pigments	129
	A. General	129
	B. Manufacture	129
	C. Properties	130
V.	Cadmium Pigments	132
	A. General	132
	B. Manufacture	134
	C. Properties	134

VI. Chromium Oxide Pigments 135
A. General 135
B. Manufacture 136
C. Properties 136

VII. Chromate and Molybdate Pigments 137
A. General 137
B. Manufacture 137
C. Properties 138

VIII. Miscellaneous Pigments 139
A. Iron blue pigments 139
B. Ultramarine pigments 139
C. Transparent pigments 139
D. Lustrous pigments 140
E. Magnetic pigments 141

References 142

I. INTRODUCTION

The modern definition of the umbrella term "colorant" encompasses not only dyestuffs and pigments but also extenders [1]. A *pigment* is a colorant that is virtually insoluble in the medium in which it is applied. It can be organic or inorganic, chromatic or achromatic. A *dyestuff* is an organic colorant that is usually soluble in the medium in which it is applied.

Natural inorganic pigments have been used as colorants since prehistoric times. Evidence of the use of these pigments more than 20,000 years ago can still be seen today in the cave paintings of southern France, Spain, and North Africa [2]. Around 4000 years ago, the Chaldeans were manufacturing red and violet pigments by roasting natural ocher. Arsenic sulfide was the first pure yellow pigment, lapis lazuli and synthetic ultramarine the first blues, green earth and malachite the first greens. Painting, enameling, glass manufacturing, and dyeing techniques reached their first peak in ancient Egypt. A synthetically produced ultramarine (a copper-calcium silicate) developed at that time is known to this day as Egyptian blue. Antimony sulfide and galena were often used to produce black, cinnabar to produce red, and ground cobalt glass to produce blue. Although dyeing was highly valued by both the Greeks and the Romans, the two cultures contributed virtually nothing to the development of new pigments. From the time of the great migration of the peoples to the close of the late Middle Ages there was no significant enrichment of the range of colors available. It was not until the start of the Renaissance period that new pigments were developed.

A true synthetic pigments industry only grew in the eighteenth century with the development of Berlin blue (1704), cobalt blue (1777), and chrome yellow (1778). In the nineteenth century, ultramarine and Guignet's green as well as cobalt, iron, and cadmium pigments were all discovered in rapid succession. The twentieth century has seen a more scientific approach to pigment development, leading to the introduction of synthetic color pigments such as cadmium red, manganese blue, molybdate red, and white titanium dioxide in anatase and rutile form.

Inorganic pigments can be classified according to a number of different criteria, three of which are chemical structure, crystal form, and color (see Table 1).

The worldwide increase in the use of color has resulted in a dramatic rise in demand for colorants. Natural sources have long been unable to keep pace with demand, and the products no longer satisfy today's quality requirements. By 1990, the production of natural pigments, mainly iron oxides, was only around 80,000 tonnes. In the last few decades, world production capacities for synthetic inorganic pigments and carbon blacks have been expanded. By 1990, they totaled some 5 million tonnes. Table 2 provides an overview of the most important pigments [3]. The major producers are in the United States and western Europe.

Figure 1 shows the main fields of application for the most important inorganic pigments. There are concentrations of use within these fields [4].

II. CHEMICAL AND PHYSICAL PROPERTIES

A. General Composition

With a few exceptions, inorganic pigments are oxides, sulfides, oxide hydroxides, silicates, sulfates, and carbonates. They are generally made up of individual particles of a uniform chemical composition.

Extended pigments and lakes do not have a uniform composition. Extended pigments are mixtures of pigments, possibly with extenders, that are formed in a dry process. For example, chrome green pigments are mixtures of lead chromate and iron blue pigments. Such mixtures may separate when used because of the differing particle sizes and shapes.

In the case of lakes, at least one further pigment component is precipitated onto a substrate (extender, pigment), usually in a wet process. Drying or calcination produces forces of varying strength between the solid particles, which has the effect that the components do not separate during application.

B. Crystallography

Inorganic pigments are based mainly on the following crystal classes:

TABLE 1 List of Color Pigments

Pigment	Chemical formula	Crystal structure	Color
Iron hydroxide			
Goethite	α-FeOOH	Diaspore	Green yellow to orange yellow
Lepidokrokite	γ-FeOOH	Boehmite	Yellow to orange
Iron oxide			
Hematite	α-Fe_2O_3	Corundum	Red yellow to red blue
Maghemite	γ-Fe_2O_3	Spinel	Brown to red brown
Magnetite	Fe_3O_4	Spinel	Black
Chromium oxide			
Chrome oxide green	α-Cr_2O_3	Corundum	Green
Chromium hydrate green	γ-CrOOH	Boehmite	Blue green
Lead chromium pigment			
Chrome yellow	$Pb(Cr,S)O_4$	Anglesite (orthorhombic)	Greenish yellow
		Crocoite (monoclinic)	Light yellow to medium yellow
Molybdate orange	$Pb(Cr,Mo,S)O_4$	Scheelite (tetragonal)	Orange to red
Molybdate red	$Pb(Cr,Mo,S)O_4$	Scheelite (tetragonal)	Orange to red
Chrome orange	$PbCrO_4 \cdot PbO$	Monoclinic	Orange to red
Chrome red	$PbCrO_4 \cdot PbO$	Monoclinic	Orange to red

Mixed-phase pigment			
Cobalt blue	$CoAl_2O_4$	Spinel	Reddish blue
Cobalt blue	$Co(Al,Cr)_2O_4$	Spinel	Greenish blue
Cobalt green	$(Co,Ni,Zn)_2\,(Ti,Al)O_4$	Spinel	Green
Zinc iron brown	$ZnFe_2O_4$	Spinel	Light brown to medium brown
Zinc iron black	$Cu(Fe,Cr)_2O_4$	Spinel	Black
Chrome rutile yellow	$(Ti,Cr,Sb)O_2$	Rutile	Ocher yellow
Chrome rutile yellow	$(Ti,Cr,Nb)O_2$	Rutile	Ocher yellow
Nickel rutile yellow	$(Ti,Ni,Sb)O_2$	Rutile	Lemon yellow
Chrome rutile brown	$(Ti,Mn,Sb)O_2$	Rutile	Light brown to dark brown
Bismuth vanadate/ molybdate yellow	$BiVO_4 \cdot 0.75\ Bi_2MoO_6$	Scheelite (tetragonal/ orthorhombic)	Greenish yellow
Iron manganese black	$(Fe,Mn)_2O_3$	Bixblite	Black
Iron manganese brown	$(Fe,Mn)_2O_3$	Corundum	Light brown to red brown
Iron chrome brown	$(Fe,Cr)_2O_3$	Corundum	Light brown to red brown
Manganese blue	$BaSO_4 \cdot Ba_3(MnO_4)_2$	Barite	Blue
Cadmium pigment			
Cadmium yellow	$(Cd,Zn)S$	Wurtzite	Greenish yellow to gold yellow
Cadmium orange	$Cd(S,Se)$	Wurtzite	Light orange to dark orange
Cadmium red	$Cd(S,Se)$	Wurtzite	Light red to dark red
Ultramarine pigment			
Ultramarine blue	$Na_8[Al_6Si_6O_{24}] \cdot S_x;\ x = 4$	Sodalite	Greenish blue to reddish blue
Iron blue	$M^{(I)}Fe^{(II)}Fe^{(III)}(CN)_6 \cdot x\ H_2O;\ M = Na, K, NH_4$	NaCl, variety (cubic)	Light blue to dark blue

Cubic	Sphalerite lattice (e.g., CdS)
	Spinel lattice (e.g., Fe_3O_4, $CoAl_2O_4$)
Tetragonal	Rutile lattice (e.g., TiO_2, SnO_2)
Rhombic	Goethite lattice (e.g., α-FeOOH, α-AlOOH)
Hexagonal	Corundum lattice (e.g., α-Fe_2O_3, α-Cr_2O_3, α-Al_2O_3)
Monoclinic	Monazite lattice (e.g., $PbCrO_4$)

C. Color and Crystal Chemistry

Color is an important aspect of expression throughout nature. In the pigmented systems under discussion here, as is the case with all non-self-luminous pigments, the perception of color is based on the subtractive color mixture. Color pigments absorb certain sections of the spectrum from the spectral energy distribution $S(\lambda)$ of incident light, which means that the eye receives only the spectral fraction $S(\lambda) \times \beta(\lambda)$, where $\beta(\lambda)$ is the spectral reflection factor. In contrast, white pigments are characterized by nonselective scattering, and black pigments by nonselective absorption, over the entire spectrum of visible light [5].

The selective light absorption of color pigments results from the interaction of visible light with the valence electrons in the solid substance. Depending on the elements (cations and anions, their oxidation number and electron configuration) and the effects of the crystal lattice, certain electronic structures and energy balances form in the solid substances via transition mechanisms such as intervalence transfers between ions with different

TABLE 2 World Production Capacities of Inorganic Pigments—1990

Pigment	Tons	%
1. Titanium dioxide	3,389,000	66.46
2. Iron oxide	610,000	11.96
3. Zinc oxide	600,000	11.76
4. Lithopone	275,000	5.39
5. Lead chrome yellow	90,000	1.76
6. Chrome oxide green	40,000	0.78
7. Iron blue	30,000	0.59
8. Molybdate red	25,000	0.49
9. Ultramarine pigments	24,000	0.47
10. Mixed-phase pigments	11,000	0.22
11. Cadmium pigments	6,000	0.12
	5,100,000	100.00

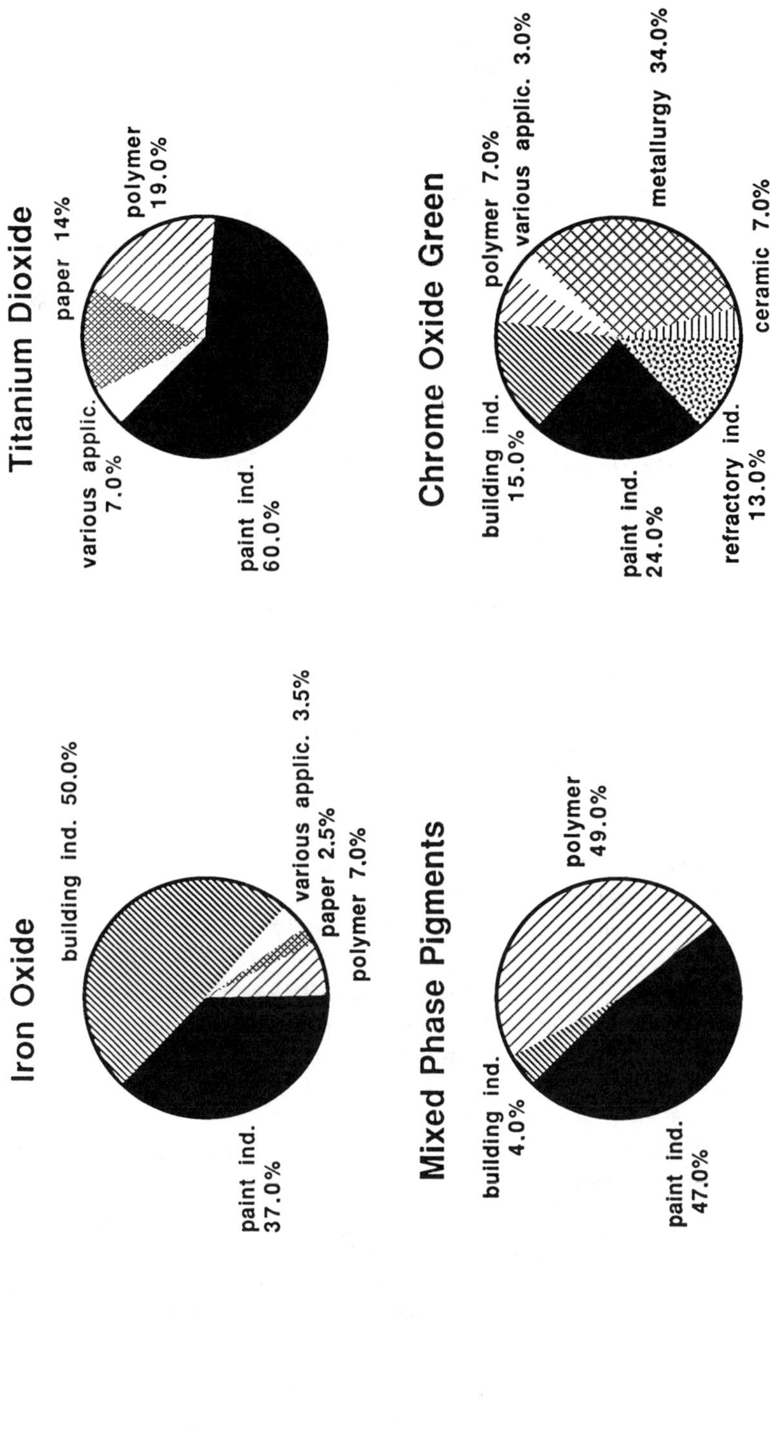

FIG. 1 Markets for inorganic pigments.

valency numbers, transfers between molecular orbitals in nonmetals, and transitions from the valence band to the conductivity band. These electronic transitions result in selective light absorption of varying intensity and produce the color of the pigment if the absorption bands are within the visible spectral range or on the boundaries of the UV or IR ranges.

D. General Chemical Properties

The general chemical properties of inorganic pigments include resistance to water, alkalis, acids, and solvents. The acid value (or alkali value) indicates how many milliliters of a 0.1 normal alkaline (or acidic) solution are required to neutralize an aqueous extract of 100 g of pigment under given conditions.

The pH of a pigment is the pH of an aqueous suspension of the pigment that is produced in a given way (ISO 787/9). Electrical conductivity and resistivity are also determined in aqueous pigment extracts.

E. Particle Characteristics

The important physical characteristics of a pigment are particle size and shape, particle size distribution, specific surface, sieve residue, hardness, density, bulk density, and oil absorption [6]. Table 3 shows these values for a number of inorganic pigments.

The primary particles of inorganic pigments are individual crystals, which can be cuboid, spherical, rodlike, or plate-shaped, depending on the chemical type of pigment and its preparative history. These can be seen in electron micrographs and, if large enough, in optical micrographs. In some circumstances, however, the optically effective particles can be segments within the crystal (crystallites) that scatter light.

The primary particles have a high surface area per unit weight (specific surface area) and tend to form aggregates. These may be compact (face-to-face contacts of crystals) or open-structured (edge or point contacts).

Specific surface area determinations (usually by nitrogen gas adsorption and the BET method of calculation) are useful in distinguishing these types of aggregate structures. The gas molecules can penetrate between particles in open (macroporous) structures and thus gain access to particle surfaces within. In such cases the measured specific surface tends to be comparable to the specific surface of the primary particles. With compact aggregates the gas molecules cannot readily penetrate between the crystals, and so the measured area is much less than the specific surface of the primary particles. The Carman flow resistance method is also of some importance in determining the specific surface area of pigment powders.

A *flocculate* is a loosely coherent aggregate that tends to form in suspen-

TABLE 3 General Technical Data of Inorganic Pigments

Pigment	Particle shape	Predominant particle size (μm)	Sieve residue on 0.045 mm sieve (%)	Oil absorption (g/100 g pigment)	Density (g/cm^3)	Tamp density (g/cm^3)
Iron oxide red	Spherical	0.1–0.7	≤0.05	15–28	4.8–5.1	0.9–1.5
Iron oxide yellow	Acicular	0.1 × 0.3 to 0.2 × 0.8	≤0.05	30–65	4.0–4.2	0.4–0.8
Cadmium yellow	Cubic to spherical	0.2–0.3	≤0.01	18–20	4.5–5.0	0.6–0.7
Cadmium orange	Spherical	0.2–0.3	≤0.01	17–18	4.5–5.0	0.8–1.0
Cadmium red	Spherical	0.3–0.4	≤0.01	14–16	5.0–5.5	1.0–1.1
Chrome oxide green	Spherical	0.3–0.4	≤0.02	10–12	5.1–5.2	1.0–1.3
Nickel rutile yellow	Spherical	0.3–0.4	≤0.03	16–17	4.5–4.6	0.9–1.4
Chrome rutile yellow	Spherical	0.3–0.5	≤0.03	15–20	4.3–4.4	0.7–1.2

sion in liquid media (e.g., a pigment–binder system, such as an ink or paint) and can be broken down by very weak shearing forces. Flocculates also occur in coated and dried paint and ink layers.

The nature of the particulate species that exist in dispersion in an application system is primarily dependent upon the preparative history of the pigment and the type of dispersion medium. The factors involved are the wettability of the pigment, the polarity of the dispersion medium, the method of dispersion used to break down the aggregates in the pigment powder, and the tendency of the separated particles to reaggregate or form flocculates.

There is a distribution of particle sizes that can be determined by various means such as microscopy, centrifugal sedimentation, and laser light scattering, the most useful parameters being the mean size and the distribution range. The basis of the size distribution (e.g., whether volume or number distribution) and the type of distribution curve used (size frequency or cumulative), and hence the numerical value of the mean, depends upon the application property being considered. Mean particle sizes of inorganic pigments are between 0.01 and 10 μm, preferably between 0.1 and 1 μm.

F. General Physical Properties

Determination of the *sieve residue* establishes the grit content of a pigment. The grit consists of pigment particles that are considerably larger and possibly harder than the normal particles. It may also include impurities or strongly coherent pigment aggregates.

The *density* of a substance is its mass in relation to its volume, and the unit of measurement is g/cm^3. Density is measured in accordance with, for example, ISO 787/10. The *bulk density* (ISO 787/11) of a powdered or granulated product is the mass in grams of 1 cm^3 of the product measured after tamping in a bulk volumeter under given conditions. The *bulk volume* is equivalent to the volume in cubic centimeters of 1 g of the product. Bulk volume and bulk density are largely dependent on the density, particle shape, and particle size of the pigment and are helpful in, for example, selecting the packaging dimensions.

Hardness is generally understood to be the resistance of a solid to mechanical influences (scratching, sanding, penetration). In inorganic pigment technology and allied technologies, however, the term "hardness" is used not only for scratch resistance (Mohs hardness) but also for abrasiveness of particles and dispersion hardness. Whereas the Mohs scale merely shows that a substance with a low rating can be scratched by a substance with a higher rating, abrasiveness is a more complex parameter and can be determined only by means of an abrasion test. The mechanical or inherent hardness of the pigment crystals does not always correlate with their abrasiveness; in the latter, sharp edges and the size of the pigment particles must also be taken into account. Dispersion hardness is a measure of the dispersibility of a pigment in binders and is therefore dealt with in the section on dispersion.

G. Optical Properties

The optical properties of a pigmented system are governed by pigment properties such as the constants of scattering and absorption, particle size and shape, and particle size distribution, by binder properties such as the refractive index and inherent absorption, and by the pigment volume concentration.

When a light quantum hits a pigmented layer, three processes are possible: it hits a pigment particle and is absorbed; it hits a pigment particle and is reflected; or it passes through unhindered (binder assumed to be nonabsorbent). This indicates that the most important physico-optical properties of a pigment are its absorption and scattering power. If the absorption over the entire visible spectral range is very small compared with the scattering, then the pigment is white. If the absorption over the same range is very large compared with the scattering, the pigment is black. If the absorption is selective, that is, dependent on wavelength, the pigment is colored.

Pigments and pigmented coatings can be clearly characterized by their spectral reflection curves $\beta(\lambda)$. Reflection spectra and therefore the optical properties of a pigment can be exclusively attributed to primary physical parameters [7]. The following relationships exist.

1. Colorimetry correlates the perceived color quality with the color stimulus and this in turn with the reflection spectrum $\beta(\lambda)$.
2. The Kubelka-Munk theory relates $\beta(\lambda)$ to scattering, absorption, and film thickness (scattering coefficient s, absorption coefficient k, and film thickness h).
3. The theory of multiple scattering (scattering interaction) relates the scattering coefficient s to the pigment concentration c and the scattering cross section Q_S of the individual particles. The scattering coefficient k has a simple proportional relationship to the absorption cross section Q_A and the concentration c.
4. The Mie theory correlates Q_S and Q_A with the particle size D, the wavelength λ, and the optical material constants of refractive index n and absorption index κ.

As mentioned, the Kubelka-Munk theory [8] links the $\beta(\lambda)$ of a pigmented system to the absorption and scattering coefficients k and s and film thickness h. Most often, the simple equation valid for nontransparent (infinitely thick) layers is used:

$$\frac{k}{s} = \frac{(1 - \beta)^2}{2\beta} \tag{1}$$

With the aid of this equation and the theory of multiple scattering, reflection measurements can be used to draw the following conclusions about the absorption and scattering of pigmented systems.

1. The absorption coefficient k follows Beer's law up to high pigment concentrations. It increases linearly with the pigment volume concentration (pvc).
2. If the individual particles are densely packed, scattering is disrupted with increasing pvc. If the distance between two primary particles falls below $\lambda/2$ of the incident light, then they no longer act as two separate entities but as one larger one. The scattering coefficient s therefore increases linearly with pvc only if the pvc is low (Beer range). At higher pvc values it remains lower than the values anticipated by Beer's law (Fig. 2).

Further conclusions can be based on the Mie theory [9]. This supplements the Fresnel theory for very large particles and the Rayleigh theory for very small particles and is valid for medium-sized particles such as are found in inorganic pigments. The Mie theory correlates the scattering cross section Q_S and the absorption cross section Q_A (i.e., the relationship be-

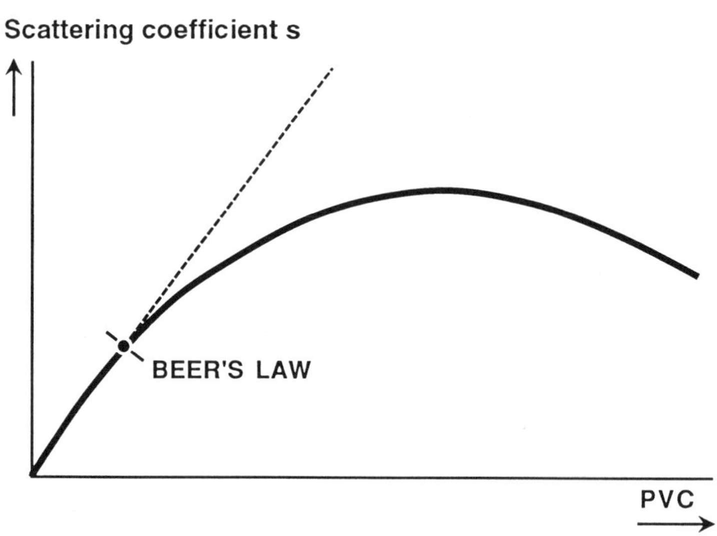

FIG. 2 Light scattering in relation to pigment volume concentration.

TABLE 4 Refractive Index of Some Inorganic Color Pigments and Mineral Fillers

	Chemical formula	Refractive index
Hematite	α-Fe_2O_3	2.988
Greenockite	CdS	2.506
Chromium oxide	α-Cr_2O_3	2.5
Hausmannite	Mn_3O_4	2.45
Iron blue	$MFe^{II}\,Fe^{III}\,(CN)_6 \cdot x\,H_2O$	2.43
Magnetite	Fe_3O_4	2.42
Crocoite	$PbCrO_4$	2.31
Goethite	α-FeOOH	2.275
Cobalt blue	$CoAl_2O_4$	1.74
Dolomite	$CaMg(CO_3)_2$	1.68
Barite	$BaSO_4$	1.636
Kaolinite	$Al_4(OH)_8Si_4O_{10}$	1.563
Talc	$Mg_3(OH)_2Si_4O_{10}$	1.54

tween the optically effective and geometric cross sections) to the mean particle size D, the wavelength of the light λ, and the optical material data such as the refractive index n and the absorption index κ. The theory shows that in inorganically pigmented systems where the particle size is constant, the degree of reflection governing the optical effectiveness usually increases dramatically with the relative refractive index m and thus also with the refractive index n of the pigment ($m = n/n_B$, where n_B is the refractive index of the binder). Table 4 shows the refractive indices of a number of important colored inorganic pigments, extenders, and dispersion agents [10].

III. IRON OXIDE PIGMENTS

A. General

Among the synthetic colored inorganic pigments, iron oxides account for a high proportion of the market in volume terms. The colors yielded by these pigments range from yellow to orange, red, brown, and black. Mixed crystal series based on the compounds described in this section and a number of other elements extend this range of colors. Browns can be created by mixing yellow, red, and black pigments. In addition, natural iron oxides are still used in certain applications.

Natural iron oxide pigments are still used to a limited extent. Colored

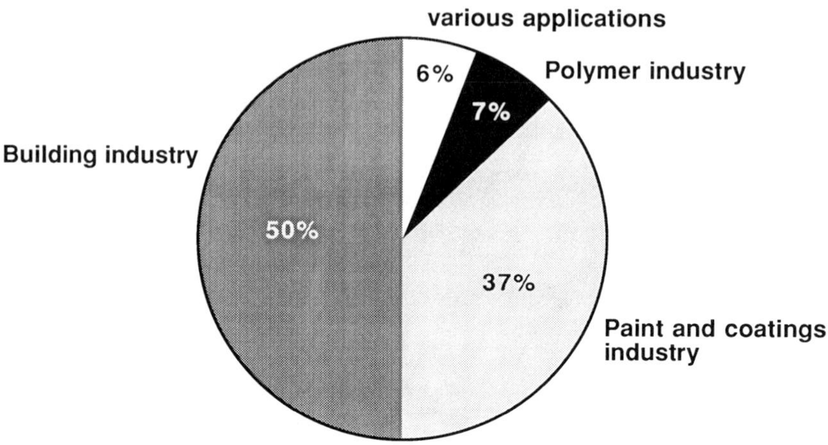

FIG. 3 Consumption of synthetic iron oxides.

earths such as ochers (α-FeOOH), red ochers (α-Fe_2O_3), and burnt ochers have been in use for a very long time. However, their market share has been declining constantly for years and, in volume terms, they now account for only about 12% of the world market. The bulk of this amount consists of Persian red, Spanish red, and burnt sienna pigments. Their tinting strength increases with Fe_2O_3 content. Less significant are the yellow ochers, which are formed by the weathering of iron ores. Their colorant component is α-FeOOH, and they also contain minerals rich in SiO_2 and Al_2O_3. Sienna pigments still have a certain significance. They consist of yellow α-FeOOH and colloidal silica and have translucent properties in the systems in which they are used.

Other natural iron oxide pigments include micaceous iron oxides (MIO), which are platelet-shaped iron oxides with the structure of an α-Fe_2O_3 and are mainly used in anticorrosion coatings. Green earth is a mica-like iron silicate produced by weathering. Earths containing vivianite [$Fe_3(PO_4)_2 \cdot 8\ H_2O$] are known as blue iron earths.

Synthetic types are now used to a vastly greater extent than natural types. In 1990, for example, world production of synthetic iron oxide pigments was approximately 610,000 tonnes, the main centers of production being in Germany, the United States, Japan, and the United Kingdom. Figure 3 shows the world consumption of iron oxide pigments in 1991. The following discussion is concerned primarily with synthetic pigments.

B. Manufacture

Synthetic iron oxide pigments can be manufactured in three basic colors—red, black, and yellow. Their structure resembles that of hematite, magnetite, and goethite or lepidokrokite, respectively. Generally speaking, brown pigments are obtained by mixing the three basic colors. However, homogeneous browns also exist. There are many different ways of manufacturing iron oxide pigments. The method chosen depends on the type of pigment and on the raw materials available. The most important manufacturing processes are briefly described here.

1. Roasting Process [3]

It is usually red iron oxide pigments that are manufactured by roasting and calcination processes. Technical development today is primarily concerned with reducing energy consumption and improving product properties. The raw materials used, such as iron(II) sulfate and iron(II) chloride from the steel industry and scrap iron from the metal processing industries, are largely unavoidable by-products. Oxidative roasting produces α-Fe_2O_3 pigments from all iron oxides, hydroxides, carbonates, and other decomposable iron compounds.

Figure 4 shows a simplified roasting process for iron(II) sulfate heptahydrate. After dehydration to $FeSO_4 \cdot H_2O$, it is heated to more than 650°C

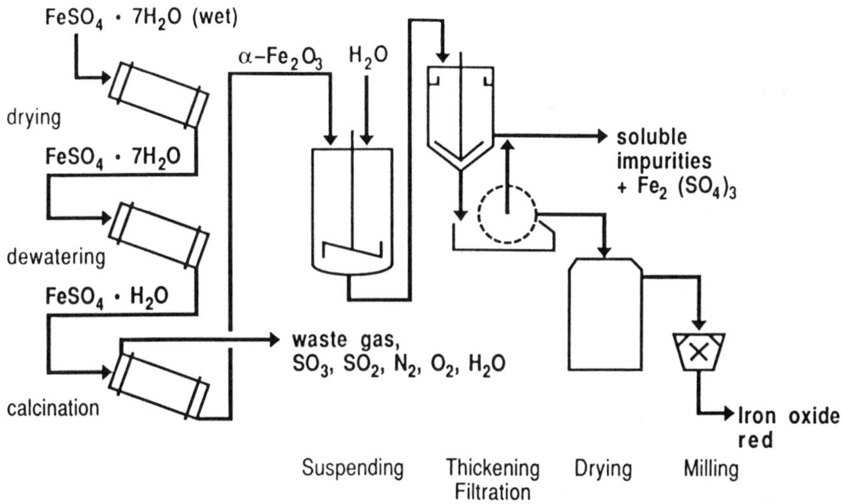

FIG. 4 Manufacture of iron oxide red by roasting $FeSO_4 \cdot H_2O$.

in a rotary kiln to form red a-Fe_2O_3 pigments as shown by Eqs. (2) and (3) [11].

$$6\,FeSO_4 \cdot H_2O + 1\tfrac{1}{2}\,O_2 \rightarrow Fe_2O_3 + 2\,Fe_2(SO_4)_3 + 6\,H_2O \qquad (2)$$

$$2\,Fe_2(SO_4)_3 \rightarrow 2\,Fe_2O_3 + 6\,SO_3 \qquad (3)$$

The sulfur oxides formed are converted into sulfuric acid. The soluble components and unreacted iron sulfate are separated by suspending the calcinate in water, then filtering, washing, and drying it.

If trivalent iron compounds are calcinated with reducing agents or bivalent iron slats in a slightly oxidative atmosphere, iron oxide black pigments are produced [12].

2. Precipitation Process [3]

Yellow FeOOH, black Fe_3O_4, and red Fe_2O_3 pigments can be produced by oxidizing iron(II) sulfate solutions with air under specific conditions. Taking α-FeOOH as an example, Figures 5 and 6 show the process in greater detail. First basic iron(II) sulfate is precipitated from an $FeSO_4$ solution with a sodium hydroxide solution. The iron(II) sulfate is converted into a suspension of α-FeOOH nuclei by oxidation with air. The crystalline α-FeOOH nuclei are very fine and can be used as transparent pigments. To obtain pigments with opaque properties, the α-FeOOH nuclei must be grown to a predetermined size for the shade required [13] by the use of

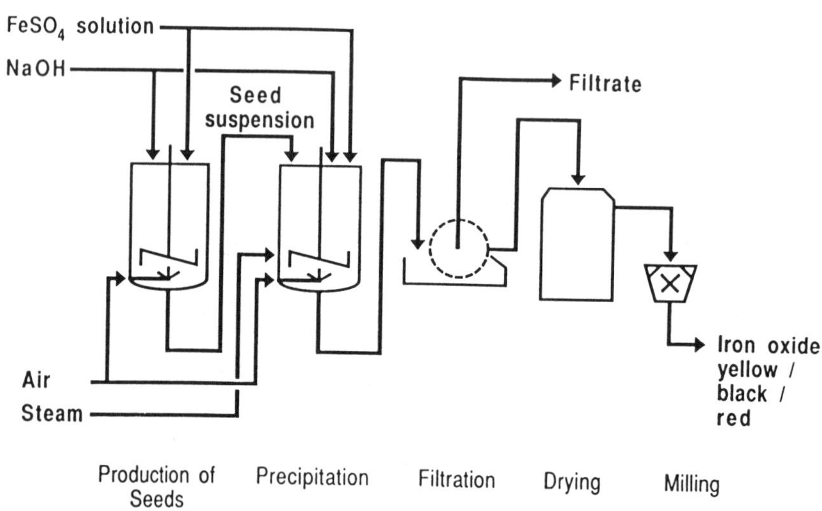

FIG. 5 Manufacture of iron oxide yellow pigments by the precipitation process.

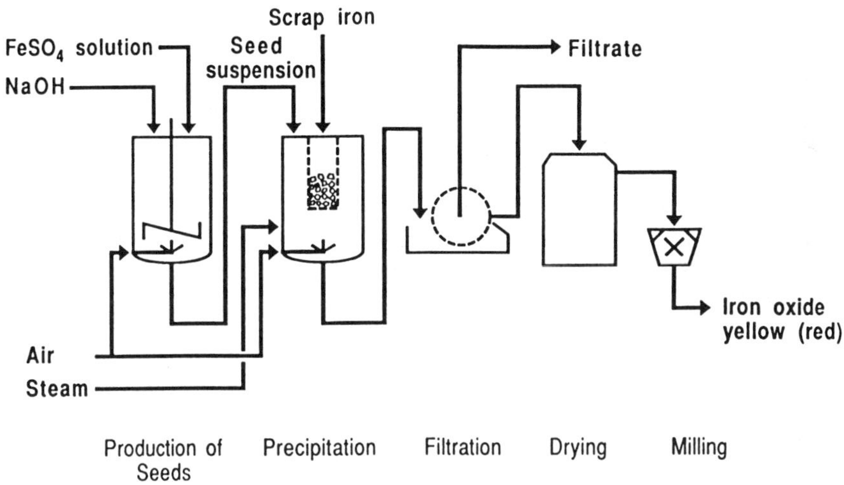

FIG. 6 Manufacture of iron oxide pigments by the Penniman/Zoph process.

further iron sulfate solution, sodium hydroxide solution, and air (precipitation process, Fig. 5. An elegant method of increasing the particle size without the use of further chemicals is by the addition of scrap iron followed by oxidation with air (Penniman/Zoph process, Fig. 6) [14]. This process does not increase the salt load in the wastewater. This process can be accelerated by maintaining the temperature at 70–90°C with steam. The formation of the nuclei is described by Eq. (4) and the formation of the pigment by Eqs. (5) and (6).

$$2\,FeSO_4 + 4\,NaOH + \tfrac{1}{2}\,O_2 \rightarrow 2\,FeOOH + 2\,Na_2SO_4 + H_2O \qquad (4)$$

$$2\,FeSO_4 + 3\,H_2O + \tfrac{1}{2}O_2 \rightarrow 2\,FeOOH + 2\,H_2SO_4 \qquad (5)$$

$$Fe + H_2SO_4 \rightarrow FeSO_4 + H_2 \qquad (6)$$

The acid formed during oxidation dissolves the scrap iron, forming $FeSO_4$.

If $FeCl_2$ solutions are reacted with just enough ammonia that an acid pH is maintained and are then rapidly oxidized under cold conditions with air, a suspension of α-FeOOH nuclei is formed. If the nuclei are then grown, an orange pigment is obtained [15].

If a neutralized bivalent iron salt solution with an Fe(II)/Fe(III) ratio of 0.5 is heated in air at roughly 90–100°C, a black Fe_3O_4 pigment with high tinting strength is formed. Given suitable nucleation conditions, the same equipment can be used to produce very fine α-Fe_2O_3 nuclei that can be grown to form readily dispersed red pigments with high tinting strength [16].

3. Laux Process [3]

The nitrobenzene reduction process with metallic iron for manufacturing aniline in acid solution was modified by Laux so that what had been an unusable residue could be turned into an Fe_3O_4 pigment with high tinting strength [17]. By adding aluminum chloride to the reaction mix, it is possible to create yellow α-FeOOH pigments [18]. The addition of other hydrolyzable trivalent and tetravalent metal salts as well as other auxiliary chemicals results in pale yellow α-FeOOH, orange α-FeOOH, or red α-Fe_2O_3 pigments. Ground and sieved, oil-free iron filings are used as the raw material for the Laux process (Fig. 7). The particle size, metallurgical structure, and metering rate of the iron are important factors for the quality of the pigments produced. The extreme heat of reaction keeps the mixture of water, nitro compounds, and amine boiling until the following reduction has been completed.

$$C_6H_5NO_2 + 2\,Fe + 2\,H_2O \xrightarrow[FeCl_2]{AlCl_3} C_6H_5NH_2 + 2\,FeOOH \qquad (7)$$

Not all details of the reaction mechanism have been fully explained. It is probable that the nitrobenzene oxidizes the bivalent iron salts to trivalent iron salts. These then hydrolyze to form the pigment and liberate acid, which is then used to dissolve metallic iron. The pigment is formed without the use of sodium hydroxide solution and requires only 3% of the theoretical amount of acid needed to transform the iron into $FeCl_2$. Because of the high capital costs involved, this significant advantage of the process can be

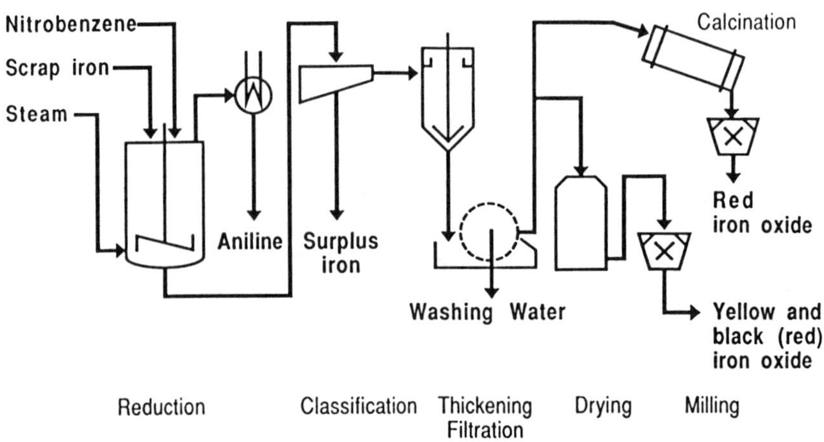

FIG. 7 Manufacture of iron oxide pigments by the aniline (Laux) process.

exploited only if a large plant is constructed and if there is sufficient demand for the aniline produced.

Once the reaction is complete, the aniline is separated. Any residual aniline is removed by water vapor distillation. The pigment suspension is separated from the surplus iron filings by sedimentation. The surplus of iron filings is necessary to ensure that all the nitro compound reacts. The pigment slurries are washed in slurry thickeners and rotary vacuum filters to remove the salts. They are then dried in conveyor or drum dryers or in more energy-efficient spray or fluidized-bed dryers. The final step is grinding or micronizing. All pigments produced by the Laux process can be calcined (e.g., in rotary kilns) to form α-Fe_2O_3 pigments.

C. Properties

The leading position held by the iron oxides among colored inorganic pigments today is due not only to their low cost but also to their excellent fastness properties. They are absolutely insoluble in and resistant to water and organic or inorganic binders such as cement, lime, and sodium silicate. They are also extremely lightfast and weather-stable and have excellent hiding power and tinting strength.

Like iron oxide black pigments, iron oxide red pigments produced by the Laux process via a black slurry, by roasting iron salts, or by the hydrolysis process have cubic to spherical particles. The only exceptions are the iron oxide red pigments produced by the dehydration of needle-shaped iron oxide yellow. These retain the needle shape of the base product.

In the case of the almost cubic to spherical pigments, the relationships between particle size and optical and other properties can be particularly clearly illustrated.

1. Iron Oxide Red Pigments

If the angle of hue of red iron oxide pigments is plotted against their particle size, a virtually straight line is obtained (see Fig. 8). This decreases with increasing particle size; that is, the hue changes from a yellowish red to a bluish red.

Not all the optically effective particles in a pigment powder are of the same size; a certain range of particle sizes will be present. This is evident in Fig. 9, which shows electron micrographs of various iron oxide reds. The larger the mean particle size, the more difficult it is to produce very uniform particle sizes.

As the hue of a pigment is dependent on its particle size, a pigment is in practice always a mixture of particles of slightly different colors. Iron oxide red, for example, is a natural mixture of yellowish and bluish components. The proportion of the various fractions in a pigment, as determined by one

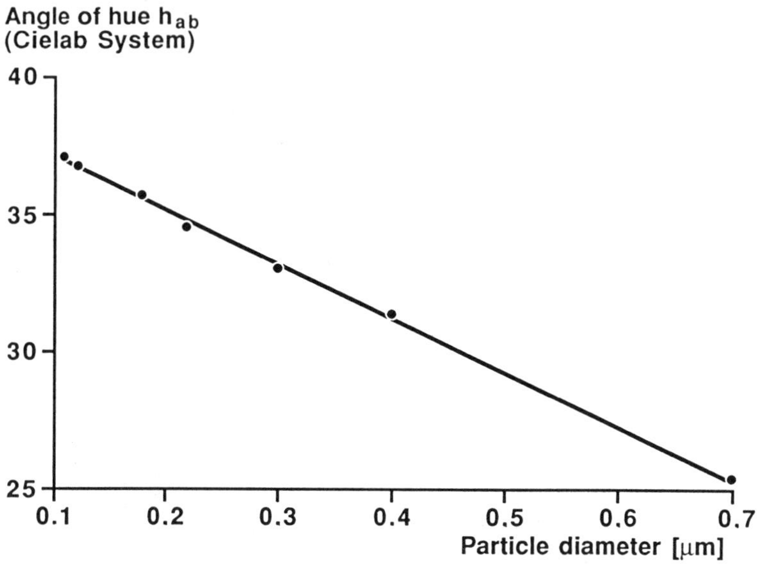

FIG. 8 Shade in relation to particle size in iron oxide red pigments.

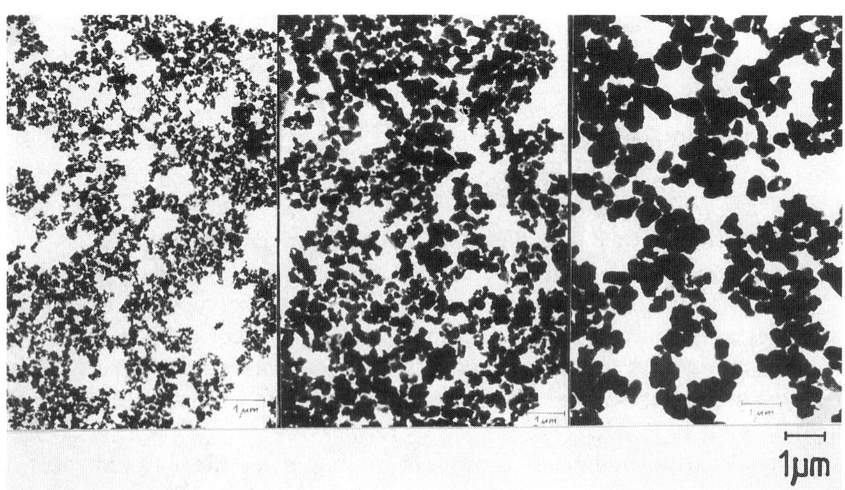

FIG. 9 Electron micrographs of three iron oxide red pigments to exemplify the differences in crystal size encountered.

or another of the methods of particle size analysis, is defined by the particle size distribution. The simplest way to display this is to plot proportion (frequency) versus particle size (Fig. 10). The hue of the pigment will appear stronger and purer the higher the frequency maximum and the narrower the distribution. This is of practical significance as, generally speaking, more uniform particles can be produced by synthesis than by grinding natural products. This is why, as a rule, natural iron oxide reds have a broader particle size distribution and therefore a weaker chroma (are less intensely colored) than synthetic iron oxide reds. Figure 10 shows the particle size distribution curves of two iron oxide reds [6]. They have the same angle of hue because their maxima are at the same particle size. However, the pigment with the broader distribution curve has a weaker chroma.

Such relationships between optical performance and particle size apply only to particle sizes of 0.1–1.0 μm, the range in which the scattering constant passes through a maximum. Only those pigments with a maximum scattering constant roughly equal to half the wavelength of the incident light (approximately 0.2 μm) have high opacity. Very yellowish reds with a particle size of approximately 0.1 μm and, above all, the violet shades with a particle size of approximately 1.0 μm result in significantly reduced opacity in pigmented media. Iron oxides with a particle size of about 0.01 μm have virtually no opacity in organic binders and yield transparent colored

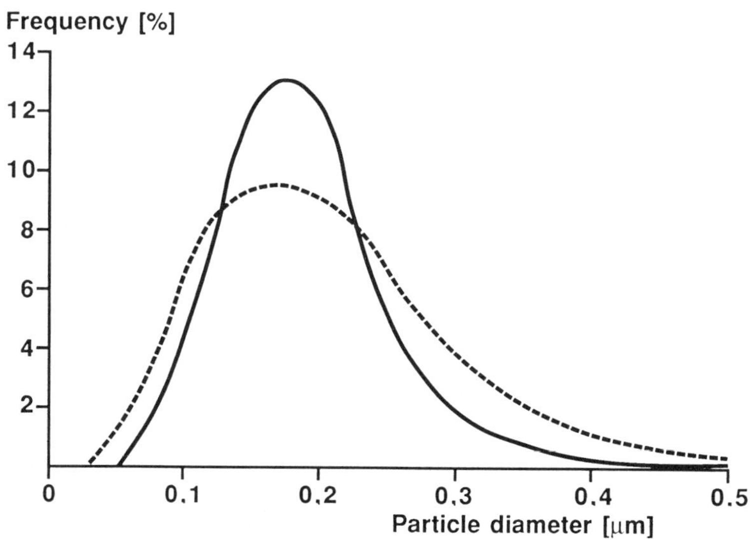

FIG. 10 Particle size distribution curves of two iron oxide red pigments.

films. Such pigments are known as transparent iron oxides and are available in red, yellow, and orange. At the other extreme are the micaceous iron oxides, mentioned briefly earlier, which have a particle size of 10–25 μm. Chemically speaking, transparent iron oxide red and micaceous iron oxide are both α-Fe_2O_3 pigments. The properties they possess over and above those of normal pigments are due entirely to their particle size and shape.

The relationships between particle size and properties in iron oxide reds are again illustrated in Fig. 11 [19].

Iron oxide red produced by calcination of an iron oxide yellow largely retains the needle-shaped particles of the base product, provided the calcination temperature is not too high. The saturation of the shade of such iron oxide reds very much depends on the calcination temperature. However, in reductions (i.e., in admixture with an excess of white pigment such as titanium dioxide), particularly in oleoresinous binders, these pigments usually have a higher tinting strength than spherical reds of a similar hue. This is due to the greater scattering power of the needle-shaped particles.

The relationship between particle size and tinting strength of opaque red pigments is especially complex because both the hue and the tinting strength vary considerably with particle size and because the tinting strength can be evaluated by various criteria depending on the hue. If the tinting strength is measured on the basis of lightness, as the color chemist usually does visually, then the maximum tinting strength produced by the virtually spherical particles of the iron oxide reds is found at a particle size of approximately 0.2 μm (as in Fig. 11). If depth of color determined instrumentally is taken as the measure of tinting strength, then, in the case of opaque pigments, the tinting strength decreases steadily with increasing particle size; that is, the pigment with the smallest particles has the greatest tinting strength.

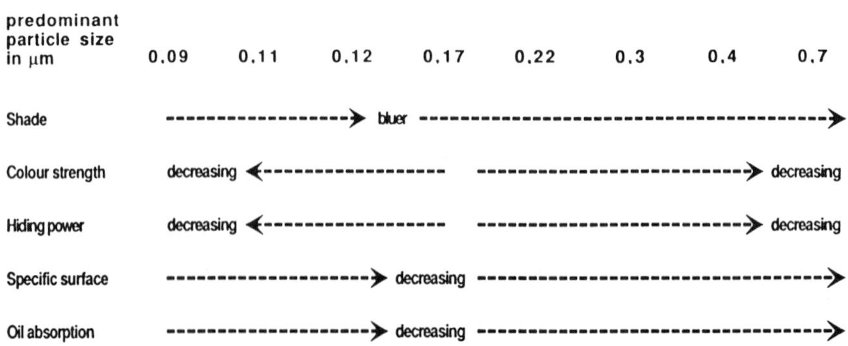

FIG. 11 Properties of α-iron oxide red pigments in relation to particle size.

2. Iron Oxide Black Pigments

In the case of iron oxide blacks, the shade varies much less with particle size. The results obtained by the various methods of evaluating tinting strength (by depth of color or lightness) therefore differ far less. The tinting strength increases with decreasing particle size. At the same time, the shade becomes browner in both full shade and reduction. This relationship is valid for Fe_3O_4 blacks. Pigments of a different chemical composition [e.g., bixbyite $(Mn, Fe)_2O_3$] cannot be compared directly with such products.

3. Iron Oxide Yellow Pigments

The particles of iron oxide yellow pigments have a fairly pronounced needle-shaped structure. It is this that determines the coloristic properties of these pigments. Iron oxide yellows with very fine needles yield a pure shade with a greenish undertone. As the cross section of the needles increases, the hue becomes more orange. At the same time, it appears to be less saturated. The length and cross section of the needles in an iron oxide yellow must therefore be known in order to characterize the optical properties of the pigment. Figure 12 shows an electron micrograph of an iron oxide yellow pigment.

In contrast to the spherical iron oxide reds and blacks, the needle-shaped primary particles of the iron oxide yellows (with the exception of very large particles) are almost always sensitive to grinding. Intense grinding breaks some of the needles and makes the shade a more achromatic yellow.

FIG. 12 Electron micrograph of an iron oxide yellow.

4. Iron Oxide Brown Pigments

Most of the commercially available iron oxide browns are mixtures of iron oxide yellow, red, and black. The pale shades have a higher proportion of yellow, whereas black dominates in the dark shades. The properties of the brown pigments are therefore dependent on the properties of their components.

5. Thermal Stability

Because iron oxide pigments react variously at elevated temperatures, some of them have only limited thermal stability [19]. Iron oxide red changes little up to temperatures of approximately 1000°C. In contrast, at temperatures above around 180°C and in the presence of oxygen, iron oxide black oxidizes to form brown γ-Fe_2O_3, which in turn is transformed into red α-Fe_2O_3 at temperatures above 350°C. Iron oxide blacks containing other metal ions are the exception and, because of their different chemical structure, are stable up to around 1000°C [e.g., bixbyite $(Fe,Mn)_2O_3$].

At temperatures above 180°C, iron oxide yellows release their chemically bound water and become iron oxide reds. Iron oxide browns, which are mixtures of iron oxide black and yellow, have the same maximum thermal stability. Exceptions are those iron oxide brown pigments that contain other metal ions and those based on γ-Fe_2O_3. The maximum thermal stability of 180°C given for iron oxide black and yellow pigments is merely a guideline value observed in the pure dry pigment. If the pigment is dispersed in a medium such as a paint or plastic, the thermal stability may be significantly higher, particularly if exposure to high temperatures during processing is only brief. The medium would seem to either protect the pigment against exposure to oxygen or have the effect that a higher vapor pressure is required to release the water.

A distinction must be made between the limitation of colorfastness resulting from chemical changes and color differences caused by an increase in particle size at elevated temperatures. Generally speaking, the size of the larger particles increases at temperatures above 800–900°C at the expense of the smaller particles. Because shade is dependent on particle size, this results in a darkening of the shade. The increase in particle size is dependent on the pretreatment of the pigment (i.e., whether it is a merely dried or an already calcined red) and on the presence of mineralizers such as alkali metal ions, which promote an increase in particle size. The length of time for which the pigment is exposed to high temperatures is also a significant factor.

6. Micronized Iron Oxide Pigments

Iron oxide pigments are usually ground in beater mills, bowl mills, or disk mills. A small proportion of the pigment remains as aggregates and cannot be dispersed finely enough in organic binders, particularly by modern grinding equipment such as dissolvers, sand mills, or bead mills. Micronizing, a special grinding method using superheated steam, breaks down the aggregate, almost completely but leaves the primary particles intact. There is little difference between the shade of a micronized pigment (red or black) and that of conventionally ground pigments. The tinting strength may increase by up to 5%. The features of a micronized pigment include a much lower residue on a sieve with 0.045 mm mesh—around 0.002% compared with approximately 0.06% for a conventionally ground pigment. The ease of dispersion of micronized pigments is also considerably better, particularly when the modern grinding equipment mentioned above is used.

Micronized iron oxide yellows not only have better dispersibility, but their oil absorption is almost half the relatively high value for conventional iron oxide yellows. The reason for this is the sensitivity of the needle-shaped primary particles to grinding, as already described. Micronizing breaks some of the needles, resulting in denser packing of the pigment in a paint film. This denser packing would seem to have a greater effect on reducing the oil absorption than the slight increase in specific surface caused by the creation of additional surfaces through fragmentation. The lower oil absorption is the main advantage of all micronized iron oxide yellows as it considerably improves the gloss and gloss retention of coatings pigmented with these products.

IV. OXIDIC MIXED-PHASE PIGMENTS

A. General

Strictly speaking, mixed-phase pigments also include the chromate and cadmium pigments in all their various forms, which will be discussed in later sections of this chapter. The mixed-phase pigments discussed in this section are those that, unlike the chromate and cadmium pigments, consist entirely of metal oxides. Guest ions are incorporated into host lattices, which, depending on the type and quantity of the guest ions, permits the creation of certain color effects. The basic structure of the host lattice is unaltered by the incorporation of the guest ions. However, the dimensions of the elementary cell (the lattice constants) may increase or decrease.

To manufacture stable pigments, highly stable host structures such as rutiles or spinels should be used. The partial or total substitution of the metal ions in a TiO_2 rutile or a $MgAl_2O_4$ spinel with coloring guest ions such as nickel, chrome, cobalt, or iron allows numerous variations [6,19].

B. Manufacture

The manufacture of mixed-phase pigments is based on the oxides and/or hydroxides of the components, which, if necessary, can be produced in an intermediate process by precipitation with alkalis from the appropriate metal salt solutions. The raw materials are homogenized in the wet phase. The actual formation of the mixed phase is a solids reaction requiring a great deal of heat. This results in the more than adequate thermal stability of these pigments for all applications in the coatings industry. Calcination can be carried out in either a continuous or discontinuous process. The products must then be washed or wet-milled intensively as the solids reaction at high temperature results in the formation of a large number of aggregates. Grinding to break down the aggregates caused by drying is carried out directly after drying in standard equipment.

C. Properties

Because of their structure and the method of manufacture, oxidic mixed-phase pigments have excellent resistance properties including lightfastness, weather stability, and resistance to acids, alkalis, and other chemicals. The method of manufacture also results in extremely high thermal stability up to approximately 1000 °C.

Of all the mixed-phase pigments shown in Table 5, nickel and chrome rutile yellows are of the greatest significance in the coatings industry. Nickel rutile yellow has a pure lemon yellow hue whereas chrome rutile yellow is much redder and more like an ocher in its hue. Compared with the conventional iron oxide yellows, chrome titanium yellow is much purer and redder. The opacity and tinting strength of nickel rutile yellow are relatively low, whereas the opacity of chrome rutile yellow is one and a half times to twice as high as that of a standard rutile.

Up until now, the excellent fastness properties of cobalt blue and cobalt green pigments have been little exploited by the coatings industry. This is because they compete with phthalocyanine blue and green, organic pigments that also have excellent fastness properties and a much higher tinting strength than the inorganic blue and green pigments. However, the excellent weather stability of the mixed-phase pigments, even in the palest of pastel shades, is of some significance in certain applications. As discoloration can be caused by weathering when cobalt blue is used in chlorinated binders such as coatings based on PVC or chlorinated rubber, trials must be carried out first.

Generally speaking, spinel black pigments are characterized by their good wettability and show little tendency to flocculate in most binder systems. This means that, compared with other combinations, there is often little or no rub-out effect in gray coatings (i.e., no increase in intensity of grayness on rubbing while "wet") based on spinel black and titanium dioxide. Depending on the manufacturing conditions, coarse spinels can be produced that have a tinting strength roughly equivalent to that of iron oxide blacks.

TABLE 5 Designation and Composition of Mixed-Phase Pigments

Trivial name	Color	Formula	Crystal structure	Color index (CI)	Density (g/cm^3)	Oil absorption (g/100 g)
Nickel rutile yellow	Lemon yellow	$(Ti,Ni,Sb)O_2$	Rutile	Pigment Yellow 53, No. 77788	4.5–5.5	15–20
Chromium rutile yellow	Ocher yellow	$(Ti,Cr,Sb)O_2$	Rutile	Pigment Yellow 24, No. 77310	4.0–4.5	15–20
Cobalt green	Green	$(Co,Ni,Zn)_2(Ti,Al)O_4$	Inverse spinel	Pigment Green 50, No. 77335	4.0–5.0	15–20
Cobalt blue	Blue	$Co(Al,Cr)_2O_4$	Spinel	Pigment Blue 36, No. 77343	4.0–4.5	20–30
Cobalt blue	Reddish blue	$CoAl_2O_4$	Spinel	Pigment Blue 28, No. 77346	4.0–4.5	25–35
Spinel black	Black	$Cu(Cr,Fe)_2O_4$	Spinel	Pigment Black 22, No. 77429	4.5–5.0	15–30
Zinc iron brown	Brown	$ZnFe_2O_4$	Spinel	Pigment Yellow 119, No. 77496	5.0–5.5	10–15
Bismuth vanadate molybdate yellow	Yellow	$BiVO_4 \cdot 0.75\ Bi_2MoO_6$	Scheelite	Pigment 184	8.0–9.0	8–10

Zinc iron brown pigments produce a yellow-brown to mid-brown color. The hue of these mixed-phase pigments is generally purer than that of the browns produced by mixing various iron oxide pigments. Furthermore, apart from special pigments, they have higher thermal stability than the normal iron oxide browns. The fact that these browns are single-phase pigments rather than mechanically produced mixtures can also be an advantage in various applications [6].

Because of their excellent resistance properties, oxidic mixed-phase pigments are suitable for use in all types of coatings, particularly those where extremely high levels of lightfastness, weather stability, or resistance to chemicals are required.

Mixed-phase yellow pigments already play a significant role in the coatings industry. Alone or in combination with high quality organic pigments, they can be used as alternatives to lead-based pigments to produce saturated and resistant shades in many coatings. Their excellent lightfastness and weather stability in coatings based on alkyd melamine resins or silicone polyesters, for example, even in high reductions (wool scale value 8), make these pigments ideal for use in high quality coatings. They have proved suitable for coloring powder coatings, not only because of their high thermal stability but also because they are very easy to disperse in binders under normal processing conditions. This guarantees a constant hue, which is particularly important in the production of powder coatings.

V. CADMIUM PIGMENTS

A. General

Cadmium pigments largely consist of cadmium sulfide, which is crystallized in a hexagonal wurtzite lattice. Wurtzite is the mineralogical designation for the hexagonal crystal modification of zinc sulfide, which is isomorphous with cadmium sulfide. Cadmium sulfide and the variations possible through the formation of mixed phases with zinc sulfide and cadmium selenide are valued as high quality, brilliant inorganic pigments.

The isomorphism with zinc sulfide is exploited to alter the golden yellow shade of cadmium sulfide. The partial substitution of cadmium with zinc in the wurtzite lattice makes the shade greener with increasing zinc content. A similar substitution is possible with selenium replacing anionic sulfur. The shade shifts from orange red to wine red with increasing selenium content. In wine reds containing the most selenium, around half the sulfide has been replaced with selenide. This is equivalent to a selenium content of about 25%. In this combined mixed-phase formation, substitution with zinc causes the lattice to contract whereas substitution with selenium has

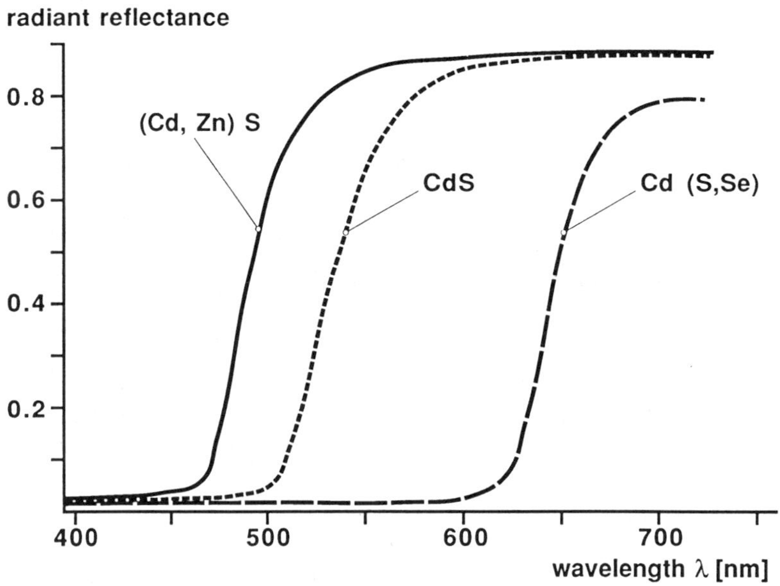

FIG. 13 Reflection curves of (Cd,Zn)S, CdS, and Cd(S,Se) pigments.

the opposite effect. The absorption limit is thereby shifted from the smaller to the larger wavelengths, as shown in Fig. 13, resulting in a wide range of shades [6, 19].

Table 6 shows the mixed-phase pigments in the most important hue classes with their composition, structure, and Color Index numbers. Every conceivable intermediate variant is possible as both the substitution of cad-

TABLE 6 Color and Chemical Composition of Cadmium Pigments

Pigment	Chemical formula	Crystal structure	Color Index (CI)
Cadmium yellow (greenish yellow)	(Cd,Zn)S	Wurtzite, hexagonal	Pigment Yellow 37, No. 77199
Cadmium yellow (reddish yellow)	CdS	Wurtzite, hexagonal	Pigment Orange 20, No. 77199
Cadmium orange	Cd(S,Se), Se $<$ 10%	Wurtzite, hexagonal	Pigment Red 108, No. 77202
Cadmium red	Cd(S,Se), Se $>$ 10%	Wurtzite hexagonal	Pigment Red 108, No. 77202

mium with zinc and the substitution of sulfur with selenium produces an endless series of mixed crystals.

B. Manufacture

The main raw materials used in the manufacture of cadmium pigments are cadmium metal, selenium, and sodium sulfides [20]. The cadmium metal is dissolved in acids to form a cadmium salt solution. Any other elements present are removed by special processes. Cadmium yellow pigments are produced by adding a given amount of a zinc salt depending on the hue required. The cadmium is then precipitated with sodium sulfide to form cadmium sulfide or cadmium zinc sulfide. The resulting product is a very fine precipitate that does not at this stage possess pigmentary properties. One variation of the process is to precipitate the cadmium with a soda solution of form cadmium carbonate. This is then reacted with sodium sulfide.

Cadmium red pigments are produced by reacting the cadmium salt solution with sodium sulfide that, in contrast to the cadmium yellow process, contains the appropriate amount of dissolved selenium to yield the shade required. Another variation of the process is to precipitate the cadmium with a soda solution to form cadmium carbonate. This is then reacted with the sodium sulfide solution containing selenium.

The very fine raw pigments produced by the precipitation process are washed and then calcined at 500–700°C. This procedure is usually carried out in air. During the calcination process, the particle size increases, with the optical properties such as hue, opacity, and tinting strength developing to their optimum level.

C. Properties

The cadmium pigments combine a number of the advantages of both organic and inorganic pigments. On the one hand, they have the good thermal stability (approximately 600°C), high opacity, good tinting strength, and absolute migration resistance common to most inorganic pigments. On the other hand, they have a very brilliant hue, similar to that found in organic pigments.

Cadmium pigments have very good lightfastness in both full shade and reductions. Depending on the medium in which they are used, this is between 7 and 8 on the wool scale. However, there are greater differences as far as weather stability is concerned. Depending on the type of pigment and on the medium, exposure to water may cause the oxidation of cadmium sulfide to cadmium sulfate with the result that the hue weakens. Cadmium yellow pigments oxidize more readily than cadmium orange and cadmium red pigments, which are stabilized against oxidation by the incorporation of selenium. Reduction with titanium dioxide also reduces the weather stability. The water absorbency of the binder also plays a significant role. All this means that only cadmium red pigments can satisfy the requirement for high weather stability in stoving coatings with low water absorbency.

Generally speaking, cadmium pigments are easily dispersed in application media. However, they are sensitive to grinding. Severe mechanical stress can result in a marked decrease in the chroma. Prolonged grinding in ball, sand, or bead mills should therefore be avoided. The most suitable equipment for gentle grinding of cadmium pigments is the three-roll mill.

In light of what has been stated in the previous paragraphs, cadmium red pigments in particular are important in the formulation of high quality industrial coatings. They are used to produce medium to dark reds.

Because of their high thermal stability, cadmium pigments are the only products that can be used trouble-free to produce brilliant yellow, orange, and red shades in binder systems that must be cured at very high temperatures, for example, those based on polytetrafluoroethylene, polyimide-polyamide, or silicone resins.

With their brilliance, high opacity, and lightfastness, all the cadmium pigments have proved suitable for the formulation of artists' colors. They are also used in significant quantities in the manufacture of printing inks. In addition to the pigment properties already described in some detail, cadmium pigments have a relatively low inherent hardness for inorganic pigments, and this means that the printing inks in which they are used generally have very good abrasion properties.

VI. CHROMIUM OXIDE PIGMENTS

A. General

The chromium oxide pigments include chemically pure chromium(III) oxide and the oxide hydrates of trivalent chromium. Because of its outstanding thermal stability, Cr_2O_3 was used as early as 1802 by the porcelain makers of Sèvres as a green colorant. More widespread use was long prohibited by its high price. The unparalleled fastness properties of chromium oxide have led to its use in coatings and cement-bonded building materials. By modifying the particle size (0.3–0.5 μm), it is possible to produce a series of very fine variations in color. The hardness and abrasiveness of chromium oxide should be borne in mind in certain applications [21].

The significance of chromium oxide hydrate green pigments declined considerably with the introduction of phthalocyanine pigments. There are no natural chromium oxide compounds that can be used as pigments.

B. Manufacture

Chromium oxide green pigments are manufactured by reducing alkali chromates to chromium(III)oxide. The following three processes are technically important [1].

(a) Reduction in the Solid Phase. The manufacture of chromium(III)oxide from solid dichromate is carried out at high temperature and in the presence of a reducing agent such as charcoal or sulfur. The reaction is highly exothermic. Sulfur is the preferred reducing agent because the alkali sulfates formed as by-products can be easily dissolved in water. The Cr_2O_3 remaining is filtered off, washed, and dried. The chromium oxide green pigment is then ground. Depending on the quality, it may also be micronized in jet mills.

(b) Reduction of Dichromate Solutions. Sulfur, thiosulfate, or organic compounds are normally used as reducing agents with dichromatic solutions. If sulfur is used, the reduction is carried out in the presence of a sodium hydroxide solution at boiling temperature. The separated reaction product is calcined. After extraction of the calcination product with water, pure chromium(III) oxide remains.

(c) Decomposition of Ammonium Dichromates. Ammonium dichromate decomposes at temperatures of 200°C and above to form nitrogen, water, and Cr_2O_3, which at this stage does not possess any pigmentary properties. These are created by subsequent calcination. The separate manufacture of ammonium dichromate can be avoided by calcining a homogeneous mixture of sodium dichromate and ammonium sulfate. If a shortfall of ammonium sulfate is used, the chromium(III)oxide produced has a particularly low sulfur content. Such a pigment is primarily required for metallurgical applications.

C. Properties

Because they consist of pure chromium(III)oxide (Cr_2O_3), chrome oxide green pigments are largely inert and therefore have excellent resistance properties. They are insoluble in water, acids, and alkalis and therefore have excellent resistance to So_2 and cement. Media pigmented with chrome oxide green have outstanding lightfastness and weather stability. Chrome oxide green pigments are stable to about 1000°C. Only above this temperature does the shade change as a result of an increase in particle size.

The chromium(III) oxide on which the chrome oxide green pigments are based crystallizes rhombohedrally in the corundum lattice and has a relatively high inherent hardness (8–9 on the Mohs scale). This should be borne in mind in certain applications. It does not mean, however, that chrome oxide green pigments are difficult to disperse, as the inherent hardness of the pigment crystals is a different property from the dispersion hardness (see earlier). The primary particles of the chrome oxide green pigments are largely isometric; that is, they are virtually spherical.

As with most colored inorganic pigments, the pigmentary properties of chrome oxide green depend upon the relative contributions of light absorption and scattering. Chromium oxide has a refractive index of around 2.5, which means that in a normal binder the optimum particle size of a chrome oxide green pigment needed to yield adequate tinting strength and scattering power is 0.3 μm. If the mean optimum particle size is increased, the ability of the pigment to scatter light decreases more than its ability to absorb light. The shade therefore becomes darker, bluer, and less saturated. The scattering power and other pigmentary properties are also dependent on the pigment concentration and the uniformity with which the pigment is distributed in the medium. Increasing pvc and flocculation also makes the shade darker, bluer, and less saturated.

Because of their high light scattering ability and relatively high absorption, chrome oxide green pigments have excellent opacity in a wide variety of media.

The outstanding fastness properties of chrome oxide green pigments allow their universal application in coatings ranging from masonry paints to automotive topcoats. However, there are certain limitations on their use in printing inks, particularly photogravure inks, resulting from the high inherent hardness and associated abrasiveness.

VII. CHROMATE AND MOLYBDATE PIGMENTS

A. General

This pigment group includes chrome yellow, molybdate orange and red, and chrome orange and red. Chrome yellow pigments consist of $Pb(Cr,S)O_4$ mixed phases or pure $PbCrO_4$. A variety of shades ranging from a greenish yellow through pale and medium yellow to golden yellow can be obtained by modifying the crystal (orthorhombic or monoclinic) or by varying the lead sulfate content.

B. Manufacture

The raw materials used for chromate and molybdate pigments include litharge, sodium dichromate, sodium sulfate, sulfuric acid, and sodium molybdate. The reaction is carried out only by precipitation in the aqueous phase. The starting solutions are placed in separate reaction vessels with temperature controls. The lead salt solution is formed by dissolving litharge in nitric or acetic acid. $Na_2Cr_2O_7$ is reacted with alkali hydroxides or sodium hydroxide to form Na_2CrO_4 and then, depending on the type of chrome yellow required, with the appropriate amount of Na_2SO_4 or H_2SO_4. In the case of the molybdate pigments, the required amount of $Na_2MoO_4 \cdot 2\,H_2O$ is added to the chromate solution. The pigments are precipitated by mixing the starting solutions. The type of pigment formed is governed by the concentration of the solutions, the pH, the temperature, and the aftertreatment of the aqueous pigment suspensions.

Aftertreatment is required following precipitation to improve the technical properties of chromate and molybdate pigments. Precipitating certain colorless oxides and hydroxides onto the pigments increases their lightfastness, thermal stability, and resistance to chemicals. Options include the precipitation of the phosphates and polyphosphates of Zn, Ce, and Mg [22]; the precipitation of zirconium dioxide hydrates and coating with amorphous SiO_2 [23]; the precipitation of B_2O_3, aluminum hydroxide, and titanium dioxide hydrates; and thick coating with SiO_2 [24].

C. Properties

Table 7 summarizes some of the physical data of lead chromate and molybdate pigments. High tinting strength and opacity, good lightfastness, and brilliant hues make lead chromate and lead molybdate pigments ideal for the use in printing inks, plastics, and, especially, coatings. Depending on the type and extent of stabilization, a distinction is made between standard types, lightfast, highly lightfast, and SO_2-resistant products, and pigments with a low content of lead soluble in hydrochloric acid. By mixing chrome yellow with iron cyanine blue (chrome green) and phthalocyanine blue (fast chrome green), it is possible to produce brilliant green pigments for use in coatings and printing inks.

TABLE 7 Properties of Chromate and Molybdate Pigments

Pigment	Chemical formula	Crystal structure	Density (g/cm^3)	Particle size (μm)
Chrome yellow (orthorhombic)	$Pb(Cr,S)O_4$	Anglesite	5.5–5.7	0.2–0.6
Chrome yellow (monoclinic)	$Pb(Cr,S)O_4$ $PbCrO_4$	Crocoite	5.4–6.0	0.1–0.8
Molybdate orange or molybdate red (tetragonal)	$Pb(Cr,Mo,S)O_4$	Scheelite	5.5–6.2	0.15–0.75

VIII. MISCELLANEOUS PIGMENTS

A. Iron Blue Pigments

Iron blue pigments have the composition $M^{(I)}Fe^{(II)}Fe^{(III)}(CN)_6 \cdot x\,H_2O$, with $M^{(I)} = Na^+, K^+, NH_4^+$. They were formerly known as Prussian blue, Paris blue, Milori blue, or Turnbull's blue.

The iron blue pigments consist of 82–88% ferrocyanide components, 6–12% alkali components, and 5–6% water [3]. Dark variants have a primary particle size of 0.05–0.2 μm. The BET surface of the virtually spherical particles is between 30 and 120 m^2/g. By varying the alkali metal ion and the pH during oxidation, green or reddish-violet shades can be produced as well as bronzing and nonbronzing types. Iron blue pigments have an extremely high tinting strength, but, because of their tendency to aggregate, they are difficult to disperse. Micronized pigments perform better.

Iron blue pigments are thermally stable up to 150°C. The pigments are decomposed by concentrated acids and alkalis. The main applications for iron blue pigments are printing inks (particularly photogravure inks), automotive coatings, and colored paper. Mixtures of iron blue pigments with chrome yellow (chrome green) and with zinc yellow (zinc green) are also widely used for coatings and printing inks.

B. Ultramarine Pigments

Ultramarine pigments are sodium aluminum silicates with the composition $Na_8Al_6Si_6O_{24} \cdot S_x$ (sodium-rich) or $Na_{8-y}Al_{6-y}Si_{6+y}O_{24} \cdot S_x$ (silicon-rich). Depending on the structure of the coloring S_x used, ultramarines with a blue, green, red, or violet undertone can be produced [3, 25, 26]. The most important of these pigments are the ultramarine blues with their brilliant hues ranging from greenish blue to reddish blue. Because the refractive index is low ($n \approx 1.5$), their opacity is relatively low. The sulfur content varies from 8 to 12%, and they are thermally stable to over 400°C.

Ultramarine pigments are used in a variety of applications including plastics, coatings, emulsion paints, printing inks, artists' colors, leather finishes, paper, textiles, and cosmetics. The pigments are physiologically harmless.

C. Transparent Pigments

If the particle size of a pigment is significantly smaller than the wavelength of the light or its refractive index differs only slightly from that of the binder, then the pigment is transparent (nonopaque). Its optical effect is based on absorption.

Colored inorganic pigments can be manufactured in a wide range of transparent shades. These include α- and γ-FeOOH (yellow), α-Fe_2O_3 (red), α-CrOOH and $NaCr(Si_2O_6)$ (cosmochlore) [27] (green), and $CoAl_2O_4$ (blue). The manufacturing processes are like those used for conventional pigments but are aimed at achieving the necessary fineness.

Transparent pigments can be used in a wide variety of applications, including decorative coatings (metallics), printing inks, plastics, and wood stains, in which their high UV absorption and lightfastness are exploited to protect the dispersion medium or the substrate on which the dispersion is coated. The main problem with transparent pigments lies in achieving a high level of dispersion in application media. Because they are so fine, they tend to flocculate and form aggregates during their manufacture even when they are wet-ground. This is why they are often used in a predispersed form in coatings applications (either as fine pastes or as pigmented binder chips).

D. Lustrous Pigments

Metallic and nacreous pigments are known collectively as lustrous pigments. Their luster effects are based on specific reflection from planar particles in a parallel arrangement. In nacreous pigments, each light ray is partially reflected many times from many pigment particles at various depths in the coating.

Metallic pigments consist of very glossy platelets or flakes of soft, ductile metals such as aluminum, copper, or α-brass [28]. They are cold-ground in special ball mills from metal powders produced by melt atomization, sheets, foils, and wire segments. Fatty acids or alkylamines are used as lubricants and cover the freshly formed surfaces, thus preventing the cold aggregation of the pigments. Depending on the grinding method used and the duration of grinding, the particles produced range in size from several millimeters (flakes) to a few micrometers (bronze powder) with a thickness/diameter ratio between 1 : 50 and 1 : 250. To prevent metal dust explosions, metallic pigments are often manufactured in an inert gas atmosphere and sold in paste form. In the case of copper and gold bronze pigments, oxidation can be used to modify the inherent colors and produce a range of variations resulting from light interference at thin oxide layers on the surface of the pigment particles. Metallic pigments can be used alone or in combination with transparent colorants in paints, printing inks, and plastics.

Nacreous pigments must have a high refractive index, a flat, even shape, and a minimal thickness in order to keep the number of diffusely scattering side surfaces as low as possible. The particle size is usually between 10 and 30 μm, and the thickness of the platelets or the reflective layer on a carrier material around 40–50 μm. Fish scale, a nacreous pigment produced from whitefish scales, is manufactured by a complex method with a low yield and is not available in sufficient quantities. Of the synthetic products, nacreous titanium dioxide pigments are the most important because of their low price and high versatility. They are manufactured by hydrolyzing titanium sulfate or titanium tetrachloride onto colorless mica platelets of uniform size. The pigments are then calcined.

Nacreous bismuth oxide chloride pigments have largely been superseded in technical applications by cheaper products with better resistance properties although they continue to be used in the cosmetics industry. They are manufactured by hydrolyzing bismuth salts in the presence of chloride ions in the form of quadragonal platelets with a refractive index of 2.1. Bismuth oxide chloride can also be deposited onto mica. Basic lead carbonate [$Pb(OH)_2 \cdot 2\,PbCO_3$] yields nacreous pigments with the highest gloss values because of their very uniform and exactly formed hexagonal crystal platelets. These are used in the manufacture of imitation pearls and fashion accessories made from cast resins.

Interference pigments are manufactured in the same way as the colorless nacreous pigments. In order to obtain interference colors, the optical phase difference (the product of the thickness of the crystal platelets or the oxide layer and the difference in the refractive indices) should be between 250 and 700 nm [29]. The mean platelet thickness must be controlled to within a few nanometers. These pigments, which create an intense display of colors and transparency, are used in the manufacture of, for example, cosmetics and fashion jewelry.

E. Magnetic Pigments

The data storage layer in movable, magnetic recording media (tapes and disks) is created using fine magnetic pigments dispersed in a binder and applied on the tapes or disks. The important properties of such pigments are saturation magnetization, remanence, coercive forces, and particle size and shape. The coercive force (H_c) determines the application. High H_c values are required for high-density storage media. Values of 25–60 A/m are standard for audiotapes and up to 80 A/m for videotapes.

The most important magnetic pigments are the magnetic iron oxides such as γ-Fe_2O_3, Fe_3O_4, and mixed phases, possibly doped with cobalt or coated with cobalt ferrite. The usually needle-shaped pigments are manufactured by dehydrating α- or γ-FeOOH pigments of a suitable particle size and then reducing them with hydrogen or solid reducing agents at temperatures between 300 and 600°C. The addition of steam prevents an overreduction of the Fe_3O_4, which itself can be used for recording media or is first oxidized with air at 250–400°C to form γ-Fe_2O_3.

The metal pigments made from iron or its alloys with cobalt or nickel, which have H_c values up to 140 A/m, can also be manufactured by reducing suitably doped α- or γ-FeOOH with hydrogen at 300–600°C. Metal pigments ignite spontaneously and must therefore be desensitized before use or used under inert conditions.

Like the magnetic iron oxides, chromium dioxide (CrO_2), is an important compound used in the manufacture of magnetic storage media.

REFERENCES

1. *DIN Catalogue 1980*, Beuth Verlag, Berlin, 1980.
2. N. Noll, *Fortschr. Mineral. 57*: 203 (1979).
3. K. Winnacker/L. Küchler, *Chemische Technologie*, Vol. 3, *Anorganische Technologie II*, 4th ed., Hanser, Munich, 1983.
4. L. Schmidt, *Defazet 33*: 198 (1980).
5. G. Kämpf, *Defazet 33*: 185 (1980).
6. H. Kittel, *Lehrbuch der Lacke und Beschichtungen*, Vol. II, W.A. Colomb, Berlin, 1974.
7. H. G. Völz, *Angew. Chem. 87*: 721, (1975).
8. P. Kubelka and F. Munk, *Z. Techn. Phys., 12*: 593 (1931).
9. G. Mie, *Ann. Phys. (Leipzig), 4th Ser. 25*: 377 (1908).
10. Landoldt-Börnstein, *Zahlenwerte und Funktionen*, 6th ed., Vol. 2, Part 8, Springer, Berlin, 1962.
11. Pfizer, U.S. Patent 2,394,579 (1946).
12. I.G. Farbenindustrie, U.S. Patent 1,813,649 (1929).
13. Cities Service, U.S. Patent 2,939,767 (1960).
14. National Ferrite, U.S. Patent 1,366,748 (1920).
15. Columbian Carbon. U.K. Patent 643 303 (1948).
16. Pfizer, U.S. Patent 2,620,261 (1952).
17. I.G. Farbenindustrie, Ger. Patent 463 773 (1925).
18. I.G. Farbenindustrie, Ger. Patent 515 778 (1925).
19. Bayer AG, Inorganic Pigments for the Coatings Industry, (ring folder).
20. *Ullmanns Enzyklopädie der technische*, Verlag Chemie GmbH, Weinheim, Vol. 12, p. 776.
21. S. Keifer and A. Wingen, *Farbe Lack 79*: 866 (1973).
22. Siegle, Ger. Patent Appl. 1 288 714 (1966).
23. Nippon Chemical Industries, Ger. Patent Appl. 1 2 325 307 (1972).
24. DuPont, U.S. Patent 3,929,500 (1975).
25. A. Weiss and M. Giesen, *XI FATIPEC Congress*, 1972, p. 155.
26. A. Winterling, *Chem.-Ztg. 85*: 230 (1961).
27. Bayer AG, Eur. Patent 14408 (1980).
28. Th. Rehner, *Defazet 27*: 113 (1973).
29. W. Bäumer, *Farbe Lack 79*: 530, 638, 747 (1973).

4
Organic Pigments

ROBERT B. MCKAY Ciba Pigments, Paisley, Scotland

ABUL IQBAL and BERNHARD MEDINGER Research Pigments Division, Ciba-Geigy Ltd., Marly, Switzerland

I. Introduction 144

II. Characteristics of Organic Pigments and Their Effect on Dispersion Properties 145
- A. Chemical constitution 145
- B. Crystal lattice properties 148
- C. Importance of physical properties 151
- D. Nature of organic pigment particles 154

III. Control of Physical Properties of Organic Pigments and Improvement of Dispersion Properties 156
- A. Control of crystal size 156
- B. Control of crystal size and lattice type in polymorphic pigments 157
- C. Control of aggregate structure and dispersibility 158
- D. Control of flocculation of pigments in paint and low viscosity ink media 164
- E. Control of the rheology of pigment dispersions in lithographic ink media 169
- F. Effect of crystal morphology on the rheology of pigment dispersions 171

IV. Particle Size Analysis of Pigment Dispersions in Application Media 172

V. Concluding Remarks 174

References 174

I. INTRODUCTION

Organic pigments are finely divided crystalline solids that are used to impart color to printing inks, paints, surface coatings generally, plastics, artificial fibers, paper coatings, and various other materials. They are sold in the form of powders, granules, and dispersions in aqueous or nonaqueous media. They are essentially insoluble in the media in which they are applied and have to be dispersed therein by mechanical means. The resulting dispersions may be of liquid consistency, such as inks and paints, or solid, such as plastics and artificial fibers. In final application the crystalline pigment particles exist in a *dispersed state* within a solid material (bulk solid, filament, or surface coating). This is their essential difference from dyestuffs, which are generally physically or chemically adsorbed from solution or dispersion onto the surface of a solid substrate, usually filaments of a fiber, or in some instances are distributed within application media via dissolution.

Although their colors depend primarily upon their chemical constitutions and to some extent on their solid-state properties, the effectiveness of organic pigments in imparting color depends upon how well they are dispersed in the application medium. The level of dispersion achieved depends not only upon factors controllable by the pigment user (composition of the dispersion medium, type of dispersion equipment, dispersion procedure) but also upon factors controllable by the pigment manufacturer (e.g., primary particle size, particle surface character, aggregate structure, dispersibility, resistance to flocculation). Indeed, pigments designed to give optimum performance in specific types of applications may be unsatisfactory in others.

Organic pigments have a marked influence on the rheological properties of dispersions in liquid media. This is important, for example, in the manufacture of printing ink and paint concentrates where rheological factors determine the upper limit of pigment concentration manageable, and hence the throughput rate. It is a major factor in printing, since ink rheology has a controlling influence on the distribution and transfer of ink on the press rollers. In the case of paints, rheology can have a specific influence on the leveling of the surface of a wet paint layer and hence on the gloss of the final dried paint coat. Organic pigments can also have undesirable effects on other mechanical properties of application media, for example, distortion of molded plastic items, structural defects in artificial fibers and blown plastic films, and abrasion of printing plates by inks. Measures have to be taken to control these effects.

A multitude of pigments with distinctive application properties can be prepared from any given chemical type by modification of solid-state and surface properties. The key factors are the size and shape of crystals, surface character and state of aggregation of crystals, and, in the case of pigments that exhibit polymorphism, crystal lattice type. Ability to control these factors leads to a continuing development of new products in response to changing applicational requirements. The technological properties of organic pigments have been the subject of many reviews; Refs. 1–4 are among the more recent.

This chapter has two aims. The first is to give some insight into how the physical character of organic pigments affects the properties of dispersions in application media. The second is to exemplify how this physical character can be controlled to give products that disperse readily into appropriate application media and give satisfactory performance.

II. CHARACTERISTICS OF ORGANIC PIGMENTS AND THEIR EFFECT ON DISPERSION PROPERTIES

A. Chemical Constitution

A wide variety of chemical types of organic pigments are available commercially. They have become established on the basis of a range of criteria, principally intensity of color; cleanliness of shade; fastness to light, weathering, heat, solvents, and chemical attack; ecological and toxicological considerations; availability of raw materials; and cost. The diversity of chemical structure coupled with differing performance levels has led to a broad classification into so-called classical pigments on the one hand and high performance pigments on the other. Details of chemical constitutions of organic pigments are given in the *Colour Index* published by the Society of Dyers and Colourists [5]. Each chemical type is given a number, for example CI Pigment Yellow 13. This nomenclature is universally accepted and is used here ("CI Pigment" being omitted for brevity). The chemistry of organic pigments is comprehensively described in Ref. 1 (in German) and in Ref. 2.

Classical organic pigments, as their name implies, have been established for many decades. They include a wide variety of monoazo and disazo pigments (mainly yellows, oranges, and reds), copper phthalocyanine pigments (blue and green), and basic dye complexes. Examples of molecular structure are given in Fig. 1. Note that the so-called azo pigments tend to exist as the hydrazone tautomer (as shown in Fig. 1) stabilized by intramolecular hydrogen bonding [6–8].

The copper phthalocyanine pigments have excellent all-around fastness properties and are the most widely used blue and green pigments for all applications, however demanding. Copper phthalocyanine exhibits poly-

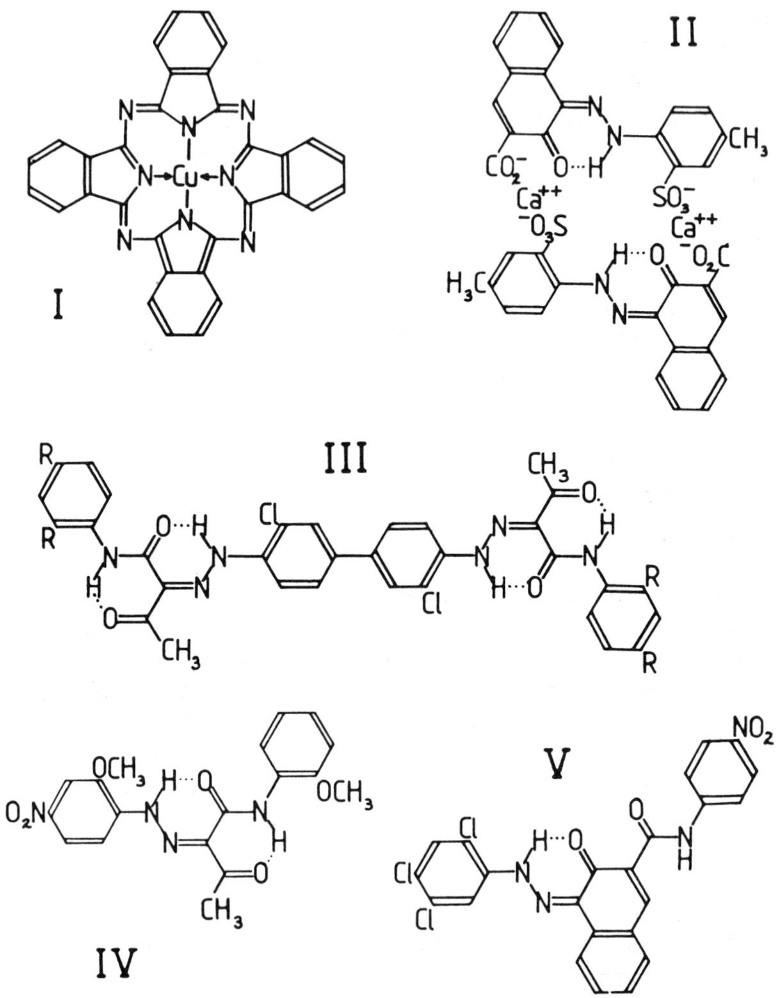

FIG. 1 Chemical structures of some "classical" organic pigments. (I) CI Pigment Blue 15; (II) CI Pigment Red 57:1; (III) R = CH_3 is CI Pigment Yellow 13, R = H is CI Pigment Yellow 12; (IV) CI Pigment Yellow 74; (V) CI Pigment Red 112.

morphism, the two commercially important forms being the cyan β form (Blue 15:3) and the reddish-blue α form (Blue 15). The α form reverts to the β form at elevated temperatures and on exposure to many organic liquids and has limited usefulness. Slightly chlorinated copper phthalocyanine (Blue 15:1) with 0.5–1 chlorine atom per molecule is isomorphous with the α form and has a similar shade. It cannot exist in the β form and is commonly called the stabilized α form; its stability is important for paints and plastics. Trichloro copper phthalocyanine is used as a cyan in metallic paints as an alternative to Blue 15:3 because of its unusual blue to dark blue flip effect (cf. blue to red with Blue 15:3); fully chlorinated copper phthalocyanine (14–16 chlorine atoms per molecule) gives bright green pigments (Green 7).

Monoazo and disazo pigments have generally less all-around fastness and are more limited in applicability. A wide variety of yellow and red monoazo pigments and some yellow and orange disazos are used in paints and plastics. The most important pigment types used in printing inks are Yellow 12 and 13 (disazos), Red 57:1 and 53:1 (metal salts of sulfonated monoazos; commonly called calcium 4B toners and Lake Red C, respectively), and β-form copper phthalocyanine (Blue 15:3).

Classical pigments are most commonly supplied as powders or granules. Predispersed aqueous pastes, however, are supplied primarily for use in emulsion paints, viscose rayon, and paper products. In the United States, flushed pigments (dispersions in viscous, aliphatic hydrocarbon-based media prepared by shear-induced phase transfer from aqueous filtercake) are commonly supplied for use in lithographic ink; in Japan, aqueous presscakes are commonly supplied for flushing by lithographic ink makers as the first stage of ink manufacture.

High performance pigments tend to be of more recent origin and have excellent all-around fastness properties. They are more costly to manufacture than classical pigments and are required for the most demanding applications, such as automotive paints and constructional plastics, where exceptional light, weather, and heat fastness are essential. Examples of molecular structure are given in Fig. 2.

High performance pigments derive primarily from two major chemical classes: azo and polycyclic pigments. The most important azo pigments are yellow to red disazo condensation pigments (e.g., Red 144), yellow azo metal salts (e.g., Yellow 183), and yellow to red pigments based on benzimidazolone (e.g., Orange 36). A number of azomethine-type pigments are commercially important; typical examples are yellow isoindolinones (e.g., Yellow 110) and isoindolines (e.g., Yellow 139).

Among the polycyclic pigments, the most important types by far are quinacridones, including the various polymorphic forms (e.g., Violet 19), perylenes (e.g., Red 179 and 224), and, of more recent origin, the diketopyrrolopyrroles (Red 254). In addition, two other types deserve mention, the anthraquinoids (e.g., Red 177 and Blue 60) and dioxazines (e.g., Violet 23).

Overall, azo pigments constitute about 50% of world volume of organic pigments, copper phthalocyanines and polycyclic pigments about 25% each.

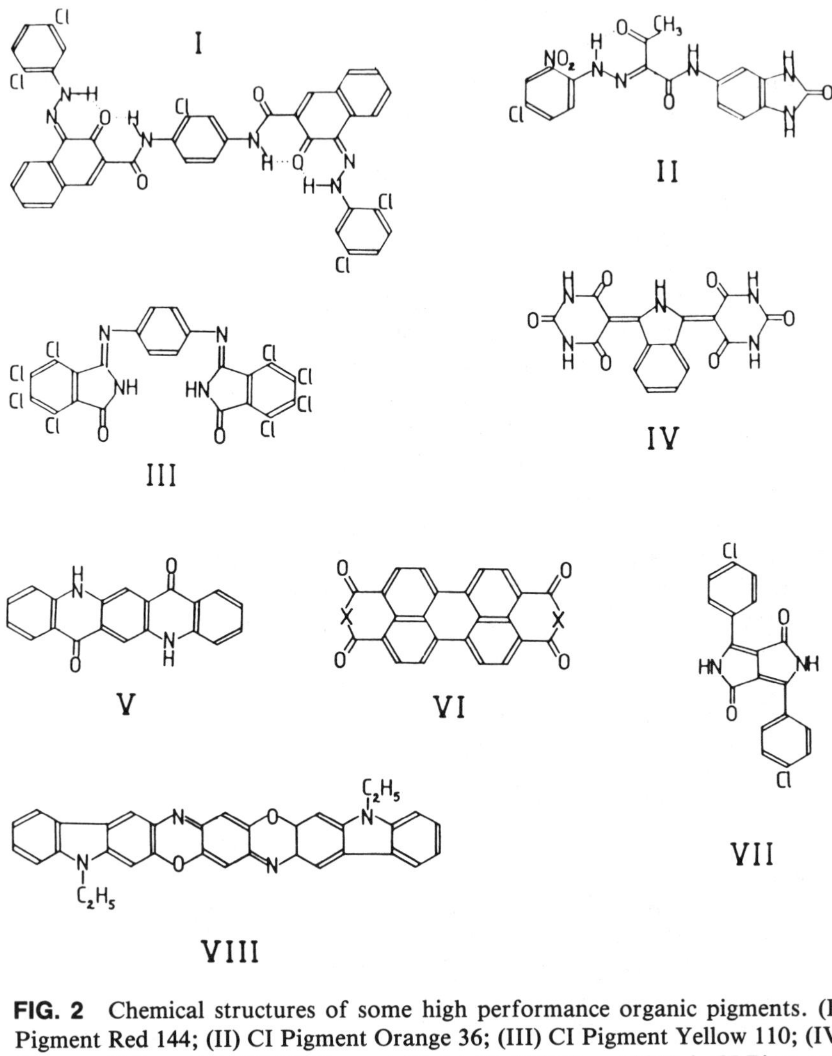

FIG. 2 Chemical structures of some high performance organic pigments. (I) CI Pigment Red 144; (II) CI Pigment Orange 36; (III) CI Pigment Yellow 110; (IV) CI Pigment Yellow 139; (V) CI Pigment Violet 19; (VI) X = NCH_3 is CI Pigment Red 179, X = O is CI Pigment Red 224; (VII) CI Pigment Red 254; (VIII) CI Pigment Violet 23.

B. Crystal Lattice Properties

The crystalline nature of organic pigments is shown by their characteristic powder X-ray diffraction patterns. High resolution electron microscopy has enabled the crystal lattice planes to be resolved in suitably oriented crystals of copper phthalocyanine pigments [9–11] and Red 57:1 pigments [12,13] (Fig. 3 provides examples), and even resolution of molecules in Green 7 [14]. X-ray diffraction studies have established the lattice structures of copper phthalocyanine blues [15,16], quinacridones [17,18], perylenes [19], diketopyrrolopyrrole [20], and various monoazo pigments [6,7]. The crystals have a molecular lattice. In most of the reported studies the intermolecular forces are van der Waals type. However, the lattice structures of Yellows 3, 65, and 74 [7], quinacridone [17,18], indigo [17], and diketopyrrolopyrrole pigments [20] feature intermolecular hydrogen bonding.

The molecules of organic pigments are essentially planar, even though the constituent atoms are not always coplanar in the strictest sense. In cases where crystal lattice structure has been determined, the molecules are generally packed face to face in columnar stacks (like coins), but with the molecular plane tilted at an angle from perpendicular to the stack axis. The stacks are packed side by side (like logs). The lattice structure of copper phthalocyanine provides the best known example (Fig. 4). In the α form the molecules in adjacent stacks are in parallel array, resulting in a layer-type lattice [15]; in the β form alternate stacks are rotated about their axis through 180°, thus giving a herringbone array [16] when viewed as in Fig. 4. There is also some interlocking of molecules in the stacks of the β form [11]; this and the herringbone array restrict plane slippage and hence deformability, so that the β form is more rigid than the α form. It is also more compact (denser). Red 57:1 pigments are a notable case in that the

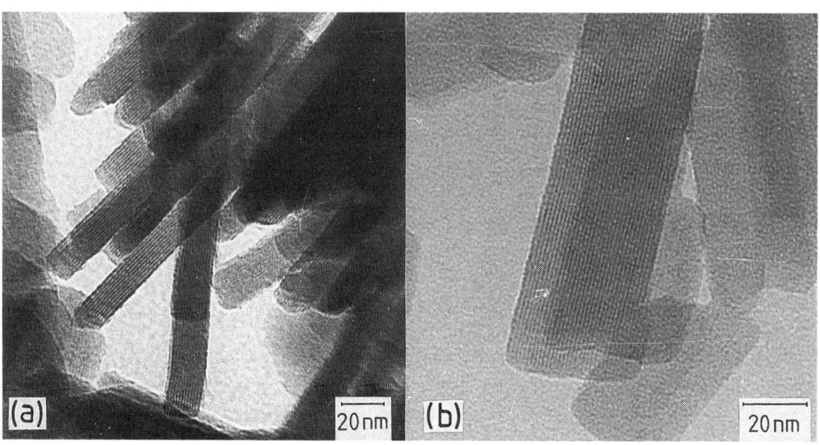

FIG. 3 High resolution electron micrographs of (a) Red 57:1 and (b) Blue 15:3 pigments showing crystal lattice planes.

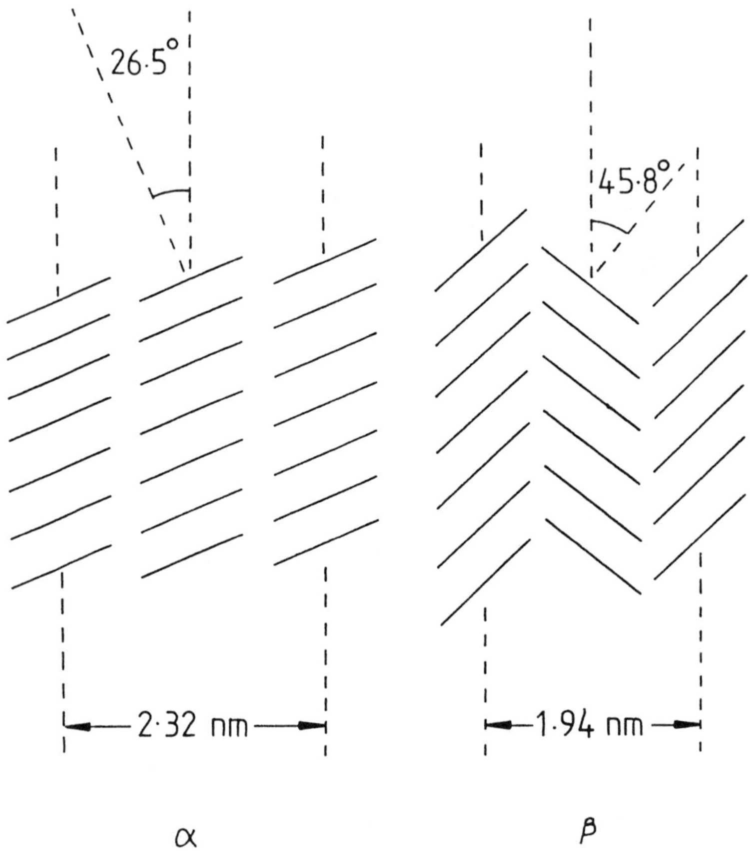

FIG. 4 Representation of the crystal lattice structures of α- and β-form copper phthalocyanine (CI Pigments Blue 15 and 15:3, respectively).

lattice consists of stacks of planar molecular dimers, the color residues being linked by divalent calcium atoms [12]; water molecules (two per dimer) are an integral part of the lattice.

In the known cases where intermolecular hydrogen bonding occurs, this bonding is between molecules in adjacent stacks [7,18,20]. It distorts the structure in monoazo yellows to some extent, resulting in puckering of the stacks in the case of Yellow 74, and in the case of Yellow 3 giving a non-columnar, layer-type lattice structure [7].

The stack character of the lattice leads to anisotropy, manifest in the brick, rod, and plate shapes of crystals. In most cases established so far

[6,7,15,16], the stack axis is parallel to the long axis of the pigment crystal. This is not so, however, in the case of diketopyrrolopyrrole; the strongest crystal growth is observed along the axis determined by the intermolecular hydrogen bonding rather than the stack axis [20].

One consequence of the anisotropy is a tendency of different faces of a crystal to have different polarity. For example, the ends of β-form copper phthalocyanine rods are more polar and more hydrophilic than the sides; as a result, cubic crystals are significantly more polar than rods [21]. This effect appears to be common to many types of pigments and can have important implications for their performance in application [22]. Another consequence is pleochroism, which is well exemplified by dichroism in copper phthalocyanine pigments [23]. All the crystal forms show different intensities of light absorption when viewed along and at right angles to the main axis of the crystals. The β form also shows a significant shade difference, being greener when viewed at right angles to the main crystal axis than when viewed along the axis. With this form, crystal shape influences not only color strength in application but also shade, cubes being perceptibly greener than rods.

The more rigid β-form copper phthalocyanine lattice is less susceptible to plastic deformation than that of the α form. Consequently the β form is less prone to strongly coherent aggregation than the α form [11]. This leads to marked differences in the technologies of the two forms. The same may apply to pigments with lattices featuring intermolecular hydrogen bonding between stacks. Such bonding also tends to enhance fastness properties such as heat resistance and resistance to bleed in organic liquids.

Overall, the crystal lattice structure is a function of the chemical constitution of the pigment. It is not generally amenable to modification except in cases that naturally exhibit polymorphism.

C. Importance of Physical Properties

The characteristic color of an organic pigment is determined primarily by its chemical constitution but is modified to a perceptible extent by intermolecular interactions in the crystal lattice [e.g., 24]. This is most clearly evident in the few cases that exhibit polymorphism, the different crystal forms having quite different shades. For example, α-form copper phthalocyanine is blue whereas the β form is cyan (greener); γ-form quinacridone is red and the β form is violet. Generally the maximum in the light absorption spectrum of an organic pigment is at a longer wavelength than that of its molecules in solution.

Although the intensity of color imparted by a pigment is dependent ultimately on its chemical constitution, the intensity actually achieved in practice is highly dependent upon the level of dispersion. Color strength increases with decrease in mean particle size. This has been firmly established by theory [25–28], experiment [27,29,30], and usage and is a fundamental principle of organic pigment technology. Figure 5 illustrates. The dependence is most marked at the smallest particle sizes, those of the order of 0.1 μm and less. In cases where particles are elongated it is the minor dimension that matters most [23,31]. This is also borne out in practice. In effect, light does not reach molecules many molecular layers below the surfaces of large particles, and the efficiency of absorption is reduced. Application of light scattering theory to pigment dispersions suggests that color should pass through a maximum with decrease in particle size at very small sizes. This, however, has not been unequivocally demonstrated in practice. Multiple scattering at practical pigment concentrations limits the

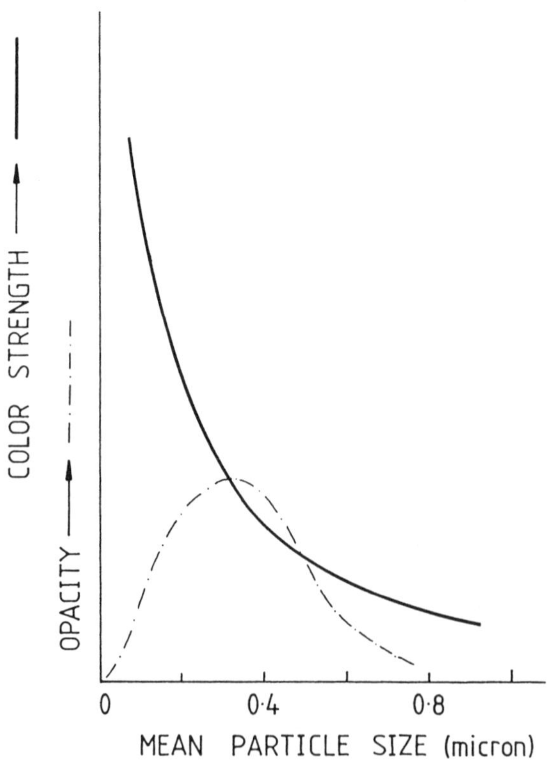

FIG. 5 Diagramatic representation of the relationships between color strength or opacity of pigmented application media and mean particle size of the pigment.

reliability of theory; quantitative assessment of particle size (especially such small sizes) in final application is problematical. Furthermore, the high interfacial energy of highly dispersed systems is likely to impose a lower limit to particle size achievable.

Increase in particle size has a small but significant effect on shade of dispersions. For example, Yellow 13 pigments become redder; Red 57:1 pigments become yellower. Copper phthalocyanine blues become redder [1,23], although the opposite tends to be observed with β-form pigments. In this latter case the situation may be complicated by their dichroic character (which makes crystal shape important) and by residual α-form crystals, which have a redder shade. Generally a wide particle size disribution gives dullness or dirtiness of shade. In practice this is usually associated with incomplete dispersion.

Particle size distribution of the pigment also affects the opacity/transparency of a dispersion. The opacity of a dispersion depends on the ability of the particles to scatter light and requires a significant difference in refractive index between the particles and the dispersion medium. It is optimal at particle sizes about half the wavelength of light. So, as illustrated in Fig. 4, opacity in a given medium passes through a maximum when mean particle size is in the region of 0.2–0.5 μm [32]; the narrower the size distribution is, the greater the opacity at a given mean size. Dispersions with mean particle sizes above and below this region are more transparent.

High color strength and transparency go together at very small particle sizes; this is ideal in inks for multicolor process printing, where successive thin ink layers (typically 1–3 μm thick) of different color are superimposed and light has to pass through the top layers to reach those below. For example, a transparent yellow ink printed on a black substrate is invisible, even though it is intensely yellow when printed on a white substrate. The light passes through the ink and is totally absorbed by the black substrate; no yellow color is visible unless there is some backscattering of light from the ink layer due to the presence of some large enough particles. With a white substrate the light is reflected back through the ink layer to be seen (minus the absorbed wavelengths) as yellow.

High transparency is also a prerequisite in the case of paints for special effects (metallic, pearlescent, etc.). In these applications the organic pigments provide the color without obstructing the unique light-reflecting and light-scattering effects of the underlying aluminum and mica platelets. Highly transparent pigments are often used as shading components in plastics applications. Here, in combination with highly opaque but dull pigments, they yield more saturated shades. In the coloration of synthetic fibers, the use of transparent pigments results in brilliance and gloss of the intrinsically opaque fibers.

The highest degrees of opacity often desirable in paints cannot in most cases be achieved using an organic pigment alone, the refractive index difference between pigment and paint medium being insufficient to give enough light scattering even at optimum pigment particle size. Admixture with an excess of a white inorganic opacifying pigment with high refractive index (usually titanium dioxide) is normally required. In the case of red pigments, admixture with a white opacifier produces a pink color rather than red. Therefore, the achievement of opaque full-shade red paints requires admixture with an inorganic red opacifier pigment (e.g., a molybdate orange). Moderate degrees of opacity can be achieved in paints using certain metal-free monoazo red and yellow pigments with optimized crystal size, but here the opacity is achieved at the expense of color strength. This latter disadvantage can be offset to some extent by certain polycyclic pigments with high extinction coefficients (perylenes, isoindolines, diketopyrrolopyrroles).

Obviously, the key to controlling the optical performance of any given type of organic pigment is control of particle size (distribution). So what, then, is an organic pigment particle?

D. Nature of Organic Pigment Particles

The primary particles of organic pigments are tiny crystals that in many cases are amenable to examination by transmission electron microscopy. They vary in shape from brick to rod to plate. Figure 6 shows some examples. Although crystal shape depends on the chemical type and lattice structure, in some important cases it can be modified by adjustment of manufacturing conditions and by the use of additives that adsorb to crystal surfaces and inhibit growth. The minor dimension of crystals varies on average from about 0.015 to 0.025 μm in pigments for transparent printing inks and paints up to about 0.2–0.4 μm in many pigments for opaque paints.

In suitable cases, crystal sizes can be determined from electron micrographs. For example, dimensions have been determined for brick- to rod-shaped crystals of β-form copper phthalocyanine [33] and metal-free monoazo pigments with large crystal size (e.g., Reds 1, 3, and 112 [34] and many others [1]), breadth being assumed to equal thickness. Also, successful determinations have been made of the area of the main faces of plate-shaped crystals of monoazo metal salt pigments (e.g., Red 57:1 [13]). Unfortunately, metal-free azo pigments with very small crystals (e.g., brick-shaped crystals of Yellow 13-type pigments for transparent inks (breadth <0.025 μm) are susceptible to serious damage in the electron beam, and an alternative means of sizing has had to be employed. Powder X-ray diffraction (XRD) has been useful in this context, breadths of lines in the diffraction line profile yielding average values of specific crystallite dimensions. Its useful-

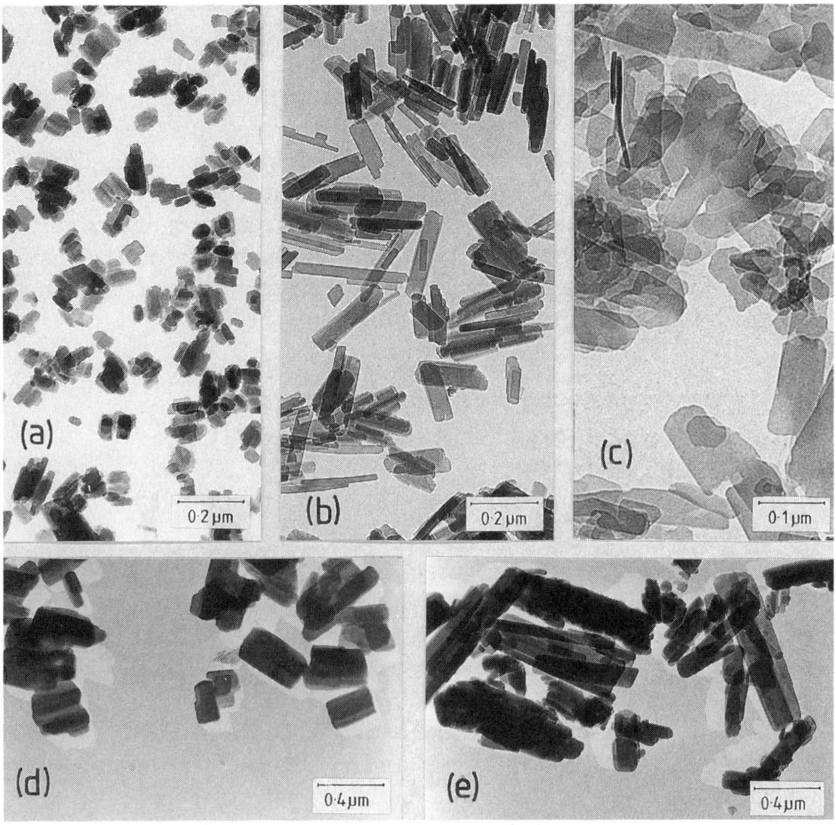

FIG. 6 Transmission electron micrographs of organic pigments showing brick-, rod-, and plate-shaped crystals. (a) and (b) Blue 15:3; (c) Red 57:1; (d) Violet 19; (e) Yellow 110.

ness depends on how well these dimensions can be related to crystal length, breadth, etc. Crystallite and crystal dimensions are the same when crystal dimensions are very small. For example, line breadths in powder XRD patterns have been used to determine the average breadths of brick-shaped Yellow 13-type pigment crystals [35,36] where, for example, a breadth of 0.02 μm corresponds to only 25 molecular breadths; the average thicknesses of the plate-shaped crystals of Red 57:1 pigment powders [13] have also been determined, 0.02 μm corresponding to only 11 molecular breadths.

Ideally, the particles of organic pigments dispersed in application media should be individual crystals. A few cases appear to approach this ideal; examples are Yellow 13-type pigments [37,38] and some calcium 4B toners

(Red 57.1) [13] extensively three-roll-milled in (highly viscous) lithographic inks. In such cases, color strength (after correction for colorless additive content of the pigments) correlates quite well with mean crystal thickness or breadth determined from X-ray diffraction line broadening. However, such simple correlation with crystal size is the exception rather than the rule. Electron microscopy of ultrathin sections of dried films of liquid inks, gravure inks, and paints has shown that the pigment particles commonly tend to be a mixture of individual crystals, small aggregates of a few crystals, and a few larger aggregates, sometimes detectable by optical microscopy. Thus control of particle size in dispersion requires control of both crystal size and the state of aggregation of the crystals.

III. CONTROL OF PHYSICAL PROPERTIES OF ORGANIC PIGMENTS AND IMPROVEMENT OF DISPERSION PROPERTIES

A. Control of Crystal Size

Monoazo and disazo pigments are prepared in aqueous media and precipitate out as pigmentary size crystals at the synthesis stage. Others, such as copper phthalocyanine and quinacridone pigments, are synthesized as large nonpigmentary crystals, which have to be subsequently reduced in size either by dissolution/precipitation in suitable media, notably concentrated sulfuric acid, or by grinding. Crystal size can be controlled within desired limits by control of process conditions and by the use of certain additives that adsorb to the crystal surfaces and inhibit crystal growth.

In some cases, notably disazos of the Yellow 13 type, minor amounts of compounds of slightly modified chemical constitution are incorporated into the pigment by cosynthesis (commonly referred to as mixed coupling) and result in smaller crystals resistant to growth. This is well established in practice.

Incorporation of Yellow 83 with bulky substituent groups is especially effective. The mechanism of crystal size control by this means has not been clearly identified; the composite material still has the crystal lattice structure of Yellow 13 even though the chemical composition is sufficiently different for the pigments to have different CI numbers. Yellows 13 + 63, 14, 83, or 12 are designated Yellows 127, 174, 176, or 188, respectively. Recent studies of cosynthesized mixtures of Yellows 12 + 14, 13 + 14, and 14 + 63 suggest that solid solution formation is an important factor leading to smaller crystal size [8], molecules of the minor component replacing some of those of the major component in the crystal lattice. Similar effects have been reported with monoazo pigments, minor amounts of Yellows 5 or 6 cosynthesized with Yellow 1 leading to much smaller crystals than those of either component synthesized alone [39]. The Yellow 1–type pigments produced appeared to be solid solutions of the minor component in the Yellow 1 crystal lattice. The same principle is used in Red 57:1 pigments, where coprecipitation of a minor amount of closely related Tobias acid–Bona coupling is used to produce a bluer shade [40]. Solid solution may be the reason for effectiveness in reducing crystal size and may be partly responsible for the bluer shade [40]. Small-particle diketopyrrolopyrrole pigments can also be obtained via a mixed synthesis technique [41].

Further control of crystal size can be achieved by treatment subsequent to precipitation or grinding. In some cases this merely requires heating in aqueous media. For example, the crystals of Yellow 13 pigments in the aqueous slurry formed in the synthesis stage are extremely small, compactly aggregated, and dull yellow. Heating to over 90°C promotes crystal growth by a two-stage process entailing annealing of crystals within aggregates followed by sintering of the annealed crystals [36] and produces the desired bright yellow color. The annealing is associated with a subtle but distinct change in the X-ray diffraction pattern (*d* spacing of 3.42 Å reducing to 3.28 Å, indicative of slightly closer face-to-face packing of molecules in stacks in the crystal lattice) [35]. This has enabled monitoring of factors influencing annealing.

The annealing process is accelerated by the presence of abietyl resin, which may be operating as a scavanger for impurities on crystal surfaces within the aggregates [36]. Conversely, it is retarded by the presence of mixed-coupled species and/or of minor proportions of sulfonated disazo derivatives ($<5\%$ w/w), which adsorb to the pigment crystal surfaces. When resin is present, however, the effect of resin dominates. At resin levels corresponding to more than about 10–15% of the final pigment powder weight, the annealing still proceeds rapidly, but the subsequent sintering stage is retarded, adsorbed resin on the annealed crystal surfaces acting as an inhibitor. At this stage, however, the sulfonated disazo compounds or certain mixed-coupled species when present are more effective inhibitors than is resin. In these latter circumstances the benefit of resin is in enabling preparation of well-annealed but still very small crystals. This is important in the manufacture of transparent Yellow 13–type pigments. In subsequent dispersion in lithographic ink media by high energy bead milling, the sulfonated disazos and/or mixed-coupled species confer resistance to crystal sintering (hence retention of color strength and transparency) at the elevated temperatures (>80°C) generated.

B. Control of Crystal Size and Lattice Type in Polymorphic Pigments

Copper phthalocyanine and quinacridone pigments are the best known examples of pigments exhibiting polymorphism. Copper phthalocyanine as synthesized (so-called crude blue) consists of rod-shaped β-form crystals of length from about 0.5 to 100 μm, size depending on the process of synthesis. Pigmentary size crystals of α or β form can be prepared by a variety of processes.

Pigmentary α-form crystals are precipitated by carefully adding a solution of crude blue in concentrated sulfuric acid to an excess of water (acid pasting). Grinding dry crude blue powder with steel balls in the presence or absence of inorganic salt (as grinding aid) produces α, β-form mixtures of brick-shaped crystals with breadth around 0.02 μm. The relative proportions of α and β forms depend on the grinding conditions and can be up to 98–99% β form. Grinding crude blue powder with salts in the presence of a trace of a powerful crystallizing liquid (but still a dry process) produces brick-shaped crystals of pigmentary 100% β-form copper phthalocyanine directly. So too does grinding or kneading with an excess of mildly crystallizing liquid. The processes and mechanisms involved have been reviewed elsewhere [11,42].

Grinding crude blue powder to produce an α, β-form mixture with >50% α form followed by heating in a suitable organic liquid produces pigmentary rod-shaped β-form crystals. In the latter stage, crystal growth occurs by a complex Ostwald ripening process [42] in which the α to β form conversion is a key factor. Optimization of processing normally yields >95% β-form crystals with axial ratio in the range of 4–10, depending on conditions, but with crystal breadth not much greater than in the ground material. Continued treatment after α to β form conversion has been completed induces a second mechanism of crystal growth involving side-by-side fusion of crystals with marked increase in crystal breadth [33]. This is undesirable from the point of view of color strength, so careful optimization of the process is essential. Stabilized α-form blues (Blue 15:1) are usually prepared by either dry grinding or acid pasting. α-Form, stabilized α-form, or β-form pigments prepared by grinding tend to have brick-shaped crystals.

Quinacridone pigments can also be prepared in different polymorphic forms by broadly similar types of processes.

C. Control of Aggregate Structure and Dispersibility

Terminology used in pigment technology to describe states of aggregation [e.g., 43] is not consistent with that recommended by IUPAC [44] and generally accepted in surface and colloid science. Most seriously, the terms "aggregate" and "agglomerate" are transposed. In this chapter the term "aggregate" is used in the general sense recommended by IUPAC to mean a group of particles held together in any way, and the strength of cohesion is indicated where appropriate. In this way the term "agglomerate" is avoided.

The smaller pigment crystals are made, the more prone they become to immediate and strongly coherent aggregation that is difficult to reverse. The reason is that not only do very small crystals have large surface area and hence large surface energy per unit weight, but also they have enhanced surface energy per unit area due to the higher number of corners and edges. Furthermore, when the pigment is isolated from the preparative medium by filtration and drying, the tiny crystals (in the size range of smokes) have to be aggregated in a controlled way to form grains large enough to give a powder or granules. This has to be achieved in such a way that the crystals can be redispersed into the application medium as easily and completely as possible. Care has to be taken to counteract difficultly reversible aggregation caused by compaction on filter presses, by capillary attraction during drying, and by compaction during subsequent blending and handling processes. Otherwise the potential benefits of small crystal size cannot be realized. Ease and completeness of dispersion are embraced in the term "dispersibility."

Wettability of accessible pigment surfaces by the application medium is a prerequisite for effective dispersion but is usually easily achieved, especially in organic media. It is not, however, a sufficient criterion for good dispersibility. It has been shown, for example, that organic liquids (including *n*-dodecane) penetrate into the interior of strongly coherent aggregates of α, β-form copper phthalocyanine mixtures prepared by dry milling and wet out the constituent crystal surfaces virtually completely [45]. This indicates that the crystal surfaces are not irreversibly linked, yet the mixtures have extremely poor dispersibility in lithographic inks (which contain high boiling aliphatic hydrocarbons). The wetting is not sufficient to overcome the strength of cohesion of the aggregate structure.

The strength of cohesion of the aggregate structure is the factor that tends to control dispersibility in practice and has to be minimized. It is dependent upon two factors: the area of contact between crystals and the strength of cohesion per unit area. The former depends on the mode of packing of crystals in the aggregate structure, face-to-face packing tending to give stronger cohesion than edge-to-edge or edge-to-face, other factors being equal. The deformability or softness of the crystals is important here, since plastic deformation of crystals forced together increases area of contact. The second factor, strength of cohesion per unit area, increases with increases in the polarity of the crystal surfaces. The importance of these two factors is demonstrated in the following examples of the main process colors for lithographic printing inks.

Openness of the aggregate structure of an organic pigment can be assessed by comparing the specific surface area of the crystals S with the BET

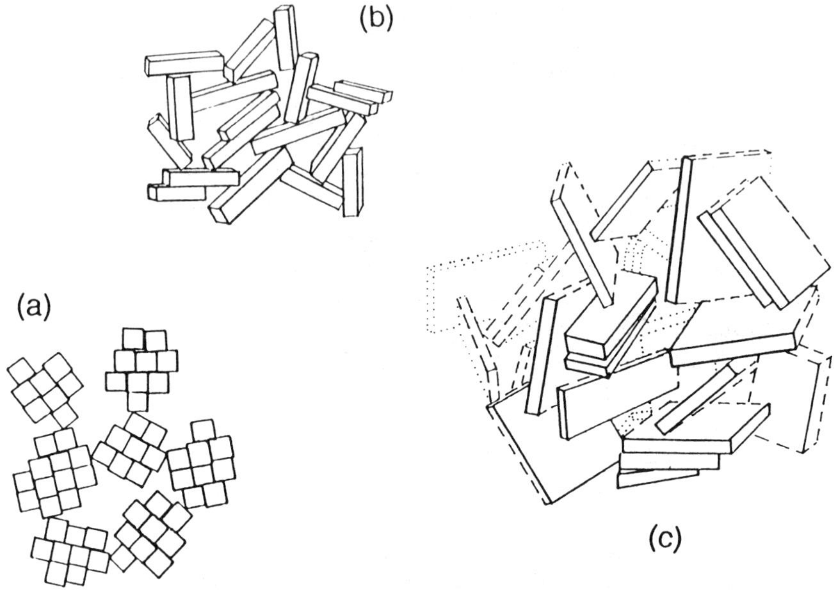

FIG. 7 Diagramatic representation of compact and open aggregate structures. (a) Compact; (b) and (c) open.

specific surface area of the pigment powder S_{BET}, determined by nitrogen gas adsorption [13,33]. At the low temperature ($-196°C$) and pressures (pressure less than 0.4 times the saturation pressure of nitrogen gas) used in the BET method, the nitrogen molecules do not penetrate between crystals in face-to-face contact [13,46]. $S_{BET}/S << 1$ indicates a high degree of face-to-face packing (compact structure); $S_{BET}/S -> 1$ indicates little face-to-face packing (open structure). Figure 7 provides diagramatic illustrations.

Rod-shaped β-form copper phthalocyanine pigments have low surface polarity [21,45] and exemplify how dispersibility is improved predominantly by minimization of face-to-face contact of crystals [4,45]. Mixtures of α, β-form crystals produced by milling crude copper phthalocyanine powder are compactly aggregated (S_{BET}/S typically <0.2; <0.05 in the absence of salts) (cf. Fig. 7a) and have very poor dispersibility. However, the rod-shaped β-form crystals produced by heating in an organic liquid have an open aggregate structure with little face-to-face packing ($S_{BET} -> S$) (cf. Fig 7b). Optimization of this type of process gives pigments with large crystal volume (but small crystal breadth) and good dispersibility in lithographic ink media. The improvement in dispersibility is related to the extent of crystal growth and is predominantly due to the opening of structure, although reduction in surface polarity resulting from the change in crystal shape may contribute. Addition of up to about 5% of abietyl resin (derivatives of wood rosin or tall oil) induces side-by-side alignment of the rod-shaped crystals into sheaves with some reduction in openness of structure [33]. Yet it produces significant further improvement in dispersibility. Immersion calorimetry has shown that the resin-treated crystal aggregates are readily and completely penetrated by ink media; the reduction in openness of structure is apparently more than offset by the reduction in strength of edge-to-edge or edge-to-face contacts of crystals [33].

Abietyl resins are closely related to abietic acid, the chemical constitution of which is shown in Fig. 8. Note the terpenoid hydrocarbon residue, which confers low polarity, and the carboxylic acid group, which confers solubility in aqueous alkali and provides a means of adhesion to the pigment crystal surfaces.

A more striking example of the beneficial effect of abietyl resin is provided by Red 57:1 pigments. These pigments consist of rectangular plate-shaped crystals that naturally form an open aggregate structure with edge-to-edge and/or edge-to-face contacts predominating (S_{BET}/S = 0.7–0.8) (cf. Fig. 7c). Nevertheless, since the crystal surfaces are polar [13,45], the aggregate structure is strongly coherent, and untreated pigments have very poor dispersibility. Resination, however, reduces the surface polarity, hence the cohesiveness, and gives pigments with excellent dispersibility in lithographic ink media [13,45]. The resin is adsorbed on the crystal surface, usually predominantly as the calcium salt. Its presence there is shown by the fact that it inhibits crystal growth during pigment manufacture so that treated Red 57:1 pigments have much thinner crystals than untreated pigments [40]. This fortuitous circumstance means that although final pigments contain up to 20% resin per gram of powder, the diluent effect of

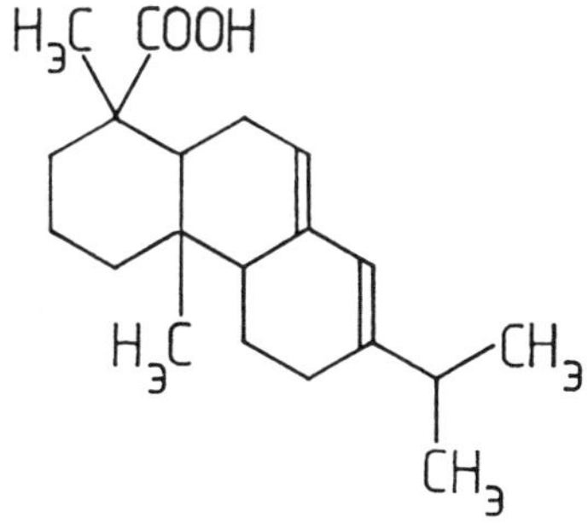

FIG. 8 Chemical structure of abietic acid.

the colorless resin on color strength is offset by the enhancement due to the smaller crystal size.

Yellow 13–type pigments have cubic to brick-shaped crystals with low polarity surfaces [37]. Additive-free pigments with relatively large crystals (breadth around 0.035 μm) have quite good dispersibility in lithographic ink media. Those with the extremely small crystal size required for transparent lithographic inks (breadth <0.025 μm) tend to form strongly coherent aggregates and have poor dispersibility. In the latter case, good dispersibility is achieved by treatment of freshly precipitated pigment slurry with sufficient abietyl resin to give up to 45% by weight in the final pigment powder. There the resin is usually present in the free-acid form as an adsorbed layer one to two molecules thick on the pigment crystal surfaces [36]. In the most highly resinated pigments (35–45% w/w), there is some resin in excess of that required to form the adsorbed layer; this tends to be present as a separate crystalline phase, detectable as a fine superstructure in the XRD pattern of the pigment powder [36]. Resination produces a compact aggregate structure (cf. Fig. 7a) in which face-to-face packing of the resin-coated crystal surfaces predominates. S_{BET} values of the highly resinated pigments are generally less than 25 m^2/g, almost an order of magnitude less than the geometric specific surface area of the resin-coated crystals [36]. The resin adsorbed on the crystal surfaces is believed to remain there after dispersion of the pigment in lithographic ink medium [37].

A few commercially available Yellow 13–type pigments contain resin at least partially in the form of calcium or zinc resinate. Unlike the free-acid form, the metal resinate softens only at temperatures well in excess of 100°C and therefore not during aqueous processing of the pigments. As a result the pigments have a more open aggregate structure than pigments containing only free resin acid, as indicated by higher S_{BET} values (e.g., 73 m^2/g for a pigment with resin completely as calcium resinate [36]), and tend to disperse more rapidly in preliminary stages of dispersion. There is no advantage, however, in the final level of dispersion.

These examples illustrate the main principles of dispersibility of organic pigments in lithographic inks. Various alternatives to abietyl resin have been reported in the patent literature and discussed elsewhere [47]. Their precise mode of action has not been fully established, but presumably the same principles apply.

The broad principles outlined above also apply to dispersibility of pigments in other low polarity media, such as hydrocarbon-based paints and gravure inks, although for such media abietyl resin levels required in azo pigments tend to be much lower. The resin dissolves off the pigment crystal surfaces virtually completely in media rich in aromatic hydrocarbons, thus facilitating dispersion; in aliphiatic media, however, only partial desorption occurs and dispersion tends to be less complete [48]. Gravure ink media have low viscosity, and the shearing forces developed during dispersion of pigments are much less than with lithographic inks. Consequently, suitable control of the cohesiveness of aggregates of pigment crystals is even more demanding. With Yellow 12 pigments (the most important yellows for gravure inks), treatment of crystal surfaces with suitable aliphatic amines results in enhanced dispersibility in aromatic-rich gravure inks. A proportion of the amine forms a Schiff's base with the surface pigment molecules to give a highly hydrophobic aliphatic surface. Capillary attraction on drying aqueous pigment presscakes is thus minimized to give less strongly cohesive aggregation [49].

It is more difficult to achieve pigments with good dispersibility in polar media such as ester- or alcohol-rich or water-based liquid inks. The low surface polarity required to avoid strongly cohesive aggregation of crystals gives poor compatibility with, for example, aqueous media. Furthermore, the low viscosity of such media limits the level of shearing achievable during dispersion. Consequently, the levels of dispersion attainable with pigments in powder form are much lower than can be achieved in lithographic inks, and the initial state of aggregation of crystals in the pigment powder is even more important.

Various types of additives can be used to improve the dispersibility of pigments in aqueous media. These range from conventional surfactants to combinations of polymers and have been reviewed recently [22,47]. Their mode of action is still subject to speculation. However, the highest levels of dispersion in polar alcohol-based or ester-based liquid inks are achieved by a quite different approach. This features pigments predispersed in molten binder resin, such as nitrocellulose, under high shear conditions to produce what are known as "chips." When contacted with solvent, the binder resin in the chips dissolves, thereby releasing the highly dispersed pigment particles. Pigments are also sold predispersed as aqueous pastes for use in emulsion paints and aqueous inks.

The deformability or softness of the crystal lattice can lead to problems during dispersion under conditions where compaction of the pigment occurs. For example, α-form copper phthalocyanine pigments are prone to irreversible aggregation by compaction when premixed with polymer granules by tumbling, prior to dispersion in polyethylene by extrusion molding at elevated temperature [11]. β-Form pigments with a more rigid lattice structure are much less susceptible.

Generally, organic pigment powders are susceptible to some degree of compactive aggregation during the initial stages of dispersion in plastics, with loss of some potential color strength. Polymer and pigment particles are forced together under pressure before the polymer warms up, melts, and "wets" the pigment. Even after the polymer has melted and wetted the pigment, the shear forces generated may be insufficient to completely reverse the effect of compaction. Prolonging the dispersion time often produces little improvement. Residual aggregates can, in severe cases, lead to mechanical defects in colored plastic materials, such as the formation of holes in plastic films and the rupture of synthetic fiber filaments.

Such difficulties can be avoided by producing predispersed pigments in the form of so-called masterbatches and preparations. The latter consist of 30–70% of pigment in specific carrier resins, with not only good pigment wetting properties but also sufficient compatibility with the final application plastic. Unfortunately, such compatibility requirements strongly restrict the range of applications. The good wetting properties of the carrier resins in the elaborations and the high shear equipment used in their preparation (e.g., high shear kneaders, double-screw extruders, and three-roll mills) result in minimal compactive aggregation. Furthermore, the high shear forces are capable of breaking down not only aggregates of pigment crystals but also large crystals, thereby leading to a narrower particle size distribution than can be achieved by conventional dispersion processes. Such preparations have the additional advantages of being dust-free and meterable.

D. Control of Flocculation of Pigments in Paint and Low Viscosity Ink Media

Pigment particles dispersed in ink and paint media tend to flocculate due to residual van der Waals attraction. This leads to the formation of three-dimensional structure, which enmeshes and immobilizes the liquid phase and causes thickening. At sufficiently high pigment concentrations, all the liquid phase is immobilized and the dispersion becomes a gel. Gel formation in some instances can occur at organic pigment concentrations of less than 5% by weight. The flocculation is mechanically reversible, thus giving rise to shear-thinning character and thixotropy, and is evidently of the weak (secondary minimum) type.

The thickening effect of flocculation on the rheology of dispersions is generally detrimental. It determines the upper limit of pigment concentration in the manufacture of ink and paint concentrates and thereby limits throughput rate. It adversely affects ease of handling of dispersions. Flocculation affects indirectly the color strength of paints that contain inorganic opacifier pigments through its role in flooding and floating (see Chapter 1). It may have a direct optical effect in inks and paints by increasing the optically effective particle size. It can lead to defective gloss of ink or paint films. Control of flocculation of organic pigments is therefore a topic of major importance, especially in dispersions with low viscosity, such as gravure and liquid inks and paints. This applies especially to modern paint systems such as medium to high solids nonaqueous dispersions and waterborne coatings, which have appeared on the market in response to the increasingly important issue of environmental protection. Not only has the content of volatile organic compounds been drastically reduced in such systems, but also the nature of currently employed solvents is much less conducive to achieving stabilized dispersions.

Control of flocculation can be considered in terms of reduction of either strength or rate of interaction of particles. The former is the more important for most practical purposes involving organic pigments, its successful achievement reducing the significance of the latter. It is not necessary to eliminate flocculation altogether, substantial reduction in the strength of structure usually being sufficient. Modification of the surface chemical character of the particles, and hence of the pigment–medium interface, can achieve this aim.

Effective modification of the interfacial properties can be achieved by the introduction of certain types of additives that adsorb to the pigment particle surfaces and anchor functional groups there. The additives can in some cases be added just prior to dispersion of pigment in the application medium. Incorporation, however, is often more effectively achieved during pigment manufacture at stages where the pigment crystal surfaces are accessible, such as during milling, or by adsorption from solution during aftertreatment. There are at least four broad categories of surface treatments effective with organic pigments. All have in common the fact that they promote the formation of an adsorbed layer containing well-solvated, extended-chain molecules. The physicochemical mechanism involved is probably steric stabilization, although electric charge stabilization may play some part in aqueous dispersions. Details of these types of mechanism are given elsewhere [e.g., 50].

The first and so far most widely used category is well exemplified in copper phthalocyanine pigments for gravure inks and paints based on aromatic hydrocarbon liquids and has been discussed at length elsewhere [4,48]. The principle is to promote interaction of the pigment particle surfaces with the binder resin in the ink or paint medium via polar anchoring groups adsorbed on the pigment surface—and in such a way as to form an effective adsorbed resin layer. For example, a substituted copper phthalocyanine derivative of composition $CuPc(CH_2NR_2)_n$ adsorbs strongly on the pigment particle surfaces. R is typically a small aliphatic hydrocarbon residue. The amine groups interact with binder resin in the paint or ink medium to form an adsorbed layer of resin with the properties required for steric stabilization. This layer is necessarily different in composition and properties from the insufficiently effective layer formed in the absence of the additive. The essential requirements are extended-chain molecules firmly anchored at the interface but well solvated and with the solvated parts extending out into the liquid phase to give a thick layer. Figure 9a illustrates the principle. It applies to both stabilized α-form pigments (Blue 15:1) and β-form pigments (Blue 15:3). The mechanism of interaction has been established by gravity sedimentation methods [51,52], by centrifugal sedimentation supported by rheological methods [48,53], and by an electrophoretic method [54]. Analogous surface treatments have been developed with other types of pigments [47,55,56].

Disadvantages are that the effectiveness of the additive is dependent on the nature of the binder resin and is limited at high pigment concentrations.

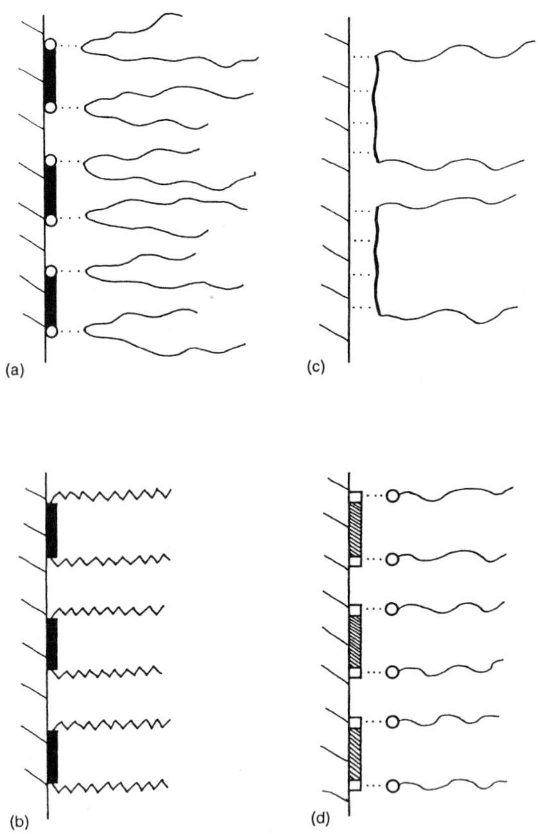

FIG. 9 Ways in which additives induce resistance to flocculation of dispersions by formation of adsorbed layers of extended chain molecules at the pigment–medium interface. (a) Additives that interact with binder resin in ink or paint media; (b) low molecular weight additives effective in their own right; (c) copolymeric hyperdispersants; (d) synergist–hyperdispersant combinations.

This limitation can be attributed to the heterogeneity of chemical constitution and/or molecular size of binder resin, exhaustion of the effective fractions leading to failure of stabilization at high pigment concentrations.

A second category of surface treatments features additive molecules that are amphipathic. They have both an insoluble anchor residue (similar in chemical structure to the pigment) and an extended chain (largely aliphatic hydrocarbon) with high affinity for the solvent in the dispersion medium. The molecules adsorb on the pigment crystal surfaces, with the chain residues well solvated and the molecules themselves forming an effective adsorbed layer, interaction with binder resin being unnecessary. Comparatively short chains are effective. For example, additives of the type CuPc-$(CH_2OOCR)_{2.5}$ and CuPc$(SO_2NHR)_{2.5}$, where R = $(C_6H_4)OC_{16}H_{33}$, are very effective with copper phthalocyanine pigments in toluene-based gravure inks and xylene-based paints [48,53]. The insoluble copper phthalocyanine residue is the anchor group, with the well-solvated substituent chains extending into the dispersion medium. (Figure 9b illustrates.) Dispersions with significant stability can be prepared in pure toluene. Colored anchor groups and short chains give an advantage over polymeric additives when the additive is incorporated in the pigment in that they dilute the color strength per unit weight of pigment less. Additives that do not rely on interaction with binder resin have the advantages that their effectiveness is not highly dependent on the nature of the binder resin and extends to high pigment concentrations.

A third category of surface treatments is similar in principle to the previous category, but the amphipathic additives are of higher molecular weight. They are copolymeric, an example being an ABA block copolymer [57] consisting of an insoluble segment B that adheres to the pigment surfaces by multipoint attachment and soluble segments A that are well solvated and form the effective part of the adsorbed layer (Fig. 9c). Such copolymers are referred to as hyperdispersants.

A fourth category of surface treatment combines the principles of the first and third categories. It features a combination of two additives. One, known as a "synergist," with insoluble anchor residue (chemically similar to the pigment) and suitable (polar) functional groups adsorbs on the surface of the pigment particle. The polar groups interact with the other additive, which is a hyperdispersant with a polar end group, the combination forming an effective adsorbed layer [57–59]. (Figure 9d illustrates the principle.) The hyperdispersant is used instead of relying on interaction with binder resin. The amount and type of hyperdispersant can be varied, giving effectiveness to much higher pigment concentrations than is possible with the first category of surface treatments.

These are the main principles. Many types of additives have been reported in this general context in the literature. Examples are given elsewhere [22,48]. In many cases their mode of action has not been clearly established, although presumably they operate by one or the other of the above mechanisms, depending on their molecular structure. Moreover, much of the literature on chemical structures of additives for surface treatment of organic pigments does not clearly differentiate between effectiveness in controlling flocculation and other aspects of dispersion.

Additives of type $CuPc(SO_3^- \ {}^+NHMeRR')_n$ for copper phthalocyanine pigments are important in that they can be effective in hydrocarbon-based or polar organic media, depending on the nature of the R and R′ residues and the value of n. One or more aliphatic groups up to C_{18} in length favor hydrocarbon-based systems; small aliphatic groups or groups with polar substituents favor polar organic media. Additives of this general type are also used as synergists with copper phthalocyanine pigments.

Flocculation of organic pigment dispersions gives rise to shear thinning properties, which can be characterized by means of a rotational viscometer [48,53,60] in terms of strength of structure, indicated by τ_o, and plastic viscosity (Fig. 10a). Effective surface treatment of the pigment reduces τ_o virtually to zero and gives almost Newtonian behavior (Fig. 10b) [48,53]. τ_o can also be markedly reduced by treatments that inhibit pigment dispersion, so care has to be taken to support interpretation by independent evidence from, for example, color strength assessments, poor dispersion giving low color strength.

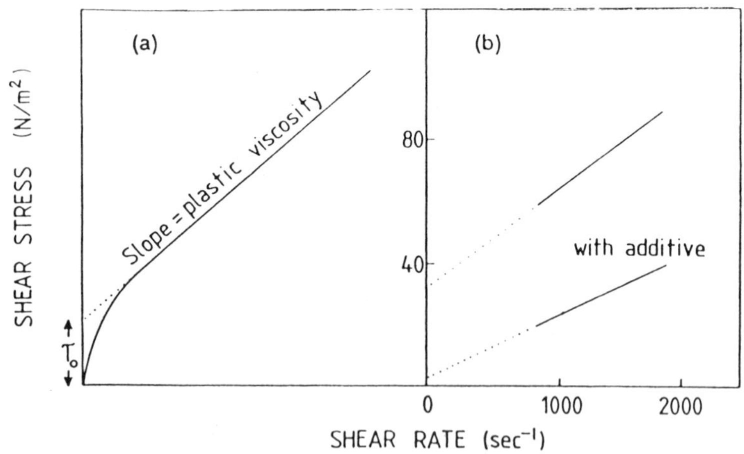

FIG. 10 Rheological curves obtained with organic pigment dispersions. (a) Interpretation; (b) effect of additive that reduces the strength of flocculated structure.

E. Control of the Rheology of Pigment Dispersions in Lithographic Ink Media

Lithographic ink media are highly viscous (e.g., 2–10 Pa.s at room temperature) and of low polarity. Thcy arc polymeric in character, the principal ingredients being usually phenolic-type resin with minor amounts of alkyd resin and high boiling aliphatic hydrocarbon. Lithographic inks contain typically 10–20% w/w pigment and exhibit shear thinning character and thixotropy indicative of flocculation of pigment particles. They are prepared via concentrated millbases containing up to 40% w/w pigment, the concentration being limited by lack of fluidity. In such solvent-deficient media, flocculation cannot be controlled effectively by surface treatment to reduce attraction between pigment particles. Instead, reduction of the strength of flocculate structure has to depend upon reduction of the number concentration of particles.

Pumpable dispersions with more than 50% w/w pigment can be prepared in resin-free aliphatic hydrocarbon containing hyperdispersants of the types illustrated in Figs. 9c and 9d and based on polyhydroxystearate [61]. Such dispersions and flushed pastes (containing media rich in aliphatic hydrocarbon) provide alternative ways to increase throughput in lithographic ink manufacture, the ink binder resins being added on dilution to the final ink concentration. However, a major disadvantage of these approaches is that the high concentration of hydrocarbon restricts the flexibility of formulation of the final inks.

Recent studies have been reported on lithographic inks containing Yellow 13–type pigments at up to 25% w/w and well dispersed by extensive three-roll milling [37,38,62]. Color strength (after correction for colorless abietyl resin content of the pigment) correlated well with mean crystal breadth, thereby indicating dispersion substantially as individual crystals. The strength of flocculate structure in the inks (indicated by τ_0) increased in an approximately linear manner with increase in the number of crystals per unit volume of ink, m (Fig. 11). At a given pigment concentration, it is possible to reduce m (and hence the strength of structure) in two ways — by increasing pigment crystal volume and by increasing the weight fraction of abietyl resin in the pigment. The latter is the preferred option; the former is less desirable in that it decreases transparency of the ink. This is one reason Yellow 13–type pigments for transparent inks tend to have very high resin contents, substantially in excess of that required to just cover the crystal surfaces.

The number concentration of particles of Red 57:1 pigments in lithographic inks and hence the strength of structure can be reduced by introducing a controlled degree of face-to-face aggregation of the plate-shaped crystals [62]. Although these pigments tend to form an open aggregate structure

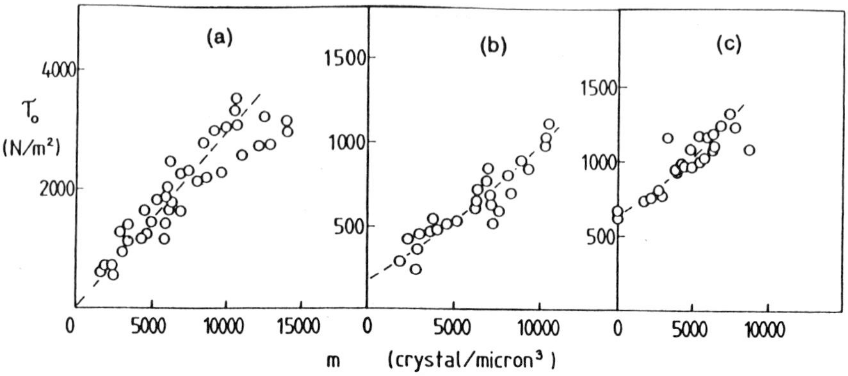

FIG. 11 Relationship between strength of flocculate structure (indicated by τ_o) and number of Yellow 13-type crystals per unit volume of lithographic ink (m). (a) Inks based on a varnish, which is free from gel structure; (b) and (c) inks based on heat-set varnishes, each with gel structure (hence positive intercept) stronger in (c) than in (b).

as powders (cf. Fig. 7c), there is a residual degree of face-to-face aggregation that survives on dispersion in the inks. The degree increases with increase in the proportion of abietyl resin present as free acid, rather than calcium salt, and thereby can be controlled. The aggregates are thought to have a sandwich-type structure with pigment crystals held together by abietyl resin acid. Such a structure featuring translucent resin layers might explain the observed fact that despite the aggregation the optically effective particles are the individual pigment crystals [13,62].

Another important rheological property of lithographic inks is the plastic viscosity. This is the viscosity under high shear conditions where flocculated structure has been broken down. It is important in relation to performance of the ink on the printing press and is profoundly affected by the pigment. In the absence of flocculation, the viscosity of a dispersion, η, is determined by the viscosity of the medium, η_m, and the volume concentration of the dispersed solid, c. This is exemplified by the well-known equations

$$\eta/\eta_m = 1 + 2.5c + 6.2c^2 + kc^3 + \cdots \tag{1}$$

[63] and

$$2.5/\ln(\eta/\eta_m) = 1/c - Q. \tag{2}$$

[64], where k and Q are constants. It follows that all dispersions with the same concentration in the same medium should have the same viscosity. This, however, is not so in practice with organic pigment dispersions. At

least part of the reason is that with particles as small as those of organic pigments, medium adsorbed at the pigment–medium interface markedly increases the effective pigment volume concentration and hence the viscosity [60]. This can be appreciated from the following simple example. A layer 30 Å thick adsorbed on the surfaces of a cube 0.03 μm in size (large for a Yellow 13–type pigment crystal) increases the effective volume of the cube by 73%. It follows that the plastic viscosity of an ink (see Fig. 10a) should increase with increase in the area of the pigment–medium interface, which indicates the capacity to adsorb medium. Indeed, this has been observed in practice with lithographic inks containing Yellow 13–type [38] or Red 57:1 pigments [62], in the latter case after allowance for face-to-face packing of crystals.

Yellow 13–type pigments with calcium or zinc resinate tend to give exceptionally high plastic viscosity in relation to their interfacial area in lithographic inks [38]. So too do unresinated Yellow 13–type pigments in heatset inks containing extensively gelled phenolic resin. These results indicate that differences in surface character from the usual resinated pigments can be important in lithographic inks. They also indicate that plastic viscosity measurements provide a means of investigating interfacial phenomena in these very difficult systems.

F. Effect of Crystal Morphology on the Rheology of Pigment Dispersions

Crystal shape can have a significant influence on the rheology of pigment dispersions. The amount of liquid phase immobilized by flocculation of pigment particles tends to be greater the more elongated the pigment particles. This is partly a geometric effect, since the void space in randomly packed rods that stick on first collision increases with axial ratio of the rods [65]. As an interesting example, rod-shaped β-form copper phthalocyanine crystals in phenolic/toluene gravure inks are especially prone to gel formation at low pigment concentrations, sometimes less than 5% w/w. Polar interaction has been shown to occur between ends (electron-rich) and sides (electron-deficient) of the neighboring (anisotropic) crystals [54]. Likewise, with lithographic inks it is well known in practice that strength of structure increases with increase in the axial ratio of rod-shaped β-form copper phthalocyanine crystals.

Another example is provided by decorative paint millbases (long-oil alkyd resin in white spirit) containing metal-free monoazo pigments with large crystals [34,62]. Strength of flocculate structure increased slowly with increase in pigment concentration at low pigment concentrations, but rapidly above a critical concentration. This concentration corresponded to the point at which the spheres of rotation of crystals of average size began to overlap. Whereas at lower concentrations there could only be isolated units of structure, at higher concentrations a coherent gel structure could form. The more elongated the crystal, the larger the sphere of rotation and the lower the critical concentration. Red 112 pigments with relatively thin plate-shaped crystals had much lower critical concentrations than the other pigments (Yellow 1 and 3 and Red 3), all with brick-shaped crystals. This explains why they have to be milled at much lower concentrations than the other pigments.

These examples indicate that attention to crystal shape is a means of obtaining desirable rheological behavior of dispersion millbases.

IV. PARTICLE SIZE ANALYSIS OF PIGMENT DISPERSIONS IN APPLICATION MEDIA

There have been few recent publications on particle size analysis of dispersions of organic pigments. This may reflect the severity of the technical difficulties encountered. Techniques are reviewed in Refs. 66 and 67. All currently available methods of size analysis for dispersions in liquid media require starting from much lower pigment concentrations than the 5–40% by weight encountered in practice, so dilution is essential. Some methods involve further dilution inherent in the analysis procedure itself, as the pigment particles migrate into or are eluted by additional liquid, examples being disk centrifugal sedimentation (using line start) [e.g., 68,69] and field flow fractionation (FFF), respectively [70]. Dilution tends to alter adsorption–solvation equilibria at the pigment–medium interface even when the diluent is further medium itself. These changes may alter the state of dispersion of the pigment and the amount of adsorbed medium. In particular, the tendency of pigment dispersions to flocculate has an indeterminable effect on particle size analysis, and flocculation has to be eliminated by use of additives [71]. Dilution, control of flocculation, and associated artifacts restrict direct correlation of size analysis data with performance in many practical situations.

Photon correlation spectroscopy (PCS) [72], which determines diffusion rates from the time dependency of light scattering by particles in Brownian movement, and FFF are two methods of special interest for organic pigment dispersions in liquid media in that they have the potential to cover the desired particle size range (0.015–1 μm). Other established methods do not. The mean particle sizes of organic pigment dispersions in liquid media are below the resolution of the optical microscope, which is therefore useful only as a qualitative means of detecting large residual aggregates that have not been dispersed. It is extensively used in this capacity in quality control. The resolutions of conventional laser light-scattering techniques (even with enhanced resolution) and the disk centrifuge method (the most commonly used method in the past) are about 0.1 μm and 0.05 μm, respectively—insufficient to detect a significant proportion of particles in many types of organic pigment dispersions.

Recent developments in the use of fiber optics with PCS equipment may alleviate to some extent the problem of the massive dilution required to render pigment dispersions transparent enough to the laser beam. The length of the light path is greatly reduced, since only the region of dispersion in the immediate vicinity of a light probe is examined.

Particle sizes from methods that involve movement of particles through a liquid medium (e.g., centrifugal sedimentation, PCS, and FFF) include liquid phase associated with the particle surfaces as adsorbed layers or within the hydrodynamic shear plane. This complicates correlation of particle sizes with properties such as color strength, on which associated liquid phase has no direct effect. Indeed, the measured sizes are more relevant to rheological properties where associated medium is highly relevant. Moreover, medium immobilized in adsorbed layers and within pores in aggregates reduces the effective density of particles and hence complicates size analysis by sedimentation [73].

A further problem with these methods is that with elongated particles the size measured is biased toward the largest dimension. This has been demonstrated in model experiments using PCS applied to dispersions of rod-shaped β-form copper phthalocyanine crystals in alkyd/toluene [71]. Furthermore, the applicability of the methods to dispersions in viscous media such as lithographic inks is not practicable. The difficulty is the high viscosity, which drastically reduces Brownian movement and retards migration in a centrifugal field to impracticably low rates. Dilution with low viscosity media introduces serious artifacts.

Some success has been achieved with aqueous dispersions. One study has shown good agreement between results of disk centrifugation and PCS applied to dispersions of β-form copper phthalocyanine [67]. PCS has also been applied successfully to extensively sonicated dispersions of Yellows 12, 13, 14, 17, and 83 [47]. It would appear from the published information that sizes for Yellows 12, 17, and 83 (with rod-shaped crystals) are more akin to the lengths rather than the breadths of the crystals, as judged from transmission electron micrographs.

As yet there is no established method for sizing particles effectively and conveniently in dried ink and paint films, plastics, etc. The most attractive method may seem to be transmission electron microscopy of ultrathin (<0.1 μm) sections of ink and paint films. In practice, however, it is laborious, time-consuming, and highly dependent upon skill in preparing sections. Care is required in interpretation, since, for example, what appear as small particles may be the truncated tips of larger particles. The method provides a qualitative impression of the state of dispersion and disposition of the pigment particles. It is best with sections of thick films containing organic pigment at low concentration and in the absence of opacifying pigments such as titanium dioxide. The much higher contrast of the latter makes discernment of the organic pigment difficult.

The fact that organic pigment dispersions are colored gives the advantage that information on the state of dispersion of the particles in ink and paint films, plastics, and artificial fibers, etc. can be obtained through color strength measurements. Such measurements can be readily made using either a densitometer or spectrophotometer. A given dispersion is always compared to a reference standard, precision being sufficient to reliably discriminate down to 5% difference. Visual comparisons can be even more discriminating, though qualitative. Such methods are used routinely in the pigments industry and by pigment users to assess levels of dispersion in quality control and technical service work. There are some drawbacks, however, in using this approach for investigating relationships between physical character of pigments and optical performance in application. For example, correction has to be made for the diluent effect of colorless additives in pigments. Also, in thin printed ink films, pigments that give opacity tend to give greater color strength than expected. The reason is that the light path through the ink layer is effectively increased due to multiple scattering from the pigment particles. Transparent dispersions must be compared at equal film thicknesses.

V. CONCLUDING REMARKS

Organic pigment dispersions impact on the quality of life by adding color. They are ever present in the home, at work and recreation, in textiles, personal possessions, decor, buildings, transport, paper, printed literature and artwork, packaging materials, and elsewhere. They have become an important part of modern life. The emergence of a new chemical class of pigments is now a very rare occurrence. However, as this chapter has endeavored to illustrate, the physical properties of pigments have a profound influence on performance in application. The ability to respond to changing application requirements with new and improved products depends increasingly on understanding how these properties influence application behavior and how they may themselves be controlled.

REFERENCES

1. W. Herbst and K. Hunger, *Industrielle Organische Pigmente*, VCH, Weinheim, 1987; English version, 1993.
2. H. Zollinger, *Color Chemistry: Synthesis, Properties and Applications of Organic Dyes and Pigments*, 2nd rev. ed., VCH, Weinheim, 1991.

3. P. Günthert, P. Hauser, and V. Radtke, *Rev. Prog. Colouration, 19*: 41 (1990).
4. R.B. McKay, *J. Oil Colour Chem. Assoc., 72*: 89 (1989).
5. Society of Dyers and Colourists, *Colour Index*, 3rd ed., (1982).
6. A. Whittaker, *J. Soc. Dyers Colourists 94*: 431 (1978).
7. A. Whittaker, *J. Soc. Dyers Colourists 104*: 294 (1988).
8. Z.S. Shenmin, L. Dongzhi, and R. Shengwu, *Dyes Pigm. 18*: 137 (1992).
9. J.W. Menter, *Proc. Roy. Soc. (Lond.) A 236*: 119 (1956).
10. J. Stabenow, *Ber. Bunsenges. Phys. Chem. 72*: 374 (1968).
11. J.R. Fryer, R.R. Mather, R.B. McKay, and K.S.W. Sing, *J. Chem. Technol. Biotechnol. 31*: 371 (1981).
12. G.R. Duckett, J.R. Fryer, and T. Baird, *Inst. Phys. Conf. Ser. 78*: 437 (1985).
13. R.B. McKay, *XVIII FATIPEC Congr. 2/B*: 405 (1986).
14. N. Uyeda, *Dyes Pigm. 6*: 115 (1985).
15. B. Honigmann, H.-U. Lenné, and R. Schrödel, *Z. Kristallogr. 122*: 185 (1965).
16. C.J. Brown, *J. Chem. Soc. A*, 2488 (1968).
17. H. Koyama, H.-J. Scheel, and F. Laves, *Naturwissenschaften 24*: 200 (1966).
18. G. Lincke, *Chem.-Zeit. 106*: 231 (1982).
19. G. Klebe, F. Grazer, E. Heidicke, and J. Blond, *Acta Cryst. B45*: 69 (1979).
20. A. Iqbal, et al., *XVIII FATIPEC Congr. 2/A*: 53 (1986).
21. R. Sappok and B. Honigmann, in *Characterisation of Powder Surfaces* (G.D. Parfitt and K.S.W. Sing, Eds.), Academic, New York, 1975, Chap. 6, p. 231.
22. B.G. Hays, *Am. Ink Maker* Oct. 1986, p. 13.
23. P. Hauser, D. Horn, and R. Sappok, *XII FATIPEC Congr.*, 1974, p. 191.
24. R.M. Christie, P.N. Strandring, and J. Griffiths, *Dyes Pigm. 9*: 37 (1988).
25. A.R. Hanke, Report of a lecture, *J. Opt. Soc. Am. 56*: 713 (1966).
26. M.A. Maikowski, *Ber. Bunsenges. Phys. Chem. 71*: 313 (1967).
27. P. Hauser, M. Herrmann, and B. Honigmann, *Farbe Lack 76*: 545 (1970).
28. P. Hauser and B. Honigmann, *Farbe Lack 81*: 1005 (1975).
29. P. Hauser, M. Herrmann, and B. Honigmann, *Farbe Lack 77*: 1097 (1971).
30. W. Carr, *J. Oil Colour Chem. Assoc. 54*: 1093 (1971).
31. B. Felder, *Helv. Chim. Acta 15*: 1224 (1968); *J. Color Appearance 1*: 9 (1971).
32. O. Hafner, *J. Paint Technol. 47*: 64 (1975).
33. R.B. McKay and R. R. Mather, *Colloids Surf. 27*: 175 (1987).
34. R.B. McKay, *Surf. Coating Int., JOCCA* 75: 177 (1992).
35. R.B. McKay, *Farbe Lack 95*: 471 (1989).
36. R.B. McKay, *Farbe Lack 96*: 336 (1990); *Surf. Coatings Int., JOCCA 74*: 169 (1991).
37. R.B. McKay, *XX FATIPEC Congr.* 1990, p. 139.
38. R.B. McKay, *Farbe Lack 97*: 576 (1991).
39. L. Dongzhi and R. Shengwu, *Dyes Pigm. 18*: 69 (1992).
40. W. Kurtz, *Am. Ink Maker* Dec. 1986, p. 20.
41. A. Iqbal, L. Cassar, A. C. Rochat, J. Pfenninger, and O. Wallquist, *J. Coatings Technol. 60* Mar.: 37 (1988).
42. B. Honigmann and D. Horn, in *Particle Growth in Suspensions* (A. L. Smith, Ed.), Academic, New York, 1973, Chap. 18, p. 283.
43. B. Honigmann and J. Stabenow, *VI FATIPEC Congr.*, 1961, p. 89; B. Honigmann, *Ber. Bunsenges. Phys. Chem. 71*: 239 (1967).

44. D.H. Everett, *Pure Appl. Chem. 31*: 579 (1972).
45. R.B. McKay, *J. Oil Colour Chem. Assoc. 71*: 7 (1988).
46. R.R. Mather and K.S.W. Sing, *J. Colloid Interface Sci. 60*: 60 (1977).
47. B. Hayes, *Am. Ink Maker*, Nov. 1990, p. 28.
48. R.B. McKay, in *Interfacial Phenomena in Apolar Media* (H.-F. Eicke and G.D. Parfitt, Eds.), Marcel Dekker, New York, 1987, Chap. 9.
49. J. Moilliet and D. A. Plant, *J. Oil Colour Chem. Assoc. 52*: 289 (1969).
50. D.H. Napper, *Polymeric Stabilization of Colloidal Dispersions*, Academic, New York, 1983.
51. W. Black, F. T. Hesselink, and A. Topham, *Kolloid-Z. Z. Polym. 213*: 150 (1966).
52. R.B. McKay, *J. Appl. Chem. Biotechnol. 26*: 55 (1976).
53. R.B. McKay, in *Proc. Sixth Int. Conf. Org. Coatings Sci. Technol.* (1980) (G.D. Parfitt and A.V. Patsis, Eds.), Technomic, Westport, Conn., 1982, p. 173.
54. W. Ditter, and D. Horn, *Fourth Int. Conf. Surf. Coatings Sci. Technol.*, Athens, 1978, p. 251.
55. O. Wallquist, SAE Tech. Paper Ser., No. 900530, Int. Congr. Expo., Detroit, Mich., 1990.
56. E. Dietz (to Hoechst AG), Eur. Patent 0 321 919 (1988).
57. A.C.D. Cowley, *J. Oil Colour Chem. Assoc. 70*: 207 (1987).
58. J.D. Schofield, *Surf. Coatings Int., JOCCA 74*: 204 (1991).
59. H.L. Jakubauskas, *XI Int. Conf. Org. Coatings Sci. Technol.*, p. 77, 1985.
60. R.B. McKay, *XIII FATIPEC Congr.*, 1976, p. 428; *Br. Ink Maker 19*: 59 (1977).
61. W. Kurtz, *Am. Ink Maker*, June 1987, p. 21.
62. R.B. McKay, *XVIII Int. Conf. Org. Coatings Sci. Technol.*, 1992, p. 249; *Progr. Org. Coat. 22*: 211 (1993).
63. G.K. Batchelor, *J. Fluid Mech. 83*: 97 (1977).
64. M. Mooney, *J. Colloid Sci. 6*: 162 (1951).
65. M.J. Vold, *J. Phys. Chem. 63*: 1608 (1959).
66. S.G. Lawrence, in *Dispersion of Powders in Liquids*, 3rd ed. (G.D. Parfitt, Ed.), Applied Science, London, 1981, Chap. 8, p. 363.
67. J. Schröder, *Farbe Lack 90*: 5 (1984).
68. P. Hauser and B. Honigmann, *Farbe Lack 83*: 886 (1977).
69. K. Brugger, *Powder Technol. 13*: 215 (1976).
70. T. Schauer and L. Dulog, *Farbe Lack 97*: 765 (1991).
71. S. Kawasaki and M. Hikosaka, *J. Jpn. Soc. Color Mater. 62*: 587 (1989).
72. D. Horn, H. Autweter, W. Ditter, and J. Eisenlauer, *J. Org. Coatings 8*: 251 (1986).
73. R.B. McKay, in *Particle Size Analysis* (M. J. Groves, Ed.), Heyden, London, 1978, p. 421.

5

Dispersion of Pigments into Printing Inks

GRAHAM BATTERSBY Coates Lorilleux Research,
Coates Lorilleux International, Orpington, Kent, England

I.	Introduction	178
II.	Liquid Inks	179
	A. Solvent-based inks	180
	B. Manufacture of solvent-based inks by bead milling	182
	C. Manufacture of solvent-based inks from chips	184
	D. Water-based liquid inks	185
	E. Properties of liquid inks related to pigment dispersion	187
	F. Publication gravure inks	190
III.	Paste Inks	191
	A. Flushed versus powder pigments	191
	B. Manufacture using three-roll mills	193
	C. Properties related to pigment dispersion	195
IV.	Newsink	195
	A. Cold-set web offset inks	195
	B. Water-based flexographic newsink	198
V.	Screen Inks	199
VI.	Radiation-Curable Inks	199
	A. Effects of pigments on curing	200
	B. Effects of pigments on shelf life	201
	C. Manufacture	201
VII.	Millbase Optimization	202
	References	203

I. INTRODUCTION

The aim of this chapter is to outline the current industrial practices for dispersing pigments into printing inks. The nature of the industry is such that it is technological rather than scientific. Therefore many of the reasons for operating in a particular way have developed empirically and are not always fully understood; opinions regularly take the place of facts.

Five different ink types are considered: liquid, paste, news, screen, and radiation-curable. Each section deals with basic formulation style, manufacture, and properties related to pigment dispersion.

Figure 1 compares the approximate volumes of the various types of ink sold in western Europe in 1992. It can be seen that publication printing; the packaging market, which, overall, uses approximately the same volume of ink as publication printing; and newsink are the three large-volume markets. These volumes in turn reflect the consumption of pigment as shown by Fig. 2. Carbon black for newsink and titanium dioxide for flexible packaging ink are major pigments, and, surprisingly, all the colored organic pigments account for only 36% by volume of pigments consumed. However, by value, organic pigments account for over 75% of pigment purchases, which illustrates the importance of maximizing the dispersion of organic pigments in printing inks. Printing inks consume over 50% of all the organic pigments produced but less than 4% of pigmentary titanium dioxide. For detailed information, see Refs. 1–3.

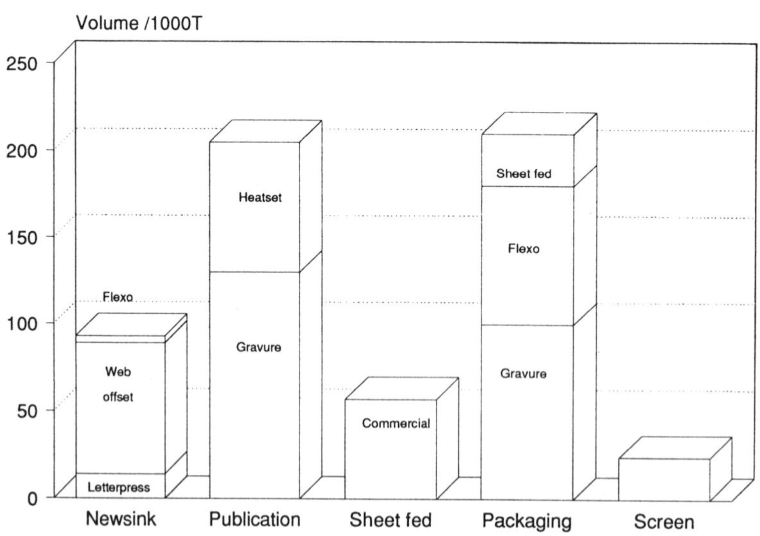

FIG. 1 Western European ink volumes, 1992.

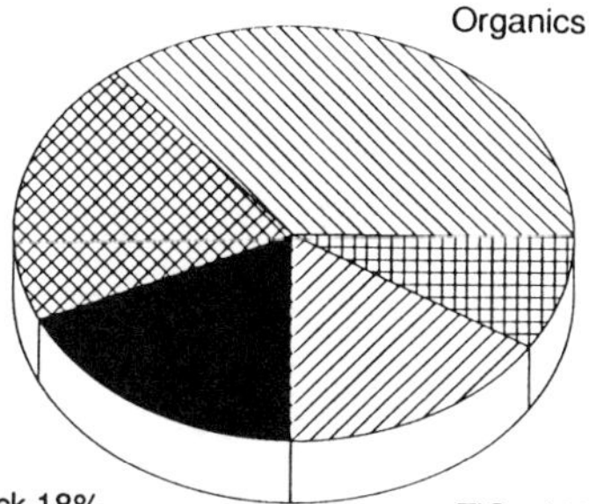

FIG. 2 Western European usage of pigments in ink, 1992.

Printing inks are dispersions of pigments in a liquid medium commonly referred to as the *varnish*. The function of the pigment is to confer color, whiteness, blackness, and, if desired, opacity. The varnish is a solution or sometimes partly a dispersion of polymer or polymer mixtures in a liquid. The liquid is essentially removed on printing either by evaporation and/or by preferential adsorption into the substrate. The function of the polymer (binder resin) is to bind the pigment particles at the surface of the substrate. The suitability of an ink for a particular type of application is determined primarily by the nature of the varnish, and it is largely on this basis that various types of ink are classified.

Liquid inks are free-flowing inks used primarily for flexographic or gravure printing on packaging materials, most commonly paper (including corrugated cartons) and polymer films. Specialized liquid inks referred to as *publication gravure inks* are used for high volume publication printing on paper (e.g., mail order catalogs, holiday brochures). Paste inks are very much more viscous. They are used for general printing on paper substrates, with special types used for metal decorating. Newsink is a special form of paste ink used for newspapers.

Screen printing inks are used for specialized printing on all sorts of surfaces, for example, plastic, metal, glass, and especially nonplanar surfaces, on a wide range of objects from computer hardware to car dashboards to cosmetic bottles. Ultraviolet-curable inks usually consist of pigment in oligomer or oligomer–monomer mixtures free of added solvent and curable by UV radiation.

II. LIQUID INKS

Liquid inks are used on flexographic and gravure printing presses. More detail on these printing techniques can be found in Refs. 4–6. Table 1 compares the main differences between flexographic and gravure inks and

TABLE 1 Comparison of Flexographic and Gravure Inks

Property	Flexography	Gravure
Application viscosity[a]	20–24	17–20
Pigmentation (at printing viscosity)		
White	29	24
Color	13	7
Dried film weight	2–3 μm	3–6 μm
Solvent restriction	—[b]	No
Quality	Medium	High
Polyethylene printing	Yes	Difficult
Water-based	Yes	Difficult
Setup costs	Low	High

[a]Seconds Zahn 2 cup.
[b]Esters attack Photopolymer plates.

print. The most important differences are solvent restrictions, film weight, and print quality.

A. Solvent-Based Inks

Typical formulations are shown in Table 2. The main resin used is low viscosity nitrocellulose (NC), which features in more than 70% of all solvent-based liquid inks. Liquid media, based on NC, disperse pigment reasonably well, have a good balance between viscosity and solids content, have low odor, print well, and are compatible with a large number of media based on other polymers. Blending of the nitrocellulose with other polymers is required to optimize properties such as adhesion, heat resistance, and

TABLE 2 Typical Flexographic and Gravure Ink Formulations at Print Viscosity, for Surface Print Coextruded Polypropylene

	Flexography		Gravure	
	White (wt %)	Color (wt %)	White (wt %)	Color (wt %)
Pigment	29	13	24	7
Nitrocellulose	9	12	7	9
Polyurethane	7	5	6	6
Solvent	56	67	56	74
Additives	8	3	7	4

flexibility, and the modifying resins are usually added in the form of varnishes. This leads to the "blending" method of producing liquid inks from a highly pigmented nitrocellulose concentrate (Fig. 3). This method is widely used by ink manufacturers, and its use by major printers is increasing.

By mixing the desired pigment concentrates with different varnishes it is possible to produce a wide variety of colors for many different applications. Typical varnishes could include the following polymer types.

Polyamide for printing polyethylene
Urethane for printing polypropylene
Maleic for printing cartons
Acrylic for coated films

The highly pigmented concentrates should (1) have maximum formulating flexibility, which entails high pigmentation, high pigment-to-binder ratio, solvency suitable for flexography and gravure, and no additives; (2) have storage stability, meaning long shelf life, nonflaking character, and lacking a tendency to coat the sides of containers; (3) demonstrate good flow for ease of dispensing under gravity; and (4) be nonflocculating, with no pigment shocking, resin systems compatible with diluting media, and no major changes in solvency.

The concentrates are typically 35% (w/w) pigment, 10% nitrocellulose,

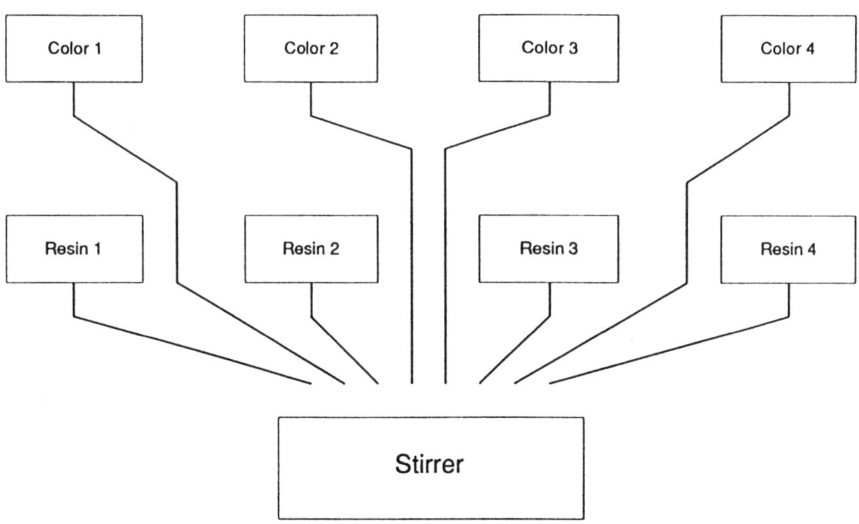

FIG. 3 Schematic diagram of a liquid ink blend scheme.

and 55% solvent. Very little is known of the colloid science behind these concentrated dispersions, and this is an area needing research. The volatility of the solvents used causes difficulty for many types of investigative work and is one of the reasons for the dearth of published information on the evaporation of solvents from thin films. Before moving on to look at how these pigment concentrates are produced, other types of resins used in liquid inks should be considered briefly.

Polyvinyl butyral (PVB) is used in inks for coldseal and laminate packaging. It is often compounded with polyethyleneimine (PEI) to improve adhesion to substrates. PVB is notorious in the ink industry as a poor pigment "wetter"; because of this characteristic, PVB pigment concentrates are difficult to manufacture and are not as highly pigmented as nitrocellulose dispersions. A further complication is caused by PEI "gelling" with some pigments. The mechanism of this effect is not known, but it does cause problems in selecting suitable pigments. The poor flow properties of PVB inks, coupled with their low pigmentation, can cause problems in achieving good color strengths when printed flexographically.

Cellulose acetate propionate (CAP) is the other major liquid ink resin. It is used in combination with acrylic polymers for printing onto PVdC (polyvinylidene dichloride) or acrylic coated films. CAP-based media are very difficult to disperse pigments into, and this, as with PVB, causes manufacturing problems; the inks have poor flow and often have low color strength due to low achievable pigmentation levels.

B. Manufacture of Solvent-Based Inks by Bead Milling

The most popular method of dispersing organic pigments into liquid inks is by bead milling. Details of bead mills are contained in Refs. 7–11. More recent developments in stirrer bead mills [12,13] allow high energy inputs with efficient cooling. This allows high quality dispersion and high throughput rates and allows bead mill dispersion to compete with chip dispersions (see later) for quality.

A recent development by Dispermat [14] is a laboratory bead mill with the capacity to monitor or control (1) constant input energy via controlling the torque of the shaft and changing the rotational speed accordingly and (2) constant rotational speed, with the instrumentation displaying the values for torque and mechanical power being used. Figure 4 shows typical data obtained from this equipment. Having the ability to measure these variables allows the work on millbase optimization, pioneered by Farkas [15] using the Brabender Plastograph, to be continued in more detail. Perhaps in the very near future, ink makers will be able to quantify the dispersibility of a

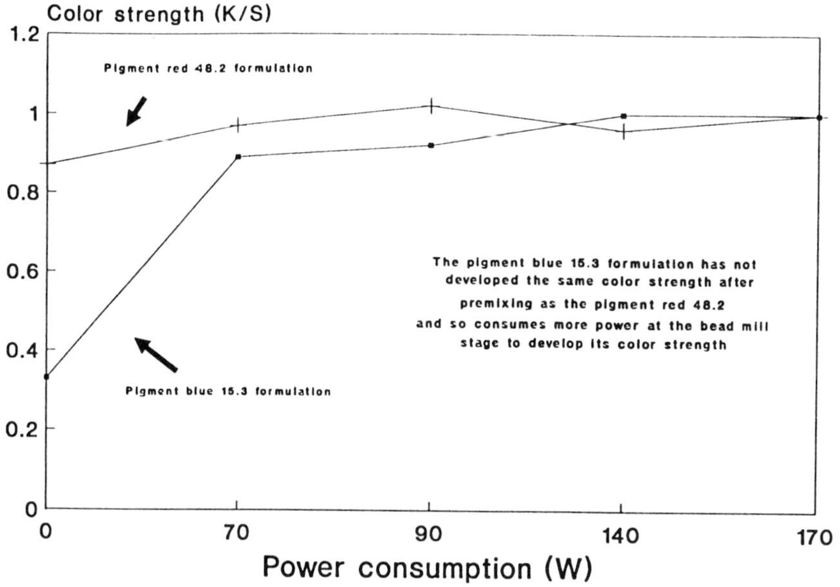

FIG. 4 Color strength versus bead mill power consumption.

pigment, the efficiency of pigment dispersants, millbase optimization, energy input versus particle size reduction and optical properties, etc. How this will translate to full-scale production machines also needs to be studied.

The Mirodur turbo mill [16] is an interesting development that could be considered to be a "reverse" bead mill. The beads are located in a cage at the end of a stirrer, and the millbase is fixed in a container below it. As the basket of beads rotates around a fixed plate that creates turbulence, the beads disperse the pigment by impaction. The machine seems to offer an advantage in ease of use.

It is essential to premix the pigment and medium efficiently prior to milling. Not only does the pigment–air interface have to be replaced by a pigment–medium interface, but also some of the larger pigment aggregates have to be broken. The more dispersion that can be achieved at the premix stage, the more efficient subsequent bead milling is. Figure 5 shows the reduction in pigment particle size during premixing. A milled sample from point B had much better gloss and transparency than a milled sample from point A.

There has been some activity in the study of bead mills. Interesting papers include work on modeling bead mills [17] and on on-line color measurements [18].

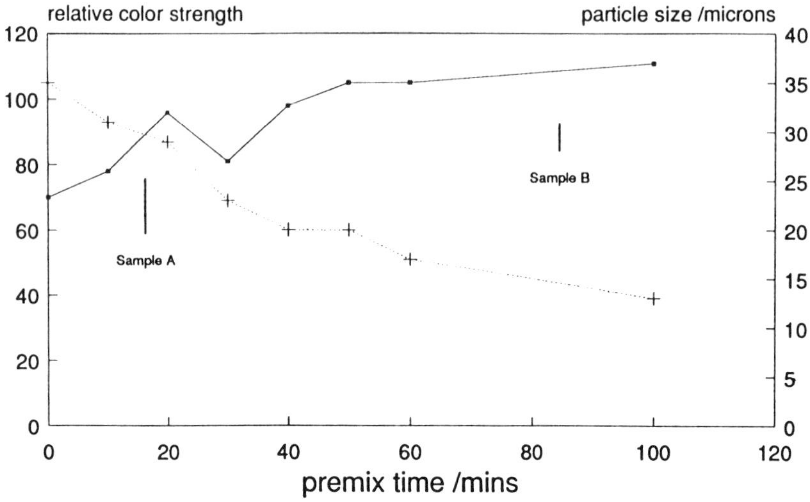

FIG. 5 Effect of mixing prior to bead milling. (·) Tinctorial strength; (+) particle size.

C. Manufacture of Solvent-Based Inks from Chips [4][1]

Chips are produced by dispersing pigment into a molten resin under high shear, high pressure, and elevated temperatures, usually on a two-roll mill. Solid pigment dispersions are produced, which have the advantage of ease of transportation but the disadvantage that they have to be dissolved before use. It is not easy to get the chips into solution, particularly when the resin component is not easily dissolved; the rate of dissolution of a resin is greatly reduced when it is pigmented. To increase the rate of solution, chips are usually "kibbled," after milling, to reduce their particle size.

Within the chips the pigment is very well dispersed. Consequently, chips produce the highest gloss and transparency achievable by any dispersion technique, although elaborated grades of pigments (specially treated, easily dispersed forms), with efficient premixing and bead milling, can approach chip quality dispersion. Although chips lead to these good optical properties, they suffer from low pigment-to-binder ratios and high plasticizer content, which in turn reduces formulating flexibility. Figure 6 shows the particle size of a pigment dispersed in an ink by bead milling and by the use of chip. The overall particle size of the pigment from chips is lower, giving

[1] For detailed information on chip manufacture see Ref. 4.

higher gloss and transparency, but the increase in the number of smaller particles causes the ink to have poor flow—hence high reduction factors (dilution) are required.

A patent issued to KVK [19] describes an alternative route to producing chip quality dispersions, in a more finely divided form commonly known by their trade name Predisol, that are much easier to dissolve than chips.

An interesting modification of chips and Predisol is powder inks [20]. The finished ink is supplied to customers in powder form; users just add solvent or water to produce a ready-to-use product. This form of supply offers advantages of ease of transportation and storage over conventional liquid concentrates but the disadvantage of requiring dissolution prior to use.

D. Water-Based Liquid Inks

There are two types of water-based liquid inks, those for porous substrates and those for impervious substrates.

Inks for porous substrates are normally manufactured much like the solvent-based products discussed previously. A typical concentrate formulation would be

Pigment	40% w/w
Acrylic resin	10%
Water	50%

The concentrates would then be blended with other resins, usually acrylics, to produce the finished ink.

The pigments used are often opaque since the ink is required to cover brown substrates. Printing corrugated cartons before corrugation (preprinting) places some unique demands on the ink, as it must be resistant to the hot humid conditions in the corrugator. The pigments also have to be resistant to those conditions and must not sublime. Inks printed after corrugation (postprinting) do not have to satisfy these requirements.

Corrugated cartons and multiwall sacks are printed almost exclusively by flexography. Preprint inks are often required to print halftone, and to compete with this quality, super postprint is being developed to print halftone work on preformed cartons.

The production of aqueous inks for impervious substrates such as plastic film is an area undergoing rapid development, and so today no definitive statement on the best available technology can be made. Pigments are required that are glossy and transparent (in contrast with properties required in pigments for corrugated cartons) and easily dispersed. Titanium dioxide,

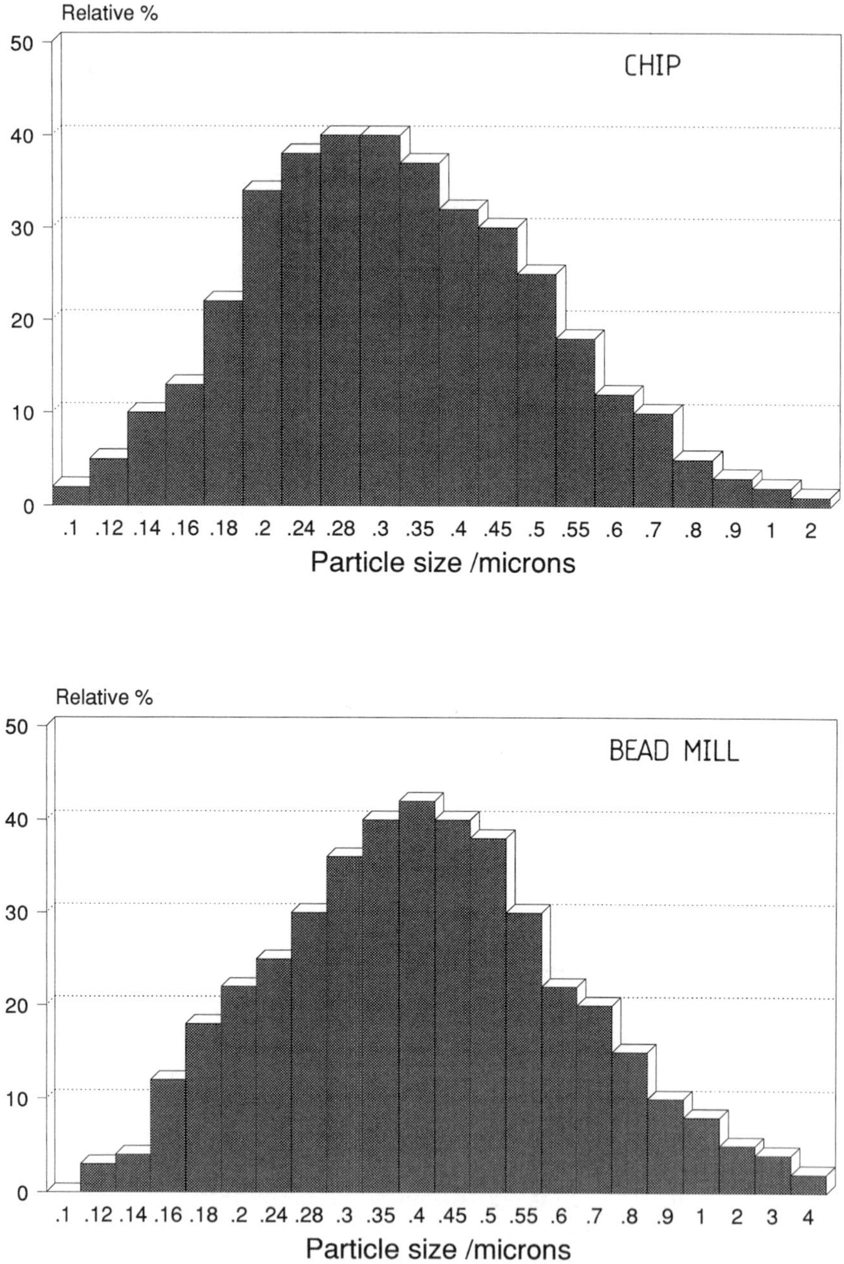

FIG. 6 Comparison of particle size distributions of a pigment (Mastersizer instrument [22]) dispersed in a liquid ink by chip (top) and bead mill (bottom) methods.

in its rutile form, is very abrasive in water, and there is a need for improved pigments if suitable gravure white inks are to be prepared. The coatings on the pigments should not detract from the resistance properties of the inks. For example, surface treatments should not reduce the water resistance of inks on polyethylene. A typical water resistance test involves assessing the amount of ink removed from a print while the print is flexed under running water, the print having previously been immersed in water for 1 h at room temperature.

The only way to achieve the resistance properties of solvent-based inks from water-based polymers is by cross-linking. The polymers are usually binary mixtures requiring mixing prior to use. An example of this type is the addition of a polyaziridine to a carboxylated acrylic polymer. However, the development of resistant one-component polymers is essential if water-based inks are to take a significant share of the flexible packaging market.

Obviously there are sections of the market where resistance properties of the materials currently available are adequate; an example is printing on carrier or shopping bags. Or it may be possible to satisfy emission regulations by combining water- and solvent-based technologies, for example, water-based inks for cold-seal packaging that are "protected" with a solvent-based cold-seal release lacquer.

There is also a need for polymeric dispersing agents for water-based inks. These dispersing agents must not reduce the resistance properties of the finished ink. The low molecular weight carboxylated acrylic resins used to disperse pigments in water-based inks for porous substrates are not suitable in inks for impervious substrates because they reduce the water resistance too much.

Other factors to be considered are gravure printability (for which optimization of inks and gravure cylinders is essential), drying speed (requiring high strength, low film weight, rapid coalescence, and optimized drying conditions), and disposal. As for the latter, solvent-based ink waste can be incinerated, but waste containing water-based inks requires treatment before disposal [21].

E. Properties of Liquid Inks Related to Pigment Dispersion

Figure 7 summarizes the effects of pigment particle size on properties of liquid inks. Transparency, gloss, color strength, and viscosity are key properties common to all liquid inks and indeed all other types of inks—a point that will be taken as understood from here on. Relating other properties to pigment particle size is not always so easy since a number of factors other than pigment dispersion are also involved.

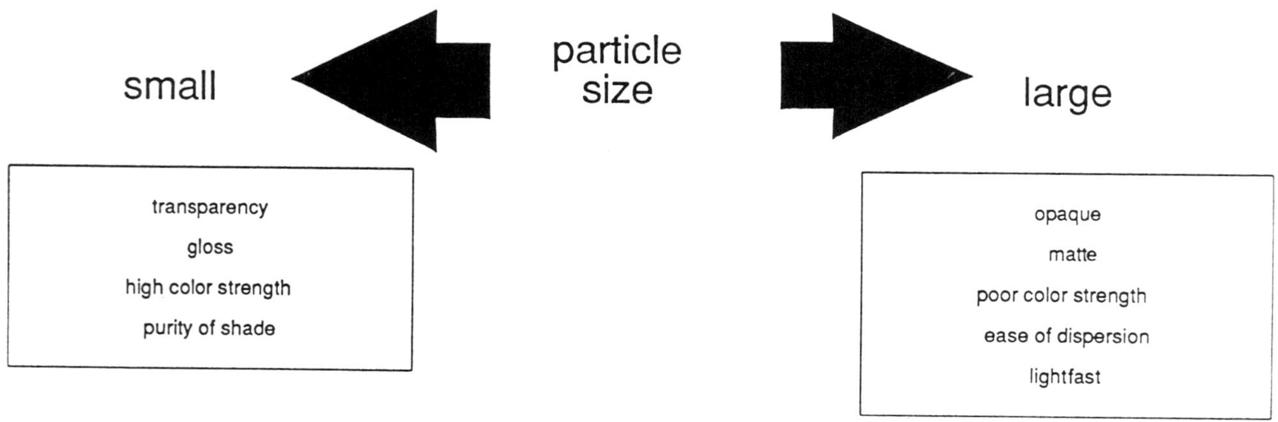

FIG. 7 The effect of pigment particle size on properties.

The actual values for large and small particle sizes depend on how the particle size is measured and on the pigment. Figure 8 shows the difference in particle size for an Orange 34 pigment dispersed by a bead mill but following different energy inputs during the premix stage. The smaller particle size ink had much better gloss and transparency and slightly more color strength. Quoting unique values for particle size does not give a full indication of the properties of the dispersion, and often the distribution of the particle size is more important (see also Fig. 6).

One final point that sometimes appears as a misunderstanding in the ink industry relates to large particles in inks. Usually inks are filtered prior to dispatch, and these filters remove large particles, typically those above 5 μm. Such particles can arise from undissolved resin, contamination from raw material containers, or undissolved chips (see Fig. 9). If not removed, large particles can often cause poor printability, particularly when printed on a gravure press. It is unusual in such cases for the problem particles to be the result of poor pigment dispersion.

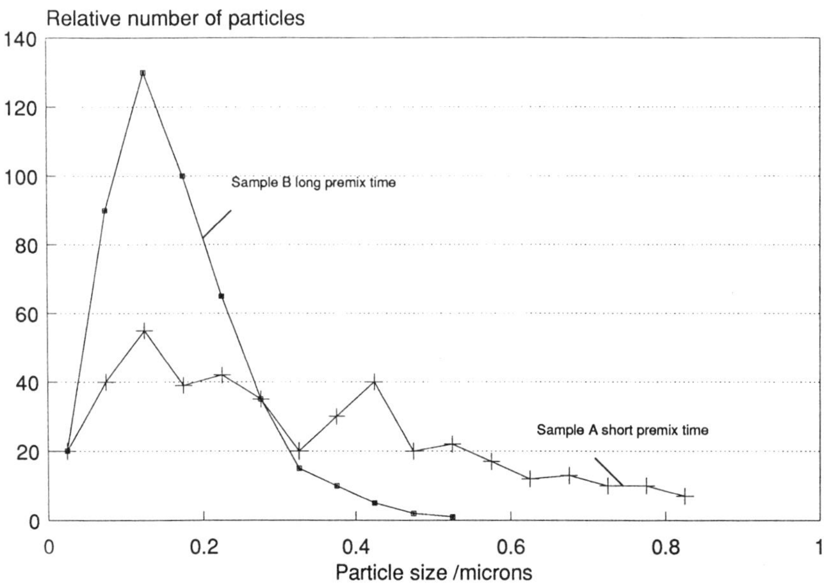

FIG. 8 Influence of premix conditions on particle size distribution of an Orange 34 pigment in liquid inks from a bead mill. Size distribution determined from electron micrographs of dried ink films.

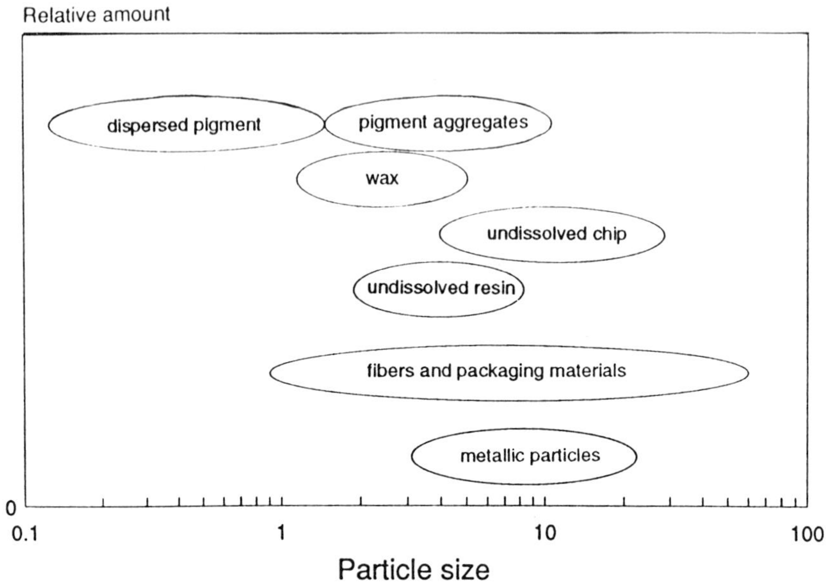

FIG. 9 Types of particles that may appear in inks.

F. Publication Gravure Inks

Publication gravure inks are probably the most price-sensitive of all printing inks and are sold in large volumes, with bulk manufacture and delivery. A basic formulation would contain

Pigment, 8% w/w
Extender, 5%
Ethylhydroxyethyl cellulose and/or phenolic resin, and/or metal resinate, 40%
Solvent, 47%

Easily dispersed pigments are frequently used to improve productivity as they require only a short residence time in the bead mill. However, if the price/benefit ratio lessens, then more conventional pigments are used. Poor pigment dispersion can have a very serious impact on profitability through decreased color strength. Abrasiveness of the ink is also a major concern with gravure inks because abrasive pigment impurities or large particles can shorten the life of the expensive gravure cylinders.

This type of ink lends itself to manufacture by a flushing process; Ref. 23 describes such a process.

It is difficult to comment on the resins used because ink manufacturers all have their own preferences. Often costs and changes in other raw materials, such as the pigment, cause changes in the proportions of resins used. Toluene and aliphatic hydrocarbons are the main solvents used, and solvent recovery systems have been in place for many years. High levels of solvent are already recovered, but work is progressing to increase this to 100% recovery.

III. PASTE INKS

Table 3 compares typical formulations for heat-set and sheet-fed paste inks. The main difference lies in the faster evaporating solvents in heat-set inks and the use of oxidation drying in sheet-fed inks. For the purpose of this chapter we can consider these inks to be similar. They are printed mainly by the lithographic printing process and to a limited extent by letterpress. Details of these printing processes are given in Refs. 4 and 24.

A. Flushed versus Powder Pigments

The pigments used in heat-set inks are mainly the process colors (yellow, magenta, cyan, and black for four-color printing), but for sheet-fed inks a wide variety of base colors are required to allow all the desired shades to be matched. This results in differences in scale of manufacturing. There is also a very interesting difference between European ink processes using mainly powder pigments, and U.S. and Japanese manufacturers using flushed pigment [25,26]. Figure 10 compares the two types of manufacturing routes.

TABLE 3 Typical Heat-Set and Sheet-Fed Paste Ink Formulations

	Heat-set (% w/w)	Sheet-fed (% w/w)
Pigment	15	15
Rosin-modified phenolic	24	30
Alkyd or drying oil	12	20
Modified hydrocarbon resin	11	0
Wax	3	3
Driers	0	2
Solvent	34.5	29.3
Chelating agent	0.5	0.5
Antioxidant	0	0.2

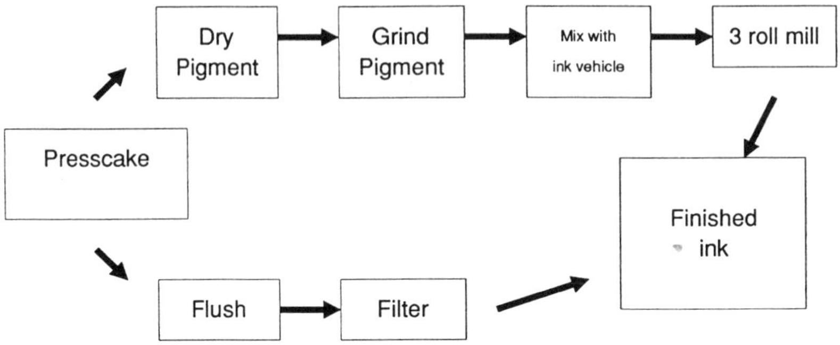

FIG. 10 Schematic diagram of paste ink production from dry powder (top route) and flushed pigment (bottom).

There has been very little published on the benefits of flushed versus powder pigments. Any comparisons are made difficult not only by opinions taking priority over scientific fact but also by the lack of any absolute methods of assessing properties—often inks are compared to a standard and referred to as being better or worse.

When comparing flushed pigment with powder pigment, what is the correct basis of comparison? Is it the optimized product from each route, or is it flush and dry powder from the same presscake? If the latter method is used, then this often results in large gains in properties via one route compared to the other. For example, 40% strength gain from a Pigment Yellow 12 when a presscake is flushed into the varnish compared to the same presscake dried and redispersed into the varnish on a three-roll mill. But if an elaborated powder Pigment Yellow 12 (i.e., a specially treated, easily dispersed form of powder or granule) is compared to a flushed pigment, the differences (if any) are very small. With these considerations in mind, the following comments can be made on flushed versus powder pigments:

1. There is little technical advantage in the use of either powder or flushed pigment. If there were, then surely one approach would be used worldwide.
2. When examined microscopically, flushed pigments are usually seen to be more finely dispersed than powder pigments. This can lead *in some instances* to cleaner shades and slightly more transparency. These effects, however, are very slight and cannot be regarded as significant technical advantages.
3. It is difficult to compare production costs, since fully automated plants

of equal capacity have to be compared. These plants should ideally be in the same factory.

4. Flushing may offer advantages to the integrated pigment and ink producer, for savings may be realized in drying and grinding of the pigment.
5. The technology for producing high quality elaborated powder pigments for lithographic inks may be restricted to a few major pigment manufacturers, whereas flushing may offer equivalent performance from a "lower technology" presscake.
6. It is difficult to produce high solids presscake of a reliable shade and strength.

A very good paper by Gomm et al. [27] applies some of the principles of colloid science to flushing. They conclude that flushing is not simply a wetting phenomenon and that the solutioning/solubilization of additives on the pigment surface is likely to be the major factor in determining the thermodynamic equilibrium. The free energy gained by solvating the pigment coating with a nonpolar solvent and vehicle is greater than the energy gained through wetting. The presscake has more influence on the flushing process than the flushing vehicle. High solids presscakes have less water to displace, less water to emulsify, and more capillary attraction between pigment particles. This causes an entropy rush. The main differences between ink varnishes are seen when resins such as alkyds are used instead of less polar types. The beneficial effect of alkyds could be attributed to better wetting but could also be due to improved solubility of the pigment coating in the varnish.

Small amounts of aromatic solvents added to the aliphatic distillate may give better wetting (benzene/water interfacial tension, $\gamma = 35$ mN/m; hexane/water, $\gamma = 51$ mN/m) but may also increase solubility of the pigment coating. Similar effects may occur with flushing aids and hyperdispersants [28].

B. Manufacture Using Three-Roll Mills

Three-roll mills are the standard machines for producing paste inks from powder pigments. Details of their operation are contained in Refs. 4, 11, and 29–31. The basic principle is to have three rollers rotating in different directions and at different speeds. As the material to be dispersed passes through the fine nips between the rollers, it is forced into a very high shear zone that imparts sufficient energy to separate pigment aggregates.

The relative speeds of the rollers vary widely, but a typical ratio is 1 : 3 : 9, with the final roll rotating at about 500 rpm. High ratio mills rely on the tack of the ink for transference. Increasing roller speed requires good temperature control and care to avoid distortion of the rollers (particularly on wider mills). The pressures used on modern three-roll mills vary depending on the cambers used on the rollers (different cambers are used for different applications). Mill pressure is the main factor in determining production quality and quantity. Most inks are heat-sensitive, and generally speaking the higher the temperature the better the dispersion but the lower the output.

If the quality of dispersion is not acceptable (usually assessed by a grind gauge or by optical microscopy), then the ink will have to be passed over the mill again. Obviously this is a time-consuming and costly operation. To reduce this problem, combination mills are sometimes used where the ink is first bead milled and then three-roll milled. Figure 11 shows a vertical bead mill in the background feeding into a three-roll mill hopper.

FIG. 11 Combination mill.

As was stressed for liquid ink manufacture, high energy input at the premix stage is a prerequisite for quality output, at a reasonable rate, even when combination mills are used. The bead mill develops approximately 90% of the color strength and the three-roll mill the remaining 10%. The three-roll mill also improves the gloss and transparency of the ink. Having two different types of dispersing mechanisms in combination inevitably means that the properties of the millbase have to be optimized to maximize the quality and throughput from the equipment overall rather than for either mill in isolation.

C. Properties Related to Pigment Dispersion

Color strength, gloss, and transparency improve as pigment dispersion improves. However, various other dispersion-related properties of paste inks are important as they can give rise to problems in printing.

Piling is the name given to a problem seen on the printing press when the ink does not transfer evenly along the inking roller chain and starts to accumulate in bands around the rollers.

Wear of plate, blanket, and rollers of the printing press can be a problem if the pigment is poorly dispersed.

Lithographic performance of an ink embraces various complex properties. One of these is that in the lithographic printing process the ink must be capable of combining some of the aqueous fountain solution to give a complex colloidal dispersion (water–oil–pigment as a stable three-phase system under conditions prevailing on the printing press). Poor pigment dispersion changes the stability of this colloid and can lead to poor press stability and print properties.

Although these problems may be associated in part with poorly dispersed pigment, it is quite possible that they are also caused by other large particulate species, especially particles of undissolved or gelled resin (Fig. 9).

IV. NEWSINK

A. Cold-Set Web Offset Inks

The main printing process for printing newspapers in western Europe is coldset web offset (Fig. 1). This is offset lithography where the inks dry by penetration into the newsprint (the paper substrate), and, like publication gravure, the emphasis is on cost, bulk manufacture, and supply (Fig. 12). Quality and consistency are essential because of the long runs and short deadlines.

The most important pigment is carbon black, and, unlike other types of

FIG. 12 Bulk storage of newsink.

inks using this pigment, it is the most expensive component in the ink. Typical formulations for a cold-set black and a cold-set color are given below.

Cold-set black		Cold-set color	
Carbon black	20	Organic pigment	14
Gilsonite resin	10	Hydrocarbon resin	10
Ink oil	68	Ink oil	74
Thixotrope	2	Thixotrope	2
	100% w/w		100% w/w

Beaded carbon blacks (spherical granules) are often used because they are of lower cost than fluffy black powders and they are also easier to handle. Beaded blacks require special premixing equipment to break the beads and fully wet the large surface area of the pigment. Bead milling without subsequent three-roll milling is normally used to disperse the pigment into a finished ink. Unlike heat-set or sheet-fed blacks, blue toner is not usually added to the formulation because of the cost.

Low rub (clean hands) black inks can be produced through the addition of rosin-modified phenolic polymers. These polymers simply act as a binder for the carbon black to prevent it from coming off onto the readers' hands.

More recent presses for printing newspapers have an anilox inking system, and the inks are required to have very good flow (very little thixotropy) yet maintain good emulsification (lithographic) properties.

The industry has moved away from ink oils based on aromatic feedstocks to oils from naphthenic and paraffinic feedstocks. This has caused problems in dissolving the gilsonite resin used to reduce the cost of black newsinks.

Carbon black properties are determined by its structure, particle size, and surface modification. Figure 13 summarizes the relationships. With black newsinks there are some properties that are especially sensitive to the degree of pigment dispersion. Carbon black has such a high specific surface area and oil absorptions capacity that any variation in particle size has a dramatic effect on ink rheology. This in turn causes difficulties in supplying inks to a constant viscosity specification. Large incompletely dispersed particles can cause poor rub resistance because, for example, they tend to protrude from the ink surface. They also increase the abrasiveness of the ink, and this leads to wear problems on plates, blankets, and rollers of the printing press.

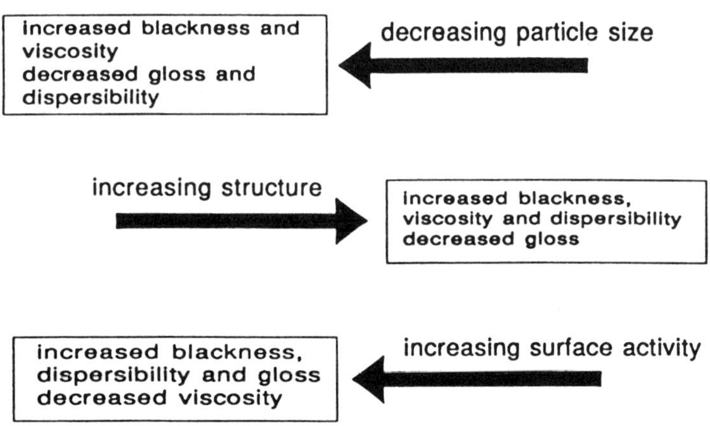

FIG. 13 Properties of carbon black.

A high print contrast ratio is desirable. The current trend is toward very dense blacks, and increasing the degree of dispersion increases print density for a given quantity of pigment.

B. Water-Based Flexographic Newsink

Flexography, although occupying a small sector of the market, is nevertheless important because it is a developing technology, with a number of advantages over cold-set web offset printing. Inks give less strikethrough (less penetration through newsprint) leading to savings on newsprint and less wastage occurs because the press can be set up for printing very easily. Good rub resistance means that the ink does not come off onto the readers' hands. More holdout ensures that colors are brighter because the inks sit on top of the newsprint rather than penetrating. The biggest disadvantage is the inherent restriction in screen size of flexography versus offset; flexography has less fidelity than cold-set web offset.

It is difficult to deink flexographically printed newspapers using the flotation process, although they can be deinked by the washing process commonly used in the United States. The alkali-soluble acrylic resins used to formulate these inks are readily soluble under the alkaline conditions of a deinking plant. However, the carbon black pigment stays suspended in the paper pulp and does not float to the surface with the bubbles used in flotation. A recent patent [32] describes the use of cationic polymers as a deinkable binder for these inks. Presumably a cationic polymer, though not soluble, can be detached from the pulp under alkaline conditions during flotation and floats to the water surface with the carbon black entrained.

There are various problems associated with pigment dispersion in such inks. Carbon blacks are not easy to disperse in aqueous media at high pigment/binder ratios and high overall pigmentation levels. It can be difficult to achieve the high print density required. The low viscosity of the inks (usually assessed by an efflux cup) is very sensitive to the degree of pigment dispersion, especially with carbon black because of its large specific surface area. The low viscosity can lead to sedimentation of pigment and stratification in the inks during storage.

Incompletely dispersed large particles can lead to problems during printing. They can cause filling in of the fine cavities on the surfaces of the anilox inking rollers of the flexographic press used in this application. They may also cause plate plugging (filling of gaps between raised image areas on the printing plate), but this has not been proved. These problems adversely affect the quality of the printed image. Rub resistance is usually so good with these inks that only very large particles would cause problems.

V. SCREEN INKS

There are such a wide variety of screen ink types that it is not possible to give a general formulation. Screen inks nevertheless have a number of typical features. They are made in small batches using small bead mills or three-roll mills, with pans and stirrers sized accordingly. Consequently their manufacture is difficult to automate. They are made via pigment concentrates or finished ink blend schemes rather than directly from pigment powders. This ensures rapid supply and better manufacturing logistics. They have a low pigmentation level (e.g., 5% w/w) since high film weight (thickness typically 25 μm) on application ensures sufficient color strength. The low pigmentation level also means that often the raw material cost of the binder resin is higher than that of the pigment—an unusual feature in inks.

Pigments with good resistance (fastness) properties are required. For example, chemically resistant pigments are used for container inks, lightfast pigments for the display market (fastness on blue wool scale of 6–7 essential), and highly resistant pigments for extended durability (blue wool scale 8). Lead and chromium pigments were frequently used in screen inks because they had good lightfastness, opacity, and chemical resistance and were easy to disperse. These have been replaced with organic equivalents, but this transition has not been trouble-free, since it is often difficult to achieve the required degree of opacity, flow, and ease of dispersion with organic replacements.

In screen inks, large particles can cause screen clogging and screen wear. When looking at the overall particle size of a pigment in a screen ink (or any ink) it is important to remember that there are various other sources of particulate species. Figure 9 indicates some of the sizes and types of particles that can appear in an ink. Particles larger than 5 μm can usually be removed by filtration or three-roll milling (three-roll mills can act as a filter).

As with inks in general, opacity increases as pigment particle size increases. In screen inks, the pigments are opaque due primarily to their large crystal size rather than their existence as poorly dispersed aggregates. Gloss increases as particle size decreases; lightfastness decreases as the crystal size decreases; perhaps chalking after extended weathering is also related to particle size.

VI. RADIATION-CURABLE INKS

This section has been included because it is very difficult to mill pigments into radiation-cured systems, and the pigment has a major influence on the curing of these inks [33–35]. The discussion here is restricted to UV curing, but many of the points will also apply to electron beam (EB)–curable inks.

At present no pigment manufacturer makes pigments specifically designed for use in radiation-cured inks, presumably because less than 1% of all the pigments produced are used in this technology. However, developments in the flexible packaging market could change this situation significantly.

A UV-curable printing ink consists of pigment, oligomer, monomer (multifunctional acrylate), and photoinitiator.

A. Effects of Pigments on Curing

The pigment can affect the curing process by either a photochemical or a physical mechanism. Assuming that the first effect is usually small, four variables can be related to cure response: the emission spectrum of the light source, the pigment absorption spectrum, the initiator absorption spectrum, and wavelengths for initiator decomposition.

The initiator must absorb light of an appropriate wavelength to start the initiation process. The more light it absorbs, the more radicals it generates and the faster the cure. If the pigment absorbs UV light, then the amount available for the photoinitiator to absorb will decrease. There is very little information published on these effects, and this is an area needing detailed research.

It is generally accepted that the influence of pigments for four-color process printing on the rate of cure of an ink is dependent upon their chemical type with the rate following the order magenta > yellow > cyan > black. Perhaps magenta ink cures most rapidly because it has a low absorption in the region of one of the mercury lamp emission lines. The radiation is therefore free to start the initiation process. Yellow pigments, although having a lower extinction coefficient than magenta pigments, have a higher absorbance in the region of the emission spectra from mercury lamps. The same applies to a greater extent with cyan pigments. Black inks are usually formulated with high levels of very active photoinitiator combinations, because very little light can penetrate due to the strong UV absorption of the black pigment.

Care has to be taken when formulating UV-curable colors containing two or more pigments, because it is possible to produce a blend of absorbances across the entire UV spectrum. For example, a green made from a yellow and a blue can give poor cure because of its high UV absorbance.

Photoinitiators are the most expensive component in a UV-cured ink, and optimization of their photoinitiation efficiency is very important. As the degree of pigment dispersion increases, the extent of UV absorption usually increases due to the larger surface area per unit volume of ink. This can reduce the cure rate. Such effects have been reported for carbon black [36], whites [33], and organic pigments [35].

B. Effects of Pigments on Shelf Life

Pigmented UV-curable inks are prone to increases in viscosity and resultant gelation as a function of time. It is not clear what causes such instability. To avoid supplying inks that are subject to this problem, it is normal practice to test inks using an accelerated method, featuring heating at elevated temperatures (typically 80°C for 10 h).

Possible causes of gelation are reaction of the binder resin through peroxide decomposition, perhaps caused by the metals in some pigments acting as catalysts. Peroxides may already be present in the oligomers used in these inks or may form during ink manufacture. The oxygen content of the ink is also important since it acts as a free radical scavenger; low levels are preferred to stop peroxide formation and high levels to stop polymerization. Peroxide scavengers can be added to the ink if peroxide formation is thought to be the cause of poor shelf life.

Chemical reaction of the acrylate group in the acrylic binder is another possible problem. If the pigments contain primary or secondary amines, perhaps added as surface treatments, then these can take part in a Michael reaction with the acrylate group.

Residual coupling components from azo pigments could be a further source of radicals. The effect of these impurities can be minimized by including an inhibitor in the ink formulation. The skill of the ink formulators lies in choosing the correct type and level of inhibitor to maximize shelf life and minimize the reduction in cure rate.

C. Manufacture

There is no solvent to evaporate from a UV-cured ink, so the pigment volume concentration remains almost the same after cure as before. There will be some small change due to shrinkage and penetration into porous substrates.

Whereas in conventional inks the presence of solvent allows formulation of low viscosity millbases, in UV-cured inks there is often insufficient monomer to formulate a low viscosity oligomer solution. As a result, the pigment wets out more slowly, and the higher millbase viscosity limits the concentration of pigment that can be incorporated, thereby decreasing production throughput rate. Sensitivity to temperature also restricts throughput.

However, it is possible to achieve better wetting with a UV-curable vehicle than with an alkyd resin if the pigments are not resinated [34] (pigment manufacturers add abietyl resin, especially to transparent yellow pigments, to improve dispersibility; see Chapter 4). This opens up the possibilities for improved pigments for radiation-cured systems if there is sufficient demand.

In the lower viscosity inks for flexographic and gravure applications, many of the previous comments apply, but in addition lower viscosity allows these inks to be manufactured by bead milling. Pigments also wet more efficiently because of the lower viscosity. Since these inks are being developed for food packaging, the cure rate must be maintained at very high levels to minimize odor and migratable materials. The optimization of pigment, pigment dispersion, and photoinitiation is therefore essential.

VII. MILLBASE OPTIMIZATION

It is interesting to consider the relative merits of producing finished ink directly or manufacturing via a highly pigmented millbase. In other words, how can output be maximized?

Kaluza [37] considers the relationship between G (the rate of dispersion of pigment) or E (ultimate color strength) and the concentrations of the pigment and the binder resin. The importance of taking dispersion quality into account when comparing throughputs is stressed. If the relationship between pigment concentration and rate of dispersion is linear, then the optimization of the amount of pigment (P_{max}) that can be dispersed per unit time, to a given color strength, is given by

$$P_{max} = \frac{G_1P_2 - G_2P_1}{2(G_1 - G_2)} \tag{1}$$

In practical terms this simply means: prepare two millbases with different pigment concentrations P_1 and P_2, but keeping the concentration of the resin in the varnish constant; then determine the throughputs (G_1 and G_2) and put the values into Eq. (1). If the decrease in the rate of dispersion is small as pigmentation increases, then P_{max} is high, and vice versa.

Determination of the optimum millbase viscosity and binder concentration is essential for good dispersion. In this context the validity of Daniel flow measurements for inks containing finely divided organic pigments has been questioned [37]. These pigments, as used in printing inks, give quite complex rheological behavior, which can be difficult to measure, especially in the case of paste inks [38]. In a three-roll mill the pigment is dispersed by shear forces, and these forces must be transmitted by the dispersion medium. The need for suitable viscosity in the material being dispersed (pigment plus medium) is therefore essential, and it is important that this viscosity prevail at high shear rates, not the low shear rates used in determining Daniel flow.

Assuming a constant energy input and continued optimized performance from the mill, the following generalizations on millbase optimization can be made. First, keep the viscosity of the dispersion medium low. Bear in mind that at high binder concentration the rate of dispersion decreases with increase in viscosity and that low binder concentration can cause flocculation. Second, have a high pigment/binder ratio. Disaggregation effects from pigment–pigment attrition cannot be ignored, and there will be an optimum pigmentation and pigment-binder ratio (again bearing in mind that flocculation will occur if sufficient binder is not present).

Since three-roll mills disperse via shearing and bead mills act more by attrition, increasing pigmentation can have different effects on the rates of dispersion in these two types of machines. In a bead mill, increasing pigmentation usually decreases the rate of dispersion; in a three-roll mill, the rate of dispersion, hence of color strength development, usually increases with increasing pigmentation until a maximum is reached. The decrease in the rate of dispersion beyond this maximum is thought to be due to flocculation and difficulties in mixing.

So, returning to the original question—Is it better to make finished ink directly or via a concentrate?—the following factors should be considered. There is a desire to maximize color strength, gloss, transparency, throughput, shelf life, handling, and tanking; there are preferences in operation of equipment, factory design, and logistics of supply; there are specific requirements of the finished ink. Each ink manufacturer will have different priorities for these factors and so no definitive answer is possible.

REFERENCES

1. W. Meuller-Kaul and C. Ratcliff, *Am. Ink Maker 70*(Nov.): 38 (1992).
2. *Chemsource '92*, Frost and Sullivan, New York, 1992.
3. *The European Market for Organic Colourants*, Frost and Sullivan, New York, 1990.
4. R.H. Leach and R.J. Pierce (Eds.), *The Printing Ink Manual*, 5th ed., Blueprint, New York, 1993.
5. P. Eno, (Ed.), *Flexography Principles and Practices*, 4th ed., Foundation of Flexographic Technical Association, New York, 1991.
6. B. Rutherford, (Ed.), *Gravure Process Technology*, Gravure Association of America, New York, 1991.
7. B. Denison, *J. Oil Colour Chem. Assoc. 73*: 256 (1990).
8. W. Tiebel and P. Hofman, *Eur. Coatings J. 1–2*: 44 (1992).
9. F. Muller and N. Stehr, *Eur. Coatings J. 5*: 418 (1989).
10. K. Engels, *Eur. Coatings J. 1*: 67 (1986).
11. I. Berg, *Polym. Paint Colour J., Supple.*, Feb. 23, 1983.
12. Fryma-Maschinen AG, Rheinfelden, Switzerland.

13. Draiswerke GmbH, Mannheim, Germany and Dr. N. Stehr short course paper at Kent State University, Kent, Ohio.
14. VMA-Getzmann GmbH, Reichshof-Heienbach, Germany.
15. F.K. Farkas, in *Surface Coatings: Theory and Practice* (R. Lambourne, Ed.), Ellis Horwood, Chicago, 1987, p. 309.
16. Mirodur SpA, Italy.
17. J. Winkler et al., *J. Coatings Technol. 59*(Nov.): 35, 45, 55 (1987).
18. A. Lust, *Farbe Lack 98*: 753 (1992).
19. KVK, Br. Patent 1 245 268 (1971).
20. Sicpa Holding Sa, Eur. Patent 281 720 (1988).
21. ETIS award to Coates Lorilleux and Staffordshire University on "The design of waterbased inks with a high efficiency of metal and colour removal by primary waste treatment" (1993).
22. Malvern Instruments, Malvern, England.
23. Sun Chemicals, Eur. Patent 341 054 (1989).
24. R.N. Blair, (Ed.), *The Lithographers Manual*, GATF, Pittsburgh, Pennsylvania.
25. A. Mercado, *Am. Ink Maker 69*(June): 19 (1991); *69*(July): 36 (1991); *69*(Aug.): 54 (1991).
26. N.J. Szalkowski, *Am. Ink Maker 69*(May): 38 (1991).
27. A.S. Gomm, G. Hull, and J. Moillet, *J. Oil Colour Chem. Assoc. 51*: 143 (1968).
28. Solsperse Hyperdispersants, ICI.
29. W. Carr, *J. Oil Colour Chem. Assoc. 63*: 438 (1980).
30. T.C. Patton, *Paint Flow and Pigment Dispersion*, 2nd ed., Wiley, New York, 1979, Chapter 18.
31. R. Maher, *Am. Ink Maker 57*(Sept.): 29 (1979); *57*(Oct.): 19 (1979).
32. M. Huber, German Patent 41 15731 Al, (1992).
33. Z.W. Wicks and S.P. Pappas, in *UV Curing: Science and Technology Vol. I* (S.P. Pappas, Ed.), Technology Marketing Corp., Connecticut, 1980, p. 79.
34. V. Kaden and P. Eckert, *J. Oil Colour Chem. Assoc. 68*: 251 (1985).
35. G. Hencken, *Am. Ink Maker 56*(Mar.): 57 (1978), C. Armstrong, Aspects of Photoinitiation, PRA Conf. Proc. (in press).
36. M.D. Garret, *Am. Ink Maker 54*(Mar.): 16 (1976).
37. U. Kaluza, *Physical/Chemical Fundamentals of Pigment Processing for Paints and Printing Inks*, Edition Lack und Chemie, Elvira Moeller GmbH, Filderstadt, Germany, 1981, p. 26.
38. T. Gregory and S. Mayers, *Surf. Coatings Int., JOCCA 2*: 82 (1993).

6

Dispersions in Plastics and Synthetic Fibers

GÉRARD HÄUPTLI Consultant, Binningen, Switzerland

ROGER C. SYKES Pigments Division, Ciba-Geigy Ltd., Basel, Switzerland

I.	Introduction	206
II.	Plastics and Synthetic Fibers	207
	A. Plastic materials	207
	B. Synthetic fibers	208
III.	Generic Composition of Dispersions	210
	A. Colorants	210
	B. Criteria for colorant selection	211
	C. Carriers	213
	D. Processing aids and auxiliaries	214
	E. Criteria for carriers, processing aids, and auxiliaries	214
IV.	Manufacturing of Dispersions for Thermoplastic Polymers	214
	A. General requirements	214
	B. Production of solid dispersions	215
	C. Production of liquid dispersions	221
V.	General Requirements and Test Methods for Dispersions	221
	A. Degree of dispersion according to end use	221
	B. Incorporation and homogenization	223
	C. Heat resistance	223
	D. Long-term compatibility with polymers	225
	E. Lightfastness	225
	F. Weather resistance	226
	G. Secondary effects on polymers and colored goods	227
	H. Physiological and ecological considerations	227

VI. Practical Incorporation of Dispersions into Polymers 228
A. Batchwise addition of solid dispersions 228
B. Continuous addition of solid dispersions 228
C. Batchwise addition of liquid dispersions 229
D. Continuous addition of liquid dispersions 230

VII. Polymer- and Application-Specific Considerations 231
A. Polyvinyl chloride 231
B. Polyolefins 234
C. Styrenic polymers and copolymers and engineering plastics 237
D. Dispersions for melt-spun fibers 238
E. Dispersions for wet- and dry-spun fibers 242

References 243

I. INTRODUCTION

A variety of additive dispersions are used for the modification and improvement of plastics and synthetic fibers, to impart light stability or flame retardancy or to achieve reinforcement, for example. Perhaps the most important role of additive dispersions, however, is to introduce the color required for product identification, safety, aesthetics, or decoration.

Many additives for plastics and synthetic fibers, particularly pigments, are insoluble in the final medium of application. They must be dispersed finely and distributed homogeneously in the polymer to ensure full effectiveness. The most efficient way to achieve this is to incorporate them already dispersed in an appropriate carrier material, which may be solid (solid dispersions) or liquid (liquid dispersions).

Examples of *solid dispersions* are ready-to-use color compounds in pellet form and color concentrates (masterbatches and pigment preparations) offered as powders, fine granules, or pellets. The advantages of granules and pellets are freedom from dust and easy volumetric metering. Dispersions in powder form are generally dusty and can be metered accurately only by gravimetric methods.

The main representatives of *liquid dispersions* for plastics are pigment pastes in plasticizer for polyvinyl chloride (PVC) or in polyols for polyurethane (PUR) systems, and “liquid colors,” which are offered mainly for extrusion and injection molding of styrenic polymers and polyolefins.

Monoconcentrates are dispersions of single pigments in suitable solid or liquid carrier materials. In practice, the end user combines several such concentrates to obtain the desired shade. Their advantage lies in reduced stock holding, as only a limited number of monoconcentrates are required to produce a number of different shades. Coloristic expertise and a profound knowledge of the general properties of the pigments used are necessary, however, to obtain consistent, high quality coloration.

Custom colors contain all necessary pigments and possibly other additives in a finely dispersed form and in the correct ratios so that the desired color is obtained simply by addition of a small amount of dispersion (usually between 1 and 5%) to the polymer. This approach allows trouble-free coloration but does require a special custom color for each shade.

Worldwide annual demand for pigment and other additive dispersions for plastics and fibers is estimated at ca. 1 million tons, of which black accounts for 27%, white for 34%, 22% is colored, and 17% is other additive masterbatches. Geographically, 40% of these concentrates is consumed in Europe, 31% in the United States, 14% in Japan, and 15% in the rest of the world [1].

II. PLASTICS AND SYNTHETIC FIBERS

A. Plastic Materials

After a slow start of the plastics industry up to the mid 1950s, the annual consumption doubled every five years until 1970, followed by slower growth rates to present times. Polyolefins—low density and linear low density polyethylenes (LDPE, LLDPE), high density polyethylene (HDPE), and polypropylene (PP)—make up the largest group of thermoplastic polymers (Fig. 1). PVC follows in importance, then the styrenic polymers and copolymers—polystyrene (PS), high impact polystyrene (HIPS or SB), sty-

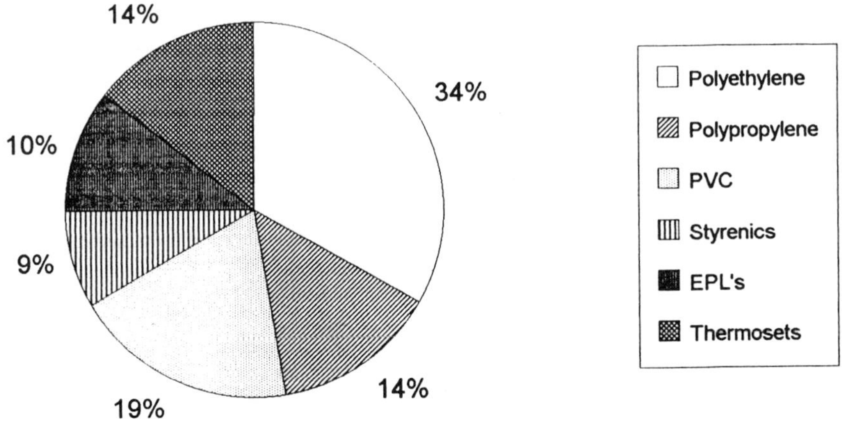

FIG. 1 World consumption of major plastics resins. Total: 105 × 10^6 tons.

rene acrylonitrile (SAN), and acrylonitrile butadiene styrene terpolymer (ABS). A variety of specialty polymers make up the group of engineering thermoplastics (EPLs).

B. Synthetic Fibers

1. Economic Significance

Almost half of the world production of textile and industrial fibers is of man-made origin. Man-made or chemical fiber production amounted in 1991 to 19.3 $\times 10^6$ tons made up of 16.2 $\times 10^6$ tons of synthetic (petrochemical-derived) fibers and 3.1 $\times 10^6$ tons of regenerated cellulose fibers. The latter group, composed of viscose and cellulose acetate/triacetate fibers, has grown little in recent years whereas synthetic fiber production has grown rapidly (Fig. 2). More than 55% of this is polyester (PES), over 20% is polyamide (PA6 and PA6.6), about 15% is acrylic (PAC) or modified acrylic, and around 7% is polyolefin fiber. The remainder includes small amounts of PVC and elastomeric PUR fibers.

Geographically, 3.7 $\times 10^6$ tons of man-made fiber was produced in the United States in 1991, 3.4 $\times 10^6$ tons in western Europe, 1.8 $\times 10^6$ tons in Japan, and around 2 $\times 10^6$ tons each in Taiwan and the People's Republic of China. India, Korea, and eastern Europe are other major producing areas for fibers, current growth occurring mainly in Asia.

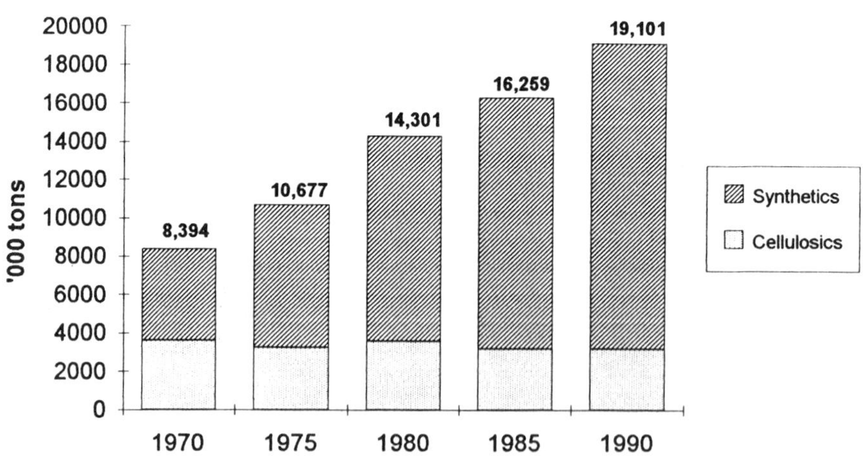

FIG. 2 World production of synthetic fibers. (Courtesy of Akzo Fibers.)

2. Mass Coloration of Synthetic Fibers

Mass coloration is an increasingly important alternative to classical methods of dyeing textiles using water-soluble or dispersed dyes applied from an aqueous bath. Mass coloration, also known as spin coloration, solution dyeing, and dope dyeing, yields colored fibers by introduction of the coloring material during or after formation of the fiber-forming polymer but prior to spinning it into filaments. The color is thus dispersed within the filament rather than adsorbed on its surface.

Both economic and technical considerations favor this process:

Economic

- Low energy costs
- No effluent or water costs
- No dyehouse investment
- Reduced labor costs
- No dyebath auxiliaries needed

Technical

- Large batches of uniform shade
- Shade uniformity batch to batch
- Higher fastness properties possible

Polyolefin fibers are a special case in that they are hydrophobic and contain no chemical groups to which conventional textile dyes can attach themselves. This lack of dyeability leaves mass coloration as the only practicable route to coloring polyolefin fibers, though attempts have been made to introduce dye sites by blending or chemical modification of the polymer. Such approaches have brought some success for printed goods but no significant breakthrough for stock, yarn, or piece dyeing.

Mass-colored fiber is produced on the same equipment as uncolored fiber, so the scale of production of the fiber has historically determined the economic viability of the mass-coloration process. With colored fiber, however, considerable time is required for cleaning equipment when changing shades. To minimize productivity losses it has thus been necessary to produce large quantities of each shade and to maintain high stocks of colored fiber. Various approaches toward increasing production flexibility are being offered and are described later in this chapter.

Traditionally, mass coloration using pigments or special dyes has been practiced primarily to achieve higher fastness properties than are obtainable with conventional dyeing. This has been the case particularly with cellulosic fibers, but the method is now employed increasingly for polyester fibers for automotive use. It is also used where large quantities of a limited number of shades provide the necessary economy of scale. Important examples are polyester fibers for military and school uniforms, black polyamide filament yarns for umbrella fabrics, and colored polyamide yarns for contract and domestic carpeting. Practically 100% of all colored polyolefin fiber, of which polypropylene forms the major part, is mass colored using pigment dispersions. With polyester, the proportion has risen from 2% in the early 1980s to about 5% today and is still growing. Similarly with polyamide, where estimates of 5–10% are quoted. Pigmented acrylic fiber is used primarily in Europe and the United States for outdoor applications and represents about 3% of the total fiber produced.

The prospects for further penetration by mass-colored synthetic fibers at the expense of conventionally dyed materials are excellent, particularly in the case of polyester and polyamide fibers. This development is driven by increasing legislative pressures on effluent quality, on increased costs of water and water treatment in dyehouses, and insufficient availability of dyeing capacity in developing countries with rapidly increasing synthetic fiber production. The need for higher performance levels in automotive textiles and in such outdoor applications as artificial turf, sports surfaces, and indoor/outdoor carpeting has generated further demand. This scenario provides a rapidly increasing potential for sophisticated pigment and special dye dispersions for synthetic fibers.

III. GENERIC COMPOSITION OF DISPERSIONS

A. Colorants

Inorganic pigments widely used for plastics include synthetic metal oxides, chromates, molybdates, sulfides, and mixed crystals of metal oxides of rutile or spinel structure. Examples are iron oxides (yellow, brown, reddish brown), titanium dioxides (white), lead chromates (yellow), lead molybdates (orange, red), cadmium sulfides/selenides (yellow, orange, red), nickel and chromium titanates (yellow), and spinel pigments (zinc/iron brown, cobalt blue, and cobalt green).

Their main advantages are their opacity (where required), heat resistance, and economy. Limitations include their abrasivity, low color strength, and, with some exceptions (cadmium- and lead-based pigments), a lack of pure, brilliant shades. The use of pigments based on cadmium, mercury, lead, and hexavalent chromium is increasingly restricted nowadays for ecological reasons, and in many cases organic pigments and dyes combined with metal oxide pigments are preferred.

All organic pigments are synthetic. Compared to inorganic oxide pigments, they are more transparent, have a softer structure, a higher color strength, and a purer, more saturated shade. The so-called classical azo pigments (see Chapter 4) are widely used in polymers processed at relatively low temperatures (180–240°C) and in applications requiring modest lightfastness, for example, polyethylene films and containers for packaging, commercial PVC goods, disposable polystyrene articles, and natural and synthetic rubber. High performance organic pigments show sufficient all-around resistance properties for use in the most demanding plastics and synthetic fiber applications. The higher coloring costs incurred tend to be offset by the performance benefits they bring to the finished goods.

A selection of special dyestuffs of anthraquinone and azo structure are suitable for the coloration of polymers with a high glass transition temperature, below which there is no risk of colorant exuding from the polymer (blooming). Although these dyes are soluble in the polymer, they are often incorporated as monoconcentrates or together with pigments in customer colors.

B. Criteria for Colorant Selection

1. Color Strength

The higher the color strength of a colorant, the lower the amount required for coloration of the polymer. Besides economy, a further important benefit of the lower amount is the reduced risk of secondary effects of the colorant. High loadings of inorganic pigment can, for instance, lead to a weakening of the mechanical properties of the polymer. In this respect the high color strength of organic pigments and, where they can be used, special dyestuffs offer advantages over inorganic oxide pigments of only moderate color strength.

2. Opacity and Transparency

Certain applications, such as textile coatings, cable insulations, marking foils, and garden furniture require high opacity. For other uses, high transparency is required, as with advertising panels, synthetic fibers, and automotive rear light lenses. For packaging, both transparency and opacity can be required, depending on the need to protect or display the packaged goods.

It is always easy to increase opacity by adding an opacifying pigment (titanium dioxide or other opaque pigment), whereas transparent or translucent coloration can be achieved only with suitable colorants of high transparency (selected organic pigments, pigment preparations, and special dyestuffs).

3. Heat Resistance

As plastics and synthetic fibers are usually processed at 200–300°C or higher, the colorants used must be resistant to such processing temperatures. This means they should not show any color change (fading or darkening) due to thermal decomposition of the molecule, they should not go into solution (except special dyestuffs), nor should they show any visible sign of interaction with the polymer or other additives contained therein. A typical test procedure is described in Sec. V.C.

With organic pigments, heat resistance depends not only on chemical constitution but also on polymer type and colorant concentration in the polymer. At very low concentrations, heat resistance is often inferior to that at higher concentrations. Partial solubility in particular polymers, chemical interactions, and different opacity levels of the polymers can also influence the thermal stability of the colored system. This emphasizes the importance of correct pigment selection for so-called universal masterbatches, offered for a wide variety of polymers and applications, and of confirmatory testing of the dispersion in the polymer to be colored.

4. Blooming and Migration Resistance

If pigments show slight solubility in the carrier resin or polymer during processing there is a risk that they will migrate to the surface of the polymer after cooling and crystallize there (bloom). They may also transfer to adjacent surfaces during service (migration). Such problems may be aggravated by the presence of dispersion aids in the concentrate or by heat treatments such as ironing or thermofixation of textiles.

5. Lightfastness and Weather Resistance

Colored goods tend to discolor when exposed to light. Depending on end use, such changes may be tolerated (for low-cost disposable products) or must be avoided (e.g., wall coverings, leathercloth for furniture or for automotive interior trim).

The durability of a colored system depends on the lightfastness and weatherfastness of the polymer itself (some polymers tend to yellow or darken when exposed to light), on the lightfastness of pigments selected, and, with many organic pigments (as with heat resistance), on their concentration in the polymer. The lower the pigment concentration, the lower the lightfastness. Only a very limited number of pigments are suitable for highly weather resistant materials, and then only in adequately stabilized, weather-resistant polymers. For a description of testing see Sec. V.E. and V.F.

6. Abrasiveness

A number of inorganic pigments have very high hardness, which can cause excessive abrasion of processing equipment. An example is titanium dioxide in its untreated rutile crystal structure, which should be avoided in some synthetic fibers. Also, chromium and nickel titanates can be responsible for high wear rates in dispersing units, particularly extruder screws and barrels.

7. Additional Criteria for Fiber Dispersions

The previously described properties are also important for fiber dispersions, but with particular emphasis on the following. *Heat resistance* is essential because dwell times in the spinning melt may in extreme cases exceed 30 mins. *Transparency* ensures high luster and a brilliant fiber aspect. *Migration resistance* minimizes colorant bleed from the polymer, which may lead to staining of spin finish baths. *Blooming resistance* (often related to sublimation resistance of special dyes) is often essential since heat-sctting treatments of the fibers of 30 s at up to 210°C may promote colorant movement to the surface of the fiber and result in reduced *rubbing* (or crocking) *fastness*. *Abrasivity* of pigments, particularly inorganic oxides, can lead to premature wear of spinning guides.

Coloring materials for fiber application should meet three further important requirements. They should have high *fastness to textile wet treatments and processing*, specifically washing, bleaching, shampoo, perspiration, dry cleaning, and cross-dyeing. They should have *minimal influence on polymer properties*. The tendency of some organic pigments to interfere with the crystallization behavior of cooling polymer melts can lead to severe spinning problems. Fiber breakage during spinning and drawing (stretching) and impairment of physical performance parameters such as tensile strength and modulus can result. Phthalocyanine pigments in particular are known to give problems of this nature. Where alternatives are not available, spinning conditions may have to be modified, resulting in loss of productivity.

They should also have *inert behavior toward other polymer additives*. The effectiveness of light-stabilizing additives used for high performance applications, for example, should not be impaired by the colorants used.

C. Carriers

The solid carrier resin system of a color concentrate may consist largely of the type of polymer in which the dispersion is to be used, but this is not always the case. Many so-called universal carrier resin systems are offered for which broad compatibility with a range of different polymers is claimed. Pigment dispersion and rheological properties of the concentrate may be enhanced by selective use of natural and synthetic waxes, oxidized waxes, derivatives of natural resins, esters, low molecular weight polymers, particularly polyolefins, and ionomeric resins. Often a combination of components is chosen to optimize the dispersion process and the ease of incorporation of the dispersion into the polymer.

Typical liquid carriers are plasticizers such as phthalates, epoxidized oils, and other polymeric plasticizers; they form the basis of pigment plasticizer pastes used for flexible PVC coloration or as an intermediate product for transformation into solid PVC masterbatches. In "liquid colors" for polyolefins and styrenic resins, complex mixtures of fatty acid esters, polyglycols, and even such simple products as white mineral oils are known.

D. Processing Aids and Auxiliaries

During the dispersion process, carrier resins must gain access to and efficiently wet pigment particle surfaces. To facilitate this wetting process, *surfactants* are sometimes added. Good wetting facilitates the attainment of high color yield and optimum dispersion quality.

Viscosity improvers are usually a judicious combination of surface-active additives that enable the pigment loading of a dispersion to be increased without increasing its viscosity. They are used mainly for liquid dispersions of inorganic pigments.

Fillers like barium sulfate or calcium carbonate find occasional use to give more "body" to the dispersions, to prevent compaction of organic pigments during the dispersion manufacturing process, to modify the surface character of the final material, or simply to lower the apparent cost of the concentrate.

E. Criteria for Carriers, Processing Aids, and Auxiliaries

All constituents of color concentrates must ideally be compatible with the polymer to be colored, to prevent exudation or migration in service and minimize the risk of surface defects and reduced mechanical or physical properties. Exceptions, however, are known, an example being the use of incompatible polyethylene dispersions in polar polymers such as polyester and polyamide plastics and fibers.

The addition of a small amount of a dispersion should neither unduly influence the processability of the polymer nor impair such desirable characteristics of the end product as mechanical properties, welding properties, printability, or suitability for vacuum forming. This issue is discussed in more detail in Sec. VII.

The rheological properties of the dispersion are of paramount importance in achieving uniform coloration and are greatly influenced by carrier resin composition. The melt flow character of the main carrier resin is in turn altered through addition of pigment, which in general increases the viscosity of the composition, and of wetting agents and waxes, which tend to reduce viscosity. The objective is to achieve a color concentrate with melt flow properties similar to that of the polymer to be colored.

IV. MANUFACTURING OF DISPERSIONS FOR THERMOPLASTIC POLYMERS

A. General Requirements

Molten polymers have low wetting ability compared to solvent-containing liquid systems. This is especially true for such nonpolar polymers as polyolefins. On the other hand, materials to be dispersed have a very high specific surface area, which in the case of organic pigments can reach 150 or more square meters per gram. Overcoming the high surface energy of such surfaces and obtaining complete wetting with molten polymers of relatively high viscosity thus requires considerable power input.

After their synthesis, many organic pigments have a very soft and fluffy structure, comparable to that of freshly fallen, dry and virgin snow. Under the effect of high pressures and impact forces, this open structure is subject to compression to hard aggregates, which are subsequently very difficult to redisperse [2]. Such compaction is most likely to occur during the first premixing, loading, and extrusion steps of dispersion production if appropriate dispersion techniques are not employed.

The degree of pigment dispersion achieved during the process influences the color strength and, in extreme cases, the hue of the final product. Severe compaction can lead to the loss of 70% or more of color strength and to the formation of aggregates visible to the eye. Under these circumstances, batch-to-batch reproducibility of shade and dispersion quality requires specialist knowledge not available to most producers of plastic goods and synthetic fibers. Such difficulties together with increasing demand for quality have induced the majority of polymer processors to abandon in-house color formulation using undispersed pigments and to purchase from specialized dispersion manufacturers.

B. Production of Solid Dispersions

The most important steps in the production of a solid dispersion are shown schematically in Fig. 3.

1. Mixing the Ingredients

The ingredients of the dispersion are premixed to ensure homogeneous distribution of all components. As solid dispersions usually contain high amounts (10–70%) of pigment powder, a satisfactory premix is obtained only if the binder is also added in powder form. If mixed with pellets, pigment tends to accumulate at the bottom of the mixing vessel. Severe compaction is a further risk when blending high loadings of organic pigments with polymer pellets.

The most common types of equipment used for discontinuous premixing are *horizontal cylindrical paddle mixers* and *ribbon mixers*. In both, a horizontal axle is equipped with paddles having either a spiral ribbon or two counterrotating spiral helixes. These mixing tools revolve slowly (20–100 rpm) in a horizontal cylinder containing the materials to be premixed. *Fluid mixers*, as described below, are sometimes used for premixing purposes also, though preferably at low speed.

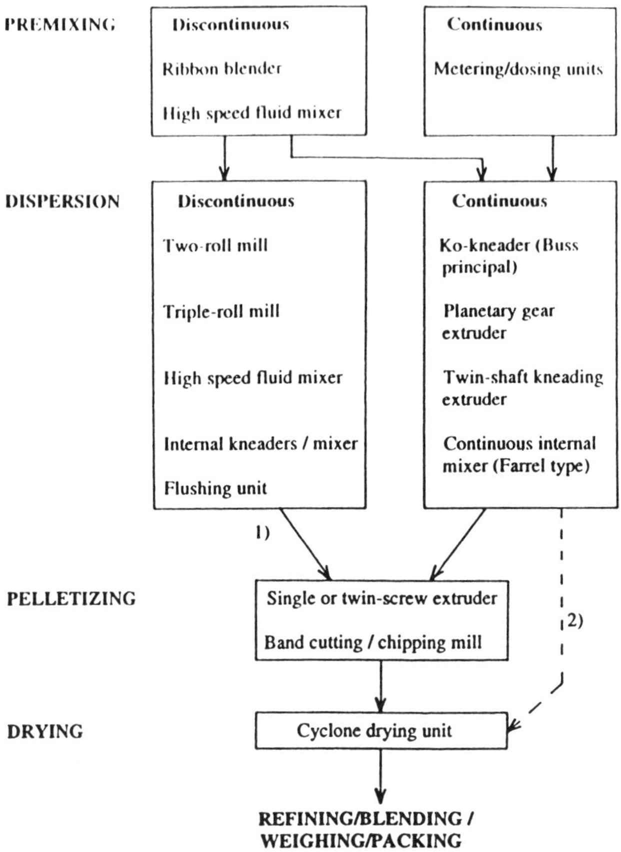

FIG. 3 Production schemes for pelletized solid dispersants. (1) Addition of principal polymer to wax-based concentrates prior to pelletizing; (2) continuously operating kneading extruders may incorporate pelletizing units.

For continuous dispersion units, the separate ingredients may also be metered continuously using appropriate devices as described in Sec. VI.B.

2. Batchwise Dispersion

Used mainly in the laboratory, *two-roll mills* still find limited use in the industry for small batch production. Two open steel rolls separated by a narrow nip are heated and rotate inwardly, the surfaces moving in the same direction in the nip but at different speeds (i.e., with "friction"). The preblended mix of resin and pigments is charged over and into the nip, is heated and melted, and forms a hide, usually on the faster roll. From time to time the hide has to be cut off and returned to the mill. The dispersion of the pigments is achieved by the high rate of shear in the nip between the two rolls. The hide is stripped off in a band, crushed, and extruded or cut into pellets.

Internal mixers with a ram and floating weight are commonly used for batchwise dispersion. Two kneading blades counterrotate with or without friction in a closed mixing chamber that can be heated or cooled. The mixing chamber is closed by a pneumatically controlled ram with a floating weight, which holds the contents under high pressure. Following the high shear dispersion step, the mass is dumped onto a two-roll mill and sheeted off into a pelletizing or cutting unit.

The other important type of dispersion equipment is the *high speed fluid mixer*. In this, impeller blades rotate vertically at high speed in a closed vessel that can be heated or cooled. Aggregates are broken down, and the pigments are distributed within the polymer. Although most of the energy input is converted to heat, a melting of the mass must be avoided, as this would destroy the fluidized bed and block the whole system. The dispersed mixture is subsequently fed into an adjacent continuous single- or twin-screw extruder for homogenization and pelletizing.

3. Continuous Dispersion with Single-Screw Kneaders

A single-screw extruder consists essentially of a screw blade that rotates in a close-fitting barrel that can be heated. The material being processed is pushed through the barrel by the action of the blade. Conventional single-screw extruders do not have the necessary kneading action for achieving adequate pigment dispersion. In the context of color concentrate production they are used primarily for homogenization and pelletizing of concentrate following the dispersion step. Specially modified screws may enhance this mixing and homogenizing action. Two developments on the single-screw principle are worth mention.

In the Buss Ko-Kneader, the dispersion is performed between a screw with interrupted helical blades, which combines rotation with a rhythmical backwards-and-forwards movement, and stationary teeth fixed in the heated barrel [3]. The planetary roll extruder is composed of a driven central screw, carrying around 6–12 intermeshing smaller planetary screws in a heated barrel [4]. The material is intensively kneaded between the central and planetary screws.

4. Continuous Dispersion with Twin-Shaft Machines

Twin-screw extruders operate with intermeshing, usually corotating screws [5], although counterrotating machines are also used [6]. The barrel is segmented in sections up to a total length of 40 times the screw diameter.

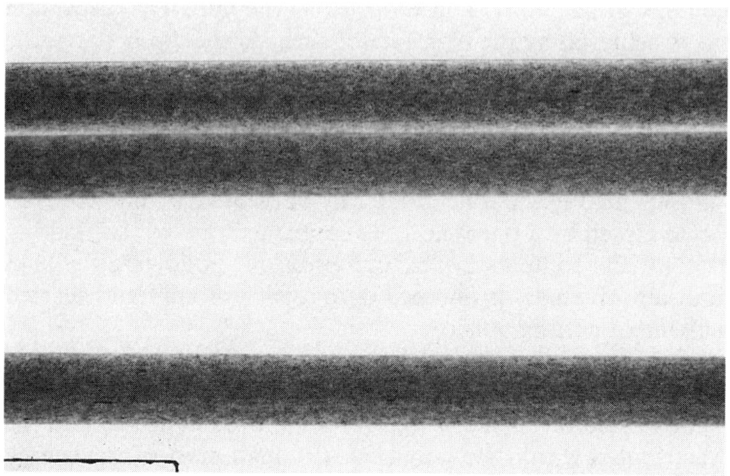

(a)

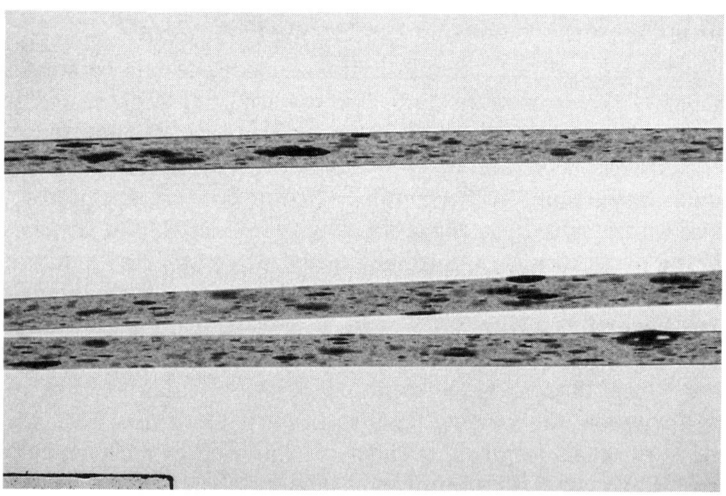

(b)

FIG. 4 Fiber filaments illustrating colorant dispersion. Scale bar = 40 μm. (a) High quality flushed dispersion of organic red pigment in PP fiber; (b) poorly dispersed organic red pigment in PA6 fiber, showing severe compaction; (c) polymer-soluble red dye in PES fiber; (d) incompletely homogenized carbon black concentrate in PP fiber.

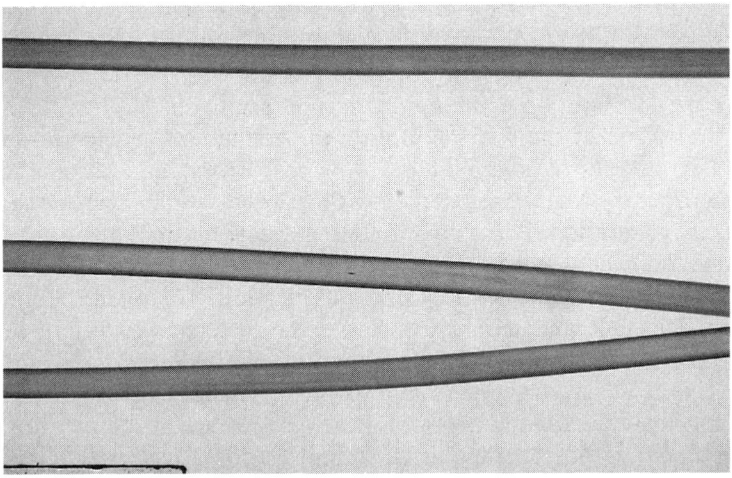

(c)

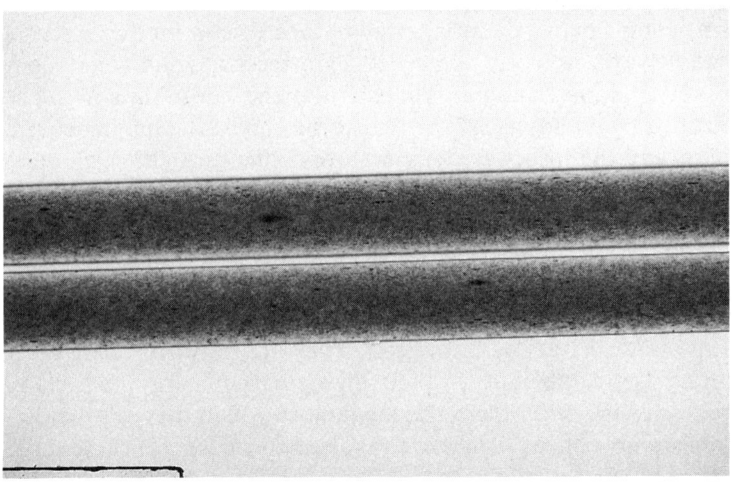

(d)

Temperature control is performed for each section by electrical resistance heaters and water cooling. A number of ports can be included in the barrel design to allow for multiple feed additions and venting.

The screws are also built up with interchangeable segments to allow flexibility and optimization of the dispersion process. A standard segmentation includes feed screws, kneading paddles, flow restrictors, and discharge screws. The configuration of the segments can thus accommodate conveying, melting, pressure generation, kneading, pressure release, venting, and downstream material addition to serve a wide range of purposes and materials.

Such machines also allow a variety of configurations for pigment and filler addition. Pigment and carrier resin may be metered separately or preblended and fed into the feed zone of the kneader. Alternatively, for pressure-sensitive or abrasive pigments and for shear-sensitive fillers, downstream metering into the molten polymer is practiced. Modular temperature control allows temperature profiles to be optimized for effective pigment dispersion—high temperature shortly after the feeding section to secure good wetting, low temperature (i.e., higher viscosity) in the kneading zone to generate high shear forces for efficacious breakdown of aggregates.

The other important type of twin-shaft equipment is the continuous screw-fed internal mixer [7]. This mixer is equipped with two counterrotating, nonintermeshing mixing paddles, similar to a discontinuous internal mixer. The paddles are, however, preceded by a feeding zone where conveyor screws press the material continuously into the kneading zone. The output is controlled by an adjustable orifice at the end of the mixing chamber. The mandrel and the rotors are temperature-controlled, although quite often the necessary melting heat is created under adiabatic conditions by the high shear forces of the kneading elements.

5. Flushing

The primary particles of organic pigments are very small crystals, which would represent an ideal state of dispersion. They are, however, extremely prone to forming aggregates due to their high surface energy, especially during the process of isolation from the medium in which they are synthesized. Very considerable shear forces are required to redisperse these large aggregates in the application medium.

The transfer of pigment particles directly from hydrophilic preparative media to a hydrophobic phase, known as flushing (see Chapter 5), is used extensively for printing ink manufacture in the United States. It avoids isolation of the pigment as a powder and associated formation of large aggregates. A related approach has also gained importance for the production of high quality pigment dispersions for mass-colored melt-spun fibers, in particular, polypropylene. Its merit lies in the extremely fine degree of dispersion attainable (Fig. 4a), which contributes to minimize filter-pack blockage and to eliminate filament breakage during spinning and drawing, these problems often being caused by residual oversize pigment aggregates (Fig. 4b).

In an adaptation of the flushing principle, pigment presscake or an aqueous dispersion of pigment containing suitable surfactants is added to a heated solvent in which a low molecular weight polyolefin is dissolved. Additional polymer, for example, fiber-grade polypropylene flake, is added. Under appropriate conditions of pH and agitation the pigment transfers to the polyolefinic phase. Following separation, the pigmented mass is dried and pelletized by extrusion [8].

In an alternative approach, pigment and low molecular weight polyolefin resin are ball-milled in water. The resulting slurry is granulated by heating under defined conditions in the presence of a water-immiscible solvent that is a partial solvent for the resin. The resulting irregular granulate is filtered, washed, dried, blended with the fiber-forming resin as carrier, and pelletized [9]. Polypropylene, polyester, and polyamides are used for this purpose.

C. Production of Liquid Dispersions

For the production of liquid dispersions, the same types of equipment and methods as featured in the paint and ink industry are generally employed, that is, three-roll mills and ball mills for batchwise processing and sand mills, pearl mills, colloid mills, and disk mills for continuous production.

A very important factor is the correct viscosity of the liquid dispersions. This must be adjusted to prevent sedimentation and floating problems during storage while not being so high that pumpability and ease of incorporation into the polymer are impaired.

The solid powders must first be premixed into the binder. Agitators, paddle mixers, planetary mixers, impellers, turbine mixers, and dissolvers are used. It may be advantageous for fine powders such as pigments to be soaked for several hours in the liquid binder to ensure efficient wetting of the surface. Wetting agents may also be used to speed up this process.

V. GENERAL REQUIREMENTS AND TEST METHODS FOR DISPERSIONS

A. Degree of Dispersion According to End Use

Increasing fineness of the produced dispersion leads to superior performance and improved color strength but may involve high costs. Production costs tend to be high for very sophisticated dispersions, such as those designed for fine melt-spun fibers where no particles over 1–2 μm can be tolerated. Where such an extremely fine dispersion is not required, and when low cost pigments are used, it may be more economical to accept a lower degree of dispersion and tolerate oversize particles up to say 10–20 μm.

A number of factors must be considered to achieve the required particle size of a dispersion. First, the coloration of any produced article should be free of specks, stripes, and streaks; that is, individual oversize particles should not be visible to the naked eye. This is more important in some cases than in others. The human eye is capable of detecting particle sizes down to 30–60 μm (depending on individual visual acuity). A dispersion free of particles larger than 20 μm would accordingly be adequate from the point of view of visual assessment of the colored article. It is indeed not uncommon to find residual aggregates of greater size in low cost plastic articles, usually those produced by injection molding. Such particle sizes cannot, however, be tolerated for films and fibers, as they tend to cause film rupture and filament breakage, resulting in intolerable production losses and inferior end-use properties.

Second, the free surface area of the dispersed phase increases with deaggregation and with any reduction of primary particle size. This means that more binder is required to cover and wet the surface of the dispersed product and may result in excessive viscosity increases. This can in practice lead to uneven coloration when such dispersions are incorporated.

A third important factor relates to the achievement of opacity and transparency. The maximum hiding power of a pigmented system is obtained when the greatest amount of light is scattered, and this occurs when the particle size of the pigment corresponds to about half the wavelength of visible light. The maximum opacity is therefore obtained in the particle size range of 0.2–0.5 μm. Below 0.1 μm, high transparency is obtained.

From these factors it can be concluded that in bulk plastics the size of the largest pigment particles should seldom be more than 20 μm, the average size being below the micrometer level. For fiber spinning, the largest particles should not be greater than 1–2 μm. Indeed, when using polymer-soluble dyes, as is common, for example, in spun-dyed polyester fibers, no colorant particles whatever should be evident (Fig. 4c).

The most convenient method of assessing the dispersion grade is the observation of a thin pressing under the optical microscope, where particle sizes down to 2–5 μm can readily be detected. This is sufficient for the assessment of pigment concentrates for injection molding, film, and general extrusion. It does not, however, indicate how well the concentrate itself will homogenize in the final polymer, and insufficiently dispersed concentrate has been mistaken for undispersed pigment on many occasions! (See Fig. 4d.) For critical applications it is therefore advisable to incorporate the concentrate in the polymer under the processing conditions to be used before assessing the degree of dispersion microscopically. This procedure is particularly advisable when testing fiber dispersions. For more fundamental investigations, electron microscopy is useful.

A practical method to judge the quality of dispersions for PVC is the observation of the "calendering effect." In this test very high shear forces are applied to a compound colored with the dispersion to check if further color development takes place. Such further development would indicate that the color potential of the pigment has not been fully exploited in the dispersion and that remaining aggregates have been dispersed in the test by the applied high shear forces. The described test is usually performed by milling the cold pigmented compound on a cool two-roll mill. A good dispersion does not show any increase of the color strength after such treatment despite the high shear forces developed on the cool mill. The principle of this method is embodied in the German DIN 53775, part 7 test for dispersion hardness.

B. Incorporation and Homogenization

As mentioned earlier, to facilitate incorporation of the dispersion into the molten polymer, the dispersion and polymer should have similar rheological behavior. A dispersion with excessive viscosity will tend to be difficult to homogenize. Visual uniformity may suffer, with streaks and stripes visible in the finished article (Fig. 5). When the viscosity of the dispersion is too low, an overlubricating effect may occur in the processing machines, leading to reduced friction between molten polymer and mixing/transporting elements such as extruder screws. Again, homogeneous distribution into the polymer will be hindered. The usual control of homogeneous distribution of the concentrate is visual inspection.

Melt-flow index measurements (ASTM D 1238, DIN 53 735, ISO 1133) to check and adjust color concentrate viscosity are widely used to ensure homogeneous results.

C. Heat Resistance

In addition to the colorants, each component of the dispersion must be stable under the highest normal polymer processing conditions. Static or dynamic test methods are used to assess the color stability or heat resistance of dispersions. In either case the test is not performed with the dispersion on its own but after incorporation of the dispersion into the polymer.

A *static test* is commonly used for rigid and plasticized PVC. Calendered strips or pressed sheets are heated for specified periods at specified temperatures (e.g., 10 min at 180°C, 10 min at 200°C, and 30 min at 180°C) in an

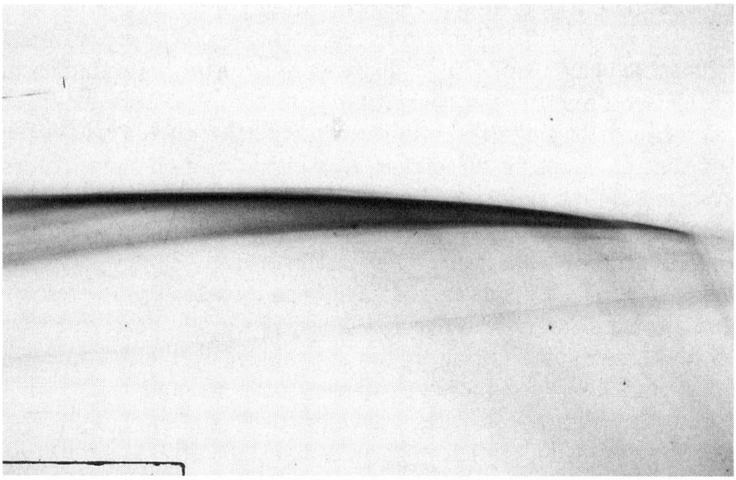

FIG. 5 Inadequately homogenized color concentrate in PET bottle preform. Scale bar = 800 μm.

air circulation oven. The specimens are then compared to untreated samples, and any color change is assessed either visually using the ISO Grey Scale for Color Change (ISO R 105/I part 2) or by colorimetric measurement.

A *dynamic heat resistance test* is used for polymers processed by extrusion or injection molding, for example, polyolefins, styrenic resins, and engineering plastics. A compound containing the dispersion at an appropriate concentration (usually corresponding to pigment loadings of 0.01–1.0%) is prepared, for example, in polypropylene (PP) at 200°C and then injection molded. In the case of PP this might be at 200° (reference), 220°, 240°, 260°, 280°, and 300°C, with a dwell time of, for example, 5 min at the test temperature. The injection molding machine used is preferably modified by using very slow screw rotation to achieve the unusually long dwell time of 5 min without actually interrupting the machine cycle. The moldings injected at the indicated test temperatures are then compared visually or by colorimetry with the reference standard molded at 200°C.

The tolerated color deviation (fading or darkening) is defined colorimetrically or by reference to the ISO Grey Scale. Individual colorants are tested in a similar way at typical concentrations between 0.01 and 1.0% in appropriate polymers. For polymers processed at higher temperatures, such as certain engineering plastics, incorporation and test temperatures are chosen accordingly.

D. Long-Term Compatibility with Polymers

The lifetime of durable plastic materials or synthetic textiles may extend to several decades. It is important that during this period no material incorporated into the basic polymer show any sign of separation, exudation, or blooming. To accelerate the appearance of such incompatibility effects, the colored material may be treated for a period of hours or even days at elevated temperatures in an oven, possibly under load. Any exudation indicates a risk of long-term incompatibility.

E. Lightfastness

Lightfastness, an essential requirement of high quality consumer goods, can be defined as the resistance of color to changes caused by the effect of dry exposure to global irradiation behind window glass, that is, to visible light (380–700 nm) plus UV-A radiation(315–380 nm), or by the corresponding radiation simulated by means of a filtered xenon lamp. Radiation in the UV-B (280–315 nm) and UV-C (below 280 nm) ranges are excluded by adequate optical filters.

Lightfastness is tested by exposure of specimens colored with the dispersion to a well-defined light source in testing cabinets. The source is usually a xenon lamp surrounded by adequate filters to simulate natural daylight (see ISO R879, DIN 53389). Fluorescent tubes and carbon arcs are also used, although their spectra do not correspond to global irradiation but show a very high energy radiation in the UV range.

Lightfastness is traditionally assessed by reference to the blue wool scale. This scale is composed of eight different wool strips dyed with standardized dyestuffs and having gradually increasing lightfastness. The test specimens are placed together with a blue wool scale into the test cabinet, exposed, and periodically inspected. As soon as a defined change in the test specimen is observed, its degree is compared with the coexposed blue wool scale samples and assessed accordingly. The highest lightfastness ("no change") is given a rating of 8, the lowest (corresponding to a visible fading after only 1 day of natural exposure or 3 h of xenon light) is given a rating of 1. Colorimetric assessment methods are also used, particularly in the United States.

F. Weather Resistance

Weather resistance may be defined as the resistance of materials colored with the dispersion to changes caused by the effect of global exposure to the elements, that is, sunshine (including UV-A and UV-B radiation), humidity (rain, dew), and ambient temperature change. Special tests are also available to simulate the effects of atmospheric pollutants.

1. Natural Weathering Testing

Test samples are fixed perpendicularly to the average position of the sun or can even follow the motion of the sun for the stipulated test period. Periodic washing and intermediate assessment of the exposed samples are usual. The limitations of natural weathering testing are the time required to obtain useful results, the poor repeatability of results, and the very limited relevance of results from a single test site. Shorter time scales can be obtained by choosing commercial exposure sites with more severe climatic conditions such as Florida and Arizona in the United States or southern France. Even at such sites, however, test durations can often exceed two or more years for particularly weather resistant materials.

2. Artificial Weathering in Test Cabinets

The weather resistance of dispersions can be evaluated more rapidly by exposing them to simulated global irradiation in test cabinets. These are similar to the cabinets described for the lightfastness tests, but in addition to the artificial light sources, which must include UV-A and UV-B radiation, they must also provide the necessary humidity (spraying cabinets, or condensation of water on the surface of the specimens from boiling water at the base of the cabinet). Test cabinets can also be equipped with devices to provide controlled SO_2 or H_2S atmospheres.

Test cabinets provide both acceleration of the weathering process and enhanced reproducibility of results. It is difficult, however, to predict from such tests the likely behavior of materials under natural weathering conditions. Their primary purpose is to allow comparison and establish a ranking order between alternative approaches to a particular problem. Another disadvantage of weathering tests in cabinets is the limitation of available space. Large articles such as garden chairs, car bumpers, or full-size plastic window frames cannot be tested in normal commercial test cabinets.

3. Accelerated Natural Weathering

One interesting new development [10] is the augmentation of natural weathering by additional artificial irradiation at night using controlled UV light. This method was developed at the main independent European weathering test station in Bandol (southern France) and is said to allow results corresponding to 2 years of natural weathering at that site to be obtained within a few months. This is also applicable to objects too large for standard test cabinets. A further advantage claimed for this method is that the test specimens are not exposed to excessive heat. An alternative approach to test acceleration uses mirrors to focus sunshine onto the samples, but with a consequent risk of excessive heat development.

4. Assessment of the Results

Initial assessment is generally based on change of color (darkening or fading) and visible change to the surface (loss of gloss, chalking, or more drastic surface degradation). This assessment may be visual, by reference to the ISO Grey Scale, or instrumental (colorimeter and goniometer). In addition, the retention of mechanical properties after exposure may be tested.

G. Secondary Effects on Polymers and Colored Goods

The design of plastics components takes account of the physical and mechanical properties of the polymer chosen. It is important that these properties not be impaired by the addition of a dispersion. Mechanical properties can suffer as a result of poor pigment dispersion. Insufficient wetting of the pigment surface or residual aggregates yield weak points that tend to rupture or fracture under excessive stress or strain. Incompatibility or inadequate melt viscosity of the binder of the dispersion can also lead to weakening of the produced material.

Severe difficulties may result from the undesirable effects of organic pigments and other additives on the crystallization rate of polymers. Of major concern is the promotion of excessive shrinkage of the polymer, causing distortion of injection-molded polyolefin articles (see Sec. VII.B.2). Further examples can be found in some of the difficulties encountered in drawing melt-spin fibers containing certain pigments during the spinning process.

H. Physiological and Ecological Considerations

Special care is necessary when formulating dispersions for food packaging and personal hygiene articles, toys, and household goods, for which legislation often regulates the choice of coloring materials. The trend recently seen in the number of local and national regulations governing disposal of waste in incineration plants and landfill sites is also increasingly influencing the selection of raw materials for dispersions, particularly as regards pigments and additives containing heavy metals.

VI. PRACTICAL INCORPORATION OF DISPERSIONS INTO POLYMERS

A. Batchwise Addition of Solid Dispersions

Drum or *barrel mixers* are the simplest type of equipment for blending concentrates with polymer. A cylindrical container (e.g., a drum) is rotated around its horizontal axis at low speed (20–50 rpm), and the materials are mixed by the rolling movement under the influence of gravity.

In *tumble mixers*, an additional end-to-end movement in the drum is achieved by an eccentric position of the axle. This allows a shortening of the mixing time from 20–40 min (drum mixer) down to 10–20 min. Such simple equipment is only suitable for mixing products of similar particle size—powders with powders, or pellets with pellets. Small quantities of powder concentrates can also be mixed with pellets, particularly if an adhesion aid is used, but quantities greater than around 1% by weight are subject to separation, with consequent risk of inhomogeneity.

High speed fluid mixers as described in Sec. IV.B.2 may be operated at lower speed for premixing concentrates with polymers of similar granulometry. *Vertical screw mixers* use a different principle. A screw rotates within a vertical silo, lifting the contents from the bottom to the top, where they are released on the bulk of the material. *Conical epicyclic screw mixers* feature a vertical conical vessel in which a screw rotates and simultaneously moves in a circular path (epicyclic rotation). The contents are mixed in three ways—from bottom to top, around the screw axle, and around the conical vessel. Both types of screw mixers are used for pelletized materials or to incorporate small amounts of powder or granular material into granulated or powder polymers.

B. Continuous Addition of Solid Dispersions

In modern processing plants, polymers are held in silos and transported to the processing machines by automated pneumatic devices. Similar methods may also be used for handling other additives such as color concentrates, fillers, and fibrous reinforcement. The main handling techniques used for adding solid dispersions to polymers are volumetric, loss-in-weight, and weight-belt feeding systems. For continuously operating fiber extrusion plants, melt injection techniques have been developed.

In *volumetric feeding* the dispersion in fine granule or pellet form is held in a hopper and discharged into the processing machine at a constant predetermined volume. This volume is controlled, for example, by a rotating punched plate or by single- or twin-screw conveyors whose rotation rate can be varied in accordance with the throughput of the processing machine.

This technique is most suitable for pelletized or granular dispersions. It is not ideal for powders with poor flow properties.

In *loss-in-weight feeding*, a discharge unit and a hopper containing the dispersion are placed on or suspended in a weighing system. The total weight is stored in a controller. The dispersion is gradually discharged into the processing machine, and the weight difference per unit time is continuously measured. The actual weight metered is compared with a desired, preset discharge rate and adjusted accordingly. When the hopper is refilled, the controller regulates the discharge on the basis of stored values.

In *weight-belt feeding*, the dispersion, again in powdered or granulated form, is transported from a hopper by a driven belt across a weigh bridge. The weight acting on the weigh bridge is measured. The feed rate is computed by a processor according to the weight and the speed of the belt, and the throughput is adjusted by varying the belt speed. Such systems are useful for metering solid dispersions of all types.

Melt injection feed systems have been developed to counteract the production flexibility limitations noted earlier in connection with fiber spinning. Such systems inject molten colorant dispersions directly into the polymer melt close to the spinning heads. They are essential when polymer is taken in molten form directly from the polymerization unit without intermediate pelletization. The use of side extruders illustrates this approach. Color concentrate is melted in the side extruder and fed into the main polymer melt stream through the side of the main spinning extruder barrel via a gear pump (Fig. 6). Homogenization of the colored melt is enhanced by the use of static mixers between screw tip and spin pump or by proprietary dynamic mixing devices. Microprocessors coupling the spinning and side extruders control dosing rates. Such approaches are said to have two principal advantages. First, they reduce significantly the time and material needed to purge the system between colors. In contrast to dosing systems for solid concentrates (Fig. 6), the main spinning extruder does not need to be purged between color changes, only the melt homogenizer and spinning manifold. A second side extruder can then be activated immediately while the first is being purged for the next color change. Second, they are claimed to yield sufficiently uniform coloration for use in continuous-filament yarns. This capability is claimed even where relatively large differences in viscosity between fiber polymer and colorant dispersion exist.

C. Batchwise Addition of Liquid Dispersions

Due to their high pigment loading, liquid dispersions usually show a high viscosity and therefore require rather slow but efficient mixers for their incorporation into liquid polymers. In planetary mixers, suitable agitators

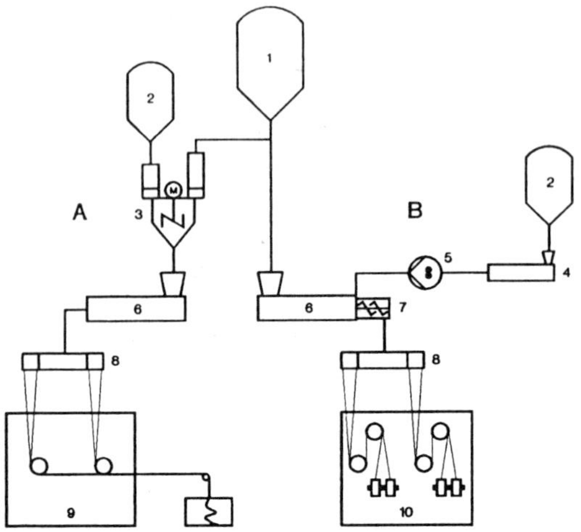

FIG. 6 Colorant dosing systems for melt-spun fibers. (A) Solid dosing system; (B) melt-injection dosing system. 1, Polymer holding silo; 2, color concentrate; 3, dosing unit; 4, side extruder; 5, gear pump; 6, spinning extruder; 7, static or dynamic melt homogenizer; 8, manifold; 9, staple (cut) fiber line; 10, continuous filament line.

rotate vertically and simultaneously move in a circular path within the mixing vessel. In double-arm kneaders, two Z-shaped blades rotate horizontally in opposite directions in a container with a divided trough or saddle. The blades can also be overlapping or run at different speeds.

Discontinuous addition of liquid dispersions to solid polymers is not of significance in practice because the high viscosity of the dispersions makes long mixing times necessary and cleaning of the mixers after use is also time-consuming.

D. Continuous Addition of Liquid Dispersions

One important advantage claimed for liquid colors is the possibility of metering them directly into the feeding zone of an extruder or injection molding machine (Fig. 7). Premixing of the dispersion into the polymer as a separated operation is avoided, with savings in cleaning time, labor, energy, and wastage costs. As the hopper is not contaminated with the dispersion, color changes are easy; and with modern metering equipment, color control and corrections are said to be practically instantaneous.

A variety of pump systems are available for metering liquid colorants. The most widely encountered types are peristaltic, diaphragm, piston, pneu-

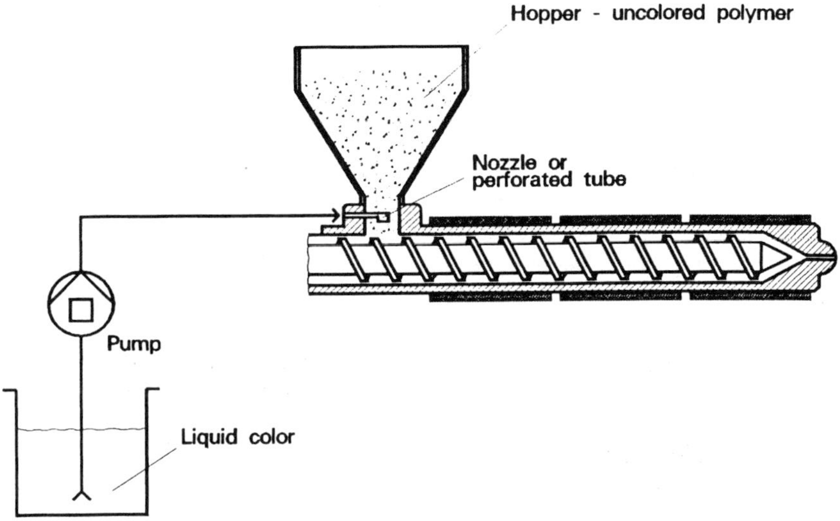

FIG. 7 Liquid color unit with injection molding machine.

matic, and gear pumps. Gear pumps can also be used to meter liquid pigment dispersions directly into the polymer melt at a point along or beyond the extruder screw. Here again, homogenization can be enhanced by attention to screw geometry or by the use of static or dynamic mixers. Such systems are also known in melt-spun fiber production. The risk of inhomogeneity, however, is greater when using liquid dispersions, the viscosity of which is often greatly different from that of the polymer melt being colored.

VII. POLYMER- AND APPLICATION-SPECIFIC CONSIDERATIONS

Whereas Sec. V covers general considerations, this section deals with specific properties, requirements, and test methods for dispersions for some important polymers and end uses.

A. Polyvinyl Chloride

1. Dispersions for Flexible PVC

Pigment plasticizer pastes are used for the coloration of flexible polyvinyl chloride (PVC) processed by spread coating, plastisol molding, and calendering (particularly in Europe). Their use in extrusion of cable insulation by the skin-coating technique has also been reported in recent years.

For calendering operations, pigment paste incorporation takes place in the compounding step where the paste is charged into the internal mixer or fluid mixer or continuously metered into an extrusion compounding unit. For the coloration of plastisols for textile coating or rotational (slush) molding, it is mixed in planetary mixers. There is, however, also a move to equip coating lines with the facility for continuous metering and mixing of pigment paste at the coating head. For continuous metering, a lower paste viscosity than that used for batchwise processing is desirable.

The main use of *solid concentrates* in flexible PVC is for injection molding of footwear and industrial goods and for extrusion of hosepipe, profiles, and particularly for coloration of cable insulation material. For the latter, very well dispersed pigment is essential to ensure satisfactory insulation properties. A limited number of custom colors, matched to the nationally or internationally standardized shades of the electrical industry, are sufficient to fulfill the color requirements of this application. Raw material selection here must take careful account of possible negative influences on electrical resistivity properties.

In the United States and Japan the calendering industry uses primarily *solid monoconcentrates*. This approach is now slowly replacing the use of pigment pastes for calendering in Europe and other markets because they afford better reproducibility of shade (no pigment sedimentation), cleaner handling, and enhanced scope for the introduction of automated feed systems.

2. Specific Requirements and Test Methods for Flexible PVC

Pigments used for calendering, whether added as a paste or as a solid concentrate, must be fully dispersed before reaching the calender. They should not exhibit what is known as a "calendering effect." This is an uncontrollable increase in color strength and shade variation due to residual undispersed pigment being dispersed by the shear forces encountered. Excellent performance in this respect is one of the reasons for the success of high quality solid monoconcentrates in the calendering industry.

With upholstery, luggage, and clothing, it is important that color does not rub off onto adjacent goods during use. *Rub* (or *crocking*) *fastness* is determined by the choice of binder and pigments and the degree of dispersion. Testing is carried out using a "crockmeter" by means of which a piece of cotton cloth drawn over a wooden peg is moved under a defined load forwards and backwards, for example, for 30 strokes, over the coated fabric. Any staining is assessed using the ISO Grey Scale for staining.

Fogging is the formation on the inside of automotive windshields and windows of a milky deposit arising from condensation of evaporated or sublimated volatile components from the interior trim of the vehicle. It reduces visibility, and the industry is investing considerable resources in an attempt to eliminate this safety risk. PVC compounds, which contain a large number of potentially volatile additives, and textile fabrics are under particular scrutiny.

There is still much discussion on the validity of the test methods used to evaluate fogging. In one approach (DIN 75201), a test specimen is placed into a glass beaker covered by a cooled glass plate. The beaker is then kept for 3 h in a bath at 100°C, thereby allowing any volatile material to deposit on the cooled glass plate. The effect of fogging is measured by the 60° gloss reflectance of the glass surface in comparison to an untreated plate.

The importance of maintaining the *volume resistivity* of electrical wire insulation compounds has been mentioned. This property not only depends on the polymer used but is also influenced by all other ingredients present in the insulation compound. Rigid PVC, for instance, has a volume resistivity of the order of 10^{16} ohm · cm, which is reduced to a value of 10^{14} ohm · cm for flexible PVC with a plasticizer content of 50%. This corresponds to a reduction of the resistivity by a factor of 100! Fillers and pigments also have a detrimental effect on volume resistivity, and careful selection of these components and plasticizers is essential in formulating dispersions for cable insulation compounds.

Volume resistivity is measured on a thin disk cut out of a foil of the compound and placed between two electrodes. The difference between the resistivity of the basic compound with and without addition of the dispersion is a measure of its influence relative to the uncolored material. Tolerances for resistivity loss are not standardized with respect to concentrates, each producer adopting an approach that satisfies major customer requirements for the finished cable.

3. Dispersions for Rigid PVC

The dry blend technique, developed for PVC, makes use of the high speed fluid mixer. PVC and all other solid additives are poured into the mixer, blended at slow speed, and then dispersed at high speed. As the temperature rises, liquid additives and inorganic pigments may be added. The peripheral velocity of the impellers reaches 40 m/s, and the temperature of the contents reaches 120°C. The compound thus produced is then transferred into an adjacent cooling mixer prior to further processing. Dry blends usually contain the compound ingredients at their final concentration and are therefore not concentrates per se.

Attempts to use this technique to prepare organic pigment concentrates tend to result in pigment aggregation under the conditions of operation of the mixer. Proprietary organic pigment concentrates are available that overcome this problem and yield reproducible, speck-free coloration.

Such monoconcentrates containing pigments very finely dispersed in a PVC/PVA copolymer carrier were introduced in the 1950s by pigment producers and are now also produced by a small number of specialized dispersion houses. The fineness of dispersion allows colored transparent rigid PVC packaging films of very low gauge to be produced free of pinholes or, in stretched film, rupturing of the film during production. Such preparations are also used for other PVC applications where highest reproducibility of shade is essential such as colored profiles for window frames and films for furniture and metal lamination.

Alternative carriers have been introduced more recently. In one development, low melting point polyester compositions are used as a basis for free-flowing granular concentrates that are metered directly into rigid pipe and profile extrusion equipment. Rigid PVC compounds themselves are also used as carriers for concentrates designed for rigid vinyl siding extrusion. These contain primarily mixed metal oxide pigments, which allow the high pigment loadings typical of such compounds to be achieved without the risk of aggregation during production that characterizes many organic pigments.

B. Polyolefins

1. Polyolefin Film and Extrusion Coating

Solid custom color concentrates based on pigments of all classes are widely used for the coloration of polyolefin films and for coatings applied to paper and other supporting substrates. Most of these materials are used for packaging purposes, but agriculture and the construction industry are also very important outlets for heavy gauge polyethylene films.

The first requirement for polyolefin film masterbatches is excellent pigment dispersion, as large aggregates may lead to collapse of the film bubble in the film blowing process (film gauges may be as low as 0.01 mm) and to rupture under load in use. High color strength ensures sufficient intensity of the coloration of thin film. Both opacity and transparency can be specified according to end use.

Heat resistance requirements are moderate for films made with low density polyethylene, stability at 200°C being sufficient. For films produced with polyolefins of higher melting range, such as linear low density polyethylene, high density polyethylene, or polypropylene, the pigments may be required to be stable at 240–260°C. Melt coating employs even higher melt temperatures, and heat resistance of up to 300°C may be specified.

It is important to avoid the use of incompatible additives in formulating masterbatches for film applications involving heat-sealing (where two films are welded together by the application of heat and pressure) or printing operations. Exudation of such materials can impair the printing or heat-sealing properties of the film surface. The same precaution is valid for the choice of pigments that may go into solution at processing temperatures and subsequently recrystallize on the film surface.

Demands on lightfastness and weather resistance of colored polyolefin films are generally modest for most packaging applications but are frequently quite high for outdoor agricultural and construction applications.

2. Dispersions for Polyolefin Injection Molding

Although in the past many processors used powder pigments directly or as in-house produced compounds for injection molded articles, the use of dispersions has become predominant for reasons of convenience, dispersion quality, industrial hygiene, and inventory control. Custom color masterbatches are by far the most important dispersion form.

The dispersion quality required of masterbatches for injection molding applications is only moderate, and a sophisticated dispersion process is often superfluous, especially when low cost pigments are used. More important for injection molded goods is uniformity of coloration. Correctly formulated solid concentrates allow typical addition rates of between 1 and 3 parts to 100 parts of uncolored polymer to be used.

Liquid color dispersions also are offered for polyolefins, usually as part of a package, which may include suitable metering pumps and control systems. Provided the processing equipment is suitable or appropriately adapted, they are said to give uniform results without significant influence on polymer properties and permit flexibility of production with the possibility of rapid color change. Market penetration, however, has not met the initial expectations of manufacturers.

Shrinkage of polymers on rapid cooling in the mold is inherent to the process and is due mainly to the rearrangement of the polymer chains into a partially crystalline matrix [11]. Shrinkage values, typically 1.5–4%, depend on molding conditions, polymer type, rate of cooling, and polymer flow pattern in the mold. This shrinkage is taken into consideration in the design of the molds. Most organic pigments interfere with the nucleation of HDPE during the cooling and recrystallization process, causing a modified crystalline structure. As a result, the shrinkage values for the pigmented polymer can differ considerably from the values of the uncolored material for which the mold was originally designed [12].

For large injection molded HDPE articles of complex shape (e.g., bottle crates) or for goods of relatively large surface with a central injection gate (e.g., lids or boxes), these different shrinkage values cause considerable internal stress forces within the molded articles. These forces can lead to distortion, warping, and even cracking of the molded part. Inorganic pigments and a limited number of organic pigments exhibit neutral behavior regarding shrinkage of HDPE. Therefore, only such pigments should be selected when formulating color concentrates for use in this process. Shrinkage-related phenomena are observed to a lesser extent with LLDPE injection molding grades. LDPE, with its mainly amorphous structure and low crystallinity, is not generally affected. Polypropylene, although of considerably higher crystallinity than HDPE, shows a very much lower crystallization rate, so that internal stress forces are not "frozen in" during the cooling process to the same extent. Nonetheless, nucleation influences of some pigments and additives have been known to lead to difficulty in maintaining dimensional and mechanical property tolerances even in this polymer.

Note that the extrusion process allows much longer polymer cooling times and is in general not significantly influenced by small differences in shrinkage rates caused by pigmentation.

To assess the influence of color concentrates on shrinkage, test specimens are injection molded under controlled and reproducible conditions. Shrinkage values are measured and compared after tempering in a hot water bath or oven. Both positive and negative shrinkage values relative to uncolored polymer are obtained. The tolerance limits for variation from uncolored polymer values are established empirically. Because of the complexity of the phenomenon, confirmation of suitability of the dispersion under actual production conditions is strongly recommended, especially where organic pigments are involved.

Durable articles such as bottle crates (where service life of up to 10 years is commonly requested) or high quality garden furniture can be colored only with selected high performance pigments of proven weather resistance. The use of optimized antioxidant and light stabilizer formulations [UV absorbers and radical scavengers, e.g., hindered amine light stabilizers (HALS)] is essential to obtain good results. Such additives can be included in the masterbatch, can be added as separate dispersions, or may be present in the polymer grade chosen.

3. Dispersions for Blow Molding Applications

Here also the use of custom color masterbatches is firmly established. Requirements regarding dispersion quality are more severe than for injection molding, especially for thin-walled articles (bottles for cosmetics), and good homogeneity is also important. High opacity is usually wanted, although transparent effects are sometimes specified.

Processing temperatures are lower than those used for injection molding (240–280°C), and there are no warping problems in this technology. The choice of suitable pigments is thus considerably wider. The demand for high lightfastness and weather resistance is limited to containers for outdoor use such as drums for chemical storage and recreational equipment. For foodstuff packaging, toys, and household goods, the masterbatches selected must conform to any relevant legislation concerning composition and nonextractability of the raw materials used.

4. Dispersions for Extruded Materials

The most important extruded polyolefin products, with the exception of film, are pipes for natural gas and drinking water and for electric cable insulation and sheathing. Solid concentrates are the preferred product form for these applications. Excellent pigment dispersion is essential for cables to ensure good insulation properties as well as for high pressure pipes to avoid weaknesses that might lead to bursting. For most other applications, dispersion is less critical. Concentrate formulation should, however, take into account the possible effects of carrier resins and other additives on the mechanical properties and life expectancy of the colored end product.

Concentrates for polyolefin telecommunication cable insulation must ensure the maintenance of good dielectric properties. Organic nonpolar pigments are in general suitable, whereas some inorganic pigments and organic metal toners may, because of their polarity, impair dielectric properties of such cables.

In addition to the excellent degree of dispersion required for long-life gas and water pipes (and for which lifetime guarantees of up to 50 years are asked), heat resistance (280°C), high weather resistance, and, for water pipes, nontoxicity and absence of migration are important elements of product specifications.

C. Styrenic Polymers and Copolymers and Engineering Plastics

Polymers of this group are characterized by second-order (glass) transition temperatures higher than the usual ambient temperature in service. This feature allows the use of special polymer-soluble organic dyestuffs that, in contrast to their behavior in olefinic polymers, do not bloom to the polymer surface at ambient temperatures. Color concentrates can therefore be based on either organic or inorganic pigments and on suitable organic dyestuffs selected mainly from the azo and anthraquinone classes. Metal complex azo dyes are also used, particularly for polyamide plastics.

The largest group of polymers here is composed of polystyrene (PS), high impact polystyrene (HIPS), styrene acrylonitrile (SAN), and acrylonitrile butadiene styrene terpolymer (ABS). As they are used extensively for food packaging and household articles, dispersions used for their coloration must conform to any relevant national legislation regulating product selection and extractability by food simulants. Additionally, ABS is an important polymer for automobile interior components where excellent lightfastness and freedom from migration into adjacent polymeric components is essential. Dispersions are here mainly pigment-based because polymer-soluble dyes, although they do not bloom, do tend to migrate into flexible PVC (widely used in interior trim) and with few exceptions do not have sufficient lightfastness.

Engineering thermoplastics have gained importance in recent years as weight-saving alternatives to metals in structural and load-bearing applications and, in the case of polyethylene terephthalate (PET), to glass and aluminum for beverage packaging. These polymers are processed variously at temperatures around 220°C for polyacetal (POM), 240–280°C for polybutylene terephthalate (PBT), 260–330°C for polyamides and PET, and from 280 to over 350°C in the case of polycarbonate. Because their utility is based on outstanding and closely specified physical and mechanical properties, the conservation of these features is of prime importance when designing dispersions for use in their coloration. Particular care is advisable in the selection of carrier resins, where the wrong choice may, for example, result in stress cracking, depolymerization, or delamination (phase separation). For food packaging, appropriately regulated colorants suitable for PET are available.

For engineering applications, however, color concentrates tend to be less widely used than in more straightforward materials. The reasons are the complexity of performance specifications and the number of additives, fillers, and reinforcing fibers that must frequently be incorporated homogeneously during the processing of such materials. Key ingredients are incorporated and dispersed at the final concentration desired during an extrusion compounding process. Nevertheless, there is a trend toward the use of monoconcentrates in such compounding operations to reduce contamination from dust and to facilitate meterability in automatic feeding systems.

D. Dispersions for Melt-Spun Fibers

1. Polypropylene and Polyolefin Fibers

Almost all polyolefin fibers are made from polypropylene, a small proportion being composed of polyethylene or blends of the two. The latter are used where the abrasive nature of polypropylene leads to excessive wear on yarn processing equipment, particularly knitting machinery. The carpet industry is the major user of polyolefin fibers in the form of staple or continuous-filament face yarns for weaving or tufting, staple fiber for nonwoven carpets, and backing yarns and fabrics produced from slit or split fibers. Other major applications, though of minor importance for coloration, are geotextiles and disposable nonwoven fabrics for personal care and hospital use. On the other hand, color is important in two interesting new uses—in pattern threads for automotive seating fabrics and in fine denier yarns for sports garments. It can be assumed that well over half the polyolefin fiber produced is colored by means of pigment dispersions and that by far the greater part will find use in the carpet industry.

Practically all fiber spinning is carried out using pelletized custom colors for reasons of uniformity. The fiber producer will often opt for in-plant compounding of his own recipes using purchased monoconcentrates but has also the option of buying ready-formulated custom colors from the dispersion manufacturer. Application is generally by automated metering devices that blend concentrate and polymer at the extruder hopper.

The pigments chosen are primarily organic, with medium to high heatfastness and lightfastness and high color strength. Most important for carpet fibers are azocondensation yellows, scarlets, and reds; isoindoline and isoindolinone yellows; and phthalocyanine blues and green. These are supplemented by perylene and quinacridone reds, quinacridone magenta and violet, and dioxazine violet. Less demanding end uses make use of organic azo reds and rubines of the 2B and 4B toner types (metal salt monoazo pigments). Once widely used, the highly economical diarylide yellow organic pigments (diazos) are being eliminated by many users because of insufficient thermal stability in polyolefin melts [13]. Selected grades of carbon black, generally produced by the so-called furnace process, are chosen for their intensity of blackness (jetness), dispersibility, and tinting strength and used both for black yarns and as shading components in custom color formulation. Titanium dioxide is the primary white pigment used, and again many types suitable for fibers are available. Other inorganic pigments of significance in olefin fibers are lead chromate yellows, which are very economical and, when appropriately treated types are chosen, very heat-, light-, and weather-resistant. Environmental considerations, however, are increasingly restricting their use. Finally, iron oxide red pigments provide a very economical basis for less saturated earth and brown tones.

Originally based on polyethylene wax–based carrier resins let down with polyethylene or preferably fiber-grade polypropylene, concentrates for polypropylene fibers based on polypropylene waxes were developed during the 1980s. These are claimed to provide better compatibility and frictional yarn properties than those achieved using polyethylene wax–based concentrates [14]. Carrier resin composition and rheological behavior also have a determinant influence on spinning properties and shade uniformity of the fiber and must be fine-tuned to facilitate rapid homogenization with the fiber polymer during the spinning process.

Fiber spinning concentrates are controlled to very tight coloristic tolerances as no correction of off-shade material is possible. They are also controlled microscopically for pigment dispersion and for their filter-blocking behavior (see below).

2. Polyester and Polyamide Fibers

Dispersions for polyester and polyamide fibers have until recent years been almost exclusively based on carbon black and used for low cost, high volume production of black polyester yarns for uniforms and geotextiles and polyamide yarns for umbrella fabrics. Such dispersions have been produced either in ethylene glycol for addition during the polymerization step in polyester fiber production or as solid pelletized concentrates. The reasons outlined in Sec. II.B have, however, recently created demand for spun-dyed fibers in a variety of colors. These are targeted first at applications in automotive interiors, where carpeting and upholstery fabrics in particular must be colored to increasingly demanding lightfastness specifications. Second, the carpet industry generally is looking for the ability to produce large quantities of fiber of uniform and repeatable shade, which, especially in the case of polyamide, is not easy to achieve. Third, the cost impact of responding to increasing environmental pressures on conventional textile dyeing processes is making the spin-dying processes increasingly attractive.

Dispersions for these colored fibers are in most cases pelletized custom colors that may be preblended with uncolored polymer pellets, metered and blended into the polymer on the extruder, or applied using melt interjection techniques. A variety of *carrier resins* are used. For polyester fibers, the fiber-grade homopolymer itself is widely used as the principal carrier. This may be combined with polyester copolymers, which serve variously to enhance wetting of pigments during dispersion or to control melt viscosity of the concentrate. For polyamide fibers, PA6 is frequently used, either alone or combined with polyamide or olefinic waxes or polyester-polyamide copolymers. The use of polyethylene concentrates for both polyamide and polyester fiber spinning is also known, notwithstanding potential compatibility-related spinning problems.

The use of liquid dispersions for injection into polyester and polyamide spinning melts via metering pumps is also known, especially in Japan. Proprietary carrier systems [15] are claimed to overcome earlier compatibility-related spinning difficulties. The colorant selection for such dispersions is similar to that for solid concentrates.

For polyester fiber dispersions, anthraquinonoid pigments and special polymer-soluble dyes make up the most important colorant group, supplemented by perinone red dyes, azocondensation reds, dichloroquinacridone magentas, and phthalocyanine blue pigments. Partial solubility of all organic pigments in polyester melts, especially when used at very low concentrations (typically below 0.2–0.3%), limits their usefulness. Not only does shade control become difficult, but lightfastness is often drastically reduced. Colorant selection is critically important for automotive use and is influenced by the widely differing fastness requirements and test methods specified by the various manufacturers [16]. Carbon blacks for use in concentrates must similarly be selected with great care to ensure optimum color, viscosity, stability, and dispersibility characteristics.

The available choice of colorants for melt-colored polyamide fibers is even more restricted than for polyester. The chemically aggressive nature of the polymer melt, which acts as a strong reducing agent, destroys most of the organic pigments and dyes that find use in polyolefin and polyester fibers. Most concentrates are based on perylene red, anthraquinone red, dichloroquinacridone magenta, and phthalocyanine blue pigments and on the yellow and red 2 : 1 metal complex azo dyes, which also find use in polyamide molding resins [17]. Selected metal oxide pigments are also used but tend to yield dull shades and abrasive fibers that accelerate wear of fiber processing machinery.

Quality control requirements are similar to those outlined above for polyolefin fibers.

3. Specific Performance Requirements for Melt-Spinning Fiber Concentrates

The presence of contaminants such as polymer gels, pigment aggregates, and inorganic salts in fiber spinning melts can lead to spinneret blockage, filament breakage during spinning and drawing, and impairment of yarn properties such as tensile strength, elongation, and modulus. Filtration of the melt is thus essential to ensure productivity and good fiber quality. Filter packs located between the extruder screw and spin pump contain wire mesh filters, generally of 325 mesh (44 μm square orifice). High productivity, however, also requires that filter pack life be as long as possible to minimize downtime when filters have to be changed. As spinning is carried out at constant throughput, the increase of back pressure in the extruder is an indication of the condition of the filter pack.

Pressure rise testing is carried out to characterize the pigment dispersion quality and filterability of color concentrates used for mass-colored fibers and is also applied in one form or another to dispersions for all melt-spun synthetic fibers [18]. For melt-spun fibers, for example, a mixture of color concentrate with uncolored polymer is passed at a constant rate through an appropriate filter pack installed in a laboratory melt-filtration extruder. Traditionally, 325 mesh screens have been specified, but more recent demands for finer filaments and higher spinning speeds have led to the specification of screens with finer mesh sizes, even below 20 μm. Pressure rise is defined and specified in terms of the increase in back pressure as a function of the quantity of pigment or pigment dispersion passing through the filter pack at a defined rate and temperature.

The achievement of fiber-grade quality pigment dispersions therefore requires the use of an efficient melt-filtration system during concentrate production. This postdispersion refining step is necessary even for an already well-dispersed pigment. The vast majority of pigment particles must be submicrometer size after dispersion, with only a few isolated aggregates over 1 μm being acceptable [19].

It must be stated, however, that the relation between pressure rise and pigment crystal geometry, particle size, and size distribution is a complex one, and low pressure rise is no guarantee that oversize particles are not present. Microscopic examination of the concentrate is advisable for this purpose.

E. Dispersions for Wet- and Dry-Spun Fibers

Wet spinning describes the process whereby a solution of the fiber-forming polymer is pumped through spinnerets into a nonsolvent where it coagulates and is then drawn into filaments. Regenerated cellulose (viscose) is the most important commercial fiber produced by this technique. Mass coloration is carried out where particularly high lightfastness and migration fastness is required and employs dispersions of organic pigments in an aqueous medium. The dispersion is facilitated and stabilized using anionic and nonionic surfactants. Pigments selected must be stable to the highly alkaline conditions encountered in the cellulose xanthate spinning solution (dope). They are added batchwise to the dope.

Dry spinning methods start similarly from a solution of the fiber-forming polymer in an appropriate solvent. On emerging from the spinnerets, however, the dope enters a column containing circulating hot air. Evaporation of the solvent yields filaments that are subsequently drawn to enhance their physical properties. Solvent is recovered. Solution dying is also practiced in the case of dry-spun fibers, the most important of which are cellulose acetate and polyacrylonitrile (acrylic) fibers. Cellulose acetate is spun from acetone solution and may be colored either with solvent-soluble dyestuffs or by means of pigment dispersions in which cellulose acetate is used as the carrier resin. Acrylic fibers possess outstanding resistance to weather and find use in outdoor applications such as tenting and awning fabrics. Colored fabrics with a correspondingly high level of weather resistance can only be achieved by the use of a small selection of high performance organic pigments. These are incorporated as liquid dispersions using polyacrylonitrile as carrier in dimethyl formamide or dimethyl acetamide solution and processed on enclosed circulating dispersion equipment such as perl mills or attritors.

Analogous requirements concerning filterability of the colored spinning solution and fineness of pigment dispersion apply to dry-spun and wet-spun fibers just as to melt-spun fibers.

REFERENCES

1. A. Reynolds, Appl. Market Information Masterbatch Conf., Basle, 1990.
2. M.J. Smith, *J. Oil Colour Chem. Assoc. 57*:36 (1974).
3. H. Thommen, *Plastverarbeiter 43*:37 (1992).
4. A. Limper and R. Stehr, *Kunststoffe 80*(1):26 (1990).
5. M.H. Mack, 25th Regional Tech. Conf. Color and Appearance Div. (CAD-RETEC) Soc. Plastics Eng. (SPE), New Orleans, 1991.
6. W. Thiele, W. Petrozelli, C. Martin, and D. Lorenc, 25th CAD-RETEC, SPE, New Orleans, 1991.
7. Anon., *Plastverarbeiter 43*(6):46 (1992).
8. W.J. Hart (to Hercules Inc.), U.S. Patent 3,755,224 (1973).
9. G.F. Jones and V.H. Ure (to Hercules Inc.), U.S. Patent 3,360,497 (1967).
10. M. Beraud and G. Häuptli, Conf. Deut. Ind. Technol., Würzburg, 1992.
11. J.M.G. Cowie, *Polymers: Chemistry and Physics of Modern Materials*, 2nd ed., Blackie, London, 1991.
12. G. Menges, R. Sarholz, and E. Krüger, *Plastverarbeiter 32*(1):59 (1981).
13. Information Notice No. 2, ETAD, Basle, 1990.
14. H.-J. Sohn, Plastics Inst. Conf. *Polypropylene Fibres and Textiles III*, York, England, 1983.
15. Dainippon Ink and Chemicals Inc., Jpn. Patent 59 075 926 (1984).
16. H. Nobs, 21st CAD-RETEC, SPE, Atlanta, Ga., 1987.
17. R.C. Sykes, 20th CAD-RETEC, SPE, Toronto, 1986; *Plast. Compounding*, March/April 1987.
18. G.W. Shore, 23rd CAD-RETEC, SPE, Huron, Ohio, 1989.
19. D. Bennett, *Int. Fiber J.* (*USA*), Aug. 1989, p. 12.

7
Paper Coatings

DEREK J. PRIEST Department of Paper Science, UMIST, Manchester, England

I. Nature of Paper Coatings 245
II. Basic Types of Paper Coating 246
III. The Paper-Coating Operation 247
IV. Composition of Coating Mixes 250
A. Pigments 251
B. Binders/adhesives 257
C. Coating and coating mix property modifiers 260
V. Coating Mix Preparation 265
A. Typical formulations 265
B. Procedures for dispersing pigment 265
VI. Structure and Properties of Pigment Coatings 266
VII. Concluding Remarks 268
Bibliography 268

I. NATURE OF PAPER COATINGS

The surfaces of paper can be treated in many ways and for many different purposes. It is important to realize at the outset that the surface of paper is complex, heterogeneous, more or less rough, and highly porous and can vary considerably from one type of paper to another. It follows that, in general, when coatings are applied to paper, there is inevitably some penetration into the sheet, a phenomenon not encountered with substrates that are hard and impenetrable. On the other hand, advantage can be taken of the nature of paper surfaces to obtain products with specific properties. For example, by controlling absorbency and surface structure in conjunction with the properties of the coating dispersion, starches can be applied either as surface treatments (to prevent disruption of surface fibers) or virtually as impregnants (to improve bulk properties of the sheet, such as stiffness). As is common practice, the term "coating" will be applied in this chapter to denote surface treatments, but the distinctive nature of the surface of the paper must not be forgotten.

With regard to supplementary sources of information, it is best in the context of this chapter to provide a general bibliography. This will readily give readers who require it the key to obtaining more detailed information from the extensive specialized literature that exists on all aspects of paper coating. The bibliography can be found at the end of the chapter.

II. BASIC TYPES OF PAPER COATING

It is possible to distinguish two distinct types of paper coating. A major category is the coated paper used primarily to obtain multicolor printed material of high quality; this normally involves the application of an aqueous dispersion of inorganic pigments such as a mixture of clay (kaolin) and chalk (calcium carbonate). Although they differ widely in detail, depending, for example, on the intended printing process, these pigment-coated papers are broadly similar in appearance. They are made in large tonnages on fast machinery and are relatively inexpensive. Their key feature is the special type of surface imparted by the pigment coating, which is able to give printed images of high resolution and clarity with high quality full color reproduction. Uncoated paper has surface structural elements that are fibers approximately 20–40 μm wide, whereas the pigment particles in the coating are typically 2 μm in their major dimension; this decrease in the size of surface structural elements is the basic reason for the enhanced printing performance of coated paper. In addition, the coating can enhance brightness and opacity and allows the production of a high gloss finish.

The other main type of coating can be described as "functional," since it serves to give the coated paper a particular function or purpose. A few examples from the large number of applications of these papers are self-adhesive labels, thermal printing paper, silicone release paper, gummed tapes, photographic papers, and paper for heat transfer printing of textiles. These coated papers tend to be comparatively expensive, are made in smaller tonnages than pigment-coated papers on slower, narrower specialized machinery, and vary enormously in composition and structure.

Within the compass of this chapter, it is only possible to deal with the generally important pigment-coated papers, since they have much in common and represent the mainstream of coating technology.

III. THE PAPER-COATING OPERATION

In outline, pigment-coated papers are produced by applying to the base paper an aqueous dispersion of pigment at a relatively high concentration (e.g., 60% w/w), usually referred to as the solids content. This dispersion is known as the coating mix or, most often in the United States, as the coating color (although they are usually colorless). In most methods of application, an excess of coating mix is applied and the superfluous mix is metered off in some way to leave the desired uniform layer of wet coating on the paper, which then passes through a suitable drying system. If a high gloss finish is required, the dried coated paper will be subjected to some kind of calendering process.

Coatings may be applied as an integral on-line part of the papermaking process, "on-machine coating," or on a separate off-line coating machine, "off-machine coating." The amount of dry coating applied, or coat weight, is expressed in terms of mass per unit area (g/m^2). This, of course, is some kind of average, depending on the area involved in the measurement; the variability that may be quoted does not reflect the microscopic variations in coat weight on a scale comparable to the roughness profile of the paper. The average thickness of the coating for a given coat weight will depend on the density of the pigment and the void volume (see below for some comments on the structure of coatings). A useful rule of thumb is that for a perfectly uniform homogeneous coating of density 1,000 kg/m^3, the coat weight in grams per square meter is numerically equal to the thickness in micrometers.

The scanning electron micrographs in Fig. 1 illustrate the change in surface characteristics brought about by coating. In this example, the coating was applied on an air-knife coater (see below) to a portion of the same making of base paper whose surface is shown in Fig. 1a. Some of the coated paper (shown in Fig. 1b) was then supercalendered, giving the surface illustrated in Fig. 1c. Thus the three micrographs give a genuine comparison, showing well the fine-grain smooth surface produced by applying the pigment particles.

High gloss coated papers for magazines are typically produced on off-machine coaters. The commonest means of metering the correct coating weight uses a flexible metal blade virtually in contact with the paper surface; the angle of the blade and the applied pressure can be precisely controlled across the web of base paper being coated. These machines are now-

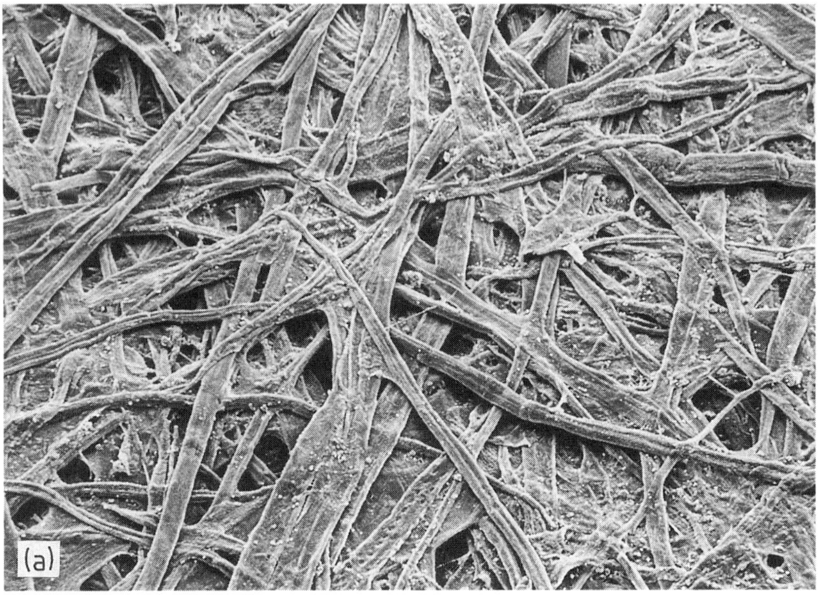

FIG. 1 Scanning electron micrographs showing the effect of applying a coating of fine pigment particles followed by supercalendering (all at the same magnification). (a) Uncoated base paper (made from a mixture of hardwood and softwood pulps). (b) The same paper coated using a pilot plant air-knife coater. (c) Some of the coated paper supercalendered on a pilot plant calender.

adays very large and fast; paper some 8 m wide can be coated at speeds of around 1200 m/min.

Coat weights applied depend on the grade of paper being made and its intended application. In recent years, manufacture of grades known as lightweight coated (LWC) has expanded greatly. A typical LWC paper has a total weight of around 60–65 g/m^2 with 4–7 g/m^2 coating on each side. Lightweight coated papers are used for mass circulation magazines and similar applications where the light weight is important in reducing the bulk and minimizing transport costs. Base paper for LWC grades normally contains a high proportion of mechanical pulp (in which wood is broken down into separate fibers principally by mechanical action), and so the grades are relatively inexpensive; they also give a high yield in terms of area per unit weight. Heavier weight glossy coated papers are also made on off-machine blade coaters; at the upper end of the market these consist of base paper made entirely from fully bleached pulps produced by chemical

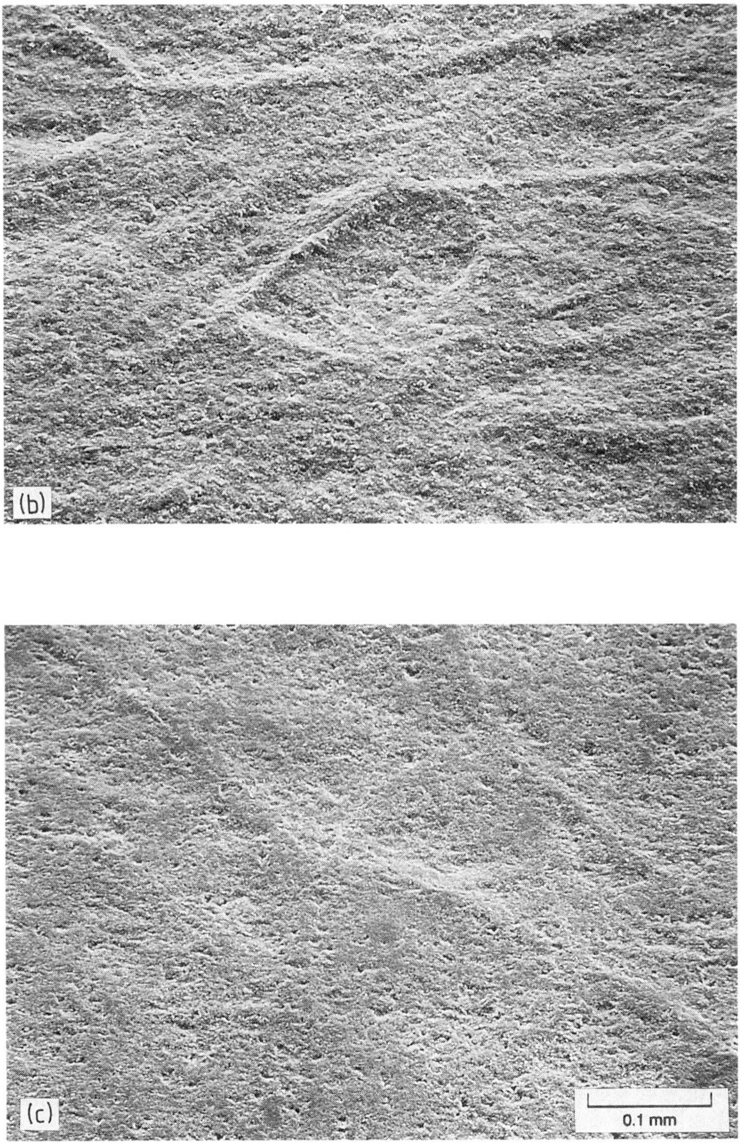
(b)
(c)
0.1 mm

breakdown of the wood (so-called wood-free grades—actually a contraction of "ground-wood-free") and have a pigment coating on each side of some 12–15 g/m^2.

For making the specialized grades of coated paper used for very high quality printing, off-machine coaters may be equipped with systems for applying two or more coats successively to each side using, perhaps, an air-knife metering system at the first coating station. In this device excess coating is removed by impinging onto the paper a narrow, well-controlled stream of air emerging from a slot located near the paper surface. Although it is something of an oversimplification, air-knife coaters can be thought of as applying a uniform thickness of coating following the microcontours of the paper surface. Application of a second coating by blade metering, which tends to fill in depressions in the surface, then gives a very smooth level coating. Air-knife coaters generally operate with coating mixes of lower solids content than for blade coating, since extra fluidity is required for efficient blow-off of excess coating. A key advantage of running any coater at high solids content is that it reduces the quantity of water to be removed in the drying stage.

A further, major use of pigment coating is in the manufacture of multilayer cardboards for packaging. The outer surfaces of cartons are often coated to obtain the high quality of print required. This is usually done by on-machine coating and may involve two or three separate coating heads of different types applying coatings successively to the same side of the board. By adjusting the composition and quantity of the coatings at each head, the cost effectiveness of the process can be optimized to produce the properties required. In total, some 20–25 g/m^2 of dry coating can be applied.

Paper machines, as opposed to board machines, are less likely to be equipped with on-machine coaters, but they are sometimes used, often adapted from size presses, to apply low coat weights as precoats or for special products such as carbonless copy papers.

For the high output glossy coated papers described above, where high speeds are involved, it is usually more cost-effective to employ off-machine coaters since running conditions of the paper machine and the coater can then be separately optimized. In the context of this book, it is not necessary to go into detail about the sophisticated application and control techniques now being employed, and the reader should consult appropriate sources from the bibliography at the end of the chapter.

IV. COMPOSITION OF COATING MIXES

Coating mixes can contain a bewildering variety and large number of different components, some having more than one function. However, constituents generally fall into four categories, to be discussed in turn below. These are pigments, adhesives or binders, coating mix property modifiers, and other miscellaneous additives.

A. Pigments

Although many substances are potentially useful as pigments, in modern coated paper production the two most common are clay and calcium carbonate.

1. Clay

Clay is a naturally occurring aluminosilicate mineral, one of a family of such minerals in which the constituent atoms are arranged in crystals that have in principle an unlimited extent in two dimensions but are very thin. The parent mineral of the clay used in coatings is known as kaolinite, and its crystalline layer structure is shown diagramatically in Fig. 2. It can be seen that a two-dimensional tetrahedral silicon–oxygen structure is combined with an octahedral aluminum–oxygen or –hydroxyl layer through shared oxygen atoms, the structure as a whole being electrically neutral. Moreover, the layers can be bonded face to face through extensive hydrogen bonding between the oxygen atoms on one surface and the hydroxyl

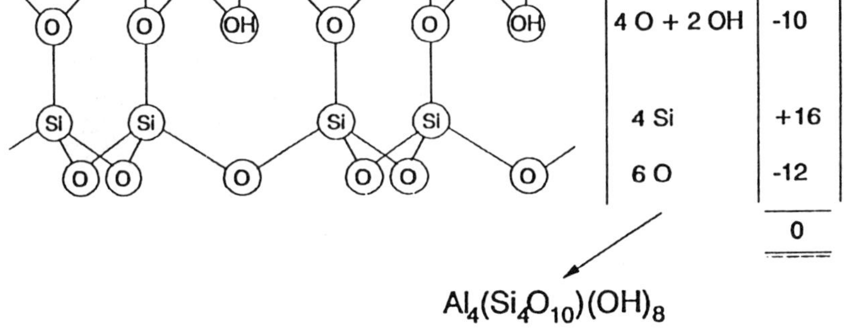

FIG. 2 Diagramatic representation of part of the laminar crystalline structure of kaolin. Layers can connect by means of hydrogen bonds. The overall charge is neutral, as shown at right.

groups on the other. The material extracted for paper coating is usually known as kaolin; it may not be entirely pure kaolinite, since there will be very small quantities of other materials present such as quartz or other metal atoms.

The crystalline structure is reflected in the macroscopic morphology of the kaolin, which typically exists as small flat platelets having a roughly hexagonal outline. In the material as prepared for use in coatings, particles are around 2 μm in width and about 0.2 μm in thickness. Clay occurs as either primary or secondary deposits. Primary clay deposits are those that have remained in the locations where the clay was originally formed. The formation process is complex and requires special conditions to have occurred over geologic time, especially to produce the white clays suitable for use in papermaking. In essence, the aluminum silicate structures have formed by chemical attack on the alkali content of feldspar, a silicate found in granite (a mixture of quartz, feldspar, and mica). The largest of such primary clay deposits are to be found in Cornwall and Devon in the southwest of England; smaller deposits are found in Europe, in parts of the United States, and in the former USSR.

Secondary deposits are those found at a distance from the location in which they were formed, having been transported by rivers to the sites where they have been deposited. The largest of these are found in the United States, principally in Georgia. In the primary deposits, as in Cornwall, which is a major source of coating grade clays, the clay is washed out of the rock by high pressure water jets and then pumped to separation plants where a complex series of separation and blending processes eventually produce a coating clay that has strictly controlled constitution and properties.

The key properties of a typical coating clay are shown in Table 1.

2. Calcium Carbonate

The second principal pigment in common use today is calcium carbonate. It is used in two forms, either a pigment obtained by refining material occurring naturally, principally as marble, chalk, or limestone deposits, or as precipitated calcium carbonate, which is the result of a chemical process usually involving the treatment of calcium hydroxide with carbon dioxide. The two forms, natural or precipitated carbonates, are distinctive in many ways and hence find broadly different kinds of applications; they are discussed briefly in turn.

There are extensive deposits of chalk and other forms of calcium carbonate, especially in Europe. The material can be recovered and treated either in the dry or wet state, dry refining being probably more common. It may be ground in sand mills to a controlled fine particle size and size distribu-

TABLE 1 Key Properties of Clay and Carbonate Coating Pigments

Property (typical values)	High grade coating clay	Calcium carbonate (from marble)
Particle size[a] data		
% by mass less than 2 μm	At least 94	At least 90
% by mass more than 10 μm	At most 0.2	At most 1.5
% by mass more than 53 μm (residue on BSS 300 mesh)	At most 0.02	At most 0.01
Optical properties		
Brightness[b] (ISO)	87.5 $\pm$ 0.7	93.0 $\pm$ 1.0
Yellowness[c]	4.2 $\pm$ 0.5	1.5 $\pm$ 0.5

[a]Particle size is in terms of "equivalent spherical diameter," the value that is given by a sedimentation method of determining particle size in which the validity of Stokes' law is assumed.
[b]Brightness is the ratio (expressed as a percentage) of light reflected from a pigment tablet to that reflected from a perfect diffuse reflector, measured at a wavelength of 457 nm.
[c]Yellowness is the difference between two "brightnesses," one measured at 570 nm and the other at the standard wavelength of 457 nm.
Source: Data from information published by ECC International.

tion; the term "ground $CaCO_3$" is used to distinguish it from precipitated $CaCO_3$. As might be expected, the individual particles of ground $CaCO_3$ are "blocky" in nature, unlike the platelets typical of clay. This affects the way in which the particles pack together at the surface of coated paper and hence the optical reflective properties of the paper, an important aspect that is discussed in more detail below. Ground $CaCO_3$ particles are somewhat more oleophilic and hydrophobic than clay, which has several consequences for coating mix properties (again, see below), but also, importantly, leads to an enhanced ability to absorb liquid organic vehicles used in printing inks. Ground $CaCO_3$ pigments can also contribute enhanced brightness and opacity in coatings. Since ground types of calcium carbonate are, particularly in Europe, less expensive than clay, manufacturers of high output commodity grades of coated paper seek to optimize the properties they require by blending appropriate proportions of clay and ground $CaCO_3$. For comparison with clay, Table 1 includes properties of a typical coating grade of ground $CaCO_3$.

Careful control of the chemical reaction resulting in the precipitation of calcium carbonate enables the production of pure material with very fine particles with a well-defined shape and size distribution. Particles usually consist of multicrystalline aggregates, which enhance light scattering and hence opacity and brightness in coated papers. In ground carbonate, problems are sometimes encountered from naturally occurring impurities that can lead, for example, to coating mixes being excessively abrasive. Such problems are avoided in the chemically prepared precipitated carbonates. Clearly, precipitated $CaCO_3$ is going to be more expensive than ground carbonate and again, particularly in Europe, its use is generally confined to applications where its special properties prove to be cost-effective. However, elsewhere in the world, notably in the United States, there is no large indigenous supply of inexpensive ground $CaCO_3$ and the use of precipitated grades has greater prominence. Given the right circumstances, costs can be reduced by operating a precipitated $CaCO_3$ plant on a papermill site, making use of cleaned power plant flue gases as a source of CO_2 and slaked lime [$Ca(OH)_2$], which can be produced by calcining grades of chalk or limestone unsuitable for direct use as pigment.

3. Titanium Dioxide

In general, this high refractive index, high whiteness, and expensive pigment is used only to a limited extent in high tonnage pigment-coated papers. This fact is linked with a very important feature of pigment coatings on paper, which is appropriately introduced at this point. These coatings are always more or less porous and contain a large number of small voids. The scattering of light that is so important to their function as opacifiers and brighteners takes place at the interfaces between these voids and the pigments (and other components). Indeed, it is now generally accepted that the light-scattering elements are in fact the voids rather than individual pigment particles, so their number, shape, and size is of great importance [see Lepoutre (1985)]. This structure is associated with the fact that in paper coatings only a relatively small proportion of binder or adhesive is used, by comparison with paint films, where, effectively, pigment particles are distributed within a coherent polymer or resin film. In paints, opacity is given by scattering at interfaces between pigment particles and polymer or resin, so a high refractive index is a great advantage and titanium dioxide finds wide application. On the other hand, the refractive index of clays and carbonates is high enough to give adequate scattering at air–pigment interfaces in paper coatings. Consequently, titanium dioxide is not widely used unless its high cost can be justified because an extremely high degree of brightness or whiteness is required; it may be added in moderate proportions to the basic clay/carbonate coating—a practice more common in the United States than in Europe. The essential difference between a paint layer and a pigment coating on paper is illustrated in Fig. 3.

The effectiveness of the titanium dioxide can sometimes be enhanced and costs reduced by adding supplementary pigments such as special silicates or

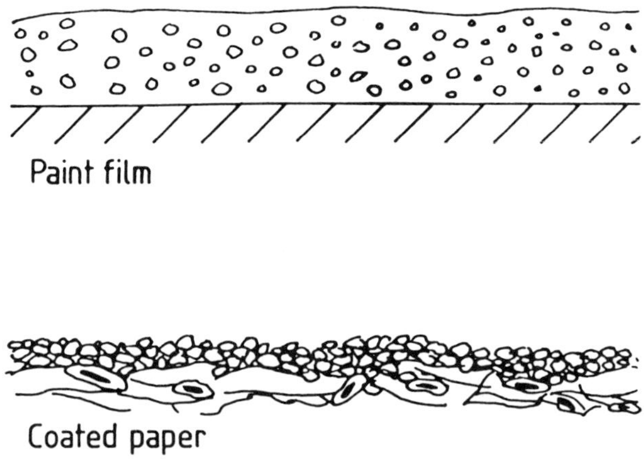

FIG. 3 Diagrams illustrating the basic difference between a paint film on a nonabsorbing substrate and a pigment coating on paper.

calcined clay, which can help to maintain the system at its optimum in terms of particle size and distribution. These materials are known as extenders.

However, it should be noted in passing that the situation can be quite different where functional coatings are concerned. An example is an opaque wax-coated paper used, say, to provide a moisture barrier. Here the polymeric wax coating needs to be coherent and continuous, and if opacity is needed it can be attained only by incorporating in the coating film pigment particles of high refractive index, such as titanium dioxide.

4. Other Inorganic Pigments

A number of materials other than clay, chalk, and titanium dioxide have been used at various times in pigment coatings either generally or for special purposes. The most notable are as follows:

(a) Calcined Clay. When kaolin is subjected to high temperatures under carefully controlled conditions, structural and chemical changes take place, yielding pigments with special properties. Partially calcined clay in which structural hydroxyl groups are removed is made at 650–700°C (to fully calcine clay, a temperature of some 1000°C is required). Brightness and opacity are enhanced, but the viscosity of useful dispersions is higher than for standard clays (see below), so their usefulness lies as minor components of coating mixes. As already mentioned, they can be used as extenders for titanium dioxide.

(b) Aluminosilicates Related to Kaolinites. Montmorillonite is a double-layer clay in which both surfaces of the layer crystal are composed of

oxygen atoms, which cannot therefore form interlayer hydrogen bonds. Consequently, these clays swell on immersion in water, severely limiting their use at high solids content because when swollen they exhibit high viscosity or even gel-like properties. However, montmorillonites often have defect sites (e.g., Mg replacing an Al atom) that give parts of the lattice an excess negative charge, usually balanced by the presence of labile ions such as Na^+. These can be replaced by acidic protons, and the acid clays so formed find use as proton donors when incorporated in kaolin-based coatings designed to interact with color-forming dyestuffs in carbonless copy paper. Such coatings are often applied at low coat weight by on-machine coaters when high solids content mixes are not required.

(c) Amorphous Silicas and Silicates. These consist of a range of artificially prepared silicas and silicates of various compositions and complexity. They can sometimes be useful adjuncts in conventional coatings—to control ink absorption, for example. They also find uses for special purposes such as matting agents in transparent varnish coats or to control surface friction in paper.

(d) Barium Sulfate (Barytes). This highly white and very pure pigment is the preferred material for coating photographic base paper of the type that then supports a gelatine-based sensitized layer. The high purity is important to avoid interference with the chemistry of photographic image formation. These papers have now been largely superseded by resin-coated types of paper, extrusion coated with various polyethylene-based formulations.

(e) Satin White. This, another high white pigment, is now mostly of historical interest. It was often made on-site in mills by mixing solutions of aluminum sulfate (papermakers' alum) and calcium hydroxide. Its composition tended to be variable, and its properties changed with time due to complex aqueous hydrolysis reactions. One possible application is in cast coating, a special type of pigment-coated printing paper or boxboard in which the wet coating is dried while in contact with a large, slowly rotating, highly glazed cylinder; the high gloss surface produced on the dry coated paper is thus essentially a replica of the glazed surface of the cylinder.

5. Plastic Pigments

A relatively new development in coating paper for print enhancement is the introduction of pigments based on hard synthetic polymers. Technology exists, based on emulsion polymerization, for preparing an aqueous suspension of polymer particles having a substantially uniform predetermined size, that is, a very narrow particle size distribution, the particles being usually spherical.

The tiny spheres (typically 0.5 μm in diameter) may even be made in such a way that in the dry coating they are hollow; the internal void then contributes to the light-scattering ability of the coating. In any case, correctly formed and uniformly sized spheres are bound to introduce suitably sized light-scattering voids into a coating. Weight for weight, the cost of synthetic hydrocarbon polymer is likely to be substantially higher than that of naturally occurring mineral pigments, although on a volume basis (i.e., giving equivalent coating thickness) the difference in cost is less because of the relatively low density of polymer (e.g., polystyrene, 1050 kg/m^3; clay, 2600 kg/m^3). However, these new materials have yet to make significant inroads into the market for conventional pigments.

B. Binders/Adhesives

The major functions of the materials known interchangeably as binders or adhesives (the term "adhesive" is used here) are to provide bonding between the pigment particles in the coating and to ensure that the internally bonded coating is in turn secured to the base paper surface. The type and strength of adhesion is dictated by the kind of printing process for which the paper is intended because the stresses imposed on the surface are different in different processes. A particularly demanding one is that known as heat-set web offset lithographic printing (HSWO) because viscous tacky inks are transferred to the surface of the paper under pressure from a rubber-covered offset roll (the blanket). High stresses in a direction normal to the plane of the paper (usually known as the z direction) are generated as contact is broken between the moving web of paper and the blanket roll; the adhesion in the coating layer must be sufficient to prevent disruption of the surface. This requirement is expressed in terms of a pick strength, typically measured in a printing simulator in which the normal stress is steadily increased by accelerating the speed of a segment representing the blanket. The pick strength is then recorded as the linear speed at which the surface begins to break up. In addition, in lithography the surface of the paper may also have been dampened at a previous printing station, so the adhesion needs to resist the effect of water, often requiring the use of wet-strengthening agents (see below). Moreover, the ink may be dried or "set" on the sheet (some chemical reaction may be involved) by passing the printed paper web at high speed through a source of high temperature; live gas jets may be employed. As the paper passes through these zones, moisture in the base paper is flashed off as steam, and the coating layer must be sufficiently porous to allow this to take place without formation of blisters or less obvious minor disturbances of the surface sometimes known as fiber puff. In simple terms, the adhesive must not form so continuous a film as to impede this transport of water vapor. Other printing processes make less stringent demands on coating adhesion. In gravure printing, for example, the inks are free-flowing and not tacky, nor is the paper surface dampened. Coated papers for gravure printing can generally tolerate a lower proportion of adhesive.

It is common practice to express quantities of adhesive as parts by weight dry (pbw) based on 100 parts by weight of dry pigment. Although there are wide variations, and detailed formulations are not generally published, an HSWO paper might require 15–20 pbw of adhesive while a gravure paper may need less than 10.

In addition to the major functions, the presence of adhesive brings about a number of other effects that show themselves (1) in the properties of coating mixes (their behavior as they are prepared, transported, and applied and their interaction with the paper surface) and (2) in the structure and properties of the dry coating itself. These topics are discussed further later.

Adhesives can be divided into two distinct types. First, there are materials such as modified starches, which give translucent viscous dispersions in water. These are often termed "solutions," and that term is used here, but although such systems are often considered to be homogeneous, on a submicrometer scale there are likely to be inhomogeneities as revealed by the translucency. The second type of adhesive is genuinely heterogeneous, being some form of synthetic polymer latex prepared by the emulsion polymerization technique.

1. Solution-Type Adhesives

The most widely used type of adhesive is some form of starch derivative designed to give the required viscosity and adhesive properties at appropriate concentrations and to be stable in properties in solution at room temperature. Unmodified starches are unsuitable because viscosity is too high in solutions of concentration high enough to be useful, and also because, on cooling, "retrogradation" takes place and the solutions gel due to partial lateral ordering of the unbranched amylose fraction of the starch. Modification can take many forms, including oxidation using sodium hypochlorite, controlled enzymatic degradation, and preparation of derivatives such as starch ethers.

Adhesives derived from proteins such as casein or soy protein can also be used. These now tend to find specialist applications such as in cast coating or coated paperboard. Soluble derivatives of cellulose can also be used in coating formulations, notably sodium carboxymethyl cellulose (CMC). However, the function of CMC is primarily in areas other than bonding (see below), although it does make some contribution to adhesive strength in the coating.

Synthetic solution-type adhesives are also used, exemplified by polyvinyl alcohol. This polymer is made by hydrolysis of polyvinyl acetate. Its solubility in water is related to the degree of hydrolysis; commercial polyvinyl alcohols are supplied either completely hydrolyzed or at a degree of hydrolysis of 88–90%. The completely hydrolyzed product is soluble only in hot water because, rather like starch on cooling, extensive ordered hydrogen bonding can take place between molecules, leading to gelation. Polyvinyl alcohol is a very powerful adhesive, and hence less is required. However, its higher price still makes it generally less cost-effective than starch. Also, great care is needed in using it in high solids mixes because its high surface activity can destabilize the dispersant system used to give manageable rheology, leading to virtual solidification.

2. Polymer Latex Adhesives

Polymers commonly employed in latex form are styrene-butadiene copolymers, polyvinyl acetate, and copolymers of various acrylic monomers. An important aspect is the nature of the adhesive film formed as water evaporates from the wet coating and the latex particles coalesce. By, for example, adjusting the ratio of styrene to butadiene, it is possible to obtain desired degrees of softness or hardness in the polymer. Adhesives that are too soft produce a limpness in coated paper, whereas if they are too hard the adhesive properties are impaired. The key parameter here is the glass transition temperature (T_g), which, for films from latex adhesive, is typically around room temperature. Latexes are normally supplied at about 50% (w/w) solids content, and these dispersions have low viscosity. It is thus possible to add required proportions of adhesive without much reducing the solids content of the final coating mix. By comparison, starch adhesives are made up at not more than 25% solids when they have high viscosity. Also starch adhesive solutions need to be prepared by solubilizing modified starch granules at temperatures around 90–95 °C, but latex adhesives can be metered directly into the coating mix. However, latex adhesives on their own may not give the required degree of water retention and rheological properties of the mix, matters discussed below.

A useful modification to the composition of the latex adhesives is to copolymerize suitable monomers which yields, for example, "carboxylated styrene-butadiene" resins. This gives several advantages.

1. The presence of anionic carboxyl groups on the surfaces of the latex particles contributes to their colloidal stability. This means that the surfactant content in the aqueous continuous phase can be reduced. Such surfactants can give adverse properties in coatings, altering the wettability of fountain solutions in lithographic printing, for example.
2. The increased emulsion stability makes the latex more readily compatible with solution adhesives such as starch, which by competitive adsorption might otherwise destabilize the latex emulsion.

3. Adhesive strength is improved by the polarity imparted to the latex particles by the carboxyl groups—that is, adhesion between particles, between particles and pigment, and between particles and the cellulose substrate. Also, water resistance in the coating is easier to obtain through interactions between the insolubilizer and the carboxylated surfaces.
4. The polar latex particle surfaces also help to retain moisture, reducing loss of water into the substrate, particularly during blade coating—a point enlarged on below.

C. Coating and Coating Mix Property Modifiers

As coating technology has developed and increasing demands have been made in terms of coating application methods and the properties of coatings, the use of a variety of supplementary materials has become more and more important. Key areas are the flow properties of coating mixes, the interaction of the mix with the base paper during the brief time elapsing between contact of mix with paper and entry into the dryer, and special properties required in the dried coating.

Before discussing some of the materials used, it is appropriate at this juncture to deal briefly with the important matter of coating mix rheology.

1. Importance of Rheology in Coating Mixes

The flow behavior of high solids content pigment dispersions in water is clearly complex, and non-Newtonian flow characteristics are commonplace; the viscosity is highly dependent on the shear rate. In addition, time-dependent features, such as thixotropy, also play an important role. A primary consideration is the preparation of coating mixes at solids contents in excess of 60% (w/w) that have sufficiently low viscosities at intermediate rates of shear to allow them to be pumped and handled as liquids. At the same time, these mixes must behave well at the very high shear rates encountered in blade coating.

Since, in general, all particles dispersed in water acquire a surface electric charge, strong interparticle interactions develop that must be modified and controlled to achieve the required flow properties. The main causes of these electric charges are the ionization of chemical groups at the surface, isomorphous substitution leading to charged groups in the crystal lattice (balanced by labile ions of opposite charge), or adsorption of cationic or anionic surfactants. In the particular case of kaolin platelets, with their underlying crystalline laminar structure, the edges carry a positive charge at low pH, while the surfaces are negatively charged. This is due to the very different nature of the surfaces and edges; at an edge, the crystal structure has to be broken, giving the possibility of a dissociation equilibrium such as

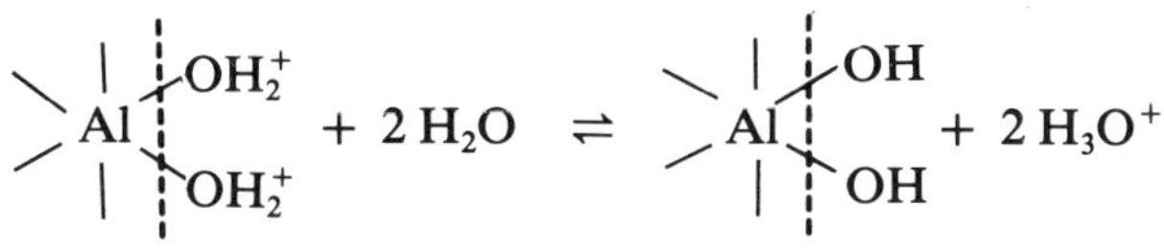

Clearly, the position of this equilibrium will be dependent on pH. However, on the platelet surface—due, for example, to isomorphous substitution of Mg for Al—the charge is usually negative and is independent of pH. Consequently, under mildly acid conditions, which usually apply when clay is dispersed in water, there can be electrostatic attraction between the edges and surfaces of platelets. This leads to formation of an extended "house of cards" structure throughout the dispersion, giving high viscosity and non-Newtonian flow behavior.

Thus for the preparation of stable high solids pigment dispersions with usable flow properties, it is vital to add an appropriate surface-active dispersing agent, which will ensure, over a sufficiently long period of time, repulsive interparticle interactions.

Also of great importance is the effect on rheological properties of the adhesive components of the coating mix. As already noted, the effect on rheology of solution-type adhesives is likely to be substantially different from that of polymer latexes. A key element in all this is the behavior at the very high rates of shear prevailing under the blade tip of a high speed coater. If excessive shear thinning takes place it becomes very difficult to apply a uniform coating of sufficient thickness; a certain amount of "body" needs to be retained in the mix under the high shear conditions. On the other hand, at high solids contents, particulate dispersions can be prone to dilatancy under high shear since particle packing can be rearranged in such a way that there is no longer adequate space to accommodate all the fluid. Very high viscosity or even solid-like behavior can result. If, on application of the coating mix, there should be rapid and excessive loss of aqueous phase into the base paper, the wet coating layer on the paper may exhibit dilatancy under the blade. This results at best in a very uneven coating, and often in scratch lines if there are small defects on the blade.

To achieve the desired rheological behavior of the coating mix, it is often necessary to use flow modifiers (which may also provide a degree of water retention) to control the rate of absorption of water by the base paper when coating is applied. The mechanism by which flow modification, or water retention, is obtained is often difficult to elucidate because there can be complex surface chemical interactions between the various constituents of a mix. Although there may be general guidelines to follow, the choice of additive and the amount to be used is generally based on an empirical approach.

There follows a brief account of the main types of additives, including some miscellaneous types not directly associated with flow properties or water retention.

2. Dispersants

Two types of dispersants are commonly used, polymeric and inorganic. Typical of polymer dispersants is sodium polyacrylate, which in solution in water dissociates to give a loosely coiled polyelectrolyte carrying a high density of negative charge. In simple terms, in the case of clay, the negatively charged polymer chains adsorb onto the positively charged platelet edges; the excess negative charges on the polymer then outnumber the positive charges, so the platelets acquire a net negative charge and are thus deflocculated. In addition, segments of the polymer extending into the aqueous phase provide a degree of steric hindrance that helps to prevent particles from adhering to one another. Inorganic dispersants are typically polyphosphates that in solution yield negatively charged ions with the general formula $P_nO_{3n+1}^{(n+2)-}$; when $n = 2$, one obtains the pyrophosphate anion,

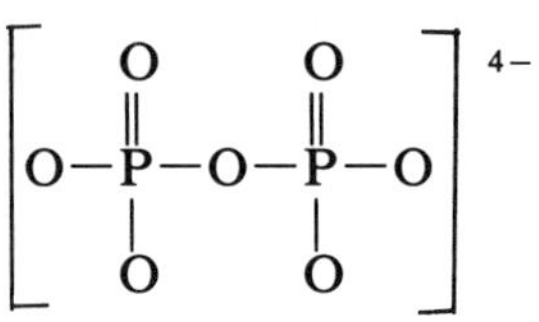

Clearly, these ions will also adsorb strongly at positively charged sites to give an overall negatively charged particle.

Due to complex hydrolysis reactions, among other things, polyphosphates tend to give a less long lasting dispersive effect than the polymers. In practice, blends of the two types of materials are found to be most cost-effective. Only small proportions are required, typically 0.3 part per 100 parts of pigment by weight, dry.

3. Flow Modifiers/Water Retention Agents

The material that best typifies this class of additive is the highly hydrophilic, water-soluble cellulose derivative sodium carboxymethyl cellulose (CMC). This can be obtained with different values of both DP (degree of polymerization) and DS (degree of substitution). A relatively small DS is sufficient to confer solubility, while the DP controls the viscosity of the aqueous solutions. Thus if only a change in viscosity is required, it may be most cost-effective to use a small quantity of a CMC having a high DP. However, in high solids pigment dispersions, the observed effects on flow properties are strongly influenced by adsorption of CMC molecules onto pigment surfaces, giving a greater increase in viscosity than might be expected from the viscosity of the CMC alone. Also, to achieve a reasonable degree of water retention it is necessary to provide a sufficient number of sites for potential hydrogen bonding and thus a sufficient quantity of CMC. Typically, about 1–2 parts CMC are employed having a DS of about 0.2.

An ingenious alternative approach to controlling flow and water retention is the use of what are known as alkali-thickenable latexes. These exploit the dissociation properties of polyelectrolytes such as sodium polyacrylate (see Chapter 8); in practice a specially designed copolymer is used. Under acid conditions, when the sodium salt is virtually undissociated, latexes of these polymers are stable and have the usual low viscosity at their commercially available solids content of about 50%. They can be added directly to coating mixes in the usual way. If the pH is then made alkaline, dissociation takes place, producing water-soluble polyanions. The latex particles swell, are highly hydrated, and dissolve, giving the required viscous coating mix with water-retaining properties. By designing suitable copolymer systems, it is possible in this way to produce so-called sole binder latexes, all the required properties being obtained by using just one adhesive component. Although these materials are used to some extent, especially in coated paper for gravure printing, where technical demands on the coating are not as high as for offset lithographic printing, coating mix formulators often prefer the flexibility afforded by separate addition of adhesive, flow control, and water retention agents.

The nature of the adhesive employed will also influence the need for flow control and water retention. In particular, if starch is used as a significant proportion of the adhesive constituent, it can provide a major degree of the required viscosity and water retention.

4. Insolubilizers

As explained above (Sec. IV.B), pigment coatings for offset lithographic printing require a resistance to wetting. The cohesion and strength of the coating layer, especially under stress in the *z* direction (perpendicular to the plane of the paper), must remain at a satisfactory level after the surface has been dampened by the aqueous fountain solution. As with water retention or flow control, the nature of the main adhesive components will affect insolubilizer requirements. If the main adhesive is a polymer latex, it will tend to be less hydrophilic than, for instance, starch. Carboxylation of the latex can improve bonding and water resistance in dry coating. However, some insolubilizer will often be required, especially where starch is used. The materials often used are basically urea formaldehyde (UF) or melamine formaldehyde (MF) resins added as prepolymer syrups to the coating mix.

Polymerization and cross-linking to give water resistance are completed during the drying stage of the coating operation. In view of current emphasis on health and safety aspects of industrial processes, intensive development work is taking place to produce cross-linking resin insolubilizers that do not require the use of formaldehyde. A completely different approach to insolubilization is the use of complexing multivalent cations such as zirconium (Zr^{4+}) added in the form of ammonium zirconium carbonate. Since this method clearly requires no formaldehyde, it may become more widely used.

5. Miscellaneous Additives

A number of other constituents are often present in coating mixes either to modify the properties of the coated paper or to assist in some way in the process. Optical properties may be influenced by the use of small proportions of dyestuffs or colored pigments to adjust the hue or shade of the paper; fluorescent whitening agents (FWAs) may be incorporated to enhance reflectivity when the paper is illuminated by light sources containing UV radiation. The interactions between FWAs and other components in the mix need to be considered carefully. The presence of TiO_2 can quench the fluorescent effect, while on the other hand it can be promoted by UF or MF resin.

Foam formation and air entrapment can be a problem in handling and pumping coating mixes. Dispersed finely divided air bubbles in high solids white opaque dispersions are difficult to detect; a useful check is to measure the specific gravity of the mix and compare the result with that calculated from the formulation. Mixes containing small bubbles give rise to flawed surfaces on the coated paper. It is therefore usually necessary to incorporate agents that act either to inhibit foam and air entrainment (antifoams) or to destabilize and break down foam already formed (defoamers). Care is needed in the use of these materials; for example, excess silicone-based additives can readily interfere with the even wetting of applicator rolls by the coating mix, thereby leading to unacceptable coated paper surface quality.

Small proportions of lubricants, for example, calcium stearate or other soaps, may be added. These are said to improve the processibility of coating mixes through applicator systems and metering blades, etc., and also to ensure a satisfactory coating surface. However, the mode of operation of these lubricants has not been clearly established.

As a final example, coating mixes containing natural colloid materials such as starch are subject to biological deterioration through the action of bacteria and fungi, and it may be necessary to add chemical preservatives.

V. COATING MIX PREPARATION

A. Typical Formulations

Because of their importance to the overall cost efficiency of a coating operation, formulations in actual use tend to be commercially confidential. In general terms, following what has already been said about lithographic and gravure printing, formulations for lithography would use 10–15 parts by weight (pbw) of dry binder per 100 pbw of pigment, while for gravure less binder would be required, say 6–10 pbw. In lithographic coating, about 2–4 pbw of insolubilizer would be used. Where a flow modifier is used, say CMC, as noted above, about 1–2 pbw would be added. A typical amount of dispersant is 0.3 pbw. Other additives, such as dyes, optical brighteners, and lubricants, are generally added in lower proportions, and this would clearly be a matter of adjustment to suit the application requirements and the production process being operated.

B. Procedures for Dispersing Pigment

The general principle of mix preparation is first to disperse the pigment in the presence of a solution of dispersant and then to blend in the adhesives and other components. The key operation is the effective dispersion of the pigments. For large operations, it is economic to purchase the pigments in bulk in the form of a high solids, predispersed, and deflocculated suspension in water. In such a case the agitators required would only be those of sufficient capacity, and suitably designed, to maintain homogeneity in the suspension and to give good mixing of the supplementary materials.

However, where pigment is purchased in powder form, special dispersing equipment is needed that is capable of providing a zone of high shear through which the entire contents of a mix is circulated. The objective is to provide interparticle friction in order to break down aggregates insofar as possible into distinct individual particles. This is best achieved at relatively high solids content. The operation can be monitored by some form of electric energy recorder, since there needs to be a high energy input and efficient energy transfer to the high shear zone until dispersion is complete and the energy input decreases. The dispersion can then be pumped into blending tanks equipped with less powerful agitators, where addition of the adhesives, etc., and also dilution, if required, can take place. At one time, when operations were generally on a small scale, Z-blade mixers were used with clay mixtures having a doughlike consistency, but most coating preparation plants (usually known in the paper industry as coating kitchens) use mixing tanks with special high power agitators and baffles. If starch is a component of the mix, equipment will be needed to prepare its colloidal solution. This normally entails preparing a slurry of the starch at room temperature, heating it to about 90°C, and holding it at this temperature for a short period (10–20 min). As heating takes place, the suspension (at first of low viscosity) passes through a relatively short range of temperature (the gelation point) at which the viscosity sharply increases to a maximum as the starch granules swell and burst. Variable-speed agitation needs to be sufficiently powerful to deal with the viscosity changes to produce a uniformly dispersed starch. Alternatively, many coating mills make use of continuous in-line starch converters, which, supplied with a slurry of raw starch, produce the required solution of colloidal starch either by thermal conversion or by the use of enzymes.

VI. STRUCTURE AND PROPERTIES OF PIGMENT COATINGS

Although at first sight it may seem an unusual concept, the dried coating on or near the surface of the paper represents a special kind of dispersion. The coating layer consists of a porous assembly of fine solid particles tacked together by adhesive and containing many small voids, all three phases having important functions as has been discussed. The properties primarily required are the following.

A smooth and uniform outer surface capable, if required, of giving high gloss on calendering

Even and controlled wetting by, and absorption of, printing inks to give a uniform and intense glossy ink film without lateral spreading of halftone dots

Even acceptance of wetting by aqueous fountain solutions in lithographic printing without disruption of the coating layer

In heat-set lithography, sufficient permeability to allow escape of water vapor from the interior of the paper

A degree of compressibility to ensure good even contact with ink applicator blankets or printing press cylinders

There is no doubt that the detailed structure of the coating layer is complex, because the systems used for applying and drying the coating can have complicated effects. A useful review has been given by Lepoutre (1989) (see Bibliography).

One of the key factors is the redistribution of components, which takes place at the moment homogeneous coating mix first contacts the base paper. The redistribution occurs because the base paper preferentially absorbs the aqueous suspension medium. It follows that the properties of the base paper, especially roughness and absorbency qualities, play a key role in determining the quality and uniformity of the finished coated paper. Roughness, it should be noted, is affected by the wetting of the hydrophilic fiber network, so, for example, if the base paper has received excessive calendering to impose an apparently smooth surface, the wet coating will tend to reverse the effect of the calendering. What this means is that the underlying structure of the base paper is the important feature, and this needs to be as even as possible. The absorbent qualities of the base are controlled, especially in paper made from chemical pulp, by the internal and surface sizing applied on the paper-making machine. Mechanical pulps used, for example, to make base papers for lightweight coated papers tend to be naturally less hydrophilic than others because of the presence of lignified fibers; such base papers are unlikely to be sized. Formulation of the coating mix therefore needs to take into account the nature of the base paper in order to give the required coating holdout (resistance to penetration) and adhesion. Although not an aspect of the basic science, a very important practical concern is to ensure that the properties of the base paper remain the same from reel to reel so that the optimum running conditions can be maintained on the coater.

The high shear regime under the blade in a modern blade coater, together with the complex flow patterns and pressure gradients, further influences the distribution of components in the surface. Then, as the surface is dried, further rearrangement can take place under the influence of temperature gradients through the coating layer. In fact, Lepoutre lists no fewer than five different processes by which dewatering can take place:

1. Pressure filtration under hydrodynamic pressure at the applicator or under the blade
2. Capillary forces, which can provide a driving force for movement of the aqueous phase remaining in the fibrous network of the base paper after external water has been removed and menisci formed
3. Diffusion of water into the hydrophilic fiber walls
4. Spreading of water over fiber surfaces, leaving solid particles behind
5. Loss of water by evaporation in the dryer

Properties of key interest are the smoothness and gloss of the finished coated paper surface. Various factors have been found to affect these properties. At one time, considering that high gloss is most easily achieved with clay coating, the main feature was thought to be the alignment of the flat platelet particles parallel to the plane of the paper to achieve a very smooth surface and high gloss. However, it is now realized that although this is a contributory factor, other matters are also important. The particle size distribution must be such that at the surface, particles fit together well to give complete and uniform coverage. Also of great importance are the dimensional changes that can take place in the coating layer as it is applied and dried. Excessive shrinkage of the surface layer will tend to cause fine scale deformation and lack of smoothness; this is another reason for coating at high solids content, since there is then less likelihood of excessive surface deformation.

Gloss is normally produced by supercalendering, a process in which the paper is passed through a nip at high pressure between a hard roll and a (relatively) soft deformable roll, which gives a microfriction polishing effect. The nature of the coating components affects the efficiency of supercalendering, and this can be related to various factors including the quantity and thermoplastic deformability of the adhesive used. In this respect synthetic polymeric adhesives are likely to give better gloss than the harder films produced by converted starches in coating. It is clearly likely to be more difficult to produce high gloss in calendering a pigment with "blocky" particles like calcium carbonate, but high degrees of smoothness and some gloss are possible if the particle size and size distribution are correctly controlled and surface deformation on drying is minimized.

VII. CONCLUDING REMARKS

Paper-coating technology has many facets, some beyond the scope of this chapter. This broad overview has aimed at giving some insight into the complexities, especially the influence of dispersion technology on achievement of coated paper with desired properties.

ACKNOWLEDGMENT

Thanks are due to Bryan Clarke and Eileen Rodgers for preparing the scanning electron micrographs.

BIBLIOGRAPHY

Readers wishing to obtain more detailed information on the coating of paper and board should find that the following selection of sources will readily give them access to the required literature.

Coating in General

Booth, G.L., and N. Millman (Eds.) (1965). *Pigmented Coating Processes*, TAPPI Press, Atlanta.

Casey, J.P. (Ed.), (1983). *Pulp and Paper: Chemistry and Chemical Technology*, 3rd ed., Vol. IV, Wiley, New York.

Landes, C.G., and L. Kroll (Eds.) (1978). *Paper Coating Additives*, TAPPI Press, Atlanta.
Lepoutre, P. (1989). *The Structure of Paper Coatings: An Update*, TAPPI Press, Atlanta.
Nadelman, A.H., and G.H. Baldauf (1966). *Coating Formulations: Principles and Practices*, Lockwood Trade Journal Co., New York.

Pigments for Coating

Garey, C.L. (Ed.) (1977). *Physical Chemistry of Pigments in Paper Coating*, TAPPI Press, Atlanta.
Hagemeyer, R.W. (Ed.) (1984). *Pigments for Paper*, TAPPI Press, Atlanta.

Polymer Latex Technology

Blackley, D.C. (1975). *Emulsion Polymerisation: Theory and Practice*, Applied Science, London.
Calvert, K.O. (1982). *Polymer Latices and Their Applications*, Applied Science, London.
Sinclair, A. R. (Ed.) (1975). *Synthetic Binders in Paper Coatings*, TAPPI Press, Atlanta, 1975.

Starches and Soluble Polymers

Davison, R.L., and M. Sittig (1962). *Water-Soluble Resins*, Reinhold, New York.
Finch, C.A., (1973). *Polyvinyl Alcohol: Properties and Applications*, Wiley, New York.
Kearney, R.L., and H.W. Maurer (Eds.) (1990). *Starch and Starch Products in Paper Coating*, TAPPI Press, Atlanta.
Whistler, R.L. (1984). *Starch: Chemistry and Technology*, 2nd ed., Academic, Orlando.

Recent Research Results

Useful starting points, in addition to the mainstream paper and board technical publications.

Abstract Bulletin of the Institute of Paper Science and Technology, published by the Institute in Atlanta, Georgia.
Proceedings of the annual TAPPI Coating Conferences, published by TAPPI Press, Atlanta.

8

Dispersions of Polymers for Special Purposes

JOHN G. LANGLEY Allied Colloids Ltd., Bradford, West Yorkshire, England

I.	Introduction	271
II.	Polymerization Processes	272
	A. Condensation polymerization	272
	B. Addition polymerization	272
III.	Aqueous Continuous (Oil-in-Water) "Emulsions"	275
	A. Water-based resins for printing inks and overlacquers	275
	B. Aqueous continuous rheology modifiers	283
	C. Aqueous continuous microcapsule dispersions	293
IV.	Organic Continuous (Water-in-Oil) Emulsions	306
	A. Polyelectrolyte flocculants	308
	B. Acrylic polymers in the textile printing industry	319
	C. Wallcovering adhesives	329
	References	330

I. INTRODUCTION

This chapter is concerned primarily with organic polymers that are manufactured or supplied as dispersions themselves and are mainly used to modify the properties of other more complex dispersions for industrial use. For convenience, the manufacture and specific characteristics of the polymeric dispersions are first described briefly and then their application in industrial practice is described in more detail.

II. POLYMERIZATION PROCESSES

Polymers can be produced by condensation polymerization or addition polymerization.

A. Condensation Polymerization

Condensation polymerization takes place when two polyfunctional monomers react together, frequently with the elimination of small molecules such as H_2O or NH_3. A well-known example of condensation polymerization is the production of nylon 66 [1]. Although condensation polymerization represents a major area of polymer chemistry it falls outside the scope of this treatise.

B. Addition Polymerization

Addition polymerization requires monomers to have active double bonds, which can be readily attacked by free radicals produced in situ [2]. Conjugated double bonds such as those in acrylic esters and amides are particularly responsive to addition polymerization reactions [3], and polymer chains can be readily built to very long chain macromolecules having molecular weights well in excess of 1 million.

Free radical addition polymerization processes consist of four distinct stages.

1. *Initiation.* Initiation is by free radical formation and its addition to the monomer.

$$R\cdot + CH_2 = \overset{\displaystyle X}{\underset{\displaystyle Y}{C}} \rightarrow RCH_2-\overset{\displaystyle X}{\underset{\displaystyle Y}{C}}\cdot$$

2. *Propagation.* Monomer units are added successively to active centers (growth step).

$$RCH_2-\overset{\displaystyle X}{\underset{\displaystyle Y}{C}}\cdot + CH_2=\overset{\displaystyle X}{\underset{\displaystyle Y}{C}} \rightarrow RCH_2-\overset{\displaystyle X}{\underset{\displaystyle Y}{C}}-CH_2-\overset{\displaystyle X}{\underset{\displaystyle Y}{C}}\cdot \rightarrow$$

$$R(CH_2-\overset{\displaystyle X}{\underset{\displaystyle Y}{C}})_{n-1}-CH_2-\overset{\displaystyle X}{\underset{\displaystyle Y}{C}}\cdot$$

This last species represents a growing radical n units in length ($M_n\cdot$).

3. *Termination.* The loss of the active center is brought about by some mechanism, that is, "dead" polymer molecules are formed. This can occur by two growing radicals combining to form a stable covalent bond (combination termination). Termination can occur by disproportionation, whereby a radical transfers from one growing chain to another, resulting in two dead polymer molecules.

$$M_n\cdot + M_m\cdot \begin{cases} \rightarrow P n + m & \text{Combination} \\ \rightarrow P_n + P_m & \text{Disproportionation} \end{cases}$$

4. *Transfer.* Transfer reactions can occur between the growing radical and another molecule, thereby transferring the activity to the molecule, which can then reinitiate monomer.

Transfer $\quad M_n\cdot + S \rightarrow P_n + S\cdot$
Reinitiation $\quad S\cdot + M \rightarrow M\cdot + S$

Addition polymerization can be carried out in various types of processes. These can be conveniently classified by the physical states of the monomer in the reaction—bulk polymerization, solution polymerization, suspension polymerization, and emulsion polymerization. Only the last two processes, in particular emulsion polymerization, are pertinent to this chapter's subject matter. However, few comments on the other two are necessary for completeness.

Bulk polymerization is carried out in the absence of any diluents. Polymethyl methacrylate is often produced by this process as a clear plastic sheet. Efficient cooling is required to dissipate the heat produced [4]. *Solution polymerization* is carried out by dissolving monomer in an appropriate solvent (water or organic solvent, depending on the solubilities of the monomer and the polymer produced) and inducing polymerization by free radical initiation. An example is the manufacture of sodium polyacrylate dipersants [5].

In *suspension polymerization* the monomer is not soluble in the continuous phase, and monomer droplets at optimum particle size are produced by gentle agitation in the presence of suitable stabilizer. Polymerization is promoted by free radical initiation, and each monomer droplet is converted to a polymeric bead. After polymerization the polymer beads can be recovered from the liquid phase by centrifugation and fluidized bed drying, and the liquid phase can be recovered and recycled.

The end product has the advantage of a narrow particle size distribution and perfectly spherical shape for ease of flow. This process is applicable to both oil-in-water and water-in-oil systems as in emulsion polymerization. Products made by an oil-in-water process include ion-exchange resins such as sulfonated polystyrene [6]. Water-in-oil examples include solid grade polyacrylamide flocculants [7].

In *emulsion polymerization*, the monomer has very limited solubility in the continuous phase. The monomer is emulsified with the aid of surfactants and/or amphipathic stabilizers, which may be assisted by high shear mixing. The polymer particle size is typically 0.1–1.0 μm in diameter, although the production of much finer microemulsions [8] is now common practice.

When the monomer is effectively water-insoluble and the continuous phase is water, the emulsion polymerization process is defined as oil in water (o/w). When the monomer is water-soluble and the continuous phase is an organic solvent, such as mineral oil, the emulsion polymerization process is defined as water in oil (w/o). In the former case, typical examples are latexes for paper coating, such as styrene-butadiene, which are used as pigment binders [9]. In the latter case, polyacrylamide flocculants can be supplied as emulsions in paraffinic mineral oils [10].

In the case of o/w emulsion polymerization, the selected monomers are mixed together in suitable proportions to form the monomer phase. This phase is mixed with an appropriate quantity of water in the presence of a suitable o/w emulsifying agent (typically an anionic surfactant) to form a monomer emulsion. The monomer emulsion is continuously fed to a reactor over a predetermined time period concurrently with an initiator solution to promote polymerization. If the selected initiators are dependent upon thermal decomposition to produce free radicals, polymerization may take place under isothermal conditions. The molecular weight of the polymer produced can be controlled by initiator concentration, temperature control, or the addition of chain transfer agents that reduce molecular weight.

This process has the advantage of producing high molecular weight polymer at a high active polymer concentration (25–50% by weight), yet maintaining the physical characteristics of an easily handled low viscosity liquid. The final product is usually a low viscosity opaque to translucent liquid with an active polymer content of around 25–40% by weight.

The incorporation of bifunctional or polyfunctional monomers into the monomer mix to promote branching and cross-linking is common practice in industry since the physical properties, in particular the rheological characteristics, can be varied to suit the specific application. The incorporation of long-chain monomers including hydrophobic and hydrophilic entities can also be used to provide interesting and beneficial rheological characteristics, particularly in aqueous systems containing high concentrations of electrolyte. These polymers, known as associative thickeners, are dealt with in more detail in Sec.III.B.

III. AQUEOUS CONTINUOUS (OIL-IN-WATER) "EMULSIONS"

This section is subdivided into three distinct subject areas, providing an up-to-date review of the latest technology. First the application of water-based resins for printing inks and overlacquers is described and compared with more traditional techniques using solvent-based systems. The second subsection deals with the use of aqueous continuous rheology modifiers for a variety of applications, with particular emphasis on associative thickeners. The third subsection discusses microencapsulation technology and its ability to provide stable aqueous continuous formulations, with very specific properties, including controlled-release characteristics.

This section is by no means exhaustive in its scope but serves to illustrate the trends away from solvent-based systems to aqueous continuous systems in line with environmental pressures.

A. Water-Based Resins for Printing Inks and Overlacquers

Over the last three decades there has been an increasing upsurge in the application of water-based inks and lacquers in the graphic arts industries. This has been stimulated by the worldwide realization that petrochemical resources must be conserved, by air pollution regulations, and by the increased pressure from occupational health and safety regulations. The expanding packaging industry has provided new outlets for water-based inks and coatings for paper and board. As the technology has improved, substrates ranging from newspapers to nonabsorbent films such as polyethylene and polypropylene are now printed using water-based technology.

Traditionally, water-based printing inks were formulated on shellac, which was solubilized into water using a suitable alkali. This type of formulation was used for inks for printing milk cartons. Shellac was subsequently replaced by maleic or fumaric resins due to the inconsistencies in the naturally occurring shellac.

During the 1960s the technology and quality of flexographic printing advanced and started to replace letterpress on corrugated boxboard and paper bag substrates. This single development was the real catalyst to the development of water-based inks. These advances together with improvements in decorative graphics and the need to increase machine speed on natural and bleached kraft prompted questions related to ink performance and formulation. Inks based on maleic and fumaric resins had been formulated to a low cost, and technical performance, particularly factors such as gloss and resistance properties, had not been considered to be important.

During the 1970s polymer emulsion technology advanced to a level that started to provide the ink manufacturer with a range of synthetic resins that were capable of being formulated into inks to provide the wide range of performance characteristics demanded by modern packaging printers. The specialized water-based dispersions of polymers formed the basis of the modern, high-tech water-based printing inks, which are themselves specialized dispersions.

1. Properties of Water-Based Inks

The basic components of a press-ready, water-based printing ink are pigment, film-forming polymer(s), liquid diluent, and additives. The term "liquid diluent" has been selected to differentiate from "solvent," which would be used in a solvent-based ink. In water-based systems the liquid phase need not be a solvent for the film-forming polymer. This is one factor that contributes substantially to the differences between solvent-based and water-based systems.

Raw material selection for use in a printing ink is determined by the performance properties required, which can be divided into two distinct groups.

(a) Liquid Ink Properties. The liquid ink has to satisfy a number of parameters. The pigment dispersion has to be optimized such that the pigment is used as cost effectively as possible in terms of pigment wetting leading to high color yield. The resultant ink concentrate must be stable to storage at ambient temperature and stable during use on the press. The dry composition has to be formulated in such a way that optimum rates of drying are achieved on the press.

(b) Dry Film Properties. Dry film properties have to be optimized to ensure that the product will perform for the intended application. If one property has to be adjusted, then the modified formulation has to be evaluated to ensure that the other properties have not been affected.

Adhesion properties are of prime importance, and various test methods have been designed to assess this property. The normal quick laboratory test is the "Scotch tape" test, which consists simply of applying a short length of Scotch tape to an area of ink and then rapidly removing it. This test gives the ink formulator a very good indication of anticipated formulation performance. A more severe adhesion test is the assessment of abrasion resistance under wet, dry, hot, and cold conditions. One of the most severe application areas is the printing of packaging films that can be used under deep freeze conditions.

Chemical resistance may be of importance, and specific tests are designed as appropriate for the end application of the product. An example of this would be the alkali resistance of inks used on soap products.

Gloss properties on permeable substrates may vary according to the substrate and the formulation. The more permeable the substrate, the less holdout and therefore the less gloss.

Odor, and health and safety aspects associated with the use of certain chemicals have become of critical importance. In the past, formulators were able to resort to chemical cross-linking agents to achieve resistance properties. Some of the most efficient cross-linking agents for example formaldehyde, exhibit undesirable handling properties that are now being discouraged by legislation.

The component that determines the performance characteristics of any ink is the resin film former. This may be a single-component system, but more usually it is a blend of resins.

2. Resins

Water-based resins divide into three distinct types: totally water-soluble polymers, aqueous alkali-soluble polymers, and aqueous dispersions of polymers.

Water-soluble polymers are defined as those that remain water-sensitive when dry (a property that has limited application in the formulation of inks). These are derived by direct polymerization of water-soluble monomers such as acrylic acid.

Aqueous alkali-soluble polymers are derived in such a way that minimum carboxylation or sulfonation is introduced into a hydrophobic resin that when neutralized with a suitable alkali, usually a volatile alkali, will solubilize into water. On drying, the alkali evaporates, leaving a water-insensitive film. Obviously, sodium hydroxide could be used as the base, but the dried film would remain water-sensitive. The selection of the base can be of critical importance and can be used to modify drying properties.

Aqueous dispersions of polymers cover a very wide range of film formers. The first products offered to the ink industry incorporated carboxylation and a complete ink could be derived from one resin. Surfactant-stabilized dispersions of this type exhibited a number of disadvantages such as poor pigment wetting, poor film formation, low gloss, and severe foaming.

As technology improved, multiresin component systems were introduced that combined aqueous alkali-soluble polymers of low molecular weight and low polydispersity as pigment wetter and dispersant vehicles together with noncarboxylated resins as let-down resins that developed the application and resultant film properties required. A further advance was made when resin-stabilized dispersions were introduced in the mid 1970s. These hybrid products moved water-based ink technology into a technical performance class that could now seriously compete with solvent-based systems. Significant improvements in the properties listed above were obtained.

The resin properties can be varied considerably by blending appropriate monomers before polymerization such that the glass transition temperature

(T_g) of the resultant polymer can be engineered. Low T_g polymers exhibit improved adhesion onto hydrophobic substrates together with good flexibility, whereas high T_g polymers exhibit heat resistance properties. I_g values of some common homopolymers are listed in Table 1. The two main extremes of this series polymethyl methacrylate, which would form a non-coalesced, powdery film at ambient temperature, and poly(2-ethyl hexyl acrylate), which would coalesce easily at ambient temperature to a film exhibiting pressure-sensitive adhesive properties. Obviously, intermediate properties can be achieved by appropriate copolymerization or by blending resins. Some of the varying requirements demanded by different application areas together with ways in which acrylic resins can be modified to satisfy those requirements are summarized in Table 2.

3. Ink Formulation

All ink formulations contain the basic ingredients of pigment, resins, and diluent. Traditionally, pigment was ball-milled into the resin solution (conventionally referred to as the varnish), using a pigment-to-varnish ratio of 15 : 85 by weight, for example. As already described, the original formulations were based on single-resin systems, which gradually advanced to multicomponent resin systems as the performance requirements became more demanding.

(a) Single-Resin Systems. To achieve optimum performance characteristics, medium molecular weight carboxylated polymer dispersions were designed that contained just sufficient carboxylation to produce alkali solubility such that the dried resin film would be water-resistant. Also, the average molecular weight and polydispersity optimization enables exceptional control of performance properties to be achieved.

For example, comparisons were made of two resins similar in composition but differing particularly in molecular weight distribution. Resin 2 contained a greater high molecular weight fraction, which gave it a higher

TABLE 1 T_g Values of Some Common Homopolymers

Polymer	T_g (°C)
Methyl methacrylate	105
Styrene	100
Acrylonitrile	105
Vinyl acetate	28
Butyl methacrylate	20
Ethyl acrylate	−24
N-Butyl acrylate	−45
2-Ethyl hexylacrylate	−60

TABLE 2 Properties of Acrylic Resins

Ink property	Resin requirement
Adhesion and flexibility	Low T_g and high emulsion-to-solution ratio
Drying rate	Faster with emulsions than with solutions, higher solids, volatile base
Resistance properties	
alkali, water	Low acid number
heat	High T_g and softening point
Heat seal	Lower T_g and softening point
Re-solubility	Higher acid number; less volatile base
High gloss	Low molecular weight and high acid number so that high solids can be achieved
Opacity	Combination of low molecular weight solution polymer as grind vehicle plus a hard resin as let-down
Flow and leveling	Increase the proportion of solution component

viscosity, particularly at increased concentration. This is graphically expressed in Fig. 1.

Figure 2 indicates the change in viscosity with respect to pH. If the alkali in the system is volatile, during a printing machine run the pH can change, resulting in a change in the viscosity of Resin 2. This could again affect printing properties. Neutralization of the dispersion to produce the varnish is carried out before the pigment is added. The mixture is subsequently passed through a suitable milling process to generate a stable dispersion. A typical formulation is as follows.

Varnish	Parts by weight
40% Solids resin dispersion	100
Water	100
Defoamer	0.2
Alkali to adjust to PH	9.0

Ink Concentrate	Parts by weight
Pigment	20
Varnish	100

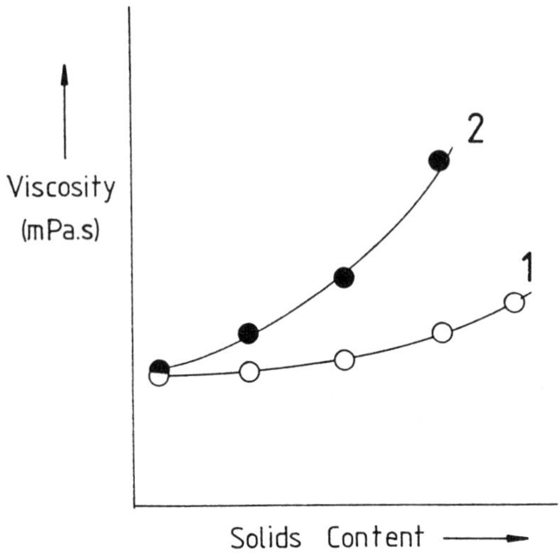

FIG. 1 Viscosity versus concentration (equivalent pH). 1, Resin 1; 2, Resin 2 with increased level of high molecular weight fraction.

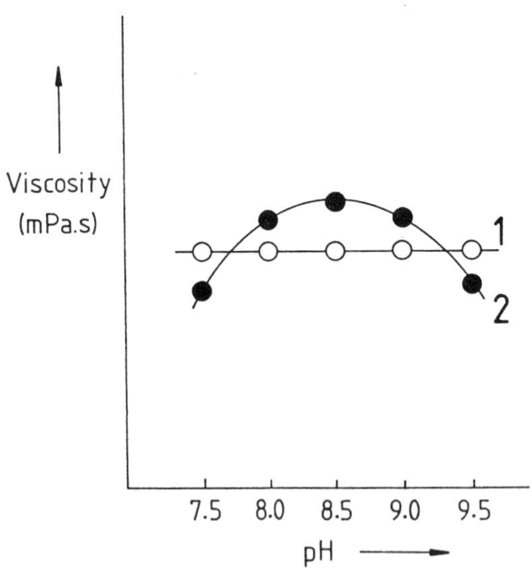

FIG. 2 Viscosity versus pH (equivalent solids content). 1, Resin 1; 2, Resin 2 with increased level of high molecular weight fraction.

The selection of alkali can affect properties such as water resistance, drying rate, resolubility, and color development. Ammonia, monoethanolamine, dimethylaminoethanol, aminomethyl propanol, morpholine, and triethanolamine can be compared in formulations containing a standard carboxylated emulsion. All the alkalis give the same water resistance properties, but the rate at which this is achieved is related to evaporation rates.

Drying rate differences are more or less in line with evaporation rates, ammonia being fastest and triethanolamine the slowest. The triethanolamine salt is markedly better for resolubility. The monoethanolamine salt is the best for color development. The best all-around properties are developed using monoethanolamine.

(b) Mixed Resin Systems. Although single-resin systems are capable of reasonable all-around performance, in practice most commercial ink formulations are based on mixed resin systems for optimum performance, typically using a grind base (ink concentrate) and a let-down varnish (dilution medium).

4. Pigment Grind Base Preparation

The most expensive component of any printing ink is the pigment, and the most important step in the manufacture of a cost-effective ink is the preparation of a dispersion that maximizes the use of the color, that is, one that produces good color development and opacity.

The principles of the process of dispersing powders in liquid media have been described by several authors [11–14]. There are many combinations of mechanical methods and chemical additives that can be employed to achieve the desirable conditions of fine particle size dispersion and resultant stability. The processes that have to be considered are wetting of the powder by liquid, separation of the particles in aggregates, and stabilization of the resulting dispersion.

Traditionally, aqueous organic pigment dispersions were produced using conventional surfactants. However, as performance demands were increased, the use of surfactants substantially decreased, and modern inks use pigment-grinding vehicles that have been specifically engineered by the polymer chemist. Pigment-grinding vehicles are essentially low molecular weight hydrophobic polymers that contain just sufficient carboxyl moiety to convert to aqueous alkali solubility. Low molecular weight polycarboxylates have been used for many years as pigment-dispersing agents for inorganic pigments in the emulsion paint and paper-coating industries. These are specifically designed with low molecular weight (1000–3000) and are highly ionized. Table 3 lists some commercially available pigment-dispersing agents.

Examples of the typical properties of a low viscosity pigment-grinding

TABLE 3 Pigment-Dispersing Agents

Pigment type	Product name	Manufacturer
Organic	Glascol LS23	Allied Colloids
	Joncryl 61	SC Johnson
	Morez 100	Morton Thiokol
Inorganic	Dispex N40	Allied Colloids
	Pigment Dispersant	BASF
	Tamol 731	Rohm & Haas

vehicle are listed in Table 4. The resin depicted has a high acid value and has been designed for good pigment wetting and color development characteristics. Ammonia has been used as the base to maximize the speed of drying. The molecular weight of the resin is low, which contributes to pigment dispersion efficiency, resolubility, and flow properties of the grinding base. Table 5 shows a number of pigment grind formulations based on this resin and using a laboratory Eiger mill at 3000 rpm.

5. Other Aspects

A wide range of let-down varnish compositions are available. These convert the bases illustrated above to commercially successful printing inks. The compositions may feature blends of resins designed to meet specific application requirements such as gloss and heat resistance.

Overlacquers are used to improve appearance or protection properties. They may be applied by gravure cylinder, roller coater, or damper application by offset. As with let-down varnishes, formulations may feature resin

TABLE 4 Typical Properties of a Pigment-Grinding Vehicle

Chemical type	Styrene/acrylic copolymer
Physical form	Clear solution
Solids	30%
Solvent	Water
Neutralizing agent	Ammonia
pH	8.2
Acid value as 100% resin	200
Viscosity	500 mPa.s
T_g	85°C
Blocking temperature	160°C
Softening point (ring and ball)	150°C
Average MW	8000

TABLE 5 Pigment Grind Formulations[a]

Component	Pigment Yellow 13	Pigment Orange 13	Pigment Red 4	Pigment Blue 15 : 3	Pigment Green 7
Pigment	40.0	40.0	40.0	45.0	40.0
Glascol LS25	19.0	16.7	16.7	21.4	29.6
Water	41.0	43.3	43.3	33.6	30.4
	100.0	100.0	100.0	100.0	100.0
Pigment/ binder ratio	7 : 1	8 : 1	8 : 1	7 : 1	45 : 1
Percent solids	45.7	45.0	45.0	51.4	48.9

[a]Quantities given in weight percent.

combinations designed to meet specific application requirements, for example, matte or high gloss, nondrip, heat resistance, or combinations of these.

Formulation details of let-down varnishes and overlacquers have been published elsewhere [15,16].

There is little doubt that as the understanding and technology of water-based systems continues to improve, there will be a continuing replacement of solvent-based systems with water-based alternatives.

B. Aqueous Continuous Rheology Modifiers

Synthetic thickeners can be prepared as polymer dispersions in water and are often referred to as alkali-soluble emulsions (ASEs) or alkali-activated thickeners (AATs). It is worth pointing out at this stage that although it is common practice to refer to products of this type as emulsion polymers (because of their method of preparation), they are in fact ultrafine (<0.5 μm) dispersions of solid polymer in aqueous continuous phase.

The end uses for products of this type are quite diverse, but one major application area in which they can be employed is as rheology modifiers. A number of variations on this basic theme are now available as commercial products, the major categories being (1) alkali-soluble types, (2) alkali-swellable types, and (3) associative alkali-soluble types. The products in each of these three categories are based on similar main component monomers. A discussion of the general principles of their preparation follows; then the more specialized monomers that are incorporated in the recipes to prepare the products in categories 2 and 3 are considered.

The concept behind these product types is to prepare a polymer that is insoluble in water in its polymerized state but can be made water-soluble by some simple chemical change, in this case a change in pH. It is therefore necessary to have present as a component in these systems some functional group whose character is strongly dependent on pH. Not surprisingly, the materials most commonly used are those that have wide commercial availability such as acrylic and methacrylic acid, although other carboxylated monomers such as maleic, itaconic, crotonic, aconitic, and other acids can be used as partial or complete replacements for the former types. The second main component is a monomer that has very much lower water solubility. Indeed, homopolymers prepared from these types would be totally water-insoluble. Examples of this type would be any of the alkyl (meth)acrylates such as methyl methacrylate, ethyl acrylate, or butyl acrylate and other monomers such as vinyl acetate. Alkali-activated polymers can be prepared from a mixture of monomers of these two types provided the correct monomer ratio is chosen prior to polymerization. The carboxylated monomer provides the hydrophilic character of the system while the non-water-soluble monomer contributes to its hydrophobic character.

The appropriate monomer mix is emulsified into water and polymerized as described earlier to form a low viscosity dispersion with an active content of 25–40% by weight.

To convert the polymer to its water-soluble form, it is necessary to neutralize the carboxylic acid groups with a suitable monovalent base. For most applications it is common to use either sodium or ammonium hydroxide or one of the alkanolamines. Once neutralized, the ionized carboxylic acid group possesses much greater hydrophilic character than its free acid analogue. This brings about a change in the overall solubility of the polymer molecule, as the non-water-soluble hydrophobic units no longer dominate the solubility characteristics. This change is obviously dependent on the degree of ionization of the carboxylic acid groups, which in turn is dependent on the pH of the system. In most commercial products the change in solubility is first seen at around pH 6.0–6.5, and this can be followed visually as a change from opaque to transparent appearance. This also corresponds to the point when significant thickening starts to occur in the system.

As the polymer particles start to dissolve, the individual polymer chains within each particle begin to disentangle and move apart from one another. Ionic groups of similar charge along the polymer chain experience charge repulsion, and this results in the polymer molecule adopting a more extended conformation than the random coil, which would be present in the free acid state, and a substantial increase in viscosity of the aqueous phase results.

Figure 3 shows chain swelling of the individual latex particles on neutralization with alkali. By careful control of molecular weight at the polym-

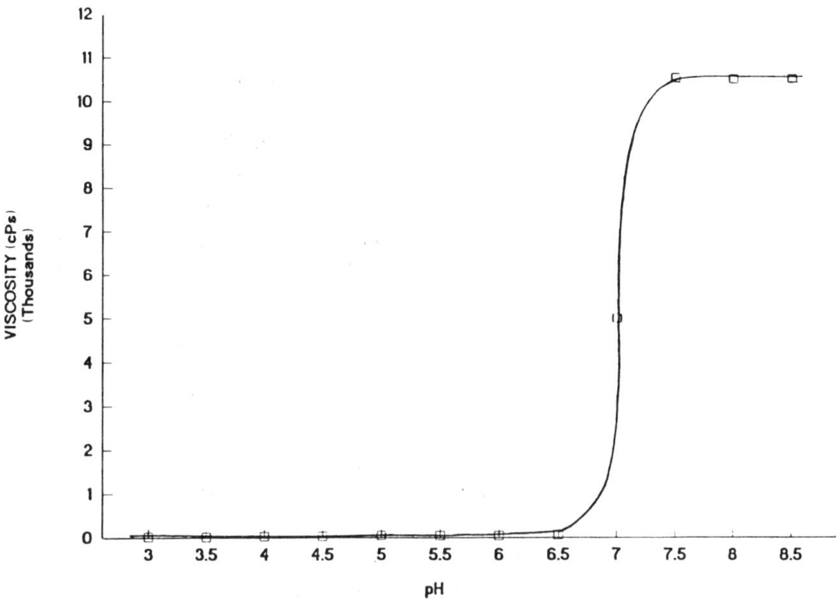

FIG. 3 Viscosity versus pH for dilute solution of alkali-activated polymer neutralized with NaOH solution.

erization stage, it is possible to produce materials that have rheological characteristics in aqueous system similar to those of certain natural and semisynthetic polymers that have been used as thickeners in a variety of industrial applications for a number of years. In particular, the paint industry consumes large volumes of thickeners for use in emulsion paints to give the finished product the correct type of rheological properties for easy application and the desired performance characteristics.

In recent years, computer-controlled rheometers have become available that allow the user to evaluate the rheological properties of a material over a wide range of stress and strain conditions. Such instruments are of particular interest in applications that use polymeric thickeners. The apparent viscosity of a system containing a polymeric thickener is often strongly dependent on the conditions under which the measurement is made. In this respect the polymer solutions are described as non-Newtonian and are shear-thinning in nature. Certain workers have suggested that all polymer solutions will follow an S-shaped flow curve, in which a Newtonian high viscosity plateau is seen in the very low shear rate region. Over the low to medium shear rate region, shear-thinning behavior is observed. When high shear rates are reached, a second Newtonian region is reached. This effect is shown diagrammatically in Fig. 4.

The Newtonian plateau in the low shear rate region is thought to arise as a consequence of the polymer chains reaching a maximum degree of entanglement. Decreasing the shear rate on the sample in this region by the application of still weaker stresses does not result in any greater chain overlap, and no further increase in apparent viscosity occurs. However, if the stress on the system is increased, the entangled polymer chain network responds to the higher shear rate by a gradual realignment of the individual molecules along the lines of flow in order to offer lower resistance to the stress being applied. This corresponds to the shear-thinning region of the curve, where the apparent viscosity decreases rapidly with increasing shear rate. Eventually, at some higher shear rate, the polymer chains reach a state of minimum entanglement, and no further decrease in viscosity occurs at still higher shear rates. (At very high shear rates, a further viscosity drop could occur if the polymer chains were mechanically degraded by the high shear environment.)

The conditions under which a formulated paint has to perform correspond to a wide range of shear rates. Examples of the shear rates that arise in particular operations are shown in Table 6. For example, at the manufacturing stage the mixture has to have a viscosity that allows it to be pumped, stirred, and filled into the can (medium shear rate operations around 10^{-1}–10^{-2} s^{-1}) without excessive energy consumption. In transport and storage (very low shear rate conditions at about 10^{-2}–10^{-4} s^{-1}), the paint should be stable against separation or settlement of the components. When the paint is applied, the user expects to find good application proper-

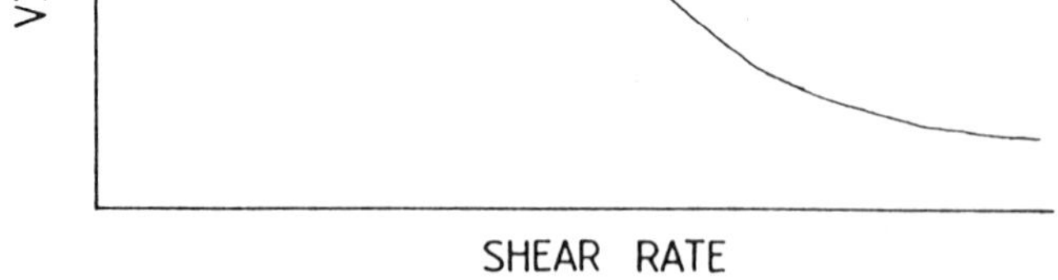

FIG. 4 Typical rheology of systems containing polymeric thickeners over a wide range of shear rates.

TABLE 6 Typical Shear Rates Found in Industrial and Everyday Situations

Situation	Typical range of shear rates (s^{-1})	Application
Sedimentation of fine powders in a suspending liquid	10^{-6}–10^{-4}	Medicines; paints; pesticides
Leveling due to surface tension	10^{-2}–10^{-1}	Paints; printing inks
Draining under gravity	10^{-1}–10^{1}	Painting and coating; toilet bleaches
Extruders	10^{0}–10^{2}	Polymers
Chewing and swallowing	10^{1}–10^{2}	Foods
Dip coating	10^{1}–10^{2}	Paints; confectionary
Mixing and stirring	10^{1}–10^{3}	Manufacturing liquids
Pipe flow	10^{0}–10^{3}	Pumping; blood flow
Spraying and brushing	10^{3}–10^{4}	Spray-drying; painting; fuel atomization
Rubbing	10^{3}–10^{5}	Application of creams and lotions to the skin
Milling pigments in fluid bases	10^{3}–10^{5}	Paints; printing inks
High speed coating	10^{5}–10^{6}	Paper
Lubrication	10^{3}–10^{7}	Gasoline engines

Source: Courtesy of H. A. Barnes, J. F. Hutton, and K. Walters, *An Introduction to Rheology*, Elsevier, New York, 1989.

ties. These can be summarized as a viscosity that, under the brush or roller (high shear rate typically $>10^3\ s^{-1}$), creates good "brush drag," which in turn provides a reasonable coating thickness. This ensures good covering power over the previous coating in one or two coats. After application, the viscosity of the film should be high enough to prevent "sag" on vertical surfaces but low enough to allow brushmarks, air bubbles, and other surface defects to disappear before the drying process is complete, leaving a smooth and even surface. In both the latter environments the shear rate is very low, typically 10^{-1}–$10^{-3}\ s^{-1}$.

A medium pvc (pigment volume concentration) emulsion paint containing a water-soluble polymer as the viscosifier can be shown to follow the rheological profile outlined earlier. This effect can be achieved with both an alkali-activated synthetic polymer (product A) or a chemically modified cellulose (product B) as shown in Fig. 5. It is worth noting that the apparent

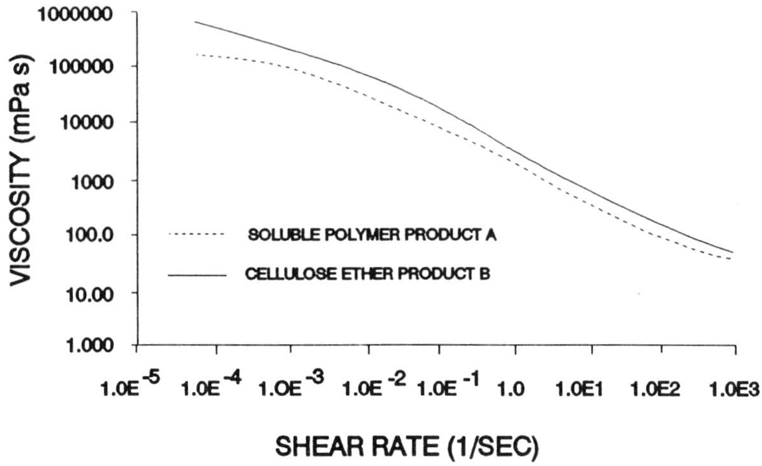

FIG. 5 Viscosity versus shear rate for medium pvc emulsion paint containing cellulose ether and Type 1 alkali-activated thickener.

viscosity changes by several orders of magnitude as the shear rate varies over a seven-decade range.

One slight drawback to the use of AATs in some emulsion paint systems is that they tend to cause a slower rate of development of water resistance in the dry paint film. Product variables that could normally be exploited to improve the efficiency of AATs as thickeners, such as increasing the polymer molecular weight or charge density, cannot be used in these systems. The maximum molecular weight of the AAT has to be kept under close control in order to prevent pigment flocculation from occurring, a phenomenon that can be common if long-chain ionic polymer molecules are used in systems that contain particulate solids. Such effects are less of a problem with nonionic modified cellulose derivatives. Increasing the charge density by using a higher proportion of free acid monomer in the recipe can result in the polymer being too hydrophilic (even in its free acid state) to be successfully prepared by emulsion polymerization.

One way in which the thickening efficiency of the product can be increased significantly is to incorporate a very low level of a third monomer, one that possesses two (or more) polymerizable double bonds, into the reaction mixture. This creates the type of product described earlier as an alkali-swellable thickener (classified as Type 2). The two vinyl groups in the difunctional monomer become incorporated in different polymer molecules during polymerization and provide a covalent cross-link between neighboring polymer chains. When the final product is diluted in water and neutralized in the usual manner, the appearance changes from opaque to transparent as before, but a significantly different type of viscosity behavior is obtained. Provided that an appropriate level of cross-linking reagent has been chosen, the final polymer will not form a true solution on neutralization but will merely swell. Because of their extremely small particle size, the swollen microgel particles are not visually apparent in solution. In this case, the viscosity of the system is not provided by soluble polymer molecules becoming entangled, but by the hydrophilic polymer network absorbing the available aqueous continuous phase. This different behavior is also reflected by the nature of the viscosity versus shear rate graph for mixtures containing polymers of this type. Figure 6 shows the flow curve data for a medium pvc emulsion paint thickened with a Type 1 alkali-soluble polymer (product A) and a Type 2 alkali-swellable polymer (product C).

A power law model provides a good representation of the flow curve data for a Type 2 polymer, and although products of this type are known to be more efficient thickeners than Type 1 polymers, such extreme shear-thinning behavior does not always suit every end use. For example, the paint formulation containing product C (Fig. 6) would not have the required application properties. The high shear environment generated when the paint is applied by brush or roller would reveal the low viscosity characteristics of the formulation under these conditions. This could result in

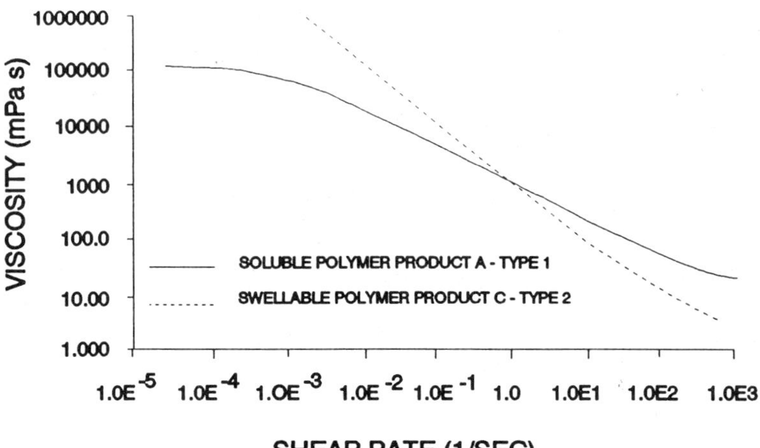

FIG. 6 Viscosity versus shear rate for medium pvc emulsion paint containing Type 1 and Type 2 alkali-activated thickeners.

overspreading of the paint, as a low level of brush drag would occur during application. Also, during the drying stage, when the effective shear rate in the system is very low, the extremely high viscosity of the formulation under these conditions would result in poor flow-out of brushmarks and other surface defects.

However, under different circumstances, such as in the thickening of a ceramic tile adhesive compound or a textured finish for walls and ceilings, product C would be much preferred over Product A. In such end uses, the low viscosity in high shear environments created by the alkali-swellable type allows easy spreading characteristics. The high viscosity in low shear conditions helps to prevent the possibility of a ceramic tile slipping down a vertical surface under the force of gravity and helps to retain the pattern in a textured finish until the coating has fully dried.

In the last few years, a new class of thickeners has become more widely available. These are the products of the Type 3 category, often referred to as associative thickeners. Products of this type possess rheological properties significantly different from those of Types 1 and 2 described so far, and this is a direct consequence of the presence of a third monomer incorporated in the reaction mix prior to polymerization. In essence, this monomer can be considered to have the structure of a nonionic surfactant as shown in Fig. 7.

Surfactants of this type can be prepared from a straight-chain or branched fatty alcohol by an addition reaction with ethylene oxide,

$$R - OH + n\,EO \rightarrow R - O(EO)_nH$$

where R is usually a C_{12}–C_{18} group, EO is ethylene oxide, and n is an integer between about 5 and 40. The overall characteristics of such a molecule vary according to the nature of R and the value of n. However, when the weight

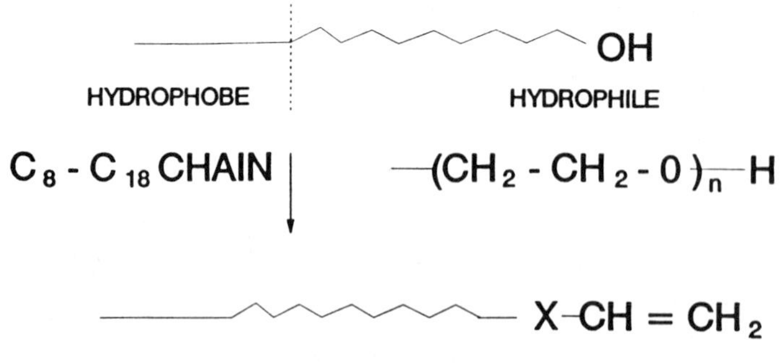

FIG. 7 Structure of nonionic surfactant monomer, X-amide, ester, ether, etc.

FIG. 8 Formation of micelles with nonionic surfactant.

fraction of ethylene oxide in such a unit is around 60% or more (i.e., when the HLB is 12 or above) the compounds are dispersible or even soluble in water. Any surfactant present in a solvent at a concentration that exceeds its critical micelle concentration will exist in the form of aggregates. These micelles provide an environment thermodynamically more favorable to the less soluble part of the molecule. Thus, in an aqueous environment the hydrocarbon chain portions of the molecule associate into a hydrophobic domain, with the hydrophilic end of the molecule remaining in contact with the aqueous phase (Fig. 8).

A number of methods have been described [17–26] whereby a polymerizable vinyl group can be attached to the terminal —OH group. Once this has been done, it is possible to use this new type of functional monomer as a component in the monomer mixture used to prepare an alkali-activated polymer. During the polymerization reaction, the vinyl group in the surfactant monomer becomes incorporated as part of the polymer chain. The long chain surfactant unit is then present as a branch, or side chain, to the polymer backbone. When diluted and neutralized, the polymer particle dissolves as in the case of a Type 1 polymer although the side chain unit can now extend out into the bulk solution. At an incorporation level of only 1–2 mol %, the nonionic surfactant side-chain groups are well spaced along the polymer backbone. The presence of charged groups along the chain results in the polymer adopting an extended configuration in solution due to electrostatic repulsion effects. However, despite the distance between the surfactant groups, there is still a strong tendency for the hydrophobic ends of the side chains to aggregate in micellar clusters. A solution of this type of polymer therefore can be regarded as a combination of the two effects. (See Fig. 9.) The structure of this solution is responsible for the rheological behavior of such polymers in water.

The micellar structures that form in aqueous solution provide a dynamic network linking individual polymer chains, with a consequent increase in thickening efficiency. The lower use levels of product of this type help to overcome the potential problems of inferior water resistance.

More important, the rheological profile is further modified in a way that enhances the performance properties of the polymer as a thickener in emulsion paint systems (see Fig. 10). Under low shear rate conditions the micellar network provided by product D (Type 3) exists in a state of dynamic equilibrium. Individual surfactant units may leave a micellar structure and be replaced from bulk solution by a previously unbound surfactant group. This allows individual polymer chains a degree of mobility and limits the maximum viscosity that is attained under low shear conditions. This in turn provides for superior flow-out, so that brushmarks and other surface defects are eliminated before drying has been completed.

At high shear rates the Type 3 associative polymer now provides the formulation with the higher apparent viscosity. This provides the effect commonly referred to as "brush drag," which results in a reasonable coating thickness being applied such that good covering or hiding power is obtained from a single application. The behavior is thought to result from the "anchor points" provided by the micelles. Under high shear conditions the micelles effectively hold the polymer chains together in a network and pre-

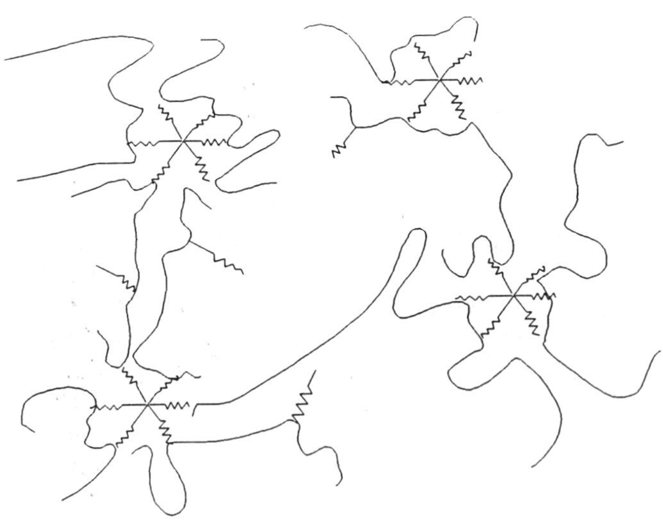

FIG. 9 Schematic diagram showing structure of associative polymer in aqueous solution.

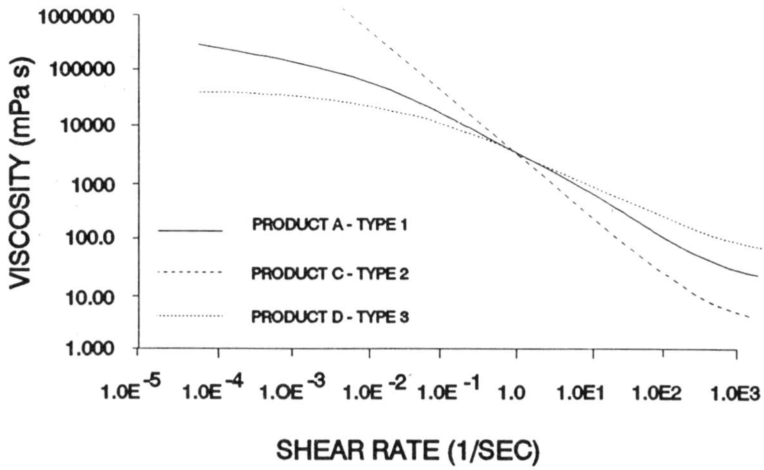

FIG. 10 Viscosity versus shear rate data for medium pvc emulsion paint containing Type 1, 2, and 3 alkali-activated thickeners.

vent the easy realignment of individual molecules against the applied stress. Any disruption or breakdown of the micelles would result in hydrophobic units returning to an aqueous environment, which would be thermodynamically unfavorable.

The technology exists for tailoring synthetic thickeners to provide the rheological characteristics specifically suited to various end purposes. The products, which are supplied as easily handled emulsions, can match the performance of natural products in most applications.

C. Aqueous Continuous Microcapsule Dispersions

Microencapsulation encompasses a broad range of processes resulting in the production of particles as small as 50 nm or as large as 1000 μm depending upon the field of application. In the pharmaceutical industry, it has become traditional to regard particles in the 100–500 μm region as microcapsules. With many other applications, however, the microcapsules are genuinely in the micrometer diameter size region. For the purpose of this review, where it is mainly colloidal aspects that are of interest, only those dispersions with a maximum dispersed phase particle size in the region of 10 μm or less will be considered, along with processes for their manufacture and their industrial applications.

1. Particle Morphology

A confusing number of terms have been used in the literature to describe encapsulated systems, and the precise meaning of the specific term is not always apparent. For instance, "microparticle" and "microcapsule" can be distinguished as having respectively homogeneous and core-shell morphologies, but "microsphere" could apply to both. Some "particles" can also be regarded as "monolithic microcapsules" as opposed to "reservoir microcapsules," which by comparison have a distinct capsular layer present. In Table 7, some descriptive terms that have appeared for encapsulated species are listed along with the inferred particle size range and morphological type. The problem of microcapsule nomenclature has been discussed by Arshady [27], who proposed a systematic procedure for naming certain types of capsules.

2. Dispersion Properties

The physical, chemical, and technological properties of a microcapsule dispersion are generally more complex than most other types of dispersions. The extra complications to be taken into account in addition to the normal colloidal and interfacial phenomena stem from the more heterogeneous

TABLE 7 Encapsulation Terminology

Size range	Morphology	Term
0.01–1.0 μm	Particulate	Nanoparticle
		Nanosphere
		Colloid
	Unilamellar	ULV vesicle
	Oligolamellar	OLV vesicle
	Multilamellar	MLV vesicle
		Liposome
		Niosome
	Capsular	Nanocapsule
		Polymerized vesicle
		Coated liposome
0.1–10 μm	Particulate	Microsphere
		Monolithic microcapsule
	Lamellar	Microsome
	Capsular	Microcapsule
		Reservoir microcapsule
		Capsular microsphere
1–50 μm	Particulate	Microbeads
	Capsular	Microcapsules, capsules

morphology of confined systems and the likelihood of more pronounced aging effects due to changes brought about by leaching, erosion, and other physical changes associated with microcapsules.

For particles or liquid droplets having a diameter in the micrometer region or above, dispersed in a liquid, settling behavior is described mathematically by Stokes' law.

$$V_m = \frac{(\rho_1 - \rho_2)d^2g}{18u}$$

where V_m is the terminal velocity, u the liquid viscosity, ρ_1 the particle density, ρ_2 the liquid density, g the acceleration due to gravity, and d the particle diameter.

For smaller particles, Brownian movement becomes significant, and this thermal effect is usually sufficient to prevent settlement or creaming.

Microcapsules having a core material with a different density than the shell material will have the characteristic of a distribution of densities varying with particle diameter (assuming wall thickness is constant). The result is that, on standing, a single dispersion might well have some particles that settle to the bottom of the container while others cream.

Like other dispersions, microcapsules are susceptible to aggregation or flocculation, with these effects being particularly problematic during preparative stages. A large part of encapsulation process development is aimed at minimizing sticky collisions between incipient capsules and coalescence caused by congealed excess wall polymer.

3. Rationale for Microencapsulation

The choice of a particular method for encapsulation and the best materials to be used for the purpose are decisions that require the practitioner to know a great deal about the intended application, manufacturing processes, and the intrinsic properties of the materials. Know-how and experience may be as important as an understanding of the fundamental principles. A general overview of the origins, underlying principles, and early process developments for microencapsulation has been published as the proceedings of a symposium held by the American Chemical Society in 1973 [28].

Some organizations [29–32] offer a specialized service for microencapsulation research and can recommend suitable materials and methods for a given application. In making even a preliminary choice, various factors have to be addressed—the reasons for encapsulation, capsule size range, release mechanism, storage stability, and activity or concentration of the active components. If the particle size of the encapsulated particle is specified to be in the range 0.1–10 μm, then physical methods of entrapment can safely be ignored and a chemical method is indicated. Chemical methods have been classified [30,31] under the headings coacervation/phase separation, interfacial polymerization, solvent evaporation, and in situ polymerization.

Many excellent reviews have been published on the general topic of microencapsulation [33–36] or on individual processes such as coacervation [37], polycondensation [38], and biodegradable microcapsule preparation [39], which all lead to entrapment of an active species. Rather less attention has been directed to colloidal aspects of the resulting products or the reaction mixtures leading to them.

4. Coacervation

Coacervation is probably the most widely studied method of encapsulation, and it is also one of the most versatile. A coacervate is the polymer-rich solution that sometimes separates from homogeneous polymer solutions when phase separation is induced by a chemical or physical agency. A list of these that have been used for microencapsulation and an assigned process code (after Kondo [33]) is given in Table 8.

The Green and Schleicher process described in a 1957 patent [40] as-

TABLE 8 Types of Coacervation Processes that Can Be Used for the Microencapsulation of Particles

Agency type (code)	Descriptor	Continuous phase
Simple		
Nonsolvent (P-S ← N)	Solvent alteration	Aqueous and nonaqueous
Ionic strength (P-S ← I)	Salting out	Aqueous
pH (P-S ← AB)	pH-sensitive polymers	Aqueous
Second polymer (P-S ← P′)	Incompatible polymers	Mainly nonaqueous
Temperature (P-S ← T)	Cooling	Mainly nonaqueous
Temperature (P,P′ − S ← T)	Thermally incompatible polymers	Mainly nonaqueous
Temperature (P-S ← T′)	Reverse-solubility polymers	Mainly aqueous
Complex		
Second polymer (P-S ← (P)	NCR process	Aqueous

Source: After Ref. 33.

signed to the National Cash Register Co. and the physical chemical studies of Bungenberg de Jong [41] form the basis from which practically the whole of microencapsulation science emanates. In the NCR process, which is a *complex coacervation* method, an oil phase is dispersed as small droplets in a solution of a colloid (sol). A second colloid of opposite charge to the first is then added as a solution, and under certain chosen conditions the polyelectrolyte complex formed between the two colloids (polymers) separates as a concentrated liquid that mostly envelops the oil droplet.

Alternatively, both polymers can be present in solution to start with, if both carry the same charge and at least one is amphoteric. The pH is adjusted to cause charge reversal on the amphoteric polymer, which then is available for complex formation. The encapsulated core is cooled to allow the liquid capsular wall to gel (sol–gel transition).

The art of microencapsulation by coacervation is to ensure that the separated phase is in reality a concentrated liquid phase and not a precipitated solid. The use of low molecular weight polymers and keeping the concentration relatively low favor this happening.

The interfacial tension between a coacervate droplet and the mother liquor is almost zero [42]. When two immiscible liquids such as oil and water are present, the continuous phase is determined by Bartell's laws of wetting:

$$S_{1\text{-}2} > S_{1\text{-}3} + S_{2\text{-}3}, \qquad \mathrm{S_3 \text{ is continuous}}$$
$$S_{1\text{-}3} > S_{1\text{-}2} + S_{2\text{-}3}, \qquad \mathrm{S_2 \text{ is continuous}}$$

where S_1, S_2, and S_3 are the dispersed phase, coacervate phase, and equilibrium liquid, respectively, and $S_{1\text{-}2}$, etc. denote the interfacial tension between phases 1 and 2, etc. The coacervate droplets wet the surface of the dispersed phase and coalesce to form a liquid membrane around it.

The second part of microencapsulation by coacervation is the conversion of the liquid membrane into a solid phase (sol–gel transition). Up to this point the process is essentially reversible, and to make capsule walls more permanent it is usual to "harden" the walls by some form of cross-linking. The microcapsules thus formed can be isolated from the continuous phase by solid–liquid separation methods, or the whole mixture can be spray-dried. For dispersions of colloidal size particles, the liquid will probably be used as such without further processing. Depending upon the method of encapsulation, a volatile internal phase can remain trapped under these conditions for years without significant loss.

For a given system, particle size is controlled by the stirring rate in the first stage and the capsule wall thickness is governed by the stirring rate in the second stage, although prolonged stirring increases the chances of aggregation.

The techniques developed by the National Cash Register Company [40] used either gelatin or a soluble formaldehyde resin or its coacervated form as the emulsifying agent. As a result of the relatively low surface activity of these materials, the emulsions formed were rather coarse and the capsules fairly large (10–200 μm commonly). The original capsule wall materials still much in evidence today are gum arabic (an anionic polysaccharide) and gelatin (an amphoteric protein). The gelatin types used are negatively charged at pH 7 and become positively charged below pH 5. Complex coacervation takes place over a surprisingly narrow pH range of about 2.5–4.5. At lower pH values the carboxylic group on the gum arabic (acacia) loses its charge, and a complex coacervate is not formed.

Since the introduction of this technique commercially for carbonless paper manufacture in the 1950s, a very large number of variations have appeared. A major development was the use of synthetic polymers in place of the natural hydrocolloids. In a process patented by Wiggins Teape Ltd [43,44], the coacervate is formed between acrylamide/sodium acrylate copolymer and a cationic urea formaldehyde precondensate. Subsequent modifications [45,46] provide various alternatives to the anionic polymer but still rely on aminoplasts for the cationic portion and wall material.

Refinements to the method include the introduction of an element of surface activity to the polyacrylates that results in them having the secondary role as shock-preventing agents [47] in addition to their function as emulsifier/wall component materials [46]. "Shock" is the term used to describe the unwanted viscosity increase or gelation that occurs during processing, especially when a polymerization or cross-linking chemical reaction is taking place.

Aminoplast walls can be formed in the absence of an anionic polymer. A process developed by 3M Company and useful for encapsulating many water-insoluble fill materials [48] uses the change of solubility with molecular weight as a method of depositing polymer. A low molecular weight precondensate is present in the medium surrounding a disperse phase, and upon polymerization the aminoplast concentrate phase separates and deposits as a capsule wall enveloping the dispersed particles.

The standard procedure in microcapsule processing is to allow polymerization to continue after the wall is formed, sometimes by addition of a highly branched reactive material like melamine/formaldehyde resin, or to make use of a cross-linking agent, thereby curing the polymer membrane into a tough impermeable film. For certain applications this final curing step may be unnecessary and may even be deleterious.

As part of a series of novel products, some major pesticides have been formulated into aqueous dispersions of micrometer-size encapsulated particles that have uncured capsule walls [49]. In spite of this the products have long shelf lives and remain well dispersed with little tendency for the active ingredients to partition into the aqueous phase and destabilize the mixture by crystallizing. The core can be a nonaqueous liquid or solution, or it can be a polymer matrix containing the active ingredient made by an emulsion evaporation technique. The wall is preferably formed out of the complex coacervate between an anionic polyacrylamide and a cationic urea formaldehyde resin. The coacervate has also been found to be a highly surface-active species, obviating the need for conventional emulsifiers, emulsion stabilizers, and polymeric dispersants.

Although complex coacervation has received a great deal of attention for microencapsulation, it is possible to use a number of other methods to phase separate the wall polymer (see Table 8).

Simple coacervation methods are characterized by the presence of only one polymer in the separated phase, unlike the polyelectrolyte complex in the NCR process. Although most physical changes can bring about phase separation in polymer solutions, some lend themselves rather more easily to practical use in microencapsulation. For example, changing the pH is one very common way of changing the solubility of certain ionic polymers (pH-sensitive or "enteric" polymers) and consequently finds use in encapsulation of pharmaceuticals. However, these products are manufactured by physical processes unsuited to colloid-size substrates. The solubility of pH-sensitive polymers usually changes dramatically over a very narrow range of pH, and it is therefore very difficult in practice to form a distinct coacervate phase by this route. Solvent alteration and salting-out procedures, on the other hand, are convenient and effective methods that apply equally well to charged polymers like gelatin or uncharged polymers like polyvinyl alcohol.

The rather unusual reverse solubility of certain acrylamide/N-substituted acrylamide copolymers can be exploited for making microcapsules [50]. These polymers are less soluble in hot water than cold. When the polymer solution is warmed, phase separation occurs, and under carefully controlled conditions a coacervate can be produced at temperatures above the lower critical solution temperature (LCST). The capsule wall can be made permanently insoluble by cross-linking with a dialdehyde at the elevated temperature. Alternatively, salt can be added to depress the LCST (Fig. 11) to below room temperature when capsule walls are wanted that dissolve on dilution from the high ionic strength storage conditions [51].

The principle underlying incompatible polymer coacervation is that with some binary mixtures of polymers in solution and at certain concentrations, phase separation into two liquids will occur, each enriched in one polymer but with a common solvent. In microencapsulation the technique is usually used in combination with a temperature change. The encapsulating polymer

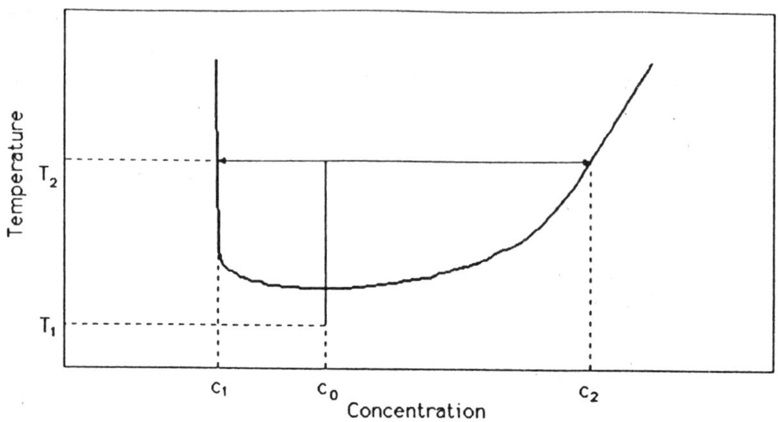

FIG. 11 Typical phase diagram for polymer solutions displaying LCST behavior. A solution at T_1 and c_0 on warming to T_2 will separate into a polymer-rich phase (c_2) and a polymer-depleted conjugate solution (c_1).

in solution is brought into contact with a second polymer known as the phase inducer. This appears to be applied mainly in nonaqueous solutions. Typical phase inducers are polyisobutylene, polybutadiene, polyethylene, and poly(dimethyl siloxane). These have generally been used with ethyl cellulose and ethylene/vinyl acetate copolymers (EVA).

5. Interfacial Polymerization

Although it is possible to generate interfacial films by polymerizing from one phase (e.g., the Fuji in situ method [52]), interfacial polymerization generally refers to the reaction of one monomer dissolved in the dispersed phase with a second oil-soluble monomer in the continuous phase. The reaction is self-limiting since the presence of the polymer film at the interface prevents monomer from diffusing into the reaction zone. The locus of polymerization depends upon the relative solubilities of the monomers in the two phases. It is preferable for the reaction to take place just inside the interface; otherwise the capsules produced may be tacky. Since the membrane formed is so thin, it is difficult to make it impermeable, and the capsules formed are often too fragile to isolate as a solid.

A commonly used method is interfacial polycondensation, which produces a thin membrane, of nylon, for example, by the reaction of a diamine or polyamine dissolved in the aqueous phase of an oil-in-water emulsion and sebacoyl chloride added to the oil phase before emulsification. Reaction is almost instantaneous. The presence of nonionic surfactants does not have an interfering effect, and particles as small as 1 μm can readily be made [33].

A variation on the idea of generating a membrane at the interface was introduced by Speaker and Lesko [53]. Microcapsules with extremely elastic walls are said to result from the formation of an insoluble "salt wall" by the interaction between a polyanion soluble in one phase and a polycation soluble in the other.

6. Emulsion Methods

The emulsification–solvent evaporation technique has wide applicability, and it is especially suited for producing microspheres as colloid-size systems for drug delivery [54]. Conceptually the technique is a simple one. The active ingredients plus encapsulating polymer are dissolved in a relatively volatile solvent. This solution is dispersed as an internal phase in a liquid–liquid emulsion, which is then subjected to evaporation or distillation to remove the volatile solvent. This leaves the active component in a microscopic particle of polymer dispersed or suspended in the second liquid.

In practice there are several difficulties. Volatile solvents for water-insoluble active ingredients tend to be flammable, toxic, or environmentally damaging. Higher boiling solvents with flash points above 41 °C are preferred industrially but require higher temperatures and/or vacuum for distillation. Under these conditions, stabilizing the emulsion can be a problem, and a polymeric emulsion stabilizer is normally required. The development of highly efficient stabilizers for water-in-oil emulsions [55] has allowed high temperature vacuum distillation recovery of water-soluble polymers dispersed in oil on a very large commercial scale.

If a water-soluble active ingredient is dissolved in the aqueous phase before emulsification into oil and subsequent distillation, the product will consist of polymeric microspheres dispersed in oil. These microspheres can be converted from hydrophilic to hydrophobic during distillation [56]. In this method the encapsulating polymer is first used in the water-soluble ammonium salt form, but on dehydration and heating it is decomposed into ammonia (which is distilled) and the insoluble polymeric free acid. This can be a useful way to entrap a material for suspension in oil under acid or neutral conditions.

Multiple emulsions are a form of encapsulation where the barrier between the internal phase and the exterior is a liquid. Alternatively, by use of the solvent evaporation technique, the inner phase can be encapsulated within a polymeric matrix. The formation of multiple emulsions can be quite difficult. A solution of the active ingredient is dispersed into a nonmiscible liquid containing an emulsifier. This emulsion is then added with gentle agitation to another liquid (usually the same solvent as the inner phase) containing a polymeric emulsion stabilizer. The process has been described in detail by Watts et al. [54].

The average diameter of the dispersed multiple emulsion droplet is usually quite large (> 10 μm) because high shear and emulsifiers in the second stage can cause unacceptable loss of internal phase into the outer liquid layer. Long-term stability is relatively poor compared with other forms of microcapsules unless the confining liquid is gelled or converted by solvent removal to a polymer matrix.

Liposomes are in effect a highly specialized form of multiple emulsion [57] (Fig. 12). The inner phase is an aqueous solution, and the "oil" barrier is a lipid bilayer. Liposomes can be made by depositing a bilayer of a phospholipid onto a solid surface and bringing this surface into contact with the aqueous solution containing the active ingredient. The bilayer detaches from the surface and wraps up into tiny spheres, trapping some of the aqueous phase inside. For laboratory experiments the liposomes can be freed of unentrapped active ingredient by techniques like gel filtration chromatography. On an industrial scale this is not practicable, and applications of liposomes are best suited to situations where it is of little consequence that some active ingredient is present in the suspending liquid or the value of the product will justify purification by, for example, diafiltration or treatment with ion-exchange resin.

Other drawbacks of liposomal systems are fragility and low encapsulation efficiency. Some modifications have been explored to improve liposome stability; these include polymerization of the bilayer and formulation of a polymer coating [58]. Liposomes can now be made on a commercial scale using sonication and a specialized high pressure apparatus [59].

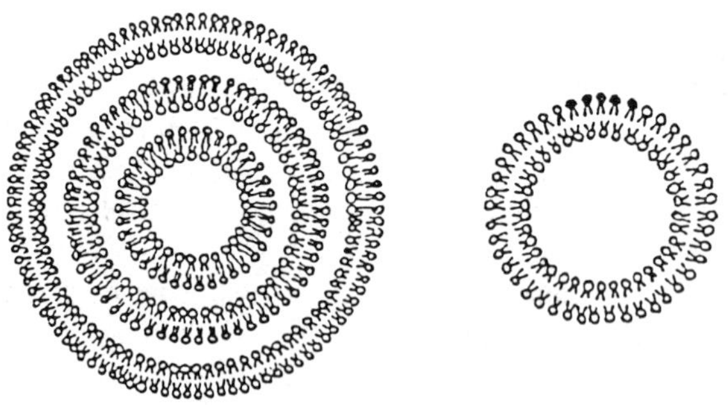

FIG. 12 Cross-sectional view of the bilipid layers of liposomes: multilamellar vesicles (left) and unilamellar vesicles (right). (From Ref. 73.)

7. Industrial Applications

Industrial applications of microencapsulation technology are expanding rapidly over a large number of domains. One university claims to have a database covering over 5000 references to microencapsulation up to 1990 [60]. The applications described here are chosen to be illustrative only and are limited in scope to colloidal dispersions.

(a) Imaging. Carbonless copying paper was the first major application for microcapsules, and it has now become a vast business area. In Europe the process based on the original gelatin coacervation method [40] accounts for about 50% of total production, but in Japan practically all pressure-recording paper is being made with synthetic polymer capsules [61].

Carbonless paper imaging relies upon the sudden rupture of microcapsules containing a dye precursor dissolved in oil when pressure is applied to the paper. The oil is forced into contact with an acid clay, and the dye is visualized by a pH change.

The microcapsules are produced as an aqueous suspension, and the capsule wall has to be strong enough to withstand the procedures of coating and drying the paper and not rupture prematurely in normal handling. Developments continue in attempts to improve sharpness and speed of response or to optimize the process. The main departures from the gelatin route have been interfacial polycondensation [62], aminoplast resin coacervation [48], complex coacervation between either a urea/formaldehyde precondensate and an anionic polyacrylamide [43,44] or cationized and sulfonated melamine/formaldehyde precondensates [45], and the Fuji in situ method [52].

Mead Corporation [63] has introduced a new concept to photosensitive imaging based on microencapsulation technology. In their method, capsules are formed as with the carbonless paper application, but the oil is replaced with polyfunctional acrylate monomers. The dye precursor is dissolved in the monomer with a photoinitiator. When capsules on a paper surface are exposed to light, the monomer polymerizes to a solid, immobilizing the dye in its colorless form. When pressure is applied, only unexposed capsules will release the dye, thus producing an image.

In reprographics, microencapsulated toner resins are being evaluated by a number of companies [64–67].

(b) Pharmaceuticals, Food, and Cosmetics. The pharmaceutical uses of microencapsulation have been a driving force behind many of the scientific investigations into the production, use, and effectiveness of microcapsules. There has been particular interest in the controlled release of drugs. This application can usually bear a higher than average cost for materials, but choice can be highly restricted by the demands of low toxicity, bioaccepta-

bility, and higher performance criteria. Polycyanoacrylate nanospheres and polylactic acid microspheres [68] are favored as drug carriers since both polymers are biodegradable.

Pharmaceutical applications of microencapsulation have an extensive literature, including references to colloid-size dispersions [69], interfacial polycondensation [70], matrix entrapments in microparticles [71], liposome-encapsulated drugs [72], and nanoparticles with adsorbed peptide drugs [73].

Liposome-encapsulated enzymes have been shown to have utility in cheese manufacture [74,75]. Other applications are anticipated for which a stage-specific or site-specific delivery system can be devised [74]. Jackson and Lee [76] have thoroughly reviewed microencapsulation in the food industry and conclude that the technology is far from being fully exploited. Large-scale production methods and the use of liposomes in foods to solve flavor and functional problems have been discussed by Vuillemard [59] and Koide and Karel [77].

Microencapsulated products are becoming of greater importance in cosmetics and toiletries. Encapsulated emollients, menthol, and antimicrobials for deodorants and absorbents are among the applications, but by far the largest market is for microencapsulated fragrances. This has been estimated to be a $2 billion (U.S.) market [78]. It is projected that in 1993 between 10–15% of the absorbent products and household product market will contain microencapsulated fragrances.

The technical demands on fragrance capsule wall materials are quite severe, since even slight permeation will be readily detectable. Any interaction must be avoided between the polymeric material and any one of the complex mixture of chemicals in a fragrance, for this could affect the crucially important organoleptic quality of the released perfume [79]. The loading of a microcapsule is also important. PFW Hercules in-Cap Fragrance is claimed to have 50% load distributed as minidroplets throughout the capsule. Another manufacturer (M-CAP Technologies International) produces microcapsules, some as small as 0.1 μm in diameter, with 75–90% oil loading for use in fragrance, pharmaceutical, and fire-retardant applications [78]. The 3M Company, which was instrumental in introducing paper coated with microencapsulated fragrances, is extending this technology into coloring materials for cosmetics [79].

The availability of functional liposomes in commercial quantities has made them a popular ingredient in cosmetics formulations [80]. Vitamins, enzymes, and moisturizers have been prepared in liposome form and appear to have enhanced activity and to be compatible with other cosmetic ingredients. However, high concentrations of surfactants will disrupt the lipid bilayer and lead to their destabilization.

(c) Agriculture. Properties associated with microencapsulation such as protection of an active material against an adverse environment and a control on release are well suited for the storage, application, and mode of action of many agricultural chemicals. It has been an active area for some years, but the market for microencapsulated products has not developed as fast as predicted [81]. An increase in regulatory and environmental pressures, however, are beginning to cause a switch away from formulations containing organic solvents. This is expected to boost demand for encapsulated materials in aqueous suspension.

Pennwalt Corporation has produced aqueous dispersions of organic active ingredients by the interfacial polymerization method with oil-in-water emulsions. A recent example [82] discloses the preparation of an aqueous formulation of fungicide microcapsules having polyurea or polyamide walls. Particle size is claimed to be in the 7–25 μm region, and fewer phytotoxic effects are observed. A number of other companies have developed variations on the interfacial polymerization method for use with agricultural chemicals and biocides [85–86]. Aminoplast microcapsules produced by simple or complex coacervation have also found use for pesticide encapsulation [87,88]. Products of this type are already on the market [89].

Monsanto Agricultural Co. is reported to be working to reduce the number of liquid herbicide preparations that contain organic solvents by use of encapsulation technology. The release of active ingredients from these products can be controlled by adjustment of the thickness of the shell wall [90].

A current trend in pest control is the increasing use of biopesticides such as the caterpillar toxin produced by *Bacillus thuringiensis* (Bt). Products like Bt have special formulation requirements because of their biological nature. Microencapsulation of Bt in a polymeric shell can lead to increased persistence on the leaf surface, Lim Technology [91] has developed a solvent alteration method for precipitating synthetic pH-sensitive polymers in the presence of Bt, entrapping the active ingredient in 5–50 μm particles. The mixture can then be applied directly to the crop or converted first to a powder by spray-drying.

(d) Miscellaneous Applications. Almost all industrial sectors provide examples of actual or potential applications for microcapsules. Even aqueous dispersions of microcapsules containing just water have been developed [92]. In the paper industry such microcapsules (2 μm in diameter) are present in a coating that is subjected to pressure in calendering. The water released at this point is a neat way to avoid the need for steaming.

The controlled release of oil field chemicals from microcapsules has been patented by Conoco Inc. [93]. In this method an active ingredient like a scale inhibitor is dissolved in an aqueous solution of hydroxyacetic acid.

This solution is next emulsified into kerosene, and a capsule wall is generated by contact at the interface with an oil-soluble initiator. The microcapsule suspension releases the active component in an oil-bearing formation because the polyhydroxyacetic acid wall slowly degrades at the high formation temperature. Particle size can be modified down to micrometers and is adjusted so that the capsules are trapped in pores of the corresponding size in the formation where treatment is required.

Lion Corporation of Japan has developed microcapsules specifically designed for use in laundry detergents. Microcapsules of a pH-sensitive polymer are made by a w/o emulsion process [94], which has been extended to include other encapsulating polymers such as polyvinyl alcohol and methyl cellulose [95]. The active ingredient in this case is a detergent enzyme that is easily damaged on storage in liquid detergents. Dewatering of the w/o emulsion is effected by addition of acetone.

As discussed above under Emulsion Methods, for large-scale manufacturing highly flammable solvents are avoided whenever possible. An improved way to manufacture encapsulated enzymes with a polymeric matrix of either pH-sensitive [96] or other water-soluble polymers [97] involves the azeotropic distillation of water from the dispersed phase of a w/o emulsion. An oil-soluble polymeric stabilizer is required to maintain stability during distillation and to keep particles well dispersed after dehydration has been completed. The oil-based dispersion of encapsulated enzyme thus formed can itself become the oil fill material of a microcapsule, this time using a coacervation process to build polymeric walls around the oil droplet (see Fig. 13) [98].

Double encapsulation procedures like the one just described lead to high protection levels of sensitive components to hostile environments, but great care is required to select the right materials and methodology to ensure that the active ingredient will be released at the appropriate time. In spite of this, systems based on these entrapment techniques do allow for considerable variation, and it is possible for a balance to be struck between impermeability and reversibility.

Microencapsulation technology is commercially very important and has widespread industrial applications. It probably represents one of the fastest growing and most sophisticated technologies within the chemical industry, as the wealth of recent literature on the subject demonstrates.

IV. ORGANIC CONTINUOUS (WATER-IN-OIL) EMULSIONS

Section III dealt with the application in industry of aqueous continuous emulsions for special purposes and identified three distinct areas for more detailed treatment. This section concentrates on organic continuous (water-

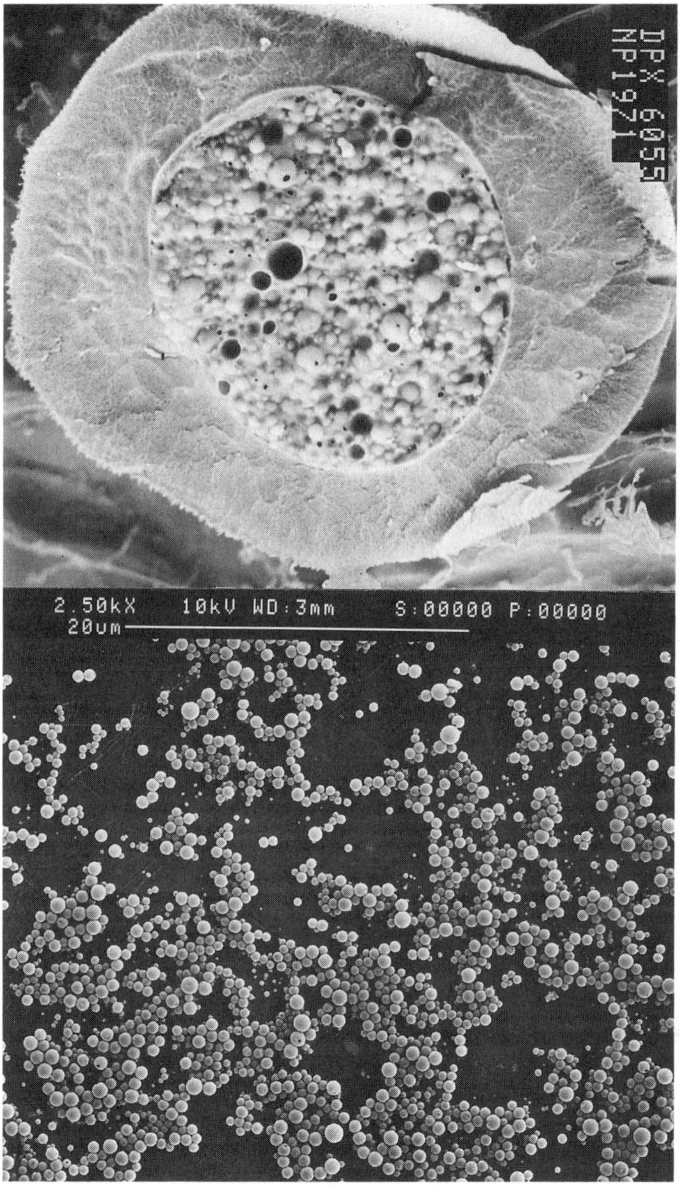

FIG. 13 (Bottom) Scanning electron micrograph showing polymer particles containing a water-soluble active component made by the method described in Ref. 47. (Top) A suspension of these particles in oil microencapsulated in a polymeric shell. Aqueous suspension of microcapsules subjected to cryostage SEM.

in-oil) emulsions and their industrial applications. Its three major subsections demonstrate the diversity of applications for these reverse-phase polymeric additives.

This section is not exhaustive in scope but serves to illustrate some very different applications of the technology.

A. Polyelectrolyte Flocculants

1. Background

Within the last two or three decades a group of water-soluble polymers have been used to supplement or replace a variety of natural substances including lime, alum, gums, and starches as solid/liquid separation aids. These compounds are high molecular weight water-soluble organic polymers prepared by a number of significantly different routes involving addition polymerization, condensation polymerization, and the modification of existing polymers [99]. They have been conveniently described as "synthetic flocculants" to distinguish them from the inorganic and natural varieties that preceded them, although those that bear charges are often described as "synthetic polyelectrolytes." The most important characteristics possessed by such compounds are, in fact, charge (or lack of it) and molecular weight, since they predominantly determine the suitability and effectiveness of the compound for a particular process requiring aided solid/liquid separation. Consequently, synthetic flocculants may be anionic, cationic, or nonionic and of low to very high molecular weight.

In application, the polyelectrolytes are first made up as dilute aqueous solutions, in which form they can be added to the suspension to be treated. They are applied at very low dose levels, and their addition results in adsorption onto dispersed solids within the suspension, which they then flocculate via processes involving charge neutralization and/or bridging. In considering the application of these products it is important to take account of the influence their characteristics will have on the solutions that have to be prepared to facilitate application. Since various levels of elevated molecular weight are involved, they will always, to a greater or lesser degree, impart viscosity to the solutions they form. With the lower molecular weight forms, this usually presents no problem; at higher molecular weight, however, the viscosity can be considerable and affect not only the handling of the polyelectrolyte solution but also its ability to disperse within the application system and thus flocculate it homogeneously. It is a fact, therefore, that some products can be made up as 50% (w/w) solutions whereas others have to be made up at 0.5% (w/w).

Leading on from the above it will be appreciated that those polyelectrolytes that can be made up at high concentration can be and are sold in solution form, a good example of these being products based on the Mannich reaction [100]. Others, however, have to be isolated as a "solid" from the manufacturing process and be put back into solution before they can be used [101]. This latter category tends to represent the technically preferred products for the industries in which they are being used, since molecular weight is an important parameter in determining the size and strength of the flocs that are formed and thus the rate of solid/liquid separation. Unfortunately, products that are in true solid form as powders or granules require special provision for make-up, and for this reason handling characteristics are not always preferred by the end user. Development has therefore been undertaken to provide a liquid form that can accommodate the requirement of high molecular weight without incurring the penalties often perceived; this has resulted in the emergence of a liquid dispersion form.

2. Liquid Dispersion Polyelectrolytes

In order to be able to provide the properties of a solid grade polyelectrolyte while at the same time fulfilling the need for a liquid, certain fundamental changes had to be made in the preparation process. To achieve the required goal, manufacturers resorted to the use of inverse emulsion polymerization [102].

It has been demonstrated that in emulsion polymerization one or more monomers are first taken up in water, and the mixture is then added to a suitable hydrocarbon in the presence of an emulsifier and high shear stirring. This causes the monomer solution to break up into very small droplets, which, following addition of a suitable water-soluble initiator, proceed to polymerize. A conversion therefore takes place that ensures the formation of very small particles of gel as a discontinuous phase suspended in a continuous phase of hydrocarbon oil. The polymer resulting from this process can have molecular weight as high as the reaction and reactants will allow but, being suspended in the hydrocarbon, can be handled as a liquid. The only problem with such a product occurs when it has to be added to water to produce the solution for application. Since the particles are protected by the oil, the polymer will not dissolve very easily; this is overcome by adding a surfactant to the make-up water or the emulsion that will strip away the oil and allow dissolution to take place. Currently the most common practice is to incorporate the "activator" in the emulsion, thus obviating the need for a two-pack system. Generally the emulsifying surfactant is of low hydrophobic lyophilic balance (HLB) whereas the activating surfactant is of high HLB.

The emulsion can be used directly as implied above, but generally emulsions as prepared have relatively low active polymer contents of the order of 25–35%. An improvement on this can be obtained by subjecting the emulsion to distillation, and at least one manufacturer does this to the extent that a substantially dehydrated suspension of polymer particles is obtained. The end product is a dispersion of very fine (less than 0.2 μm) particles in a hydrocarbon oil, which on coming into contact with aqueous medium inverts to product a solution of polymer, the oil being emulsified away. Since the dispersion is very stable within the temperature range −30 to +50°C, it can be stored and transported for many months and is hence a very practical answer to requirements for "liquid polyelectrolyte" worldwide.

3. Make-up and Handling of Polyelectrolyte Solutions

To ensure that polyelectrolytes are utilized to maximum efficiency, it is essential that they be made up and applied at solution concentrations that provide for maximum stability but at the same time provide fast dispersion within the substrate being treated. This is often catered to by making up a stock solution at relatively high concentration of 0.5–1.0%, depending on the molecular weight of the polyelectrolyte being used (bearing in mind that molecular weight determines the solution concentration). Figure 14 shows a typical viscosity versus concentration plot of a high molecular weight cationic dispersion polymer.

At these make-up concentrations a degree of stability will be maintained, thus ensuring resistance to such things as degradation and hydrolysis, which would have a very marked effect upon performance. The viscosity will, however, still be too high to allow dispersion of the solution; this is overcome by providing in-line dilution as the solution is transferred to the substrate and may result in a concentration of between 0.01 and 0.1% at the point of addition.

In order that solutions can be made up and added without difficulty, various pieces of equipment have been designed by manufacturers of polyelectrolytes. These extend from simple eductors to ensure that products can be dispersed easily in the make-up water during manual make-up [103, p. 84] to completely automated make-up units that will, with minimum supervision, provide continuous make-up and supply of solution [103, pp. 85–88]. In the case of liquid dispersion grade products, one system that accommodates the automated make-up and addition is the Allied Colloids DMU system shown in Fig. 15. This system comprises a small flash mixing tank, which is built inside a larger holding tank, and various pumps, solenoid valves, and pipework required in making up and dosing a flocculant solution. The reason for constructing the unit in such a way arises out of a need to keep polymer in the liquid dispersion form and polymer solution separate once the solution has begun to develop viscosity. A liquid dispersion polymer solution develops viscosity very rapidly; therefore, if an attempt were made to make up in one large tank simply by adding polymer

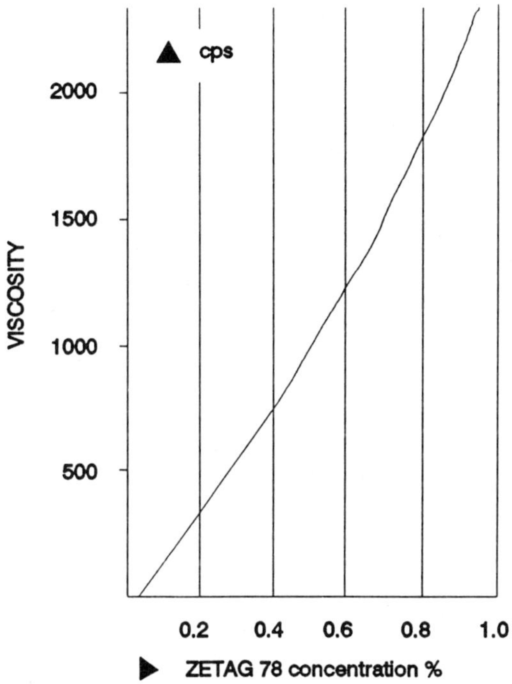

FIG. 14 Viscosity versus concentration for solutions of cationic liquid dispersion polyelectrolyte. (Courtesy of Allied Colloids Limited.)

to stirred water, there would be a risk, at some point, of viscosity impeding the dispersion of the incoming polymer. A consequence of this would be the formation of "gel lumps," which are manifest as large agglomerations of polymer, each protected by a hydrated layer and hence unable to disperse and go into solution. Such a state is obviously undesirable as it not only leads to underutilization of the polymer but also causes blockages and other adverse effects within the process to which the polymer is being applied.

4. The Need for Solid/Liquid Separation

Solid/liquid separation is a process that is widely used in many industries. It is often required where there is a need to remove either the liquid phase or the solid as a waste product, and instances of such use can be found in many areas of activity. For the present text, examples are taken from the mining, papermaking, pollution control, and water industries, although similar requirements, can be found in the brewing, pharmaceutical, and food industries, to cite but a few.

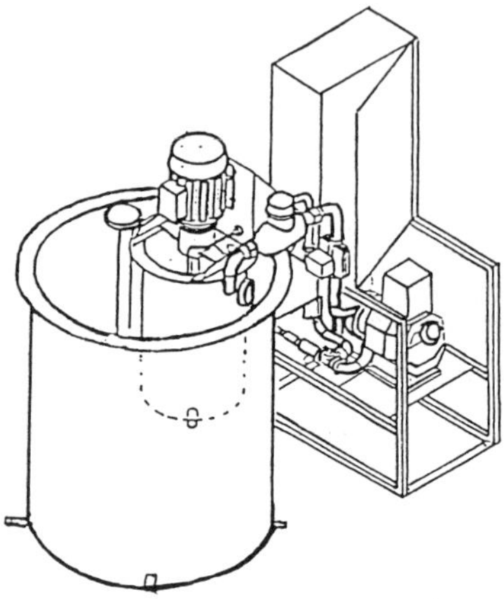

FIG. 15 Allied Colloids dispersion make-up unit. (Courtesy of Allied Colloids Limited.)

The process of solid/liquid separation and how it can be aided through the use of polyelectrolytes has been dealt with in a number of publications [103, pp. 136–165; 104] and need be given only cursory attention here. It relies on an ability for the products in question to attach by adsorption onto charged particles within the suspension being treated. They can then, because they are charged and are of high molecular weight, bring about destabilization of the suspension through charge reduction and bridging. The end result is the formation of flocs of suspension held together in a three-dimensional matrix. These flocs, comprising aggregated particles, behave as if they are large individual particles and give rise to greater ease of separation. Hence, elimination of solids by sedimentation results in flocculated particles falling through the liquid phase very rapidly to form a sludge that can be removed from the process. On the other hand, where filtration is practiced, the flocs form into a permeable mass from which the liquid can be more easily removed. These mechanisms give rise to two important areas of application for polyelectrolytes—(1) clarification and thickening and (2) underflow or sludge dewatering [105, 106].

(a) Sedimentation and Thickening. As a primary separation process, sedimentation is common to a broad range of industries. Thus we find that the first stages of treatment in sewage disposal and drinking water purification involve the use of sedimentation, as do those of water recovery in such processes as coal washing, hydrometallurgy, sand washing, and other process water treatments.

The use of polyelectrolytes as sedimentation aids sets out to achieve two objectives. The first is to accelerate the sedimentation rate and hence increase the capacity of the process; this ensures that small rather than large sedimentation tanks, or thickeners, can be employed. The second objective is to improve the efficiency of solids capture and hence the clarity of the supernatant being produced. In many cases it is possible to achieve both of these objectives with equal satisfaction; in some cases, however, it may be necessary to improve one at the expense of the other. Generally, as the molecular weight of the polyelectrolyte is increased, an ability to induce larger flocs is realized; increasing the ionic character improves the solids capture. Varying molecular weight and ionic character, and in some cases the chemical nature of polyelectrolytes, through the use of alternative monomers caters to the variation in the chemical nature of the substrate. It is therefore the case that substrates having different constituents and levels of pH may require different polyelectrolytes.

There are variations in the equipment used for sedimentation and clarification, but the general principles as far as polyelectrolyte treatment is concerned are the same. In some cases it may require a two-stage chemical treatment comprising both coagulation and flocculation. Such cases are encountered where there is a need first of all to destabilize soluble or colloidal components and thus render them susceptible to flocculation. The distinction drawn here between coagulation and flocculation is one that assumes addition of both multivalent metal salt and polyelectrolyte. In practice, a metal salt, usually alum, is added to the water first to destabilize the solid and form a microfloc. On treatment with high molecular weight polyelectrolyte, very large flocs are formed; these will settle out and compact very rapidly [107].

Suspensions with coarser solid particles can be flocculated directly. In such cases, suspensions are treated with high molecular weight polyelectrolyte and then passed to settlers, where the solids settle out and are removed in the underflow, with the clarified water being passed for reuse or disposal.

(b) Solids Dewatering. The dewatering of materials in the course of processing or after they have been produced as underflows, or sludges, represents a significant area of application of liquid dispersion polyelectrolytes. This has particularly proved to be the case where organic materials are to be dewatered. The dewatering of such material to aid a production process through the removal of water has always been a difficult operation. On their own, many organic suspensions will not dewater, and any attempt to accomplish dewatering without some form of pretreatment results in rapid filter blinding and cessation of the dewatering process. This is also true of many mineral substrates, particularly when the material to be dewatered is of a very fine or small particle size.

5. Examples of the Use of Water-Soluble Polymers in Industry

To create a clearer understanding of how liquid dispersion polymers are used within industry, three main areas of use—mineral processing, water environmental management, and paper production—are now considered.

(a) Mineral Processing. Walter-soluble polymers are used by mineral processors to enhance thickening/clarification and filtration operations. At present, by far the most important applications are for tailings and concentrates, where thickening/clarification and hydrometallurgical countercurrent decantation/filtration operations are carried out.

Enhanced solid/liquid separation is important within the mining industry for the following reasons. Fine particles must not be retained within mill circuits, which would interfere with flotation, solvent extraction, ion exchange, and electrowinning; greater economy of water must be achieved while ensuring greater efficiency of recovery of solids; throughput with existing equipment has to be improved to minimize capital expenditures.

In thickening/clarification applications, the water-soluble polymers are added to the feed launders or feed wells at concentrations of 0.01–0.05%, with doses as low as 0.01 kg/per tonne of solids being treated. In this role they promote flocculation, which leads to fast settling and greater concentration of the underflow. The efficiency of a particular operation can be controlled by varying the amount of flocculant being added, thus providing a very valuable means of operating the process.

Filtration is carried out where there is a need to further concentrate values or tailings, presented as underflows or slurries from thickening/clarification, in order to assist with recovery or disposal. Such operations are commonly carried out on belt or vacuum filters where further addition of polymer may be required to accelerate the removal of water. At this later stage, the polymers used may be very different from those used in the primary separation stage.

To satisfy the requirements of the industry as a whole, there is a need to draw on the variation that can be built into a range of water-soluble polymers. This in essence means that products have to be manufactured at various levels of ionic activity and molecular weight to provide a range of characteristics that will meet the needs of various conditions of use. Within the mineral processing industry this applies mainly to nonionic and anionic polymers, with some cationic polymers being used to satisfy a broad range of conditions as exemplified schematically in Fig. 16.

As a broad generalization it can be said that nonionic polymers are used widely under acidic conditions, with highly anionic polymers being applied substantially in situations where high alkalinity exists. Low to medium mole-

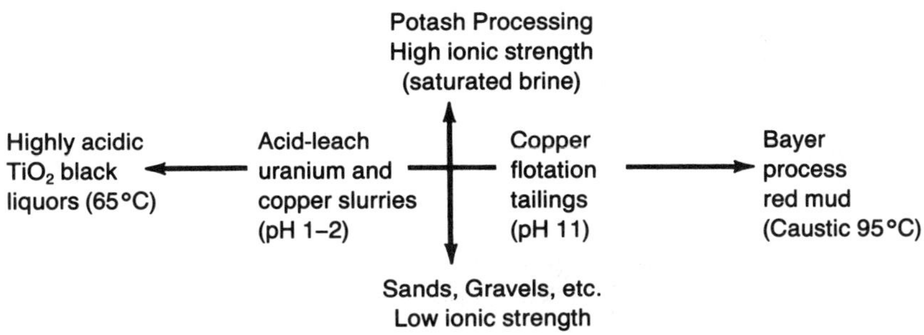

FIG. 16 Variation of ionic strength and pH within the "mineral processing" industries.

cular weight anionic and certain cationic polymers are suitable for neutral conditions. In addition to this, the type of solid/liquid separation process influences selection in terms of the molecular weight of the product, with low to medium molecular weight predominating for filtration and high molecular weight for sedimentation.

There are exceptions to some of the statements in this by no means comprehensive review of the use of water-soluble polymers in mineral processing, and Ref. 108 is recommended as a more detailed account if deeper insight is required. It is, however, an area in which liquid dispersion polymers find significant use alongside their solid-phase contemporaries.

(b) Water Pollution Control. This area of application covers municipal sewage and industrial effluent treatment. These are in effect very similar types of operations; both are very heavily dependent on solid/liquid separation and in this respect have many similarities in terms of the unit treatment processes involved. Thus sedimentation to remove suspended solids is common to both, as are the sludge dewatering processes that are used to deal with the sludges produced as a result of sedimentation. If anything, the only aspect that distinguishes the two treatments is the scale upon which they are carried out, with municipal sewage treatment being by far the larger of the two. As an example of this area of application, then, one need only consider municipal sewage treatment, bearing in mind its similarity to requirements within the industrial sector.

To better identify the areas of application within pollution control, a schematic diagram of a simple sewage treatment plant is shown in Fig. 17. This indicates how the flow of raw sewage taken into the works is subjected to primary sedimentation, biological treatment, and secondary sedimentation. Within the sedimentation tanks of a sewage treatment plant, two types of sludge are generally produced. These are identified as primary and secondary sludges. Primary sludge is collected in the primary sedimentation

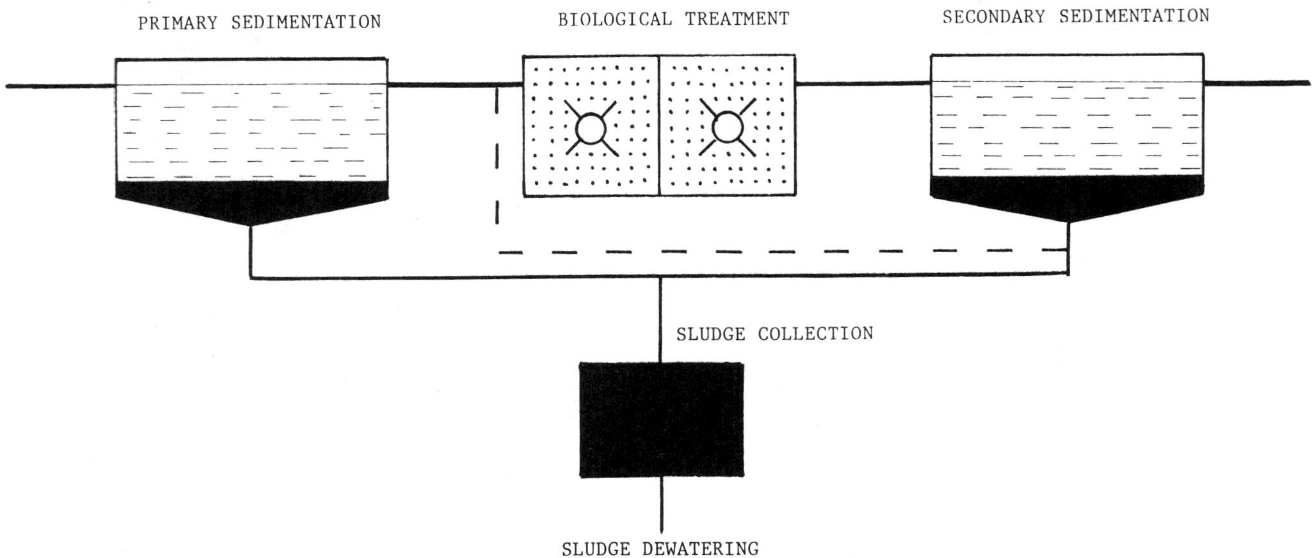

FIG. 17 Simple schematic diagram of sewage treatment plant. Polyelectrolytes are increasingly used to aid sedimentation and are essential for sludge dewatering. (Courtesy of Allied Colloids Limited.)

tanks after the sewage has passed through the inlet part of the works where screening and grit removal take place. The secondary sludge is collected following biological treatment, which converts soluble components, not taken out in the primary sedimentation tanks, into biomass and consequently settleable solids; this sludge is particularly difficult to dewater. The two sludges are mixed to provide final blend that can be presented for dewatering.

It is obvious from the discussion on mineral processing that water-soluble polymers have the capability to provide improved efficiency, at the sedimentation stages, of sewage and industrial effluent treatment and are in fact used in practice. In particular, they have in the past been applied to the inflow prior to sedimentation to improve the capacity of plants that have become overloaded as a result of increases in population and industrial activity. Increasingly, however, they are being considered at the design stage for proposed plants with a view to reducing plant size and thus the capital expenditure on new plants.

The dewatering process, which can involve filtration or centrifugation, was originally carried out with the aid of chemical treatment entailing the use of heavy inorganic chemicals such as lime and/or aluminum or iron salts. This, however, resulted in slow dewatering in the case of filtration and little or no dewatering with centrifugation. A further problem arose because of the low solubility and high dose requirement (sometimes as high as 40% on dry solids); they made a considerable contribution to the amount of material that eventually had to be disposed of. The development of cationic polyelectrolytes that need only be added at doses less than 0.5% on dry solids and produce high degrees of solid/liquid separation, and hence rapid dewatering, represented a considerable breakthrough (see Fig. 18).

In this area of activity the use of polyelectrolyte dispersions has made considerable impact; some countries use them to treat as much as half of their sludge. Unlike mineral processing, pollution control applications are largely the domain of cationic products. A similarity exists, however, in that they are prepared at varying levels of ionic activity and molecular weight to satisfy, first, the needs of the different characteristics that are to be found within sewage and sludges from different areas, and second, the different demands of processes used to deal with them throughout the world. Consequently, the product range comprises low ionic copolymers to high ionic homopolymers at low to high levels of molecular weight.

(c) Paper Production. The formation of the sheet during papermaking is basically a filtration process that over the years has been given much technical consideration in arriving at its perfection. An important consider-

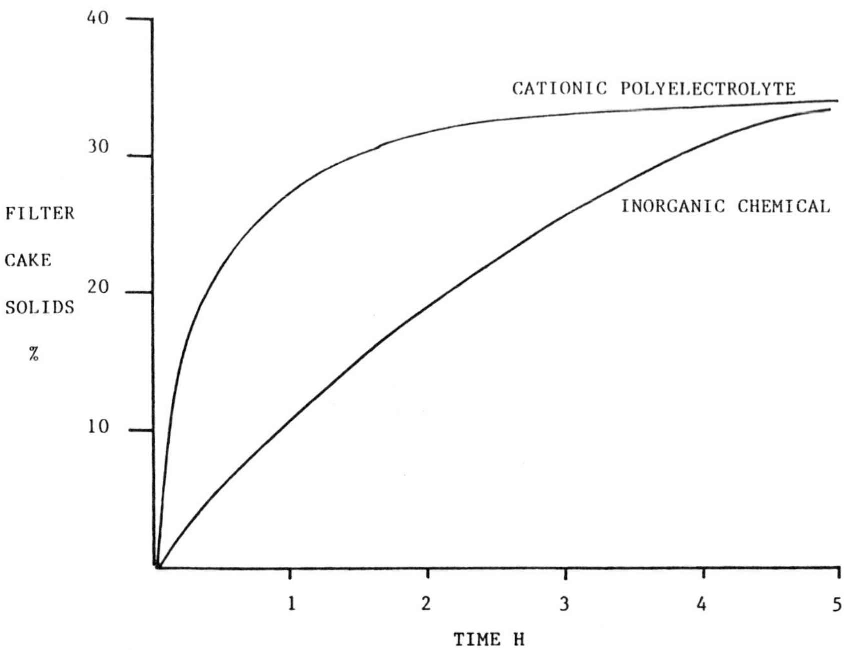

FIG. 18 Typical dewatering rates for cationic polyelectrolyte and inorganic salt on a filter press.

ation in obtaining good sheet formation is that of ensuring good retention of the fibers and fillers that are used in producing the paper, and many important factors affect these parameters [109].

The role of water-soluble polymers within the paper industry identifies and helps to answer this area of special need. The primary use of such products is as retention aids added at the wet end of the production process. In essence this involves the use of a solution of the polymer to flocculate the fibers and fillers, thus, as the description would imply, promoting retention of the flocs on the machine wire. Good retention aids help to ensure that the web of fibers formed on the machine wire can drain efficiently while retaining the maximum amount of filler, usually china clay or calcium carbonate, with low loss into the white water leaving the machine. The white water is subsequently returned for use in the preparation of further stock required to maintain continuity of production and therefore has to be as clean as possible.

It is essential that a good retention aid have a high level of solubility and that there be no extraneous insoluble components that can get into the stock. If this is not the case and solid particulate matter becomes embedded in the web of fiber used to form to the paper, imperfections will arise and/or the discontinuity in the web will cause breaks and disruption of production, which can lead to very serious loss. The liquid dispersion polymers have a considerable advantage in respect of insolubles, since the particles of polymer that make up the discontinuous phase are too small to cause problems, even in the unlikely event of some of them being insoluble.

The types of water-soluble polymers that have been used to improve retention extend across the entire spectrum of ionic character, although the most important group are the cationic polymers. Generally speaking they are required to be of low to medium ionic character and medium molecular weight [$(7-10) \times 10^6$] and are sometimes used in conjunction with more highly charged products of low molecular weight, which help to promote better drainage.

A more recent innovation that has served to greatly improve both retention and drainage uses a multicomponent system comprising cationic polymer and bentonitic clay. This is known as the Hydrocol process [110] and involves the addition of cationic polymer to the system followed by shearing down of the floc to produce a microfloc in the stock. Subsequent addition of the bentonitic clay as a slurry produces very stable flocs that dewater well and provide excellent retention and formation.

B. Acrylic Polymers in the Textile Printing Industry

1. Pigment Printing

Considering all the textile printing systems available, printing with aqueous pigment dispersions offers the cheapest and simplest means of coloration. Pigments are crystalline solids that are insoluble in the media from which they are applied and in the substrate to which they are applied; they are secured to the fiber surface simply by the use of a binder. In principle, this simple low cost, high throughput system with no subsequent washoff is obviously very attractive, particularly as the process can be applied to fabrics made of blends of fibers that can pose technical problems when substantive dyes are used.

However, relative to soluble dye systems there have always been well-recognized disadvantages to the use of pigment dispersions. These include a tendency to inferior wet-rubbing fastness, stiff handle (particularly with lightweight, flexible, or bulky materials). Problems can also arise with running properties (blocking of screens and engravings, and build-up on rollers) and environmental concerns (relating to pollution, flammability, and odor when emulsion thickeners are used).

The earliest printing pastes were thickened by simply emulsifying kerosene into the aqueous phase, which resulted in the correct rheological characteristics for printing. Levels of 70% kerosene in the paste were not uncommon. Later technology used solid grade polyacrylic acid dissolved in the aqueous phase, which made it possible to reduce the amount of kerosene to about 40% within the paste. Even these have been superseded by water-in-oil liquid dispersions as described in Sec. IV.A.2.

After the pigment, the two most important components of a pigment printing paste are the binder and the thickener. Both are now usually acrylic polymers, careful selection of which has enabled minimization of the above problems. The binder is generally supplied as a so-called o/w latex, where the discontinuous phase consists of acrylic polymer with minor amounts of dispersing agents. Modern thickeners are usually supplied as w/o liquid dispersions consisting of substantially dehydrated acrylic polymer particles suspended in a paraffinic hydrocarbon oil.

(a) Pigment Binders. The binder is a film-forming polymer that, when applied to the substrate together with pigment, coalesces and secures the pigment in place. Its binding properties are enhanced by incorporating self-cross-linking groups to produce a molecular network that insolubilizes the binder, further reduces its water sensitivity, and anchors the pigment more securely to the fiber. An ideal binder should possess all the following properties: good adhesion to the fiber, without being tacky; resistance to chemical attack and mechanical stress; good fastness; flexibility to give a soft handle; nonyellowing, during both the curing and subsequent life of the material; easy removability from screens; and shear stability.

There are a wide variety of monomers that can be copolymerized to give products of varying properties, but for reasons of economy, availability, simplicity, etc., acrylic pigment print binders are based on butyl acrylate, ethyl acrylate, methyl methacrylate, and acrylonitrile. The choice is governed to a great extent by the cost and the handle required. Comonomers other than acrylics are sometimes included to impart special properties, such as the use of styrene to harden a polymer. In general, the chemical constitution of the commonly used comonomers influences the glass transition temperature and hence the flexibility of the polymer as shown in Table 9.

Increasing the molecular weight generally gives improved wash fastness and rubbing fastness with surprisingly little adverse effect on handle.

Fastness properties are achieved by the incorporation of a cross-linking monomer into the formulation, the one most commonly used being *N*-methylol acrylamide (NMA).

$$CH_2{=}CH{-}\underset{\underset{O}{\|}}{C}{-}NHCH_2{-}OH$$

TABLE 9 Commonly Used Monomers which can be Copolymerized to Provide Ideal Film Forming Properties for Binders

Comonomer: Formula	Comonomer: Name	T_g of homopolymer (°C)	Hardness[a]
$CH_2{=}CH{-}COO{-}C_4H_9$	*n*-Butyl acrylate	−56	
$CH_2{=}CH{-}COO{-}C_2H_5$	Ethyl acrylate	−22	
$CH_2{=}CH{-}$ 	Styrene	100	
$CH_2{=}C(CH_3){-}COO{-}CH_3$	Methyl methacrylate	105	
$CH_2{=}CH{-}CN$	Acrylonitrile	110	

[a]Hardness increasing top to bottom.

Under acidic conditions the binder incorporating such a reactive group reacts with itself or with any reactive group on the substrate. For example,

$$B{-}CH_2{-}OH + HO{-}CH_2{-}B \xrightarrow{pH < 5} B{-}CH_2{-}O{-}CH_2{-}B + H_2O$$

and

$$B{-}CH_2{-}OH + HO{-}cellulose \longrightarrow B{-}CH_2{-}O{-}cellulose + H_2O$$

where B is the methylated side chain of the binder copolymer, incorporating NMA.

Increasing the amount of cross-linker increases the reactivity of the binder and improves the fastness properties, but the amount that can be used will be limited by the stability and handleability of the product, as increasing the cross-linker level increases the reactivity and viscosity of the polymer. Generally, incorporation of 1–3% of cross-linking monomer gives good cure properties. To avoid the release of formaldehyde during the manufacture of these cross-linkable polymers, *N*-isobutylmethylol acrylamide (NIBMA) is increasingly being used.

Since acrylic binders are supplied dispersed in aqueous media, their performance is influenced by the choice and quantity of stabilizing surfactant, particle size, and particle size distribution. These factors influence such properties as product stability (settlement, freeze/thaw, skinning), shear stability (runnability, screen blocking), and foaming. In general, a relatively small particle size is needed to avoid settlement problems. However, the smaller the particle size, the greater the amount of surfactant stabilizer that is needed during manufacture, which adds to the cost and can result in foaming problems.

Acrylic binders are available that satisfy all the criteria mentioned earlier—giving stable products with good running properties that gives nonyellowing prints of good fastness. Aqueous pigment print systems have been criticized for their stiff handle. However, very soft flexible prints are achievable, particularly if the manufacturer has designed the binder to be responsive to additional softeners. Silicones have a beneficial effect on handle by affecting the surface smoothness of yarns and fabrics. However, by using alternative softeners that actually affect the flexibility of the binder film, pigment prints can be achieved the softness of which is very similar to those produced using traditional emulsion thickening. This is illustrated by the effect of the softener Alcoprint PSM on the acrylic binder Alcoprint PBA. Assuming the fabric bending length to be an indication of softness (the shorter the bending length, the softer the fabric), Fig. 19 shows that Alcoprint PSM has a much more dramatic effect than the silicone. The latter cannot approach the softness achieved using the emulsion thickener irrespective of how much is used.

(b) Pigment Print Thickeners. Like all other printing systems, the aqueous pigment dispersions used as colorants must be thickened to achieve well-defined uniform prints. It is well documented that early attempts to use water-soluble natural or modified natural thickeners failed. It was only when emulsion thickeners containing kerosene were introduced that good pigment prints were achieved. Partial replacement of kerosene by solid grade polyacids provided a significant technological improvement, and commercially successful products featured high molecular weight polyacrylic acid as the active ingredient, swelling and developing viscosity when neutralized with a suitable base. Emulsion thickeners are, however, under increasing pressure on environmental and cost grounds.

A more acceptable alternative is to use w/o liquid dispersions. These consist typically of very small (about 1 μm) dehydrated polyacrylic acid particles suspended in a paraffinic hydrocarbon oil, the stability of the dispersion being ensured by the use of a stabilizer and an emulsifier. The

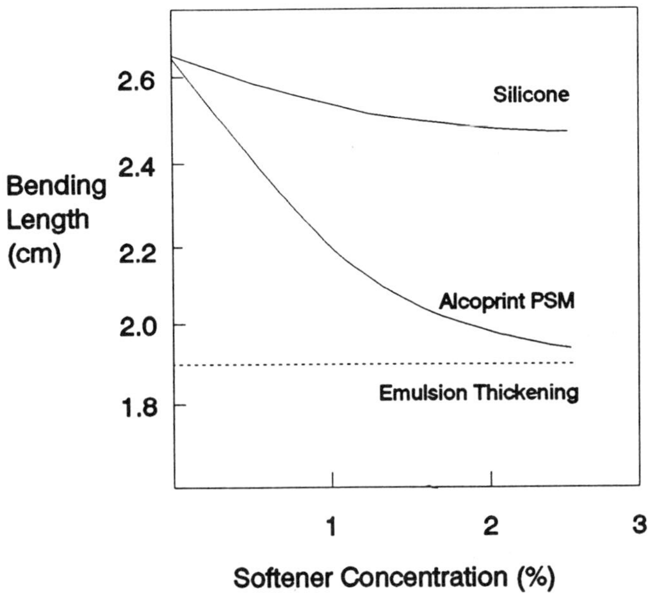

FIG. 19 Influence of softener concentration on bending length.

fine polymer particles are formed directly in the dispersion (see Sec. IV.A.2), thereby resulting in a much narrower range of particle size than could ever be achieved by grinding a solid product and dispersing it in a liquid. This uniformity and the small size of particles, besides contributing to the product stability, is the key to ease of use.

On addition to the aqueous pigment dispersion, the hydrocarbon is emulsified away, thus releasing the dispersion particles. Small particles give the dispersion a large surface area, which helps to achieve rapid swelling and a fast thickening action in the slightly alkaline aqueous medium. The particle diameter increases approximately tenfold, corresponding to a 1000-fold increase in volume. Careful control of the molecular size and cross-linking capability of the polymer molecules ensures that the neutralized polymer remains as swollen particles rather than dissolving. The flow properties of the resultant print pastes are very similar to those of emulsion-thickened pastes, as Fig. 20 depicts.

In general, increasing the molecular weight increases the swellability and hence thickening effect of the polymer. However, uncross-linked polymer is water-soluble and stringy, exhibiting rheological properties quite unsuitable for printing. Introduction of cross-links by means of such monomers

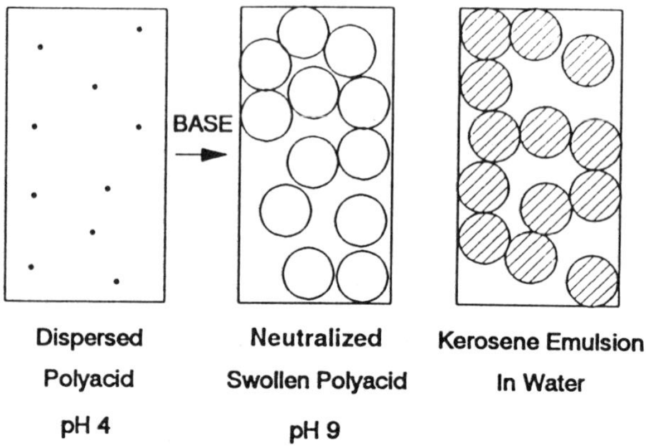

FIG. 20 Diagram demonstrating the development of viscosity on neutralization of synthetic cross-linked print thickeners.

as methylene bisacrylamide (MBA) ensures swellability rather than solubility. As the degree of cross-linking increases, the efficiency of the thickener increases as shown in Fig. 21, which compares three products of the same polyacid and molecular weight but different degrees of cross-linking.

Reducing the cross-linking reduces flushing (the diffusion of color from the printed area to the adjacent area) and penetration but can adversely affect color yield (intensity). The balance of molecular weight and degree of cross-linking also influences the overall appearance of the print and whether a smooth blotch area is achieved, as can be seen in Fig. 22.

The above considerations explain why it is not enough to simply monitor the efficiency of a thickener in terms of the viscosity created by a set concentration of polymer. For example, Fig. 23 shows that although product D (a low molecular weight version of product A) has lower viscosity than product A, the slope of the graph remains the same. If the printer were to simply compare the viscosity generated by product D to that generated by the "ideal" product B at one concentration, x, he would conclude that they were identical. In practice there could be wide variations, with product D being far more subject to flushing than product B. It is essential, therefore, to characterize a thickener by at least two points on a graph of viscosity versus concentration to measure its thickening power and also to get some indication of its effect on such properties as flushing, coverage, color yield, and penetration as dictated by molecular weight and degree of cross-linking.

Viscosity is developed when the molecular entanglements in the polyacrylic acid aggregates swell in the water. This will occur only if the poly-

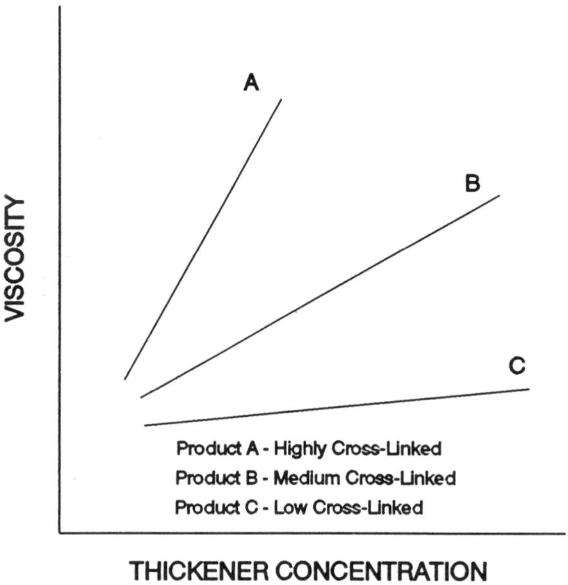

FIG. 21 Effect of cross-linking on thickening efficiency.

acrylic acid is in its ionized form, that is, if it has been neutralized with a suitable base. The most commonly used bases are caustic soda and ammonia. However, for pigment printing, caustic soda is unsuitable because there is no washoff process, the thickener remains on the fabric, and the residual water-soluble sodium polyacrylate would detract from the waterfastness properties of the print. It is essential, therefore, that a volatile base be used so that the acidic conditions necessary for cross-linking are achieved and after curing the polymer is left on the fabric as the insoluble polyacrylic acid. Ammonia, being cheap and readily available, is frequently used, but alternatives such as triethanolamine may have to be substituted if the odor of ammonia is unacceptable. Whatever base is selected, it is essential that the volatility be such that the base will be completely removed during drying. It will be seen later that with soluble dyes the situation is quite the reverse.

Liquid dispersion thickeners exhibit all the usual environmental, safety, and cost advantages of acrylics over emulsion systems. There are, however, additional benefits. They can be offered at 50–60% active ingredient compared to the usual 20–30% for gels and emulsions, thus contributing to lower packaging and transport costs and requiring smaller storage areas. They are low viscosity mobile liquids that can be tapped from drums, are

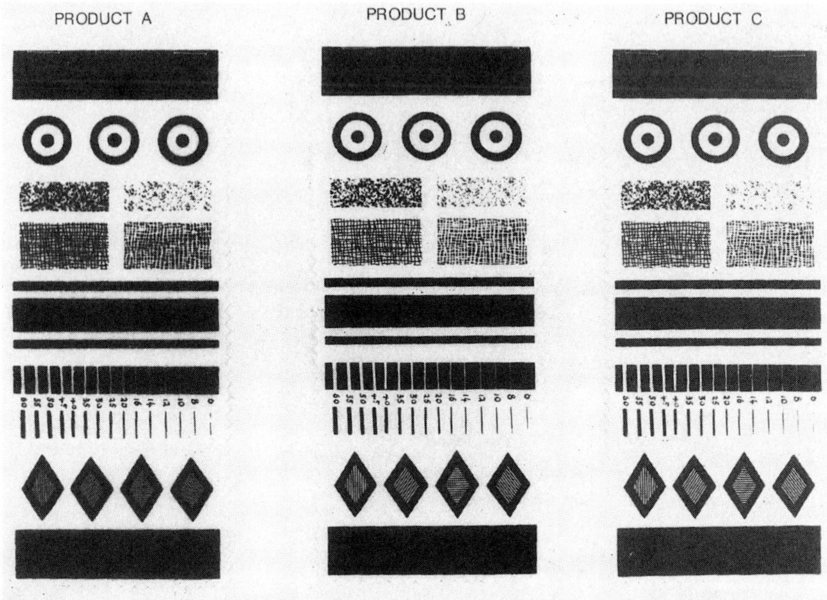

FIG. 22 Effect of cross-linking on print performance (same products as in Fig. 21).

easily metered, and can be weighed accurately. Because they are water-free they can be fully neutralized with ammonia yet retain their low viscosity and hence dispense with the need to handle ammonia within the printing plant. The small particle size together with the special stabilizer system ensures a long shelf life. Being water-free, they are insensitive to frost.

(c) Pigment Dispersants. Both binders and synthetic thickeners contain surface-active agents that will help disperse pigments. However, there are instances where the system benefits from additional dispersants, particularly when the pigment loading is high, as in the case of printing whites with titanium dioxide. An investigation into the use of a variety of different chemical types of dispersing agents has led to the conclusion that the sodium salt of low molecular weight polycarboxylic acid is the most suitable. Such products are stable over a wide range of temperatures and pH and are nonfoaming. The amount of dispersant required will depend on the amount of pigment, but a print paste containing 35% titanium dioxide would require no more than 0.5% of dispersant. In some cases it is advantageous to add the dispersant to the paste before adding the pigment.

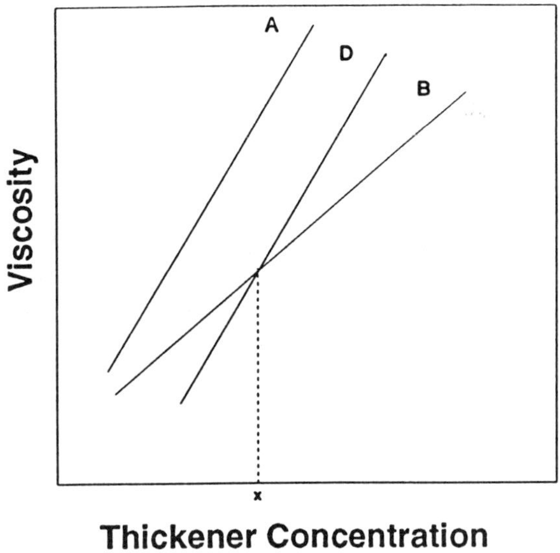

FIG. 23 Viscosity versus concentration for textile print thickeners with varying degrees of cross-linking. A and B are same products as in Fig. 21; D is a lower molecular weight version of A.

2. Water-Soluble Dye Printing

The success gained by acrylic thickeners in pigment printing during the past decade has not been matched by their progress in water-soluble dye printing. There are a number of reasons for this fact. The cost advantages of acrylic thickeners compared to emulsion thickeners are easy to demonstrate in most parts of the world, but the relative costs of acrylic versus the commonly used natural thickeners such as alginates and cellulose products are often quite close.

Efficient acrylic thickeners are usually sensitive to the high concentrations of electrolyte present in the print pastes, and there is no obvious environmental advantage such as can be demonstrated when aqueous systems based on w/o liquid dispersions are compared to emulsion systems. However, w/o dispersion thickeners do offer considerable advantages, and progress has been made, particularly in the printing of reactive and disperse dyes.

(a) Reactive Dye Printing. Because of the chemical reactivity of this class of dye, printers are very restricted in their choice of thickener. In practice they will probably use sodium alginate or a part emulsion system with kerosene. Sodium alginate is one of the more expensive natural thick-

eners, and the supply situation has come into question from time to time. It was understandable,therefore, that the swellable liquid dispersions of acrylic polymers gaining such success in pigment printing should be examined. There are a number of reasons, however, why ammonium polyacrylate-based thickeners are unsuitable for printing with reactive dyes; poor color yield, foam generation, and difficulty of removal (all associated in different ways with ammonia release). These problems can be overcome by producing the sodium salt of the polymer, which does not react with the alkali in the print paste and remains in this fully neutralized, water-soluble form throughout fixation. Thickeners of this type have been offered to the industry and have proved to be successful. Alcoprint RTA, for example, is a thickener designed as a direct replacement for sodium alginate. Since it is in liquid dispersion form, it exhibits all the additional advantages discussed previously, such as being highly concentrated and easy to handle, and gives rapid print paste preparation.

For some printers there remains one weakness with this type of synthetic thickener—its electrolyte sensitivity. The traditional method of preparing a print paste is to add the color to a prepared thickened stock that contains all the necessary chemicals. Since stock thickenings based on sodium alginate are largely insensitive to the addition of reactive dyes and other electrolytes, it is normally found that the variation in viscosity from one paste to another is small. The polymer chemist can modify the polymer to reduce the sensitivity to electrolyte, for example, by introducing higher proportions of acrylamide. However, such modifications invariably reduce the efficiency of the thickener. A better approach is to take the most technically suitable product and devise a method for using it that is practical but not governed by the traditional techniques optimized for using alginates. Such a system has been successfully promoted by Allied Colloids and is referred to as direct dosing. The thickener (e.g., Alcoprint RTA) is treated like a dye in a print recipe insofar as the printer accepts that for different shades one can expect to use different amounts of dyes; similarly, different quantities of thickener will have to be used to generate the same printing viscosity. Once the recipe is established for a certain shade, it is perfectly reproducible.

Such thickeners offer the printer several advantages over alginates. Being liquid products they are easy to dispense and use; stock thickenings are not required; the final viscosity is developed within minutes; if needed, "back thickening" is simple and accurate; after preparation, the printing paste is stable for very long periods; and a 20% advantage in color yield is consistently achieved.

(b) Disperse Dye Printing. As with reactive dye printing, liquid dispersions of swellable acrylic polymers are not available as replacements for natural polymers as thickeners in disperse dye printing. The resulting print-

ing pastes can produce excellent quality prints on very lightweight polyester fabrics. They are also simpler than formulations required with natural polymers in that there is no need for an acid or acid donor (to protect dye from oxidation) or a fixation accelerator (to improve dye retention in the print).

3. Printing Table Adhesives

When printing with more than one color it is imperative to ensure accurate registration of the design. This can only be achieved if the fabric is held firmly in contact with the blanket so that no relative movement takes place. To achieve this, it is normal practice to apply a suitable table adhesive to the blanket.

Ideally, water-based "temporary" adhesives should possess the following properties; minimum transfer from blanket to fabric, good tack and adhesion to a wide variety of different fiber types and fabric constructions; and ease of removal from the blanket by water. Polyvinyl alcohol has become generally accepted by the industry and gives excellent results with hydrophilic materials, particularly cellulosics and cellulose-rich blends. However, problems may be encountered with hydrophobic fabrics such as polyester.

The adhesion of acrylic polymers to fibers will be influenced by their own hydrophilic/hydrophobic nature. Polyacrylic acid will exhibit good adhesion to cellulose but poor adhesion to polyester. If ester comonomers such as methyl acrylate, methyl methacrylate, and butyl acrylate are introduced into the polymer chain, this situation can be reversed. It is possible, therefore, to optimize the comonomer ratio to give the best adhesion for a particular substrate, and by varying the molecular weight it is possible to produce a polymer of convenient viscosity at working concentrations.

C. Wallcovering Adhesives

Liquid dispersions of acrylic polymers have recently found application in prepasted wallcoverings. Prepasting entails coating a wet adhesive composition onto the substrate and drying. The wallcovering is simply dipped in water to swell the adhesive layer prior to application. Convenience of application is of paramount importance, since the main market is the "do it yourself" home market.

Prepasted wallpaper was introduced into the North American wallcoverings industry in the late 1950s and early 1960s. Since then a wide variety of substrates have been developed—coated paper, latex-impregnated paper, paper-vinyl laminates, nonwoven vinyl laminates, nonwoven metallized foil laminates, vinyl foil, and expanded vinyl-paper laminates. This, together with the increasing need for rapid processes of coating evenly on the substrate and user friendliness, has required the development of better adhesives. Especially important has been prepasting of paper bases for expanded vinyl products. Starch, cellulose derivatives, and water-soluble synthetic polymers have been the conventional bases of adhesives. Starch adhesives are especially problematical, tending to be sticky to handle, susceptible to oxidative cross-linking on prolonged storage (reducing water absorbency and effectiveness), and tending to cause visible blemishes when accidentally smeared on the decorative face of the wallcovering.

Liquid dispersions of alkali-swellable acrylic polymers (prepared as described in Sec. IV.8.2) [111] are coated on the wallcovering substrate by conventional printing equipment, namely, flexographic, gravure, or rotary screen printing presses or reverse roller coater. The best results have been achieved using gravure printing because the cylinder, can be accurately set to apply the required weight of material. This removes any potential of operator error to which other methods are prone. The hydrocarbon carrier medium is evaporated off.

Acrylic polymers have a number of major advantages over conventional types of adhesives. They give improved adhesion at lower coat weights (typically 6–8 g/m^2, compared to 10–20 g/m^2 for conventional) and can be applied at higher active concentrations (50–60%, compared to 20–30%). These factors and the greater volatility of the nonaqueous solvent enable faster coating, with lower thermal energy requirement for solvent removal. Also, the acrylic polymers are not susceptible to oxidative cross-linking and bacterial attack that are problematic with traditional adhesives, so antioxidants and bactericides are not required. A disadvantage, however, from the environmental point of view is the associated emission of volatile organic solvent during coating on the substrate in wallcovering manufacture.

Wallcoverings prepasted with these products are now very well established both in North America and in Europe, and as the wallcoverings markets develop so will the adhesive technology. Research and development will continue with the objective of developing prepaste adhesives that can be applied from highly concentrated, solvent-free systems.

ACKNOWLEDGMENTS

I thank John Clarke, Malcolm Hawe, Roy Seddon, Rodney Field, Ken Symes, and Michelle Stead for their invaluable contribution in the preparation of this chapter.

REFERENCES

1. I.K. Miller, and J. Zimmerman, Condensation polymerization and polymerization mechanisms. ACS Symposium 285 *Appl. Polym. Sci.* (2nd ed.), pp. 159–73 (1985).

2. W. Ray, Harmon modelling of addition polymerization processes—free radical, ionic group transfer and Ziegler-Natta kinetics, *Can. J. Chem. Eng.* *69*(3): 626–629 (1991).
3. R. Bhattacharga and S. Nandi, Addition polymerization in mixed solvent systems, *J. Indian Chem. Soc.* *48*(11): 1033–1038 (1971).
4. W. Bauer, O. Rohm, and C.T. Kautter (to Rohm & Haas AG), Ger. Patents 639 094 (1936) and 875 092 (1953).
5. D. Farrar (to Allied Colloids Ltd.), Eur. Patent 0 129 329 (1984).
6. J.K. Millington and S.V. Slovenskai (to Huntsman Chemical Corp.), U.S. Patent 4,730,027 (1988).
7. W. Heide, K. Schmitt, and H. Hartmann (to BASF AG), Eur. Patent 107 020 (1984).
8. C.A. De Fazio (to Celanese Corp.), U.S.Patent 4,656,222 (1987).
9. G.L. Mier (to Dow Chemical Co.), U.S. Patent 3,575,913 (1971).
10. T.W. Lai and B.R. Vijayendran (to Air Products & Chemicals, Inc.), Eur. Patent 264 649 (1988).
11. G.D. Parfit, *Dispersion of Powders in Liquids*, Applied Science, London, 1973.
12. Th.F. Tadros, *Solid-Liquid Dispersion*, Academic, New York, 1987.
13. J.W. Goodwin, *Colloidal Dispersions*, Royal Society of Chemistry, Special publication no. 43 (1982).
14. T.C. Patton, *Paint Flow and Pigment Dispersion*, Wiley-Interscience, New York 1978.
15. Allied Colloids, Technical and Processing Data Sheet No. 6200, 1991.
16. Allied Colloids, Technical and Processing Data Sheet No. 6216, 1991.
17. C.-J. Chang and T.E. Stevens, U.S. Patent 4,663,385 (1987).
18. L.G. Dammann, U.S. Patent 4,338,239 (1982).
19. D. Engel, H. Rinno, and K. Zimmerschied, U.S. Patent 4,569,965 (1986).
20. J. Fock and E. Esselborn, U.S. Patent 5,086,142 (1992).
21. M. Hawe and D. Farrar, Eur. Patent 216,479 (1991).
22. M.J. Karickhoff, U.S. Patent 4,464,524 (1984).
23. C.G. Ruffner and J.M. Wilkerson, U.S. Patent 4,600,761 (1986).
24. C.G. Ruffner, U.S. Patent, 4,616,074 (1986).
25. G.D. Shay, E. Eldridge, and E. Kail, U.S. Patent 4,514,522 (1985).
26. L.F. Sonnabend, U.S. Patent 4,384,096 (1983).
27. R. Ashady, *J. Microencapsulation 9*:187 (1992).
28. J.E. Vandegaer (Ed.), *Microencapsulation: Processes and Applications*, Plenum, New York, 1973.
29. M-CAP Technologies International, Wilmington, Del.
30. South-West Research Institute, *A Capability Statement for Microencapsulation*, San Antonio, Tex. (Technical brochure).
31. Washington University Technology Associates, *New Processes for Microencapsulation and Granulation*, St. Louis, Mo. (Technical brochure).
32. Ronald T. Dodge Co., *The Handling of Difficult Materials Using Microencapsulation Technology*, Dayton, Ohio. (Technical brochure).
33. A. Kondo (Ed.), *Microcapsule Processing and Technology*, Marcel Dekker, New York, 1979.

34. R.E. Sparks, in *Kirk-Othmer Encyclopedia*, Vol. 15, (M. Grayson, Ed.), Wiley-Interscience, New York, 1981.
35. P.B. Deasy, *Microencapsulation and Related Drug Processes*, Marcel Dekker, New York, 1984.
36. M.H. Gutcho, *Microcapsules and Microencapsulation Techniques*, Noyes Data Corporation, Park Ridge, N.J., 1976.
37. J.R. Nixon and M.S. Harris, in *Development of Drugs and Modern Medicines* (J.S. Gorrod, G.G. Gibson, and M. Michard, Eds.), Verlagsgesellschaft, Weinheim, 1986.
38. R. Arshady, *J. Microencapsulation 6*:1 (1989).
39. R. Jalil and J.R. Nixon, *J. Microencapsulation 7*:297 (1990).
40. B.K. Green and L.S. Schleicher (to National Cash Register), U.S. Patent 2,800,457 (1957).
41. H.G. Bungenberg de Jong, *Proc. Acad. Amsterdam 41*:646 (1938).
42. T. Kondo, (Ed.), *Microencapsulation: New Techniques and Applications*, Techno Inc., Tokyo, 1979.
43. D.J. Hasler and T.A. McGhee (to Wiggins Teape), U.K. Patent 1 507 739 (1978).
44. P. Sinclair (to Wiggins Teape), U.K. Patent 2 073 132 (1983).
45. G. Pietsch and K.-H. Schrader (to Papierfabrik August Koehler) Australian Patent 27028/88 (1989).
46. R.P. Bowman (to Appleton Papers), Australian Patent 70144/87 (1987).
47. S. Irii, H. Kumamoto, and T. Schiozaki (to Kanzaki Paper Manufacturing Co.), Jpn. Patent 01 130 727 (1989).
48. G.W. Matson (to Minnesota Mining and Manufacturing), U.S. Patent 3,516,941 (1970).
49. P. Chamberlain and J.G. Langley (to Allied Colloids), Eur. Patent Appl. 379 379 (1989).
50. P. Studt, H. Zobelein, and J. Galinke (to Henkel et Cie), U.S. Patent 3,244,640 (1966).
51. K.K. Mistry, J.G. Langley, and K.C. Symes (to Allied Colloids), U.K. Patent Appl. 9 110 407.5 (1991).
52. Fuji Photo Film Company Ltd., U.K. Patent 1 319 123 (1973).
53. T.J. Speaker and L.J. Lesko (to Temple University), U.S. Patent 3,959,457 (1976).
54. P.J. Watts, M.C. Davies, and C.D. Melia, *Crit. Rev. Therap. Drug Carrier Syst. 7*:235 (1990).
55. D.B. Anderson, J. Forder, I.M. Johnson (to Allied Colloids), U.K. Patent 1 482 515 (1977).
56. K.K. Mistry, J.G. Langley, and K.C. Symes (to Allied Colloids), U.K. Patent Appl. 9 110 408.3 (1991).
57. N. Oku, *Polymeric Drug Delivery Systems*, ACS Symp. Ser. No. 469, American Chemical Society, Washington, D.C., 1991, pp 24–33.
58. Y. Ozden and V.N. Hasirci, *Br. Polym. J. 23*:229 (1990).
59. J.C. Vuillemard, *J. Microencapsulation 8*: 547 (1991).
60. B. Boh, A. Kornhauser, E. Knez, M. Kukoric, and V. Birsa, 7th Int. Symp. Microencapsulation, Glasgow, 1990.

61. K. Nakamura, 7th Int. Symp. Microencapsulation, Glasgow, 1990.
62. Moore Business Forms Inc., U.K. Patent 1 046 409 (1966).
63. J.D. Chen, K. Kovacs, M.T. Thomas, and R.-C. Liang (to Mead Corporation), Eur. Patent Appl. 433 094 (1990).
64. K. Matsumoto and Y. Ugajin (to Toppan Printing Co.), Jpn. Patent 02 229 545 (1990).
65. S. Matsunaga and S. Doi (to Canon) Jpn. Patent 02 004 276 (1990).
66. H. Havashi (to Fuji Photo Film Co.), Jpn. Patent 02 123 357 (1990).
67. A. Sakai (to Brother Industries), Jpn. Patent 03 077 961 (1991).
68. B. Conti, F. Pavanetto, and I. Genta, *J. Microencapsulation 9*:153 (1992).
69. S.P. Li, C.R. Kowarski, K.M. Field, and W.M. Grim, *Drug Dev. Ind. Pharm. 14*:353 (1988).
70. J.W. McGinity and G.W. Cuft, *Methods Enzymol. 112*: 84–101 (1985).
71. R. Gurny, N.A. Peppas, D.D. Harrington, and G.S. Banker, *Drug Dev. Ind. Pharm. 7*:1 (1981).
72. G. Gregoriadis (Ed.), *Liposomes as Drug Carriers: Recent Trends and Progress*, Wiley, Chichester, 1988.
73. J.L. Grangier, M. Puygrenier, J.C. Gautier, and P. Couvreur, *J. Controlled Release 15*:3 (1991).
74. C.J. Kirby and B. Law, in *Chemical Aspects of Food Enzymes* (RSC Spec. Publ. No.63) (A.T. Andrew, Ed.), Royal Society of Chemistry, Cambridge, 1987, pp. 106–117.
75. W. Alkhalaf, J.-C. Piard, M. El Soda, J.-C. Gripon, M. Desmazeaud, and L. Vassal, *J. Food Sci. 53*:1674 (1988).
76. L.S.Jackson and K. Lee, *Lebensm.-Wiss. Technol. 24*:289 (1991).
77. K. Koide and M. Karel, *Int. J. Food Sci. Technol. 22*:707 (1987).
78. K. Topfer, *Chemical Marketing Reporter*, Dec. 28, 1990. Schnell, New York.
79. R.J. Versic, *Drug Cosmet. Ind. 144*:30 (1989).
80. G. Brooks, *Manuf. Chem. July 1989*: 36–79.
81. R.C. Koestler, ACS 197th Meeting, Dallas, 1989, Industrial and Engineering Chemistry, No.147.
82. J. Thenard and S.B. Advani (to Pennwalt Corp.) U.S. Patent 4,915,947 (1990).
83. A.J. Kielbania, W.D. Emmons, and G.H. Redlich (to Rohm and Haas), Eur. Patent Appl. 305 212 (1989).
84. H.W. Haesslin and W.P. Iwanzik (to Ciba-Geigy A.-G.), Eur. Patent Appl. 281 521 (1988).
85. H.B. Scher, in *Pesticide Chemistry, Human Welfare and the Environment*, Vol. 4 (J. Myamoto and P.C. Kearny, Eds.), Pergamon, Oxford, 1983, pp. 295–300.
86. M. Rodson and H.B. Scher (to ICI Americas), Eur. Patient Appl. 336 666 (1989).
87. G.W. Matson (to Minnesota Mining and Manufacturing Co.), U.S. Patent 3,516,941 (1970).
88. M. Naoki, Y. Igarishi, and S. Aoyagi, (to Kureha Chemical Industries), Eur. Patent Appl. 368 576 (1990).
89. Anon., Empire 20 insecticide brochure, Dow Elanco Europe, Valbonne, France.

90. J. Shearman, *Chem. Eng. 99*:59 (1992).
91. H.A. Bohm (to Lim Technology Laboratories), U.S. Patent 4,948,586 (1990).
92. M.S. Korpela (to Kaukas Division, Kymmene Corp.), Eur. Patent Appl. 425 975 (1991).
93. L.A. Cantu and M.E. Yost (to Conoco Inc. and E.I. Du Pont de Nemours), U.S. Patent 4,986,354 (1991).
94. T. Okabe, M. Takizawa, and H. Takahashi (to Lion Corp.), Jpn. Patent 61 028 440 (1986).
95. M. Takizawa and H. Takahashi (to Lion K.K.) U.S. Patent 4,77,089 (1988).
96. J. Langley and K.C. Symes (to Allied Colloids), Eur. Patent Appl. 316 677 (1990).
97. J. Langley, K.C. Symes, and P. Holm (to Allied Colloids), U.S. Patent 5,035,900 (1991).
98. J. Langley and K.C. Symes (to Allied Colloids), Eur. Patent Appl. 356 239 (1990).
99. W.L.K. Schwoyer, *Polyelectrolytes for Water and Wastewater Treatment*, CRC, Boca Raton, Fl., 1981, pp. 2–45.
100. J.R. Field and G. Smalley, Flocculating agents, U.S. Patent 3,897,333 (1975).
101. I.H. Hess and K.K. Kurts, U.S. Patent 3,046,259 (1962).
102. J.W. Vanderhoff and R.M. Wiley, Water-in-oil emulsion polymerization process for polymerizing water-soluble monomers, U.S. Patent 3,284,383 (1966).
103. J. Bratby, *Coagulation and Flocculation*, Uplands, Press, Croydon, England, 1980.
104. N. Moss, *Theory of Flocculation*, Vol. 7, No. 5, Mine and Quarry, 1978, pp. 57–61.
105. T.K. Hunter and M.J. Pearse, The use of flocculants and surfactants for dewatering in the mineral processing industry, *Proc. CIM XIV Int. Mineral Process. Congr.*, Toronto, 1982, pp. 1–39.
106. M.J. Pearse and A.P. Allen, *Filtration Separation 20*(1): 22–27 (1983).
107. J.M. Cohen, G.A. Rourke, and R.L. Woodward, Natural and synthetic polyelectrolytes as coagulant aids, *J. Am. Water Works Assoc.*, April 1958, pp. 463–478.
108. T.K. Hunter and M.J. Pearse, The use of flocculants and surfactants for dewatering in the mineral processing industry, *Proc. CIM XIV Int. Mineral Process. Congr.*, Toronto, 1982, pp. 8–9.
109. D. Eklund and T. Lindstrom, *Paper Chemistry: An Introduction*, D.T. Paper Science Publications, Grankulla, Finland, 1991, pp. 145–191.
110. J.G. Langley and D. Holroyd (to Allied Colloids Ltd.), (Eur. Patent 235 893) (1986).
111. P. Flesler and G. B. Clarke (to Allied Colloids Ltd.), U.S. Patent 4,361,452 (1978).

9

Strength Enhancement of Inorganic Materials Through Dispersion

KEVIN KENDALL Department of Chemistry, University of Keele, Keele, Staffordshire, England

I.	Introduction	335
II.	Cracks and Strength of Inorganic Materials	337
III.	Poor Dispersion as a Source of Natural Flaws	339
IV.	Strength Improvements in Cements Through Dispersion	341
V.	Effects of Dispersion in Ceramic Processing	343
VI.	Stronger Ceramics Through Dispersion	344
VII.	Conclusions	347
	References	347

I. INTRODUCTION

Many inorganic materials, including cements and ceramics, are made from dispersions that are subsequently dried, reacted, or fired in a furnace. The final structure of such materials depends on the state of dispersion of the solid particles and also on flocculation, which can occur during drying or reaction. Since the strength of materials depends on microstructure, it is evident that strength may be related to the dispersion of the particles during the process. This paper demonstrates the connection between dispersion and strength of inorganic cements and ceramics. First, the theory of strength of these brittle materials is outlined. Then experiments are described on hydraulic cements, showing how strength has been raised by an order of magnitude through improved dispersion. Finally, it is demonstrated that similar considerations apply to ceramics, which can be strengthened considerably through extra dispersion.

The major bulk construction materials of our civilization, concrete and bricks, are made by dispersion processes. They are less advanced then engineering metals and plastics, though far exceeding them in terms of volume production. In addition, they suffer from brittleness, porosity, and low strength. Thus they have been neglected in terms of understanding. Why should they be porous and weak? Why should they be brittle? Condoms are cast from dispersions and are neither porous nor weak nor brittle. Powder metals can be injection-molded from dispersions to form dense, strong, tough components. The objective of this chapter is to show that substantial improvements in cement and concrete and in ceramics and related materials can be obtained through study of the connection between microstructure and properties. In particular, strength can be enhanced by improving microstructure through better dispersion.

Materials such as cement and ceramic are made in dispersion form because of the ease of processing. Although some small and simple articles such as cement catalyst supports or ceramic bathroom tiles are made by pressure compaction of dry powder, most cement and ceramic materials are made by mixing the powders with water to form a high solids dispersion, typically 50–70% by volume solids, then casting, filtering, or extruding the material to form the product. For example, Portland cement is ball-milled to a fine powder, then mixed with water, sand, and aggregate before being cast into shuttering to erect a building. The cement then reacts with the water to form a hydrate that glues the structure together. Or clay is swollen by stirring with water, then filter-pressed to give a stiff paste that is vacuum-extruded to form electrical insulators. These are dried and fired in a furnace to sinter the particles together. It would be impossible to make such large objects other than by a dispersion method, because dry powders clump and cake and so cannot be molded uniformly in large volumes. Nonuniformity leads to cracks, which are the scourge of brittle solids like cement and ceramics because they weaken them.

The message of this chapter is that inorganic materials can be strengthened by paying attention to the details of these dispersion processes. First it is shown that there is a direct and simple connection between strength and crack size. Then poor dispersion is identified as a source of large cracks. Finally, by improving dispersion, stronger cements and ceramics are revealed.

II. CRACKS AND STRENGTH OF INORGANIC MATERIALS

Cracks weaken inorganic materials enormously. This can be demonstrated using the two strength tests illustrated in Fig. 1. In the tensile test (Fig. 1a) a sheet of material is gripped at its ends and pulled with a force *F*, leading to a stress *F*/*bd* in the material, where *b* is the thickness and *d* the breadth of the sheet. If the sheet contains a crack of length *c* in its edge, then the crack propagates when the stress σ is raised to

$$\sigma = 0.89\sqrt{ER/\pi c} \tag{1}$$

where *E* is the Young's modulus of elasticity of the material and *R* its fracture surface energy, that is, the energy required to break a unit area of material. Here the crack is assumed to cross the full thickness *b* of the specimen but to have no gap across the crack faces until the force is applied to open the crack. As the force is raised, the crack length *c* suddenly increases to cause catastrophic failure across the breadth *d* of the sample. Equation (1) was originally discovered by Griffith [1] and shown to describe the cracking behavior of glass bulbs and tubes stressed by internal pressure. The fracture surface energy *R* has the units of Joules/m^2 and was thought of by Griffith as being akin to thermodynamic work of cohesion of liquid glass. Now we know that *R* is a more complicated quantity than that, and *R* is often much larger than could be expected from such a theory. For typical ceramics, measurement of *R* by this approach gives values ranging from 5 to 50 J/m^2.

Equation (1) is useful because it shows how sensitive such brittle inorganic materials are to cracks. As a crack gets longer, increasing *c*, the

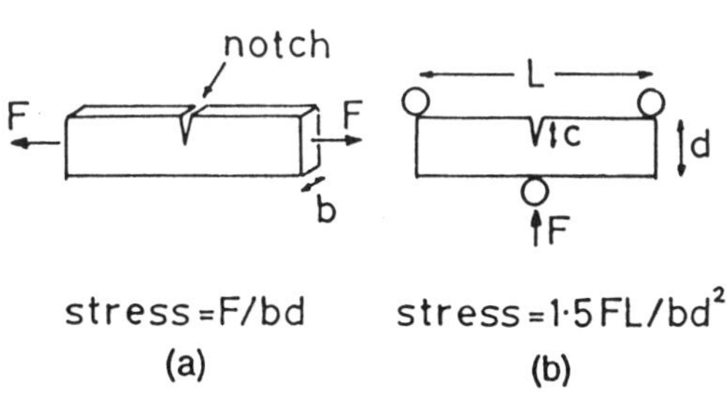

FIG. 1 (a) Tensile strength test on a notched strip of material. (b) Bending strength test on the same strip.

strength drops rapidly. This theory explains how an optical fiber can be very strong when it is free of imperfections, giving a strength of around 5000 MPa, yet very weak when it is scratched with a diamond scriber, say 10 MPa. Griffith also recognized that materials can contain other imperfections (arising from manufacture of from the environment) that act like cracks and so weaken the material as predicted by Eq. (1).

Later in this chapter, it is demonstrated that various flaws such as air bubbles and powder aggregates can give cracklike behavior. Thus, this theory provides a mechanism for the influence of dispersion on strength. Dispersion could reduce the size of air bubbles, for example, or it could break down particulate aggregates into smaller entities. Consequently, c in Eq. (1) would decrease and strength would increase as a result of dispersion.

The bending test shown in Fig. 1b is usually more convenient than the tensile test because no clamps are needed to grip the specimen. In this case, the stress distribution is more complex and is calculated from the formula

$$\sigma = 1.5FL/bd^2 \tag{2}$$

where L is the span of the sample, b its thickness, and d its depth. Once the strength has been worked out using this equation and the measured fracture load, the relation between strength and crack size can again be established through Eq. (1).

Griffith and many of his followers, for example Lawn and Wilshaw [2] and Atkins and Mai [3], have used Eq. (1) to relate strength and crack length, thereby allowing the fracture surface energy to be calculated for a brittle material. The equation can also be used to define the natural crack size in a material by the following method. Make several samples of brittle material and cut notches of known length into several specimens. Test these for strength, and plot out a Griffith curve as in Fig. 2. Next, measure the

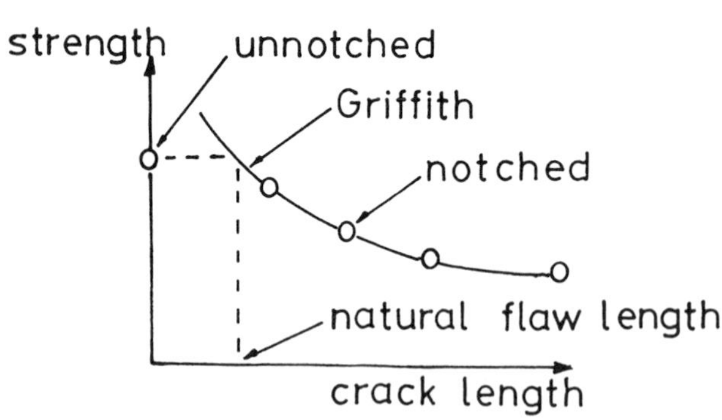

FIG. 2 Griffith curve and the natural flaw size determination.

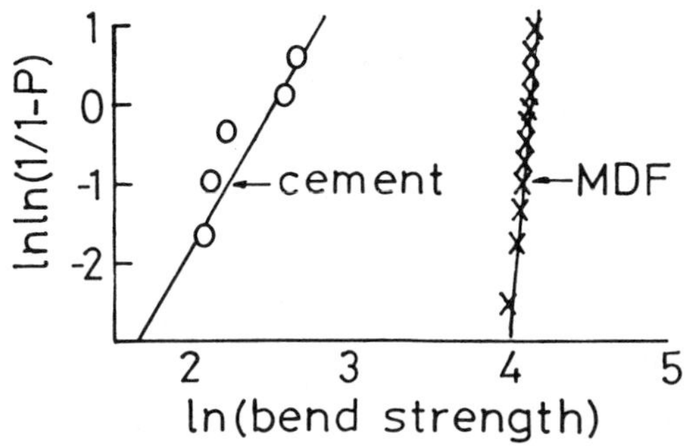

FIG. 3 Weibull plot for strength of macro-defect-free (MDF) cement samples.

strength of specimens without cutting notches into them. The mean of these points falls on the vertical axis away from the Griffith curve. However, it would be on the Griffith curve if we knew the natural crack length in the material. Therefore, by drawing a horizontal line from the axis across to the Griffith curve, the natural crack length is determined. In practice, the strength of each unnotched sample is different because the natural crack length varies from sample to sample. Thus one finds a range of natural crack lengths that are plotted statistically, usually on a Weibull plot as in Fig. 3 [4].

III. POOR DISPERSION AS A SOURCE OF NATURAL FLAWS

The key problem is finding the source of the natural flaws within the material. One source of flaws in ceramics is large grains, which grow during the sintering stage in the furnace [5]. Alumina is notorious for such flaws because some grains grow very large and cracks appear around the grain boundaries as differential shrinkage occurs on cooling. These cracks then reduce the strength of the alumina according to Eq. (1).

Another plausible source of natural flaws is poor dispersion. Consider air bubbles, for example. An air bubble trapped in a cement or ceramic casting will behave like a crack. The bigger the bubble, the bigger the crack. It follows that flocculated suspensions with high yield points will tend to trap large bubbles, which will weaken the final solid product considerably. One practical solution to this problem is to vibrate the mix so that air bubbles rise to the surface and escape. Alternatively, dispersing agents such as sodium lignosulfonate or sulfonated naphthalene resin are added to reduce the yield point, facilitating bubble removal. Applying vibration to a cement mix with added dispersing agent has a large effect on bubble removal, and strength can be raised considerably, as shown in Fig. 4 [6]. The large effect of deaerating cement mixes has been recognized for many years, but the precise source of the strength improvements has been controversial [7].

A similar effect of air removal can be found in extrudable cement mixes, which are prepared by mixing a water-soluble polymer such as hydroxypropyl methyl cellulose into a composition made of cement powder and water [8]. If this formulation is kneaded in a Z-blade mixer, then approximately 10% by volume of air becomes entrained in the dough, and these bubbles act as flaws that weaken the solidified product. Two methods have been described for removing this air. One is to use a vacuum Z-blade mixer to suck out the air before extruding the paste. The second is to press the dough flat at a pressure around 5 MPa for several hours. This squeezes the bubbles into solution and the gas then diffuses out through the sample. It was found that the strength of the cement product was doubled by these de-airing procedures [6].

The same sorts of effects have been observed in ceramics processing. For example, a clay suspension can trap air bubbles because of its non-Newtonian behavior. To avoid this problem, dispersing agents such as polyacrylic acid or its salts are added to the suspension to make it more fluid, and the mixture is vacuum degassed before being poured into molds. Another alternative is to pressure cast the slurry into the porous molds,thus compressing any bubbles. In the same way, a plastic composition of clay and water is kneaded in a screw extruder, where it picks up several percent

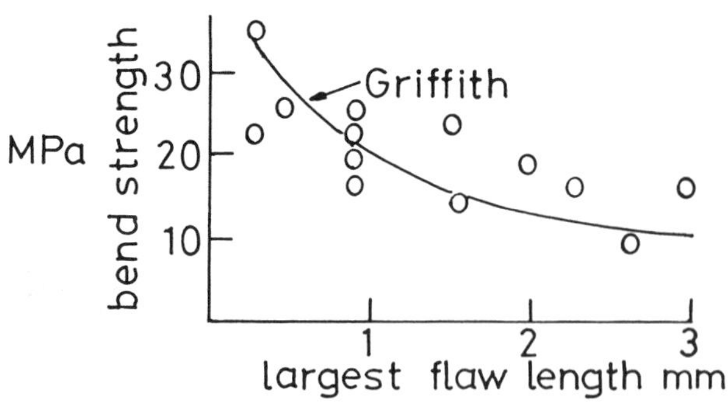

FIG. 4 Strength of cement versus largest natural flaw size after vibration with surfactant.

of entrained air. This is removed by extrusion into a vacuum chamber before screw-extruding the product through the die.

IV. STRENGTH IMPROVEMENTS IN CEMENTS THROUGH DISPERSION

Over the years, there has been a steady improvement in the properties of concrete as dispersing agents have been added to mixes, improving fluidity and reducing the water volume in the composition [9]. Ordinary cement mixes are flocculated and so require an excess of water to provide fluidity. This extra water reduces strength because it produces more hydrate phase and therefore more porosity in the hardened product. Feret [10] demonstrated this relationship in 1897, showing that the strongest cements are those made with the least water.

A mix of Portland cement powder and water becomes rather stiff and intractable above 50% volume fraction of solid particles because of the strong interparticle attractions within the dispersion. Under these conditions extra water must be added to give castability, and the resulting products show poor strength. However, the addition of 0.5% of sodium lignosulfonate or sulfonated naphthalene resin to the mixture disperses the cement grains, and so less water is needed when stirring and pouring the mix. Typically, casting can then be achieved at 60% solid volume in cement–water pastes. Consequently, strength improves considerably. As long ago as 1972, Brunauer and his colleagues [11] showed that a compressive strength of 204 MPa could be obtained with a finely ground Portland cement mixed with water and 1% calcium lignosulfonate.

The ultimate in this process of water reduction was an invention called DSP in which larger amounts of surfactant were added to the cement together with a substantial fraction of silica fume, which also aided rheology. A typical formulation contained 30% by weight of silica fume and 3–5% of sulfonated naphthalene resin [12]. The mixture was vibrocast into molds and allowed to set in the normal way. It produced a remarkably dense product with a much increased compressive strength, up to 300 MPa, but with only a modest tensile strength, about 30 MPa. This material has been tested in several applications including metal forming tools, plastic molds, and machine beds [13].

There are two problems with such fluid cement formulations. The first is that of applying shear stress to the mix to break down lumps and bubbles. The second is controlling the structure of the material when gelation occurs as the cement reacts. Flocculation can arise at this stage and cause cracks or other discontinuities in the product. One way of solving both of these problems is to add water-soluble polymer to the cement mix. This polymer thickens the paste and allows it to be worked on powerful mixing machinery such as two-roll mills, Banbury mixers, and calenders. Large aggregates are seen to be broken down by this high shear treatment. The added polymer also inhibits flocculation as pH alters during setting, giving a uniform microstructure in the product.

The high strength properties of such formulations were first reported in 1981 [14] when it was shown categorically that cement materials obeyed Eq. (1). Cement powder was mixed with water and polymer, typically in proportions of 100 : 11 : 5 by mass, using a two-roll mixer, followed by pressing overnight at 5 MPa pressure to form a deaerated flat plate around 4 mm thick. The polymer was either polyacrylamide, hydroxypropyl methyl cellulose or polyvinyl acetate/alcohol, giving an extremely viscous, almost rubbery mix. Typical solids content was up to 70% by volume in the mix. Since very high shear forces were applied in mixing such compositions, most aggregates were broken down and the microstructure was close-packed and uniform. Figure 5 shows the results obtained from notched bend tests on samples of this material, which was called macro-defect-free cement (MDF cement) because of its flawless structure. The results followed Eq. (1) over a wide range of crack lengths and gave a natural flaw size of about 100 μm, corresponding to an unnotched strength of 66 MPa. The elastic modulus (40 GPa) and fracture energy (30 J/m^2) were double the values normally found in cement, indicating superior structure.

The same processing method was applied to different kinds of cement, such as aluminate and sulfate cements, and found to give substantial improvements. Moreover, it was discovered that the polymer could be removed from the product by pyrolysis, followed by rehydration to fill the

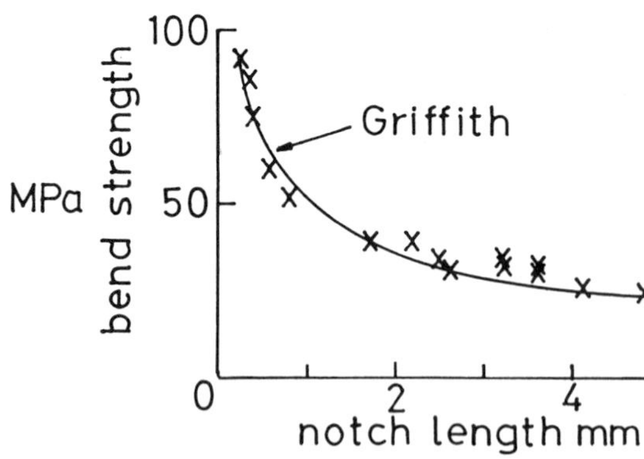

FIG. 5 Griffith plot for notched macro-defect-free cement samples.

space with cement hydrate. High strengths were obtained for these MDF cements containing no polymer, showing that the organic component was acting largely as a dispersing aid with little contribution to the mechanical performance of the product. In an experiment to find the highest solid content that could reasonably be achieved in a moldable composition, a specially graded aluminate cement was mixed on a two-roll mill with water and polyvinyl alcohol in a proportions of 100 : 6 : 2.5 by mass [15]. This represented a solid packing of 81% by volume. The polymer was removed by heating to 500°C, and the sample was rehydrated in water for 24 h at 50°C. The final sample after drying gave an open porosity of only 1% by volume, the lowest ever achieved. The Young's modulus was 75 GPa, the highest ever found for aluminate cement, and the bend strength was 80 MPa. These results proved that hydraulic cements could be made essentially nonporous, strong, and stiff through improved dispersion.

Further improvements in such cements were obtained when polymer additives were used as both dispersing aids and mechanical reinforcements. Polyvinyl alcohol was found to disperse aluminate cements and then to act as a reinforcing material that stretched across the crack faces during fracture, inhibiting propagation of the crack. A typical formulation was 100 : 10 : 7 by mass of calcium aluminate/water/polyvinyl alcohol. When mixed on a two-roll mill, this composition gave excellent rheology, and after hydration the material was much more crack-resistant than when formed with other polymers such as polyacrylamide or cellulose derivatives. Typically the bend strength was 150–200 MPa, at least double the ordinary MDF value. These compositions were good for machining and for acoustic damping applications [6,16].

V. EFFECTS OF DISPERSION IN CERAMIC PROCESSING

Dispersion plays an even larger part in the processing of ceramics because particle size is smaller, around 1 μm compared with 15 μm for a typical hydraulic cement. Particle size must be so small if the ceramic is to sinter in the furnace to give a dense product. Sintering is the mechanism that allows solid grains to pull each other together below their melting point, to densify a power compact from, say, 50% packing to 100%. The densification is driven by surface energy of the particles and, in a sense, is the ultimate coagulation step in which colloid forces fuse solid grains together, thus removing porosity. The problem lies in making the molded powder body from such fine powders because fine sinterable particles also flocculate and adhere strongly at room temperature, making them difficult to handle. Ideally, in a ceramic manufacturing process, the particles should be kept apart while flowing into the mold then gelled uniformly together to give the molded shape, which can subsequently be sintered in the furnace. In clays, hydrated surface layers provide this separation function, and drying provides the gelling force. For other ceramics, dispersing agents are used.

Consider a typical ceramic manufacturing route—for example, the making of sparkplugs. Hydrated alumina is precipitated from solution and calcined to grow α-alumina grains about 1 μm in size. This calcined powder is generally ball-milled to give a fine powder, which is then compacted to shape and fired in a furnace at around 1600°C. Dispersion plays a significant role in the milling stage and in the molding step. During milling, the alumina is dispersed in water at around 20% volume fraction by adding acid to pH 4 or by adding polyacrylic acid or ammonium polyacrylate. Other additives such as binders and lubricants—for example, 0.5% polyvinyl alcohol or polyethylene glycol—are also added at this point. Typically a ball mill or an attritor mill is used, and the fine suspension is then spray-dried to give spherical aggregates 50–100 μm in diameter. These are readily pressure-compacted into molded shapes, which are machined before sintering at 1600°C to give dense products.

In an alternative pathway, the milled dispersion is concentrated by sedimentation, then poured into porous molds that suck out the fluid, leaving a shaped body that is dried before firing at 1600°C. Or the milled slurry is filter-pressed to a thick concentrate. Polymer is then added, and the mixture is kneaded to a plastic state before being extruded, dried, and fired. Typically, methyl cellulose is used, around 1–2% by mass, to give this extrudable rheology.

It can be seen that in all these situations the particles of ceramic are held apart by the dispersing agent during comminution and during shaping. Then the dispersion is dried to bring the particles into solid contact to form the molded part before the temperature is raised to start sintering. The problem is to understand how flaws are introduced into the material during these steps. By systematically removing these flaws, the strength of the ceramic should be increased according to Eq. (1).

VI. STRONGER CERAMICS THROUGH DISPERSION

There is no doubt about two issues. The first is that ceramics obey Eq. (1) very well for cracks ranging from 100 mm down to less than 1 μm in length. The only difficulty is that the fracture surface energy R can appear to increase slightly as the crack grows in length, the so-called R-curve effect. This can arise for a number of reasons, including plastic deformation, phase change, and crack bridging [17]. But this effect is small and does not significantly affect the theoretical argument. The second issue is that ceramics have enormous potential for being stronger. Melt-formed and vapor-formed ceramics typically give strengths around 5 GPa because flaws can be readily excluded from such processes. By contrast, ceramics made from powders are usually only one-tenth as strong [18]. In other words, from Eq. (1), ceramics made from powders contain flaws 100 times as large, typically 20–100 μm in length.

Several studies have shown that such flaws originate from various causes, ranging from large grains and air bubbles through dust contamination to ball-mill chips [19]. However, even when these extraneous flaws were removed by working in clean conditions, large flaws were still observed, and these were identified with powder aggregates. One particular effect of aggregates was shown to be inhibition of sintering [20,21]. It was demonstrated that better dispersion and sedimentation of fine powders (less than 0.1 μm in diameter) led to sintering at temperatures some hundreds of degrees lower than usual. Since then, the same effect has been achieved by breaking down aggregates through high shear during plastic mixing in a polymer-based composition [22].

It has become clear that large powder aggregates can appear in two ways. The first occurs when a stable dispersion is destabilized slightly [23]. The 0.2 μm particles begin to stick together to form small flocs, but large clusters also appear near 60 μm in size. These clusters are not easily broken down by shearing and so can act as flaws in the final product when cracks appear around the clusters, as shown in Fig. 6. The second mechanism for aggregate formation is drying. During drying, destabilization must occur at some stage, and then large flocs appear. These are resistant to breakdown by ultrasonics [24]. Since all ceramics must go through a drying stage during powder production or during molding, this mechanism for producing flaws is ubiquitous.

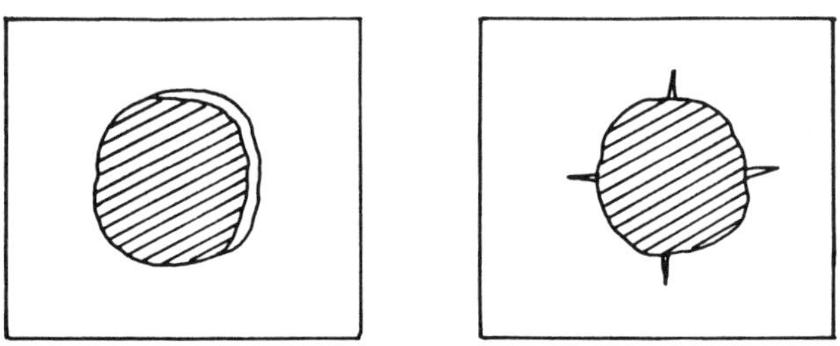

(a) (b)

FIG. 6 (a) Crack formed by shrinkage of a well-packed aggregate. (b) Cracks caused by shrinkage of a poorly packed aggregate.

Typical results for a zirconia powder are shown in Fig. 7. The powder was molded into ceramic in two ways—by die pressing and by mixing in a polymer solution using a two-roll mixer. Both methods led to a low bend strength (around 400 MPa), indicating large flaws in the 50 μm range from Eq. (1). Such large flaws were found in the samples when they were polished and then inspected with a microscope and image analyzer. The spread of observed flaws gave fair agreement with the measured strengths of a number of samples. Then attempts were made to break down powder aggregates by applying extra shear forces to the polymer mix during forming. This was achieved by increasing the viscosity of the mix and also by extruding the mix through fine orifices [25]. These samples gave a much improved microstructure, as demonstrated by microscopy, and also showed much increased bend strength (Fig. 7). A factor of 3 improvement in strength and in Weibull modulus of bend strength (reliability) was proved experimentally, as shown by the increased mean value and slope of the right-hand curve in Fig. 7.

These observations indicate that polymer mixes of high viscosity can be used to break down aggregates and that such aggregates do not regrow on drying from the viscous medium. By contrast, drying from fluid dispersions always seems to produce aggregates that weaken the dried and fired ceramic product. No one has yet succeeded in making strong ceramic moldings by

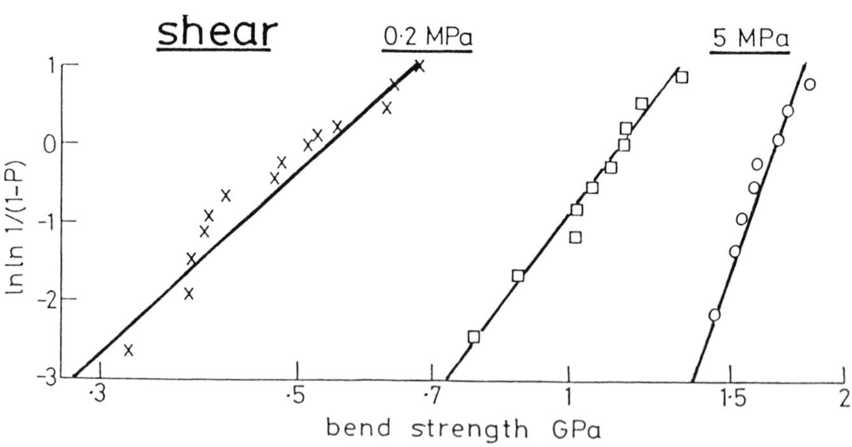

FIG. 7 Weibull plot showing the improvement in ZrO_2 strength and reliability obtained in two ways: by weakening the powder aggregates and by increasing the shear stress during mixing (x) results for strongly aggregated powder at low shear stress (0.2 MPa); (□) strongly aggregated powder at high shear stress (5 MPa); (○) weakly aggregated powder at high shear stress (5 MPa).

drying a good dispersion of ceramic particles from a pourable suspension. This is a major target that remains to be achieved in the field of ceramic powder processing.

VII. CONCLUSIONS

Dispersion has a substantial effect on the strength of inorganic materials such as cement and ceramic. This is explained theoretically through the Griffith equation, Eq. (1), which shows that the strength of brittle materials decreases rapidly as the length of flaws is increased. Flaws in these materials can be equated with gas bubbles and with lumps, both of which can be influenced by addition of dispersing agents and by shearing.

Flaws in cement are generally large air bubbles that cannot escape because of the paste rheology. On removing these millimeter scale defects by adding dispersants or by mixing with water-soluble polymers, followed by sharing, bending strength has been increased by a factor of 10.

In fine ceramics, the most persistent flaws are strongly adherent aggregates caused by flocculation during drying. Such aggregates grow remarkably big, around 70 μm in diameter for 0.2 μm alumina particles. These can be removed in certain instances by producing a highly viscous mix of the powder in a polymer and applying high shear stress to the composition. Large improvements in strength and reliability have been attained through this approach. New ways of inhibiting flocculation during drying are needed to make use of these ideas on an industrial scale.

REFERENCES

1. A.A. Griffith, *Phil. Trans. Roy. Soc. Lond. A221*:163 (1920).
2. B.R. Lawn and T.R. Wilshaw, *Fracture of Brittle Solids*, Cambridge Univ. Press, Cambridge, 1975, Chap. 1.
3. A.G. Atkins and Y.W. Mai, *Elastic and Plastic Fracture*, Ellis Horwood, Chichester, 1985, Chap. 2.
4. K. Kendall, N.McN. Alford, J.D. Birchall, and S.R. Tan, *J. Mater. Res. 1*: 120 (1986).
5. R.W. Davidge, *Mechanical Behaviour of Ceramics*, Cambridge Univ. Press, Cambridge, 1979, Chap. 6.
6. K. Kendall, A.J. Howard, and J.D. Birchall, *Phil. Trans. Roy. Soc. Lond. A310*:139 (1983).
7. N.B. Eden and J.E. Bailey, *J. Mater. Sci. 19*:150 (1984).
8. S.B. Downing, D.W.J. Osmond, M.W. Skinner, E.J. West, and D.G. Dawson, U.K. Patent 1 563 190 (1976).
9. M. Regourd, in *Very High Strength Cement Based Materials* (J.F. Young, Ed.), MRS, Pittsburgh, Penn., 1985, p. 3.

10. R. Feret, *Bull. Soc. Encour. Ind. Nat.,Paris II*: 1604 (1897).
11. M. Judenfreund, J. Skalny, R.S. Mikhail, and S. Brunauer, *Cement Concrete Res. 2*:313 (1972).
12. H.H. Bache, 2nd Int. Conf. Superplasticizers in Concrete, Ottawa, Ontario, June 1981.
13. S. Wise, J.A. Satowski, B. Scheetz, J.M. Rizer, M.L. MacKenzie, and D.D. Double, in *Very High Strength Cement Based Materials* (J.F. Young, Ed.), MRS, Pittsburgh, Penn., 1985, p. 253.
14. J.D. Birchall, A.J. Howard, and K. Kendall, *Nature 289*:388 (1981).
15. K. Kendall and J.D. Birchall, in *Very High Strength Cement Based Materials* (J.F. Young, Ed.), MRS, Pittsburgh, Penn., 1985, p. 143.
16. S.R. Tan, A.J. Howard, and J.D. Birchall, *Phil. Trans. Roy. Soc. Lond. A322*:479 (1987).
17. D. Munz, *The Energetics of Fracture*, (A. G. Atkins, Ed.), Griffith Centenary Meeting, The Institute of Materials, London, 1993, p. 36.
18. K. Kendall, *Mater. Forum 11*:61 (1988).
19. F.F. Lange, *J. Am. Ceram. Soc. 67*:83 (1984).
20. W.H. Rhodes, *J. Am. Ceram. Soc. 64*:19 (1981).
21. M.F. Yan and W.W. Rhodes, *Mater. Sci. Eng. 61*:59 (1983).
22. W.J. Clegg, N.McN. Alford, J.D. Birchall, and K. Kendall, *J. Am. Ceram. Soc. 72*:432 (1989).
23. K. Kendall, N.McN. Alford, W.J. Clegg, and J.D. Birchall, *Nature 339*:130 (1989).
24. K. Kendall, *Powder Technol. 58*:151 (1989).
25. N.McN. Alford, J.D. Birchall, and K. Kendall, *Nature 330*:51 (1987).

10

Liquid Crystal Dispersions

JOHN L. WEST Liquid Crystal Institute, Kent State University, Kent, Ohio

I.	Introduction	349
II.	Liquid Crystals	350
III.	Liquid Crystal Devices	352
	A. Cholesteric temperature indicators	352
	B. Liquid crystal displays	353
IV.	Electro-Optic Liquid Crystal Dispersions	355
	A. NCAP dispersions	358
	B. Polymer dispersed liquid crystals	358
	C. Electro-optics of PDLC materials	361
	D. Recent developments in liquid crystal/polymer dispersions	365
V.	Conclusion	369
	References	369

I. INTRODUCTION

Dispersions of liquid crystals in a solid polymer matrix have improved and expanded the performance of liquid crystals in a variety of applications. Dispersions of cholesteric liquid crystals were first used as temperature indicators to improve their ease of use and environmental stability. Over the last decade, dispersions of nematic liquid crystals have been developed for entirely new electro-optic applications such as switchable windows and high definition projection television.

This chapter provides a brief overview of the physics and chemistry of liquid crystals and introduces the principles of operation of several conventional liquid crystal devices. Techniques used to disperse liquid crystals in a solid matrix, how these dispersions affect the properties of liquid crystals, and their commercial application are discussed.

II. LIQUID CRYSTALS

Liquid crystals were discovered over a century ago and have been extensively researched since that time. A complete overview of liquid crystals and their applications can be found in several recent books [1–4].

Liquid crystal, as the name implies, is a phase of matter that is intermediate between the solid and liquid phases. There are two distinct types of liquid crystal phases: lyotropic and thermotropic. Lyotropic phases are formed by amphiphilic molecules dissolved in a solvent at an appropriate concentration. Lyotropic liquid crystals will not be discussed in this chapter; however, a number of reviews can be found elsewhere [e.g., 5,6].

Thermotropic liquid crystal phases are formed upon melting a crystalline solid, typically one composed of elongated molecules with a rigid core and flexible tail. Upon melting from the crystalline phase, the molecules lose their positional order while maintaining some degree of orientational order. The transition from the liquid crystalline to the isotropic liquid is marked by complete loss of positional and orientational order (Fig. 1). Like crystalline phases, liquid crystals are optically and electrically anisotropic. Liquid

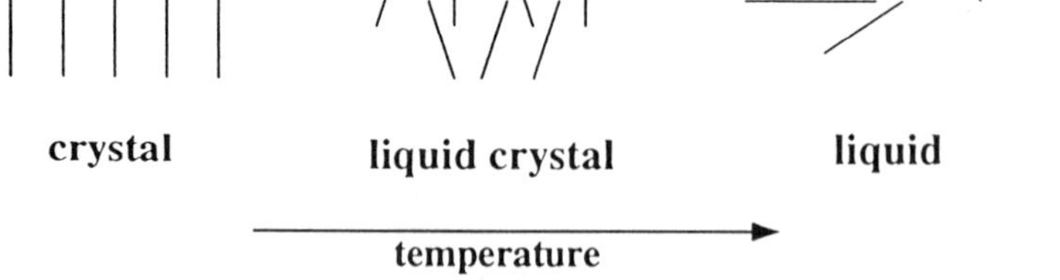

FIG. 1 Representation of the crystalline, liquid crystalline, and liquid phases of matter.

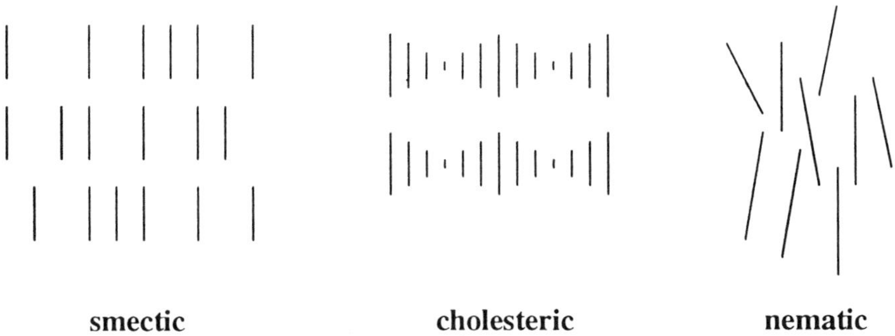

FIG. 2 Smectic, cholesteric, and nematic liquid crystal phases.

crystals have in common with liquids their ability to flow and assume the shape of a container.

A director is associated with the liquid crystal and is aligned with the average orientation of the long axis of the molecules. The liquid crystal is birefringent, with an extraordinary refractive index measured for light polarized parallel to the director and an ordinary refractive index measured for light polarized in the plane normal to the director. The phase is also electrically and magnetically anisotropic. The dielectric and magnetic anisotropies allow easy alignment of the liquid crystal director either parallel or perpendicular to an external field. The optical anisotropy and the ease of aligning the director with an external field is the basis for their electro-optic applications.

Three major categories of thermotropic liquid crystals have been identified: smectic, cholesteric, and nematic (Fig. 2). The smectic phase retains some of the positional order found in the crystalline solid, with the molecules being arranged in layers. The molecules may be tilted within the layers, and some limited positional order may be maintained. A number of smectic phases have been identified that vary by the amount and type of positional order of the molecules.

The nematic phase retains no positional order but does maintain orientational order, with the molecules aligned in generally the same direction, defined as the director. The cholesteric phase is similar to the nematic phase except that the director rotates periodically. The pitch length of the cholesteric phase is defined as the distance required for the director to rotate through 180°.

Pure materials are liquid crystalline only over a narrow temperature range, rarely exceeding 10°C. Wider liquid crystal temperature ranges have been achieved by blending materials to form a eutectic mixture. Multicomponent eutectic mixtures have been formulated with nematic ranges greater than 100°C.

III. LIQUID CRYSTAL DEVICES

Commercial development of liquid crystal devices began in the 1960s. Cholesteric liquid crystals were used for temperature indicators, while nematic liquid crystals were developed for electro-optic devices. Display devices became commercially viable in the 1970s. Their combination with the burgeoning semiconductor technologies produced a revolution in watches and calculators. Advances in liquid crystal technology continue today with displays for laptop computers and projection TV. Dispersions of liquid crystals in a solid matrix improve the mechanical properties and environmental stability of liquid crystal devices and have afforded entirely new applications of liquid crystals.

A. Cholesteric Temperature Indicators

As previously discussed, the cholesteric phase is characterized by a helical variation in the orientation of the liquid crystal director. The periodic variation in the director orientation results in a periodic variation in the refractive index in the material. This effectively produces a grating that reflects light of a wavelength on the order of the pitch length of the cholesteric material (Fig. 3).

The pitch length is temperature-dependent, generally decreasing with

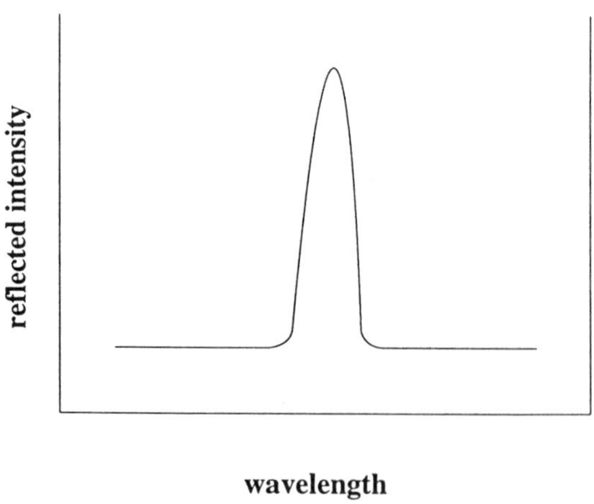

FIG. 3 Reflectivity of cholesteric films as a function of wavelength.

increasing temperature. Therefore the wavelength of light reflected by the cholesteric material is temperature-dependent. Complete color change from violet to red can occur over a temperature range as small as 0.2°C or as large as 30°C. Cholesteric materials are used as thermometers in medical diagnostics and in the nondestructive testing of electronic components.

Initial applications used pure cholesteric materials, which are thick viscous liquids, coated as thin films on the material to be analyzed. The colors are brightest with cholesteric material coated on a black background. This requires the object to be painted with a black paint or stain, making the use of cholesterics cumbersome. Also, because the pure cholesteric material is a thick liquid, its application and removal are cumbersome.

Early cholesteric materials were derivatives of cholesterol, most typically cholesterol esters. These materials oxidize easily, and therefore thin films decompose in a matter of hours, destroying their usefulness. Encapsulation of cholesteric liquid crystals in a polymer binder was used to improve their ease of use and stability [7].

Dispersions of cholesteric liquid crystals can be made in numerous polymers, greatly extending their lifetime. Microencapsulation can be achieved with polymers such as polyvinyl alcohol or gelatin. Emulsions can be formed with polyurethanes [8]. These dispersions generally exhibit a lower intensity iridescence than pure materials; however, liquid crystal aggregates smaller than 2–3 μm have an iridescence similar to that of pure materials. The intensity of the iridescence in dispersions has been related to the texture of the droplets [9].

B. Liquid Crystal Displays

Interest in liquid crystals expanded in the 1960s and led to the development of displays based on dynamic scattering [10], guest–host [11,12], and cholesteric–nematic phase change [13,14]. The advent of the twisted nematic (TN) display [15] coupled with the introduction of environmentally stable cyanobiphenyl nematic liquid crystals [16,17] led to the rapid commercialization of liquid crystal displays.

Liquid crystal devices produce a low power, lightweight display, first used extensively in the 1970s for wristwatches and calculators. The technology was improved in the 1980s with the introduction of the supertwist display and active matrix addressing. Liquid crystals are now the display of choice for laptop computers and hand-held TVs.

A schematic of a TN device is given in Fig. 4. The device consists of a nematic liquid crystal sandwiched between two conducting substrates, at least one of which is transparent. Treatment of the substrates, usually with a rubbed polyimide layer, causes the nematic director to lie parallel with

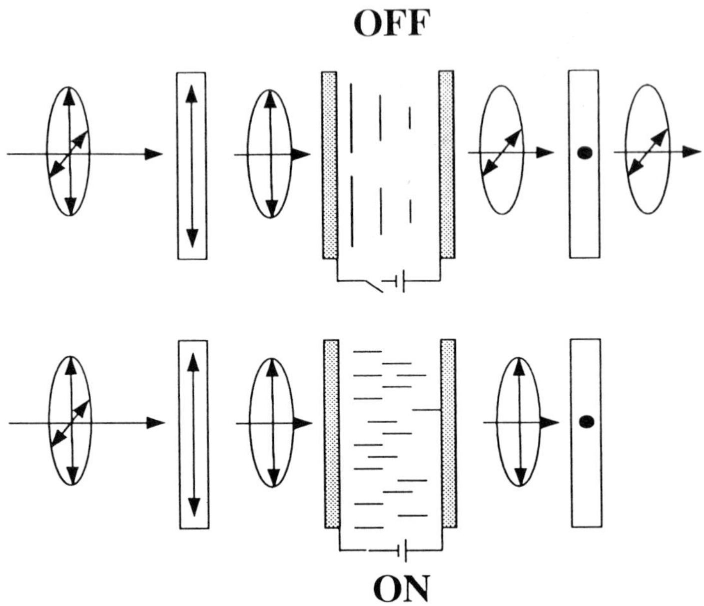

FIG. 4 Twisted nematic liquid crystal display in the ON and OFF states.

the substrate surface (homogeneous alignment) and in the rub direction. The top and bottom substrates are oriented so that the nematic director rotates through 90° from the bottom to top substrate of the cell.

In one configuration of the TN cell, crossed polarizers are adhered to the top and bottom substrates. Light incident on the shutter is linearly polarized. The direction of polarization follows the rotation of the nematic director on passing through the cell in the OFF state and is passed by the second polarizer, called the analyzer. In the ON state, an electric field is placed across the cell, and the nematic director aligns parallel to the field and perpendicular to the substrates. The linearly polarized light passes through the nematic layer without rotation and is absorbed by the analyzer.

The TN cell is switched from a light-transmitting to a light-absorbing state by application of an electric field across the cell. Images are produced using patterned transparent electrodes [18]. Direct drive addressing schemes are used for low information content displays such as wristwatches and simple calculators. Multiplexing schemes are used to produce moderate resolution displays used in calculators with simple graphics capabilities and low resolution TVs. High resolution TN displays utilize a complex active matrix addressing scheme [19].

While TN is the predominant liquid crystal display technology, other display schemes with enhanced performance have been developed. The supertwist display (where the nematic director rotates through $>180°$ passing through the cell) has a steep voltage response curve that allows for high level multiplexing and is now found in most laptop computer screens [20]. While providing higher resolution displays, the response time of the supertwist is too slow for real-time video displays. Fast bistable shutters can be produced using ferroelectric liquid crystals. The ferroelectric devices can be used to produce high resolution multiplexed displays [21]. Problems controlling surface alignment and defects have precluded commercialization of the ferroelectric displays despite their improved performance.

IV. ELECTRO-OPTIC LIQUID CRYSTAL DISPERSIONS

Conventional liquid crystal displays consist of a thin (<10 μm), uniform film of liquid material sandwiched between transparent conducting electrodes, usually glass coated with a vacuum-deposited layer of indium tin oxide. Because the active layer flows like a liquid, it is difficult to maintain uniform thickness over large areas. Also, the light efficiency of most conventional liquid crystal devices is low (typically less than 35%) because they require polarizers. Dispersions of liquid crystal in a solid material have been developed for high efficiency, large area displays. The electro-optic dispersions operate on the principle of electrically controlled light scattering. Application of an external field serves to align the liquid crystal and match its refractive index with that of the solid.

The first demonstration of a solid/liquid crystal dispersion for electro-optic applications consisted of small glass particles dispersed in a liquid crystal film [22] (Fig. 5a). Micrometer-size soda glass spheres were dispersed in a nematic liquid crystal sandwiched between two transparent conducting glass substrates. In the absence of an applied field, the nematic liquid crystal is randomly aligned. The refractive indices of the liquid crystal and glass spheres are mismatched and the device scatters light in the OFF state. The nematic director aligns parallel with an applied electric field, matching the refractive index of the liquid crystal with that of the glass spheres and substantially reducing the scattering. While these films are not self-supporting, they eliminate the need for polarizers, greatly increasing the light efficiency.

Liquid crystals have been imbibed in microporous films such as membrane filters [23,24] (Fig. 5b). As with the soda glass sphere dispersions, these materials scatter light in the OFF state and become more transparent upon application of an external electric field. Neither approach achieved

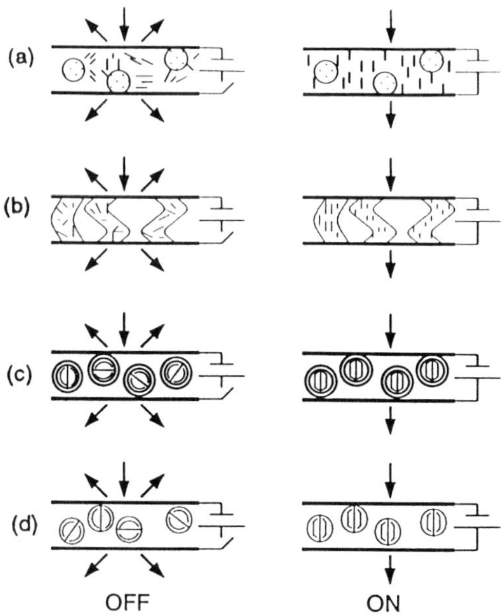

FIG. 5 Types of solid liquid crystal dispersions. (a) Glass particles dispersed in a liquid crystal film; (b) liquid crystal imbibed in a microporous film; (c) NCAP dispersions; (d) PDLC dispersions.

commercial viability due to lack of clarity in the ON state and high driving voltage, perhaps caused by the irregular structure and strong surface anchoring of the liquid crystal at the solid interface.

Two liquid crystal dispersion technologies have been extensively researched. Liquid crystals have been encapsulated in a polymer film by microencapsulation or emulsification techniques and are generically termed NCAP (nematic curvilinear aligned phase) (Fig. 5c) [25,26]. Finally, liquid crystals have been dispersed in a polymer binder by phase separation from a homogeneous solution and are known as polymer dispersed liquid crystals or PDLCs (Fig. 5d) [27,28]. Both NCAP and PDLC films consist of droplets of a nematic liquid crystal (usually with a positive dielectric anisotropy) dispersed in a polymer matrix. The droplets typically have a bipolar configuration where the director lies parallel to the droplet wall and have two point defects (Fig. 6a) [29]. A radical configuration has also been identified in which the director is normal to the droplet wall with a point defect in the center of the droplet (Fig. 6b). An axial configuration is possible in which the director is normal to the film's surface with a line defect through the

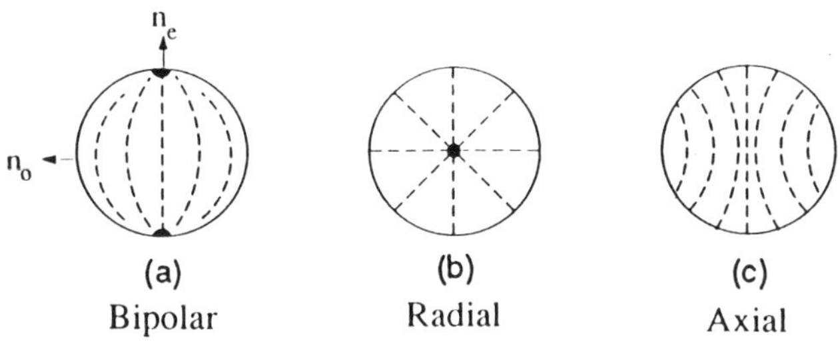

FIG. 6 NCAP and PDLC droplet configurations.

center (Fig. 6c) [30]. These droplet configurations were predicted and observed earlier for liquid crystal droplets suspended in a viscous liquid such as glycerol [31,32]. The morphology of a PDLC film is shown in Fig. 7. The droplets are relatively uniform in size and densely packed. The droplets are aspherical and randomly shaped, resulting in a random orientation of the liquid crystal droplet directors.

The bipolar droplet is used in virtually all applications of these materials.

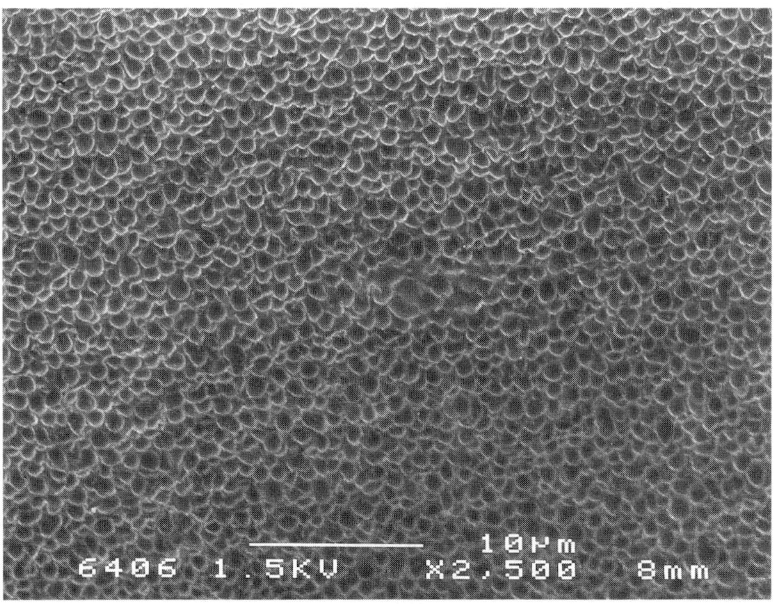

FIG. 7 Scanning electron microscopic photograph of a PDLC film.

The bipolar droplet is birefringent with extraordinary and ordinary refractive indices n_e and n_o, respectively. In the absence of an applied field the droplet directors are randomly aligned, and on average the refractive indices of the droplets are mismatched with that of the isotropic polymer binder, causing the films to scatter incident light. Application of an electric field across the film aligns the droplet parallel to the field if the liquid crystal has a positive dielectric anisotropy. The polymer is chosen to have a refractive index n_p matched with that of the droplet n_o. In this orientation the refractive indices of the droplet and polymer binder are matched for normally incident light and the film is transparent.

A. NCAP Dispersions

James Fergason first reported on NCAP dispersions in 1985 [25,33]. The NCAP material consists of a nematic liquid crystal dispersed as droplets in a polymer film. The material is formed by emulsifying a nematic liquid crystal with a waterborne polymer. The polymer can be either a colloidal suspension such as a latex or water-soluble such as gelatin or polyvinyl alcohol. The resulting emulsion is then coated on a transparent conductive substrate and allowed to dry. Finally, a second conductive substrate is laminated to the dry assembly. The resulting film consists of roughly spherical liquid crystal droplets with some interconnecting passages. Droplet sizes range from submicrometer to greater than 10 μm.

Dichroic dyes dissolved in nematic droplets of NCAP films produce color and higher contrast [34,35]. The dichroic dye is aligned with the nematic liquid crystal in the droplet. In the absence of an applied field, the droplet directors and the dichroic dye are randomly aligned in the plane of the film, and the material is colored. Alignment of the directors normal to the film's surface orients the dichroic dye into a relatively nonabsorbing orientation, switching the film to a transparent, colorless state. Dichroic dyes have been used to produce high contrast colored displays. NCAP materials have been developed for a variety of applications including switchable windows and color displays. Raychem Corporation is tailoring NCAP materials to be incorporated with active matrix substrates for high resolution displays [36].

B. Polymer Dispersed Liquid Crystals

Polymer dispersed liquid crystals (PDLCs), like NCAP materials, consist of liquid crystal dispersed in a polymer matrix; however, PDLCs are formed by phase separation of the liquid crystal from a homogeneous solution. PDLCs are made by forming a homogeneous solution of the liquid crystal with a polymer or prepolymer followed by phase separation, droplet formation, and finally polymer gelatin and solidification. PDLCs can be formed by three distinct processes: polymerization-induced phase separation (PIPS), thermally induced phase separation (TIPS), and solvent-induced phase separation (SIPS) [37]. All three processes can be used to coat PDLC films continuously on flexible, transparent conducting substrates. The resulting large area electro-optic shutters can be used as switchable windows for architectural, automotive, and aeronautic applications. Figure 8 shows a prototype PDLC window in the ON and OFF states.

The PIPS process involves dissolving the liquid crystal in a polymer, monomer, or oligomer precursor. The prepolymer is then polymerized thermally [37], photochemically [38,39], or with an electron beam. The solubility of the liquid crystal decreases as the molecular weight of the polymer increases, eventually becoming immiscible and phase-separating by nucleation and growth or spinodal decomposition. The droplets continue to grow until polymer gelatin locks in their morphology. The droplet morphology is determined in the time between droplet formation and polymer gelatin. This process is depicted in Fig. 9.

Typical polymers used in PIPs are epoxy resins such as Epon 828 and

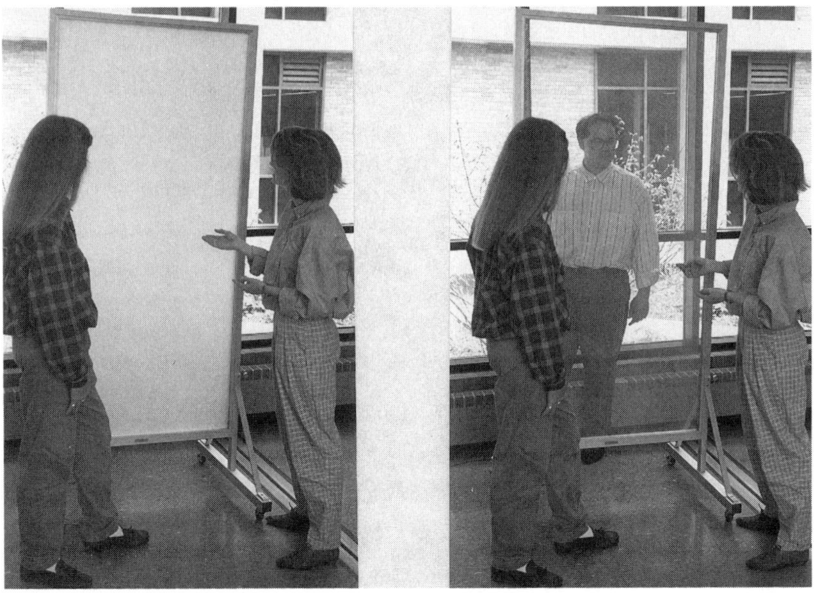

FIG. 8 A PDLC window in the scattering (OFF) and transmitting (ON) states. (Photographs courtesy of Renate Ondris-Crawford and Greg Crawford.)

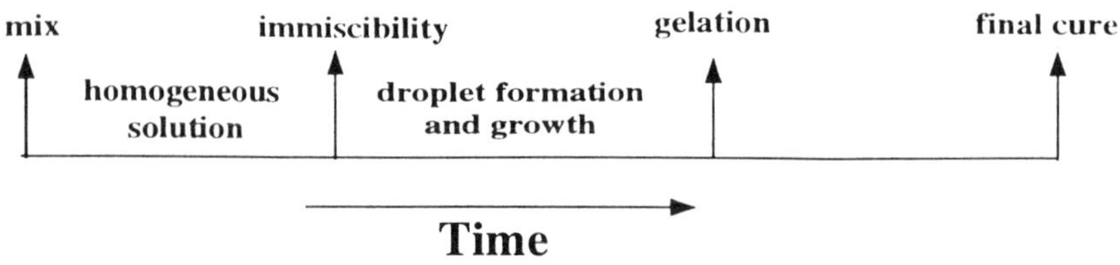

FIG. 9 Time line depicting the PIPS process.

cure agents such as Capcure 3-800 [37]. Polymerization occurs through a thermal condensation reaction with no low molecular weight by-products. The droplet density and size are controlled by the type and relative concentration of the components and by the cure temperature. Yamagishi et al. [40] report that for photopolymerization, the morphology of the PDLC is controlled by the composition and concentration of the monomer and liquid crystal, the type of photoinitiator, and the rate of polymerization. They used a variety of acrylate-based monomers and found that the morphology of the resulting film was switched from a "polymer ball" morphology to a "Swiss cheese" morphology by changing the photoreaction from a step growth to a chain reaction.

Thermoplastic polymers are used in the SIPS and TIPS processes [37]. For SIPS the liquid crystal is dissolved with the thermoplastic in a common solvent. Evaporation of the solvent leads to phase separation, droplet formation and growth, and polymer gelatin. The droplet size is controlled by the rate of solvent evaporation, with larger droplets being formed at a lower rate of evaporation. The process can be visualized with a ternary phase diagram as in Fig. 10. The original composition of the mixture is

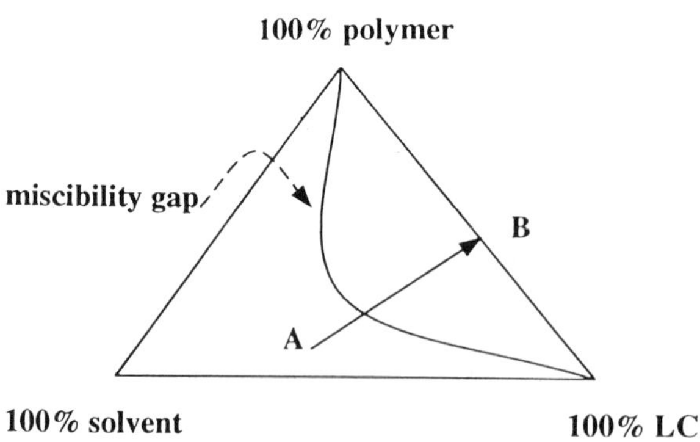

FIG. 10 Ternary phase diagram depicting the SIPS process.

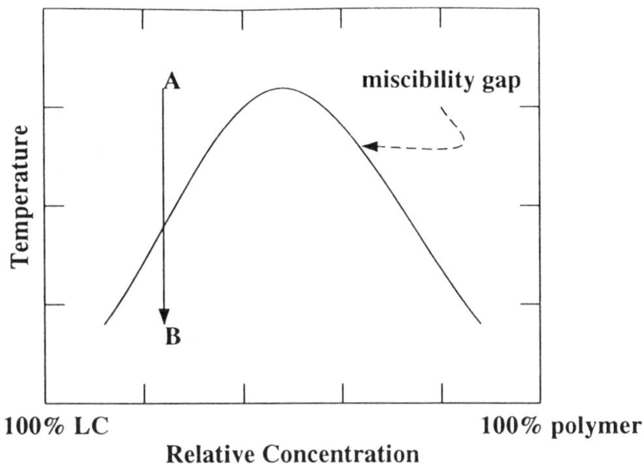

FIG. 11 Phase diagram of a liquid crystal and thermoplastic.

depicted as point A. Evaporation leads to the system following the line A-B. Phase separation and droplet formation occur when the system passes through the miscibility gap, and final evaporation of the solvent leaves the system at point B.

The TIPS process involves dissolving the liquid crystal in a thermoplastic melt. This can be achieved by using a low melting polymer such as polyvinylbutyral or a PDLC made by the SIPS process and represented as point B in Fig. 10. Heating the material forms a homogeneous solution of the liquid crystal and polymer. Subsequent cooling results in liquid crystal immiscibility, droplet formation and growth, and polymer gelatin. The process can be visualized as in Fig. 11. The droplet size is controlled by the rate of cooling for the TIPS process. The faster the cooling rate, the smaller the droplets.

Control of droplet size and density is relatively straightforward with PDLCs. Droplet size and density are maximized by increasing the liquid crystal concentration. Speeding up the droplet formation process decreases the droplet size [37]. The distribution in droplet size can be made small, often less than 10%.

C. Electro-Optics of PDLC Materials

The important electro-optic parameters of PDLC materials are determined by the intended application. For example, window applications require high ON state clarity and efficient OFF state scattering, while other parameters such as driving voltage and switching speed are relatively unimportant. This is in contrast to display applications, where low driving voltage and fast switching speeds are essential.

The electro-optic performance of PDLC shutters is dependent on the materials used, film thickness, and droplet size, shape, and density. The ways in which ON and OFF state scattering, driving voltage, and switching speed are controlled by these various material parameters are reviewed briefly in this section.

1. ON State Clarity

The ON state clarity is maximized when the refractive index of the polymer matrix, n_p, is precisely matched with the ordinary refractive index of the liquid crystal in the droplet, n_o. Because liquid crystal is dissolved in the polymer matrix and because the configuration of the liquid crystal in the droplet is distorted from the bulk configuration, the refractive indices of the polymer and liquid crystal are different from that of the bulk materials. Therefore, n_p and n_o of the PDLC can only be estimated.

It is possible to empirically determine the match of n_p and n_o by measuring the transmission through the film as a function of incident angle and applied voltage [41]. Figures 12 and 13 schematically show the shapes of the transmission curves as a function of incident angle and applied voltage, respectively. The shape of these curves can be used to determine how well n_p is matched with n_o.

If n_p is greater than n_o, the transmission of vertically polarized light through the PDLC film will initially increase as the viewing angle deviates from the normal. For both $n_p = n_o$ and $n_p < n_o$ the transmission steadily decreases as the viewing angle deviates from normal. A plot of the transmission versus applied voltage also indicates the match of n_p and n_o. For $n_p > n_o$ and $n_p = n_o$, the transmission asymptotically approaches a maximum value with increasing voltage. For $n_p < n_o$, the transmission versus voltage curve passes through a maximum value at an intermediate voltage. The match of n_p and n_o can then be determined by measuring the transmission as a function of angle and applied voltage.

2. OFF State Scattering

The OFF state scattering is maximized by using liquid crystals with large birefringence and droplet sizes on the order of visible wavelengths and a high density of droplets. The shape of the droplet can also affect the OFF state scattering by controlling the director orientation. If the director is aligned in the plane of the film, the scattering will be maximized. Scattering is enhanced by thicker film because of a greater number of scattering centers and the introduction of multiple scattering. Increasing thickness is most

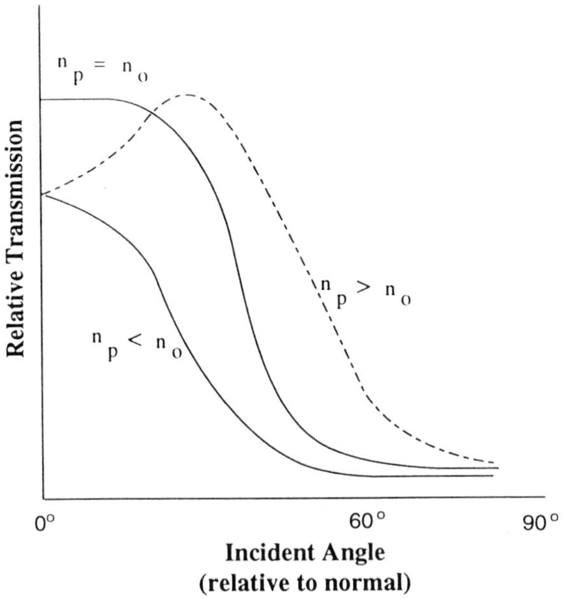

FIG. 12 Transmission of vertically polarized light through a PDLC as a function of incident angle and refractive index match.

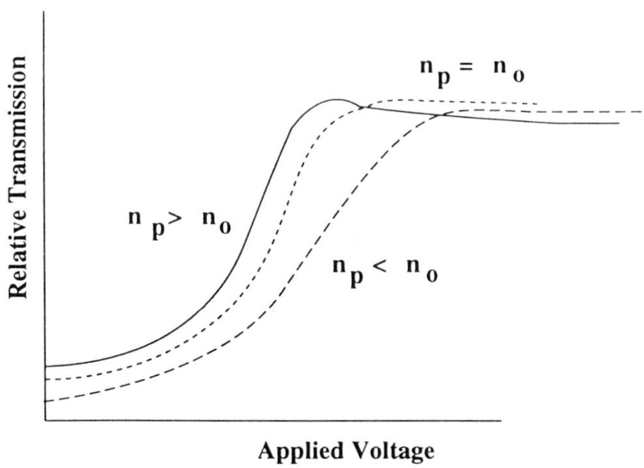

FIG. 13 Transmission as a function of applied voltage and refractive index match of a PDLC.

effective with thin films, and little improvement in scattering is achieved by increasing the thickness above 25 μm.

3. Driving Voltage

Low driving voltages are required for PDLC materials to be used for display devices. Therefore, a number of studies have been conducted on the factors controlling PDLC driving voltage [29,42,43]. The driving voltage depends on a variety of factors and can vary by more than an order of magnitude. The driving voltage can be estimated by the equation [44]

$$V_B = \frac{d}{3a}\left(\frac{\rho_\rho + 2}{\rho_{lc}}\right)\left[\frac{K(l^2 - 1)}{\Delta\epsilon\epsilon_0}\right]^{1/2} \tag{1}$$

where d is the thickness of the PDLC film; $l = a/b$ is the ratio of the length of the semimajor axis, a, to the length of the semiminor axis, b; ρ_ρ is the resistivity of the polymer; ρ_{lc} the resistivity of the liquid crystal; $K = K_{11} = K_{22} = K_{33}$ is the elastic constant of the liquid crystal; $\Delta\epsilon$ is the dielectric anisotropy of the liquid crystal; and ϵ_0 is the dielectric permittivity of free space. This formula predicts zero driving voltage for perfectly spherical droplets because it does not include a term for surface anchoring. However, it is useful for predicting trends in driving voltage.

Equation (1) predicts that thinner films containing nearly spherical droplets of a liquid crystal with a large dielectric anisotropy will have a low driving voltage. Low resistivity polymers and higher resistivity liquid crystals also produce lower driving voltage. Dielectric studies of PDLC films have shown that charge is built up both at the electrode and at the droplet–polymer interface [45]. This charge buildup shields the droplet and raises the driving voltage.

4. Switching Speed

The switching speed of the PDLC shutters is minimized when the droplets are small and aspherical. The switching speed is also minimized for liquid crystals with large elastic constants and low viscosity [46].

Equation (2) can be used to determine the effect of liquid crystal elastic constants and droplet size and shape on switching speed [46].

$$\tau_{off} = \frac{Y_1 a^2}{K(l^2 - 1)} \tag{2}$$

where Y_1 is the viscous torque[$\simeq 4 \times 10^{-2}$ kg/(m·s)]. Other values are defined for Eq. (1). Switching speed is therefore minimal for small elliptical droplets and for liquid crystals with large elastic constants.

Trade-offs must be made in the design of PDLC shutters. For example, many display applications require low driving voltage and fast switching speeds. Minimum driving voltage requires large spherical droplets, while fast switching speed requires small aspherical droplets. The OFF state scattering is also affected by the droplet shape. Clearly, compromises must be made in the design of PDLC materials.

D. Recent Developments in Liquid Crystal/Polymer Dispersions

1. Colored PDLCs

Isotropic and dichroic dyes have been incorporated in PDLCs to produce colored displays. The color contrast produced with an isotropic dye is the result of the difference in the effective path length of light passing through a PDLC in the transparent ON state and the scattering OFF state [47]. This difference can be used to estimate the scattering efficiency of a PDLC film.

Greater color contrast is achieved with dichroic dyes where not only the path length of light but also the extinction coefficient of the dye is modulated [48]. To achieve high contrast, the dichroic dye must be segregated in the droplet. The degree of segregation depends on the type of dye used and can approach 100% [49]. By proper selection of dichroic dyes, a black PDLC can be fabricated with a large contrast ratio over the entire visible region.

2. PDLC Infrared Applications

PDLCs have been tailored for operation in the infrared [50]. The materials used to fabricate the PDLC must have reasonable transmission in the infrared. The droplet size must also be on the order of the longer wavelength of infrared light. The effect of droplet size on the light scattering is clearly seen in the infrared. Films with droplet sizes on the order of 5–10 μm scatter efficiently in the 2–5 μm region, whereas 15–25 μm droplets scatter more effectively in the 8–14 μm region (Fig. 14) [51].

PDLC shutters formed with the eutectic liquid crystal, E7, and a polyvinylpyrrolidone matrix sandwiched between germanium substrates have been used as modulators for infrared video cameras. They can be used to replace mechanical shutters where maximum reliability is required and vibration must be eliminated [52].

3. Haze-Free PDLCs

The practical applications of PDLC and NCAP materials have been limited by the appearance of haze in the ON state for other than normal viewing angles [53]. The haze is caused by the increased effective refractive index of the droplets when viewed at oblique angles. The droplet refractive index is

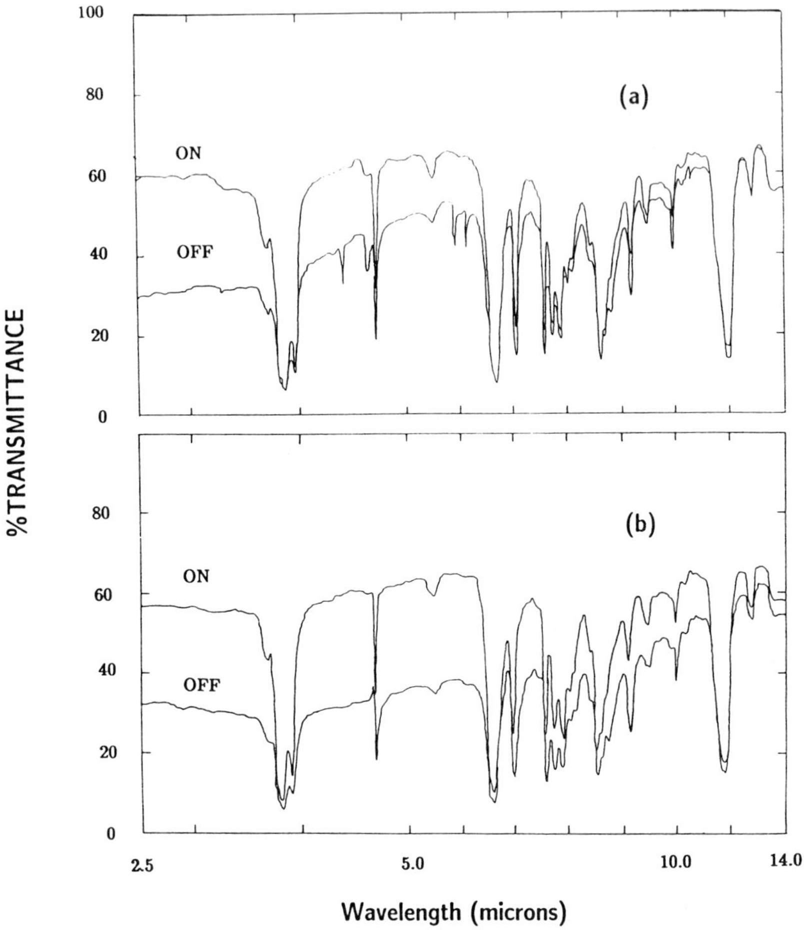

FIG. 14 Infrared spectra of a PDLC film after (a) quenched cooling, <2 μm diameter droplets; (b) cooling at −2°C/min, 10–25 μm diameter droplets. (Reproduced from McCargar et al. [51], by permission of SPIE.)

angle-dependent, whereas the isotropic polymer has a constant refractive index for all viewing angles.

Haze-free PDLC films have been reported that use a birefringent liquid crystal polymer matrix [54]. The polymer is chosen to have both the ordinary and extraordinary refractive indices of the polymer matched with those of the liquid crystal. In the ON state the refractive index of the droplet and polymer matrix are matched for all viewing angles and the film is haze-free (Fig. 15) [55].

Development of haze-free PDLC shutters has been slow because of the difficulty in achieving and controlling phase separation of the low molecular weight liquid crystal from a liquid crystal polymer matrix and controlling the alignment of the polymer matrix.

4. Polymer Network Liquid Crystals

Liquid crystal/polymer dispersions have been developed that are not based on droplets but rather consist of a polymer network supporting a continuous liquid crystal phase. This type of structure was first described by Yamagishi et al. [40]. They found that photopolymers produce the standard droplet morphology when a step growth reaction is used, but the photopolymers produce a polymer ball morphology with a continuous liquid crystal phase when a chain reaction is used. More recently, investigators from Dainippon have reported a polymer network using relatively high concentrations of liquid crystal (70–90%) [56]. The electro-optics of these materials is similar to that of standard PDLC materials.

5. Polymer Microspheres

Electro-optic materials that switch from a scattering to a transparent state have also been made from mixtures consisting of nematic liquid crystals dispersed in polymer microspheres connected by a small quantity of binder [57]. The gap between the microspheres is filled with nematic liquid crystals. Films of the material are formed by dissolving the liquid crystal and polymer binder in an organic solvent. The microspheres are added to the solution, which is then coated in a thin layer on an ITO glass substrate. The substrate is then heated to drive off the solvent. The binder separates

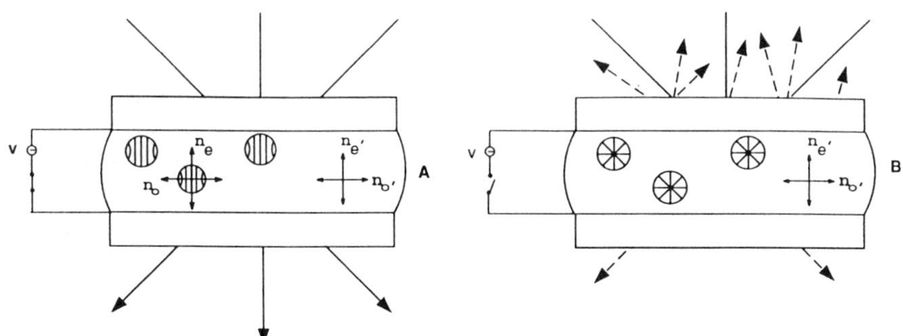

ON OFF

FIG. 15 Haze-free PDLC in the ON and OFF states.

from the liquid crystal and serves to connect the microspheres. The electro-optics of these materials is similar to that of the PDLC and NCAP materials described above.

6. Polymer Gels

Anisotropic gels have been produced by photopolymerization of liquid crystal diacrylates dissolved in an oriented conventional liquid crystal mixture [58,59]. The unpolymerized material is used to fill a cell with rubbed polyimide surface coatings. Photopolymerization results in a polymer network that maintains the original orientation. Application of an electric field across the cell switches the material to a translucent state, with the scattered light being highly dependent on the polarization and wavelength of the incident light.

The polymer structure of the anisotropic gel is hypothesized to strongly anchor the low molecular weight liquid crystal in the near vicinity. Therefore, when a field is applied across the anisotropic gels, only a fraction of the low molecular liquid crystal aligns with the field and a fraction remains oriented by the polymer network. These materials operate in a reverse mode, that is, they are clear in the OFF state and switch to a scattering state upon application of an external field.

7. Polymer-Stabilized Cholesterics

Photopolymerization of liquid crystal diacrylates dissolved in an oriented cholesteric liquid crystal have been used to produce a variety of liquid crystal shutters. The specific electro-optic properties of the shutters depend on the initial orientation of the cholesteric materials. A shutter that is scattering in the OFF state and switches to a transparent state upon application of an external field is formed by filling a cell with no surface treatment and photopolymerizing the diacrylate in the presence of an electric field. The resulting cell is highly scattering in the absence of an applied field. Application of a field serves to switch the material to a homeotropic texture, and the films are transparent for all incident angles [60]. A reverse-mode cell uses buffed polyimide alignment layers. The cell is prepared by irradiating a surface-aligned cell with UV light. The completed cells are clear in the absence of an applied field. Similar to the polymer/nematic gels, when a field is applied, only a fraction of the liquid crystal is reoriented, resulting in scattering.

Polymer-stabilized cholesteric textures have also been demonstrated to be multistable at zero field [61]. The shutters can be switched into a Grandjean reflecting state with a high voltage pulse and into a focal conic scattering state with a low voltage pulse. Pulses of intermediate voltage switch the shutter to an intermediate state that is also stable at zero field. The color of the reflecting state depends on the pitch length and birefringence of the cholesteric nematic mixture. The color is adjusted by varying the concentration of chiral additive. These materials offer the possibility of high level multiplex displays.

V. CONCLUSION

Dispersions of liquid crystal and polymers were first employed for cholesteric thermometers to improve their environmental stability and simplify their use. In the 1980s and into the 1990s, liquid crystal/polymer dispersions have been developed for electro-optic applications. In general these materials operate on the principle of electrically controlled light scattering between the liquid crystal and polymer phases. These dispersions offer improved light efficiency and make possible large area devices fabricated on flexible substrates in a continuous process. These new dispersions are being developed for use in high resolution direct-view and projection displays and for entirely new applications such as switchable windows. The recent development of entirely new liquid crystal/polymer dispersions such as polymer networks and polymer-stabilized cholesterics demonstrate that this field is in its infancy.

ACKNOWLEDGMENTS

This work was supported in part by the National Science Foundation Science and Technology Center ALCOM DMR89-20147, the National Center for Integrated Photonic Technology supported by DARPA under contract MDA972-90-C-0037, and DARPA under contract MDA972-91-J-1020.

REFERENCES

1. B. Bahadur (Ed.), *Liquid Crystals: Applications and Uses*, Vols. 1 and 2, World Scientific, Teaneck, N.J., 1990 and 1991.
2. J.L. Ericksen and D. Kinderlehrer (Eds.), *Theory and Applications of Liquid Crystals*, Springer-Verlag, New York, 1987.
3. P.J. Collings, *Liquid Crystals: Nature's Delicate Phase of Matter*, Princeton Univ. Press, Princeton, N.J., 1990.
4. L.M. Blinov (Ed.), *Electro-Optical and Magneto-Optical Properties of Liquid Crystals*, Wiley, New York, 1983.
5. S.E. Friberg, in *Liquid Crystals: Applications and Uses*, Vol. 2 (B. Bahadur, Ed.), World Scientific, Teaneck, N.J., 1991, p. 157.
6. D. Chapman, in *Liquid Crystals: Applications and Uses*, Vol. 2 (B. Bahadur, Ed.), World Scientific, Teaneck, N.J., 1991, p. 185.
7. J.L. Fergason, *Appl. Opt. 7*: 1729 (1968).

8. L.C. Scala and G.D. Dixon, *Mol. Cryst. Liq. Cryst. 10*: 411 (1970).
9. J.P. Quigley and W.J. Benton, *Mol. Cryst. Liq. Cryst. 42*: 43 (1977).
10. G.H. Heilmeier, L.A. Zanoni, and L.A. Burton, *Appl. Phys. Lett. 13*: 46 (1968).
11. G.H. Heilmeier and L.A. Zanoni, *Appl. Phys. Lett. 13*: 91 (1968).
12. G.H. Heilmeier, J.A. Castellano, and L.A. Zanoni, *Mol. Cryst. Liq. Cryst. 8*: 293 (1969).
13. J.J. Wysocki, J. Adams, and W. Haas, *Phys. Rev. Lett. 20*: 1024 (1968).
14. H. Baessler and M.M. Labes, *J. Chem. Phys. 51*: 5397 (1969).
15. M. Schadt and W. Helfrich, *Appl. Phys. Lett. 18*: 127 (1971).
16. G.W. Gray, K.J. Harrison, and J.A. Nash, *Electron. Lett. 9*: 130 (1973).
17. G.W. Gray, *J. Phys. (Paris) 36*: 337 (1975).
18. I. Sage, in *Thermotropic Liquid Crystals* (G. W. Gray, Ed.), Wiley, New York, 1987, p. 64.
19. F.C. Luo, in *Liquid Crystals: Applications and Uses*, Vol. 1 (B. Bahadur, Ed.), World Scientific, Singapore, 1990, p. 397.
20. T.J. Scheffer and J. Nehring, *Appl. Phys. Lett. 45*: 1021 (1984).
21. N.A. Clark and S.T. Lagerwall, *Appl. Phys. Lett. 36*: 899 (1980).
22. C. Hilsum, U.K. Patent 1 442 360 (1976).
23. H.G. Craighead, J. Cheng, and S. Hackwood, *Appl. Phys. Lett. 40*: 22 (1982).
24. G. Beni, H.G. Craighead, and S. Hackwood, U.S. Patent 4,411,495 (1983).
25. J.L. Fergason, U.S. Patent 4,616,903 (1986).
26. P.S. Drzaic, *J. Appl. Phys. 60*: 2142 (1986).
27. J.W. Doane, N.A. Vaz, and G. Chidichimo, U.S. Patent 4,688,900 (1987).
28. J.W. Doane, N.A. Vaz, B.-G. Wu, and S. Žumer, *Appl. Phys. Lett. 48*: 269 (1986).
29. P.S. Drzaic, *Liq. Cryst. 3*: 1543 (1988).
30. J.H. Erdmann, S. Žumer, and J.W. Doane, *Phys. Rev. Lett. 64*: 1907 (1990).
31. E. Dubois-Violette and O. Parodi, *J. Phys. (Paris) Colloq. C4*: 57 (1969).
32. S. Candau, P. LeRoy, and F. Debeauvais, *Mol. Cryst. Liq. Cryst. 23*: 283 (1973).
33. J.L. Fergason, *Soc. Inf. Displays, Int. Symp. Dig. Tech. Papers 16*: 68 (1985).
34. J.L. Fergason, A. Dalisa, S. Lu, and P.S. Drzaic, *Soc. Info. Displays Int. Symp. Dig. Tech. Papers 17*: 126 (1986).
35. P.S. Drzaic, R.C. Wiley, and J. McCoy, *Proc. SPIE 1080*: 41 (1989).
36. A.B. MacKnick, P.J. Jones, and L. White, *Proc. SPIE 1080*: 169 (1989).
37. J.L. West, *Mol. Cryst. Liq. Cryst. 157*: 427 (1988).
38. N.A. Vaz, G.W. Smith, and G.P. Montgomery, Jr., *Mol. Cryst. Liq. Cryst. 146*: 1 (1987).
39. N.A. Vaz, G.W. Smith, and G.P. Montgomery, Jr., *Mol. Cryst. Liq. Cryst. 146*: 17 (1987).
40. F.G. Yamagishi, L.J. Miller, and C. I van Ast, *Proc. SPIE 1080*: 24 (1989).
41. B.-G. Wu, J.L. West, and J.W. Doane, *J. Appl. Phys. 62*: 3925 (1987).
42. G.P. Montgomery, Jr., N.A. Vaz, and G.W. Smith, *Proc. SPIE 958*: 104 (1988).
43. A.M. Lackner, J.D. Margerum, E. Ramos, and K.-C. Kim, *Proc. SPIE 1080*: 53 (1989).

44. J. Erdmann, J.W. Doane, S. Žumer, and G. Chidichimo, *Proc. SPIE 1080*: 32 (1989).
45. D. Seekola and J.R. Kelly, *Proc. SPIE 1455*: 19 (1991).
46. B.-G. Wu, J.H. Erdmann, and J.W. Doane, *Liq. Cryst. 5*: 1453 (1989).
47. J.L. West, W. Tamura-Lis, and R. Ondris, *Proc. SPIE 1080*: 48 (1989).
48. J.L. West, R.J. Ondris, and M. Erdmann, *Proc. SPIE 1257*: 76 (1990).
49. J.L. West and R. Ondris-Crawford, *J. Appl. Phys. 70*: 3785 (1991).
50. J.L. West, J.W. Doane, Z. Domingo, and P. Ukleja, *Polym. Prepr. 30*: 530 (1989).
51. J.W. McCargar, J.W. Doane, J.L. West, and T.W. Anderson, *Proc. SPIE 1455*: 54 (1991).
52. J.W. McCargar, R. Ondris-Crawford, and J.L. West, *J. Electron. Imaging 1*: 22 (1992).
53. P. van Konynenburg, R.T. Wipfler, and J.L. Smith, *Proc. SPIE 1080*: 62 (1989).
54. J.L. West, L.-C. Chien, and W. Tamura-Lis, *Pacific Polym. Prepr. 1*: 247 (1989).
55. J.W. Doane, J.L. West, W. Tamura-Lis, and J.B. Whitehead, Jr., *Pacific Polym. Prepr. 1*: 245 (1989).
56. T. Fujisawa, H. Ogawa, and K. Maruyama, *Jpn. Display 690* (1989).
57. H. Nomura, S. Suzuki, and Y. Atarashi, *J. Appl. Phys. 68*: 2922 (1990).
58. R.A.M. Hikmet, *J. Appl. Phys. 68*: 4406 (1990).
59. R.A.M. Hikmet, *Liq. Cryst. 9*: 405 (1991).
60. D.-K. Yang, L.-C. Chien, and J.W. Doane, *Appl. Phys. Lett. 60*: 3102 (1992).
61. Y.K. Fung, D.-K. Yang, and J.W. Doane, *Proc. SPIE 1664*: to appear (1992).

11

Use of Monosize Dispersions in Silver Halide Photographic Emulsions

TREVOR MATERNAGHAN Research and Development,
ILFORD Limited,* Mobberley, Cheshire, England

I. Introduction 374

II. General Technology of Photographic Materials 376
A. General chemistry of photographic emulsions 376
B. The photographic process 378
C. Photographic materials 379

III. Photographic Processing 385
A. Development of the latent image 385
B. Processing of a holographic film and developed image structure 388

IV. Solid-State Properties 390

V. Monodisperse Photographic Emulsions 392
A. Control of size distribution 392
B. Crystal growth processes 393

VI. Emulsion Technology 394
A. Size, size spread, halide composition, and crystal habit 397
B. Preparation of silver halide emulsions 398
C. Defects, twinning, halide boundaries, and dislocations 399
D. Core-shell systems 400

---

*ILFORD Limited is a business unit of the Imaging Products Division of International Paper.

VII. Emulsion Sensitization 402
A. Chemical sensitization 402
B. Spectral sensitization 405

VIII. Emulsion Characterization 407
A. Surface chemistry and physics 407
B. Instrumental and analytical methods 408
C. Complementary instrumental studies 409

IX. Design of Tabular Emulsions 409
A. Twinning and general properties 409
B. Epitaxial methods 414
C. Applications 421
D. Core-shell tabular crystals 423
E. Nucleation and growth at high halide concentrations (T-grains) 424
F. Recent developments 427
G. ILFORD 400 and 100 DELTA materials 432

X. Future of Silver Halide Photography 433
A. Development of competing electronic systems 433
B. Potential for improvement in AgX systems 434

References 436

I. INTRODUCTION

Photographic emulsions comprise silver halide microcrystals of 0.03–3 μm diameter dispersed in aqueous gelatin. The use of the term "emulsion" to describe such systems is traditional, although in fact they are dispersions of solid particles in a polymer binder such as gelatin. In an increasingly competitive market, the performance of photographic materials depends primarily on the design of these specialized dispersions. The most sensitive silver halide crystals may become developable after an exposure of only two or three photons. This response may be called upon after years of storage of photographic products in unfavorable conditions of temperature and humidity. In another application, an X-ray product may be required to produce a stable, dry image in a hospital after a processing time of less than a minute.

Silver halide photography is a mature technology over 150 years old. Many of the current manufacturers have celebrated the centenary of their founding. However, there has been no decrease in the rate of technological change and achievement in silver halide phototechnology during the last 15 years.

The emulsion microcrystals are contained in coated papers or films in layers, each layer about 1 μm thick. There may be nearly 20 such layers in a modern color film, and a film may contain up to 100 different components or chemical compounds. The performance of the material also depends on the design and fabrication of these multilayer elements.

Much of the progress in the design of photographic emulsions during the last 15 years has been due to research on monodisperse or monosize photographic emulsions. Their exploitation as model systems initiated a much greater understanding of the fundamentals of the photographic process. In principle, the photographic requirements can be met in a more systematic fashion by suitable adaptation of crystal size, size distribution, halide composition, crystal habit (shape), doping, and other features of experimental monodisperse microcrystals, so that they now provide the image-forming elements of many recent products. This new technology, allied with established patterns of sensitization (digestion) treatments affecting the sensitivity (speed) and sensitivity distribution (contrast) of coatings, has led to "designer" grains with more efficient internal structures that would not have been achievable by conventional methods.

The most sensitive (but not yet necessarily the most efficient) microcrystals are tabular microcrystals containing stacking defects, that is, twin (mirror) planes in their structures. Crystals with parallel twin planes predominantly have flat hexagonal or triangular shapes. The problem of producing monosize crystals of this type has been solved within the last 15 years; several photographic products now contain them.

It has usually also been necessary to exploit composite crystals to achieve the highest speeds or superior image quality. These have different halide compositions in different parts of the grains, for example, composite crystals of AgCl with AgI or AgBr. In fact, it has been a long-term aim of silver halide technology, currently at least partly achieved, to provide the silver halide microcrystals with precisely engineered structures. Different parts of each grain will perform different functions, such as light capture or storage of developable image centers.

Hence, the use of monosize silver halide dispersions now extends well beyond model systems for the study of crystal growth and sensitization mechanisms to ever-widening commercial application as emulsion components in large-scale manufacturing of photographic films and papers. The photographic knowledge and methods used for control of the characteristics of the microdispersion are reviewed in this chapter, using examples of commercial applications from the ILFORD Laboratories and those of other research groups.

Silver halide photography has greatly benefited from the recent advances in modern analytical instruments and techniques. Some key problems that seemed intractable two decades ago are now much better understood and applied. Particularly useful, of course, are the various forms of microscopy; studies employing electron microscopy, scanning ion microscopy, or atomic force microscopy have been featured.

Electronic imaging systems such as video recording and solid-state cameras were seen as eventually replacing silver halide photography. While this has indeed effectively taken place for some applications (e.g., hobbyist film-making), silver halide photography has made remarkable progress in image quality in recent years and proved remarkably resilient. The limits of sensitivity for the silver halide system have almost certainly still not been achieved. Instead, an interesting phenomenon at the scientific and technological level is that many advances in silver halides (themselves with semiconductor properties) have paralleled concepts, understanding, and methods derived from electronic imaging. Systems seeking the "synergy" of the two imaging methods (the image quality of silver halides and the convenience and image processing ability of electronics) are becoming available.

An excellent review of the photographic market with an analysis of recent trends is that of Krause [1].

II. GENERAL TECHNOLOGY OF PHOTOGRAPHIC MATERIALS

A. General Chemistry of Photographic Emulsions

The silver halide particles in photographic emulsions are invariably precipitated from aqueous solutions of silver and halide salts stirred into an aqueous solution of gelatin above 35°C and under safelighting. The average final sizes of the crystals are in the range of 0.03–3 μm.

The isoionic point of silver bromide in gelatin is close to pAg 6.5 at 40°C. [pAg values are derived by analogy with pH and are formally defined as $-(\log_{10}[Ag^+])$, where $[Ag^+]$ is the silver ion activity.] The pBr for the emulsion system is defined by the *solubility product relationship* in the form

$$\text{pAg} + \text{pBr} = \text{p}S$$

where pS is the −logarithm of the solubility product $[Ag^+][Br^-]$, 11.5 for AgBr at 40°C. Analogous relations exist for other halides (and pseudohalides such as thiocyanate).

Gelatin is a polypeptide and is composed chiefly of linear chains distributed around a common unit, the α-chain with molecular weight of ca. 95,000. Hydrolysis of the gelatin from collagen derived from bone or hide leads to a peptide fraction of lower molecular weight, and residual cross-linking leads to chain molecular weights higher than this. The molecular weight distribution tends to be multimodal at multiples of 95,000. Gelatin is also a polyelectrolyte, having positively charged (amino) and negatively ionizable (carboxyl) groups on side groups on the peptide backbone.

The isoionic point of gelatins derived by alkaline or lime extractions from bone or hide are lower (ca. pH 5) than the corresponding final values for acid extraction (ca. pH 6–8). The difference in the balance of ionizable groups occurs because the alkaline hydrolysis during lime extraction causes conversion of amide groups to carboxyl groups. Thus limed ossein gelatin becomes positively charged below pH 5, and silver bromide particles in gelatin become positively charged below pAg 6.5. The relative charge densities on gelatin and silver halide can affect the conformation of adsorbed gelatin and colloid stability behavior.

The primary role of the gelatin binder here is to restrict coagulation of the initially formed particles by the rapidly increased salt concentration and to preserve stability (prevent aggregation) during further growth. The gelatin is thus a *protective colloid* for the silver halide particles and stabilizes by a mixture of steric (osmotic) chain segment interactions during particle collisions and by electrostatic forces due to the repulsion of adsorbed coils with like charges.

The silver halide content of the emulsion can be high (ca. 1 mol/L). The silver halide crystals in gelatin are desalinated by *flocculation* and/or washing. A chemically modified gelatin or anionic polyelectrolyte such as a polystyrene sulfonate derivative is commonly used so that the dispersion can be flocculated with a suitable acid addition, washed when cold to remove salts, and then redispersed by warming with additional gelatin.

The intrinsic sensitivity or *photographic speed* (expressed in some units like the ISO standard for photographic products) is normally increased by *chemical sensitization* (often a mixture of sulfur or gold salts, added in less than millimolar quantities at temperatures greater than 40°C for a few minutes). This process, still called a *digestion* process, can often increase the arithmetic speed of the emulsion by a factor of 16 or so.

Unfortunately, *fog* is also often increased during digestion. This is the visible silver density that can be developed without any exposure. It can be classified as emulsion fog or developer fog. Emulsion fog is introduced during crystal growth or sensitization and often increases with increasing crystal surface area, suggesting random silver speck formation. Developer-specific fog is triggered by reduction, normally at a crystal defect. This is counteracted by organic antifoggants such as 1-phenyl-5-mercaptotetrazole.

Most reducing agents may increase speed or, even more often, increase fog. Such a sensitization method would be termed a *reduction sensitization*, although the distinction with respect to other forms of sensitization is often rather unclear. For example, sodium thiosulfate is a commonly used sulfur sensitizer, but it is also a mild reducing agent and, as a mild silver halide solvent, can also create etch sites on the grain surface that may be active areas.

For many applications, the spectral range of the silver halide is too confined to the blue and green wavelengths, and a *spectral sensitizing dye* is also added during manufacture. The commonly used dyes are the cyanine dyes, which have a pronounced tendency to form organized molecular aggregates on the silver halide particle surfaces. The so-called J-aggregate forms, for example, consist of dye molecules stacked face to face with electronic interaction along the stack direction. They tend to absorb light of longer wavelengths than the monomers and also show peaks with relatively narrow wavelength spread.

The photographic emulsion is then coated in one or more thin layers of about 1–3 μm thickness on a *polymeric support*. For films, this is a transparent polymer film base of polyethylene terephthalate (polyester) or cellulose triacetate, varying in thickness from 50 to 250 μm. For papers, this is now invariably a rigid paper core resin-coated on both sides with water-impermeable polyethylene layers to aid processing. The front layer contains a titanium dioxide layer that ensures high reflectivity. Additional gelatin-bearing layers are applied to these supports to provide the required photographic, chemical, and physical properties. Such layers are usually hardened with an organic cross-linking agent such as formaldehyde, a triazine salt, or a vinylsulfone.

After suitable exposure of the material in a camera or enlarger, or by a cathode-ray tube or X-ray screen, the image is then developed in a reducing agent such as hydroquinone, which converts the exposed AgX to metallic silver. The remaining unexposed silver halide particles are then removed by a *fixer*, a compound such as sodium thiosulfate that forms a soluble complex with silver halide, allowing it to be washed out to produce a stable image.

B. The Photographic Process

Silver halide photography depends on the unique combination of chemical and physical properties that allow the exposure of a silver halide microcrystal by a few photons to form "latent image" specks. These centers act as catalysts for the reduction of perhaps the whole particle of AgX. Thus, as this particle could consist of perhaps 10^8–10^9 silver and halide ion pairs, an image of this size can be formed in a developer following exposure to only a few light photons. One that has received less than three or four photons is certainly not developable. Geometric and absorbance factors also dictate that normally a grain may have to be exposed by considerably more than this threshold number of photons in order to ensure developability, and in practice the quantum efficiency of photographic systems is often only about 20%.

In the case of color materials, the photographic assembly invariably consists of a tripack, at least three layers that form images corresponding to blue, green, and red exposing light. The assembly usually contains *color coupler dispersions*, which are actually oil-in-water emulsions containing the coupling agents dispersed in the oil phase.

On so-called *chromogenic development* in an agent such as a substituted phenylenediamine, the coupler reacts with the oxidized developer formed as a result of the redox reactions with the exposed AgX. A dye image is formed. The silver image then has to be bleached in a subsequent step to render the dye image visible. This technology can also be used to generate a monochrome image, as in ILFORD XP2 black-and-white film. Here the exposed imaging layers contain a blend of different coupler dispersions. On chromogenic development, a neutral dye image is formed after the bleaching step. The main advantage of this approach is that it produces an image that is much less "grainy" than that from a conventional monochrome film of similar speed.

The photographic process is rather familiar, and keen photographers will be quite aware of the extended choice of "new technology" films and papers. Some of the design problems of photographic emulsions will be similar to those encountered with many other types of particulate dispersions. However, for photographic emulsions, there are additional considerations in order to meet the product requirements, over and above the usual handling and technical requirements of industrial preparations such as flow properties and corrosion control. Also, emulsions have to be prepared, treated, and coated under restricted, safelight illumination. Any manufacturing effluent containing silver salts is both precious and potentially hazardous, requiring recovery and processing treatment.

C. Photographic Materials

The most comprehensive review of photographic science is given by James [2]. The most important features of the silver halide/gelatin photographic system are that

1. Sensitization leads to an efficient photochemical reaction.
2. The product of exposure can be concentrated to form the latent image.
3. It has an efficient amplification step (ca 10^9) between latent images and the final developed image.

4. It has a good shelf life—the n-atom threshold favors stability, where $n > 2$.
5. It gives a stable latent image.
6. It can be spectrally sensitized to all visible wavelengths (and even to the infra-red).
7. It can be designed for a wide range of applications such as from continuous tone to line reproduction as in graphic arts applications.
8. It has a high information capacity (a 35 mm slide is often quoted as having half a million bits of information).
9. The silver image has archival permanence.

The advantages of gelatin in the silver halide photographic process are that

1. It adsorbs to AgX yet allows crystal growth.
2. It provides a steric and electrostatic barrier to particle aggregation.
3. It allows desalination techniques such as flocculation by use of phthalate-modified gels and acidification, etc.
4. It sets after coating and provides a mechanical support for film.
5. It cross-links with hardening agents and produces a protective topcoat.
6. It protects against latent-image fading.
7. It swells during processing, facilitating development and washing.
8. A filamentary form of silver is promoted during development. This is an efficient light scatterer and so optical density is enhanced.
9. It ensures a high speed/fog discrimination. Fog tends to be much higher in alternative binders.

The function of gelatin as a protective colloid has been studied in a model system [3].

The laminated assemblies required for photographic products that have to be made in a very wide range of types, formats, and sizes have been discussed elsewhere [1,2,4–6]. The constructions of typical monochrome paper and color film materials are illustrated schematically in Fig. 1.

A monochrome camera film consists of one or more layers of silver halide in gelatin with a layer of gelatin coated on top—the supercoat—to prevent physical damage to the emulsion. The supercoat may contain matting agents to reduce the gloss of the film surface and antistatic agents and lubricants to avoid friction. In color materials, an ultraviolet light-absorbing dye is sometimes added.

Modern color films have at least two blue, two green, and two red recording layers. Other layers are positioned next to the appropriate emulsion layer for optical or control purposes. Each pair of layers consists generally of an upper high speed layer and an underlying layer of finer

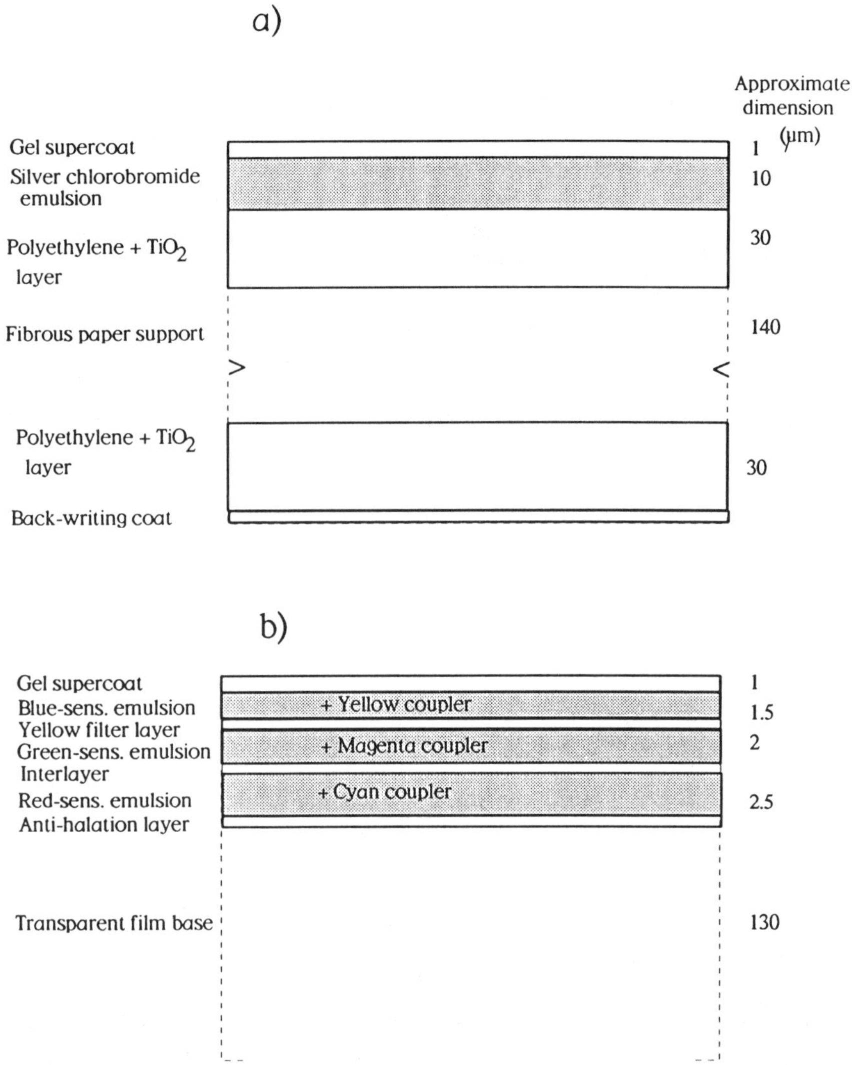

FIG. 1 Construction of typical monochrome paper and color film materials. (a) Monochrome enlarging paper; (b) color negative film. (After Ref. 6.)

grained emulsion with a higher concentration of dye-forming couplers, thus providing the material with sufficient exposure latitude and low granularity in the important midtone and highlight regions. The layer nearest the film base often contains dyes to prevent halation (diffuse imaging due to back-scattering of light). The antihalation dyes in this layer are bleached during processing. In integral masking, the unwanted blue absorption of the available magenta couplers is compensated by the contra-imagewise formation of a yellow dye in the magenta layer. The masking systems and the coupler dispersions result in some loss in sharpness.

The rapid advances and revised assemblies in color negative film materials have been reviewed by Krause [7] and by Sasai [5].

"Instant" photography requires a high complexity of the photographic assemblies to cope with the processing step. This was illustrated by Land, who outlined the achievement of "one-step photography" [8] in the Polaroid SX-70 material. In a later material, development was initiated by spreading a viscous alkaline developing composition as the film is ejected from the camera [9]. The coating of up to 17 layers may be necessary for the functioning of the material: the primary image-forming layers, layers to release the diffusible intermediates, mordanting layers to receive the final image, opacifying layers to protect the photosensitive layers from exposure to light before and after development is completed, and control layers with neutralization and "chemical" timing mechanisms to control the termination of the processing sequence.

Any imaging process based on diffusion of an intermediate may suffer a loss of sharpness of the final image by the lateral spreading of reactants; this is minimized by keeping the total thickness as small as possible. Such materials are therefore rarely used for images in which great enlargements will be required.

The wide range of addenda and of preparation schemes for the assembly components for photographic materials is cataloged in Refs. 10 and 11. Some detail on the technology of production coating of photographic materials is given in Ref. 12.

Reference 6 gives a short, general review of photographic materials, and a survey of film materials particularly designed for radiographic processing is given by Roberts and Smith [13]. The latter contains a useful review of the basic physics and also covers the structure of film emulsions including T-grains etc., spectral sensitization, grain size and distribution, gelatin properties, film types, films for specialized purposes, and manufacturing processes. "T-grains" was the name given by Eastman Kodak to highly tabular, thin grains grown at high bromide ion excess.

Reviews of some of the underlying photographic science can be found in the reviews of James [14], Locker et al. [15], and Sturmer and Marchetti [16].

As a specific example of a photographic material an enlarging monochrome paper is considered first, ILFORD Multigrade III RC [17]. This resin-coated paper features cubic silver chlorobromide microcrystals below 0.4 μm in size. Exposure of the paper to light of varying spectral composition results in a wide range of printing contrasts (Fig. 2), which is a substantial user benefit.

All chlorobromide (black-and-white) emulsions are blue-sensitive with a slight sensitivity to green light. To make an emulsion sensitive to colors other than blue, sensitizing dyes need to be added. ILFORD variable contrast papers are coated with an emulsion that is a mixture of a basic blue-sensitive emulsion and an emulsion to which is added green-sensitizing dye. Thus, part of the emulsion is sensitive mainly to blue light and the other part is sensitive to blue and green light (Fig. 3a). These two parts of the emulsion have the same contrast. They also have the same speed to blue light, but, naturally, the undyed part of the emulsion has a low speed (that is, is less sensitive) to green light.

When the paper is exposed to blue light, both parts of the emulsion respond and contribute equally to the final image. This image is of high contrast because of the additive effect produced by two emulsions with the same speed and contrast. When the paper is exposed to green light, only the dyed part of the emulsion reacts initially. This is because the two emulsions have very different sensitivities, as explained earlicr. The resultant curve has a very much wider exposure range and is thus of low contrast.

FIG. 2 Example of wide range of printing contrasts available for ILFORD Multigrade paper. (Photo by Gered Markowitz.)

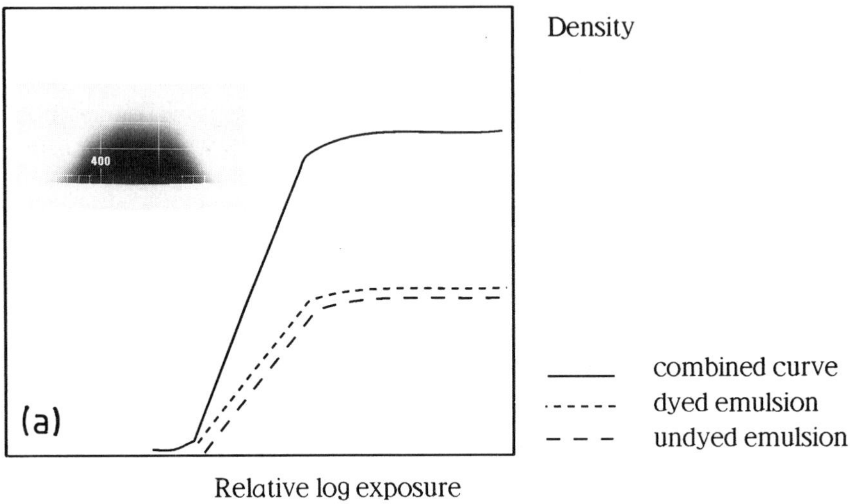

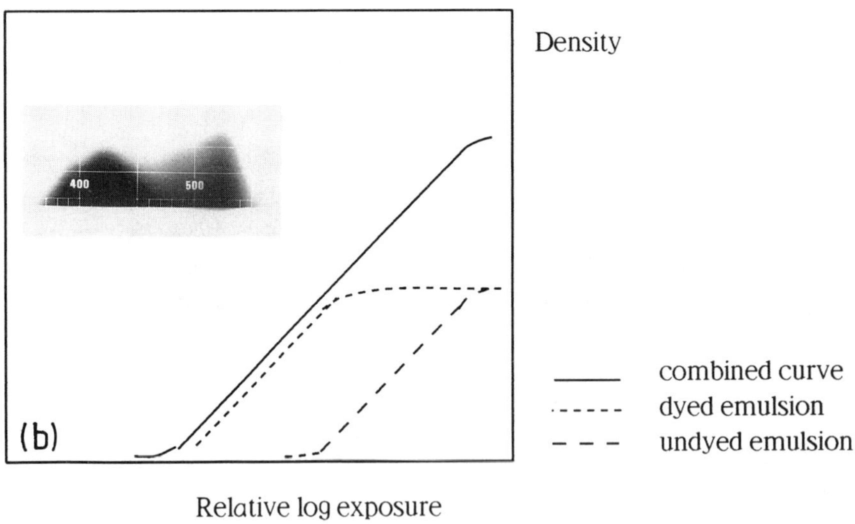

FIG. 3 Principle of ILFORD variable contrast papers. (a) High contrast response to blue light exposure; (b) low contrast response to green light. Inserts show spectral response (horizontal scale in nm) of undyed (a) and dyed (b) emulsions.

By varying the proportion of blue to green light, a contrast range between these two extremes can be obtained. The color of the light reaching the emulsion during exposure is controlled by the use of filters. A magenta filter absorbs green light and transmits blue; a yellow filter absorbs blue light and transmits green. In this way, high and low contrast images can be made (Fig. 3b).

Other variable-contrast papers now available include Kodak Polycontrast and Agfa Multicontrast products.

III. PHOTOGRAPHIC PROCESSING

A good general review of photographic characteristics and silver halide technology was compiled by Locker et al. [15]. Lohmann reviewed current trends in photography at the 1986 ICPS conference in Cologne [4].

A. Development of the Latent Image

Exposed silver halide can be reduced to silver by two pathways: chemical development in which metallic silver is formed by the reduction of the silver ions in the developing grains, and physical development, in which the supply of silver ions comes from other grains or from a silver salt added to the developer solution. Although chemical development predominates in most monochrome developers, developer components complex an appreciable amount of silver ion.

The composition of a typical monochrome developer for monochrome film and papers comprises

Developing agents—combinations of *N*-methyl-*p*-aminophenol or 1-phenyl-3-pyrazolidone and a dihydroxybenzene, known as Metol-hydroquinone and Phenidone-hydroquinone developers, respectively.

A restrainer—typically KBr or an antifoggant such as benzothiazole.

An antioxidant—sodium sulfite is commonly used to prevent aerial oxidation of the developer.

An alkali such as sodium carbonate (developer activity increases with pH) and a pH buffering agent such as sodium borate.

The fixer removes unexposed silver halide from the emulsion; it is a silver halide solvent such as sodium or ammonium thiosulfate. For reversal processing, the originally developed silver halide must be removed while the unexposed silver halide is left intact. A bleaching agent such as ferric ammonium ethylenediamine tetraacetic acid is therefore used, normally in combination with potassium bromide.

The developer may be designed for full development of all exposed grains, for example in radiographic films of graphic arts. However, for most camera films, partial development is required to provide the most suitable tone reproduction and image quality for pictorial work. The selection of the appropriate characteristic curve of developed density versus the logarithm of the relative light exposure (log E) for the photographic specification is illustrated in Fig. 4. This is the subject of frequent sensitometric testing by manufacturers and processing laboratories.

The use of monosize emulsions has allowed many of the complex interactions during development to be identified. For example, Radcliffe [18] comprehensively investigated the effects of crystal size, iodide content, and iodide distribution on the covering power (ratio of optical density to silver coating weight) and development rate at D_{max} (maximum developed density) of chemically sensitized cubic monosize emulsion types in ILFORD ID2 (PQ) and ID11 (MQ) developers. The development characteristics with I-layered AgBr emulsion crystals suggested that solution physical development was retarded by the iodide ion released by crystals that were themselves developing; the covering power was reduced relative to the corresponding pure AgBr emulsions.

The factors influencing the morphology of developed silver were set out by Farnell and Flint [19] using tabular AgBr emulsion crystals. Partially developed grains and their associated silver centers were separated from the coatings. These were then imaged by electron microscopy using the usual carbon replica/gold shadowing techniques used for silver halide emulsions. The form of the developed centers depended not only on the location of latent image centers on particular faces, but also on the ease of release of bromide ion from particular centers. It was deduced that at higher developer activity, a halide ion adjacent to a development center near a grain edge would pass more easily into solution than a similar one near the center of a grain face.

Sahyun [20] recently described some mechanistic investigations of photographic development. The initiation stage of development, when the first silver cluster is being formed, was emphasized as being mechanistically distinct from the continuation stage. Until a cluster size of 2 nm, or about 2000 atoms, is built up, the reaction depends on the slow external injection of electrons and neighboring silver ions. After this critical step and during the continuation stage for chemical or direct development, the rate of silver production was often found to vary according to the surface area of the Ag already developed. This accords with the frequently cited electrochemical model of the developed silver acting as a cathode for the redox couple made up of the developer and its oxidized form. The silver then grows autocatalytically. According to Sahyun, oxidation of the developing agent takes place at the surface of the cluster in contact with solution; reduction

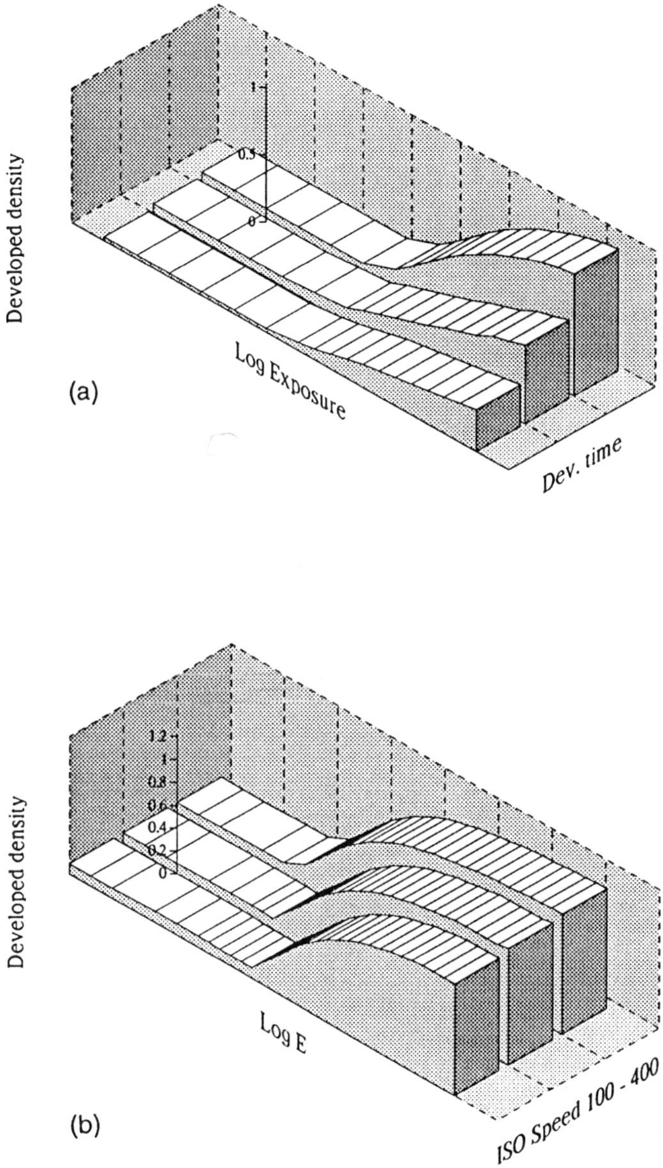

FIG. 4 Characteristic curves of (a) developed density versus the logarithm of the relative light exposure (log E) for increased development time for the same film, and (b) for various ISO speeds of photographic film with a fixed development time.

of the silver ions takes place at the interface with the silver halide lattice or elsewhere on the grain interface (solution physical development).

The latent subimage has been examined in a silver bromide model emulsion sensitized with sulfur or sulfur plus gold [21,22]. Further exploration of the nucleation and growth model of latent image formation has focused on the effect of a distribution of developable image sizes by determination of the minimum size of a latent image cluster, the latent image size developability threshold [23].

B. Processing of a Holographic Film and Developed Image Structure

The material chosen to illustrate specialized processing methods is an experimental red-sensitive ILFORD HOTEC 'R' holographic film, a holographic film assembly. The example illustrates how the normal processing routines can be adapted to produce special results. The holographic emulsion used for the recording was a silver iodobromide emulsion of extremely high resolution and uniform, low crystal size (0.03–0.04 μm diameter) grown by pAg-controlled techniques as for other monosize emulsions. Sensitizing dyes are used to extend the spectral range of the material to the ruby or helium-neon laser used to record the image, but the final image is made up by the migration during processing of part of the original coated silver halide.

The aim is to record developed density data throughout the thickness of the recording layers, but not as the color information stored in successive layers of the laminar coating in a color negative tripack. Instead the recorded data would be phase and amplitude information—the interference pattern formed from the light beams coming from the object and the reference source. After exposure, the material is put through a normal negative developer. Silver is formed where the incident intensity was highest. It is then bleached with a halide salt to produce a positive image in which the highest silver halide density corresponds to the least exposure.

The preparation of the specimen by C. McCullough and N. Ashford and the complete study have been reported elsewhere [24]. The coating was exposed, developed, and bleached to give a reflection hologram. Ultrathin sections of the coating were then prepared by ultramicrotomy. The intricate structure of the ILFORD processed holographic assembly was investigated using the scanning ion microprobe instrument at the University of Chicago [25], which uses a scanning ion beam of gallium ions to build up an elemental map of the specimen.

The construction of the product was a cellulose triacetate base, gelatin subbing layer, an emulsion layer containing extremely fine grain AgX crystals suitable for holographic recording, and a protective gelatin non-stress overcoating (in vertical order of deposition). Scans of the coating section show clearly (Fig. 5a) the emulsion layer in Ag^+ (or Br^-) scans. The vertical striations in Fig. 5b are wrinkles or shrinkage folds in the very thin emulsion slice, on the base side.

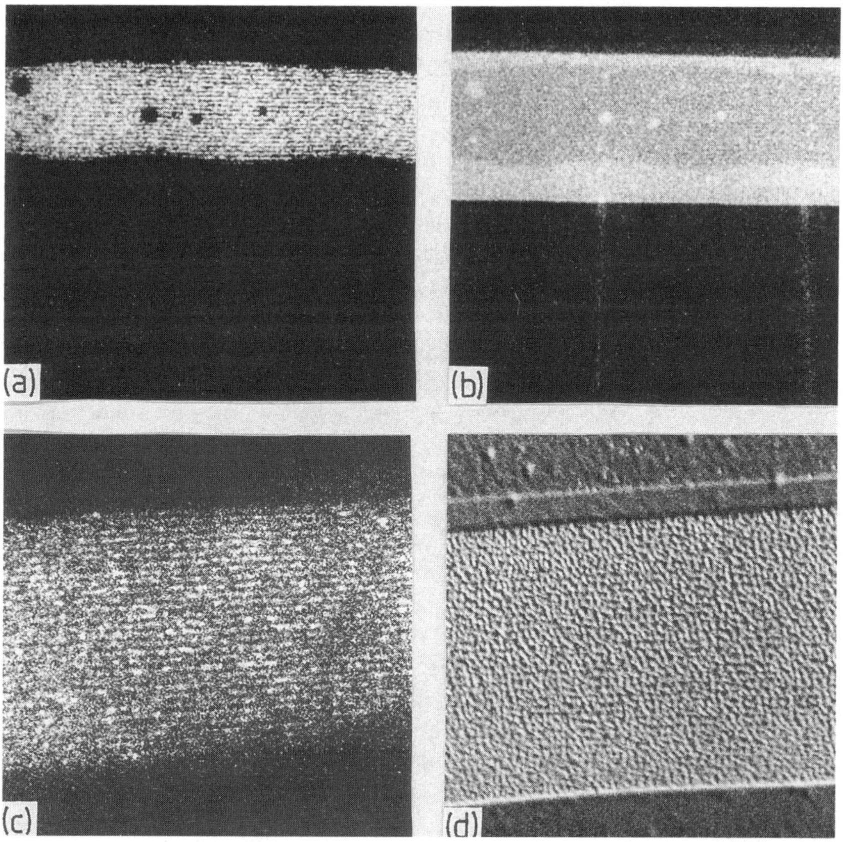

FIG. 5 Scanning SIMS images of ultramicrotomed sections of ILFORD HOTEC R holographic recording film exposed, developed, and treated with rehalogenating bleach. The holographic image is then composed of "fringes" of AgX particles lying approximately parallel to the base. The ions mapped are (a) Ag^+ (silver halide) and (b) CN^- (gelatin) at 40 μm full scale; (c) Ag^+; and (d) ISI map (all ions, overall structure) at 20 nm full scale. (a) and (b), (c) and (d) are image pairs of the same image portions.

The Ag^+ (or Br^-) scans of the processed holofilm show remarkably well the parallel fringes for the processed film. The process should give "fringes" parallel to the base, composed of layers of re-formed silver halide particles separated by layers containing little silver halide. The fringe separation should theoretically be about 300 nm if a red hologram is to be replayed. The successful imaging of these fringes (Figs. 5c and d) and their component particles (originally ca. 30 nm) is a good demonstration of the imaging capability of the Chicago instrument. The retention of the delicate fringe layering also illustrates the ability of the gelatin matrix to support and stabilize high quality images during exposure, processing, and drying.

IV. SOLID-STATE PROPERTIES

Detailed discussion of latent image formation in silver halide systems is beyond the scope of this chapter. Many details are still controversial, and a summary of the opposing mechanisms is given by Roberts and Smith [13]. It is appropriate, however, to indicate some important factors.

The levels of light exposure employed in photography create small numbers of photoelectrons and photoholes in the silver halide grains. The photoholes can be thought of simplistically as "electron vacancies" in the crystal lattices, or, for example, isolated bromine atoms missing their normal negative electronic charge. The semiconductor properties of silver halides are involved. The absorbed light raises the energy level of the photoelectrons from the valence band of the silver halide to the conduction band and thereby becomes mobilized throughout the crystals.

Interstitial silver ions are always present at a finite low concentration inside the silver halide materials used in photography, due chiefly to Frenkel defects. These interstitial ions are also mobile inside the crystal lattice. An aggregation process can then occur between the photoelectrons generated by the exposure and some of the interstitial silver ions. The product, whatever the detailed steps are, is a small cluster of silver atoms, probably concentrated at an impurity speck provided by the chemical sensitizers used. This cluster can accept electrons better or faster from a developer than the unexposed crystals and can therefore catalyze the development stage of that whole grain. By development, the original latent image comprising only a few atoms becomes potentially "amplified" to a clump of 10^8Ag atoms from the whole grain. This high amplification factor at low "noise" (that is, fog) is an outstanding feature of the silver halide imaging system.

Crystal defects play an important role in many key photographic processes. The physical properties of silver halide microcrystals have been examined comprehensively by Hamilton [26], including those features of AgX emulsions that are important in latent image formation. Some of the mechanisms involved in the photochemistry, reduction, and dissolution by complexation of AgX in which crystal defects were found to play an active role were reviewed by Sahyun [27]. The various factors involved and discussed in the literature include the limits of sensitivity of AgX systems with respect to emulsion grains [28]; the effects of crystal habit, iodide content, and location in core-shell iodobromide emulsions and gold, etc. [29,30]; the principles relating to chemical and spectrally sensitized materials; and the factors reducing the photographic sensitivity of those compounds [14,31].

Large emulsion crystals tend to absorb more photons and hence become more developable at low light levels, conferring greater photographic speed to the material. A study of the relationship between speed and grain size has shown that efficiency decreases at large sizes [32], an effect attributed chiefly to dispersal of latent image [33]. The relation between sensitivity, fog, and grain size in emulsions has also been discussed [34].

Recording, amplification, and discrimination in silver halide imaging have been tackled, using electrochemical arguments to treat the activity of what emulsion chemists might describe as defects in the grains, or what Hoffman and Weiss term the "oxidised and reduced" states [35]. The photographic sensitivity and the redox environment are related and used as the basis of a model for the photographic behavior.

Studies of the physics of the photographic latent image have indicated that the high sensitivity and efficiency of AgX materials can be traced to mesoscopic metal clusters—small groups of silver atoms subject to quantum size effects—and the influence of their immediate environment [36].

Molecular orbital theory has been applied to the silver clusters on AgBr surface sites [37]. In a different approach, the analysis of efficiency of films in terms of a signal-to-noise parameter, the DQE (the detective quantum efficiency), has shown that most photographic films are only 10% efficient in this parameter, primarily due to shortcomings in the reproduction of contrast [38].

A photograph depends on the formation of an image that has to be amplified to be seen. Yet studies of the underlying physics indicate that the image formation is in principle an improbable event [39]. That image formation occurs at all is due to a rather unlikely combination of properties [40] through which the standard AgX photographic emulsion continues to offer a unique combination of sensitivity, resolution, tone quality, convenience, and economy.

Recombination of photoholes and photoelectrons in latent image formation has been identified as the major source of inefficiency during exposure

[41] of photographic emulsions. On storage, the exposed material may suffer progression or regression effects due to redox effects in the coated layer [42].

V. MONODISPERSE PHOTOGRAPHIC EMULSIONS

Klein and Moisar [43] in Agfa-Gevaert and Berry and Skillman [44] of Eastman Kodak pioneered the laboratory preparation of monodisperse silver halide emulsions. The application of monosize and core-shell grains as models in basic photographic research and in the production of precisely doped core-shell grains has been reviewed by Moisar [45].

A. Control of Size Distribution

Methods for producing successful monodisperse silver halide dispersions, as with other colloidal systems, depend normally on the appropriate control of the supersaturation of the growth species. Cubic monosize pure AgBr emulsions have been produced in the ILFORD laboratories by controlled growth of seed crystals maintained in a supersaturated solution by double jetting at constant pAg (that is, simultaneous and continuous additions of soluble silver and bromide salt solutions) [46]. The addition rates necessary to maintain the supersaturation at just below that required for secondary nucleation to occur were determined experimentally. This enabled the experiments to be carried out at a fixed supersaturation. Under these conditions, where only growth of the seed crystals had occurred, the effects of crystal number, crystal size, and intercrystal distance (dilution of the emulsion) could be assessed.

The narrowest size distribution was obtained when the supersaturation (i.e., addition rate) was at a maximum. Under optimal conditions, emulsions of cubic crystals could be produced with a standard deviation of less than 0.02 μm on an edge length of 1.0 μm.

The maximum addition rate was such that the rate of increase in volume of the crystals was proportional to the surface area plus a small constant; that is, the rate of increase of the linear dimensions was nearly constant. From this and the effects of intercrystal distance it was concluded that growth rates were controlled mainly by multiple nucleation on the crystal surface and not by diffusion. However, these conclusions are not necessarily valid for other crystal forms such as twinned octahedra (see later).

A patent review for the period 1955–1975 has been published relating to monodisperse silver halide emulsions and methods of obtaining monodisperse (uniform) AgX emulsions [47]. Also exemplified in this review are in-vessel and external-to-vessel mixing devices desirable to ensure uniform supply of reactants to the growing monosize crystals.

B. Crystal Growth Processes

Berry and Skillman [48] defined the four basic processes of silver halide crystallization as nucleation, growth, Ostwald ripening, and recrystallization. The onset of precipitation in photographic systems is characterized by rates of addition corresponding to 10 times the equilibrium solubility of dissolved or complexed silver halide. Under these conditions it is believed that the dominant process is homogeneous rather than heterogeneous precipitation, despite the undoubted presence of extraneous particles from the gelatin. X-ray evidence showed that the initial cubic crystals precipitated had various lattice defects in the early stages, but these were largely "annealed" out after further growth.

Growth of crystals was stated to occur by two processes: "pure" growth and Ostwald ripening. Ostwald ripening is the preferential growth of the larger crystals with lower excess Gibbs surface energy at the expense of small crystals. This is a dominant process when the equilibrium solubility is high (i.e., in the presence of a silver halide solvent such as ammonia) or when the addition rates are much lower than those required to initiate a secondary nucleation as discussed above. The latter conditions of cubic monosize growth described probably correspond to almost "pure" growth. Ostwald ripening was used to determine the effects of crystal size distribution, crystal habit, and halide composition [49].

Growth caused by coalescence ripening (failure of colloid protection by gelatin) has sometimes been invoked as a growth mechanism, although gelatin is normally extremely effective for particles greater than 0.1 μm in diameter. Although the driving force for Ostwald ripening forces in silver halides is essentially enthalpic, the recrystallization of two pure halides to a mixed halide has a significant entropic component. The entropy gain of forming a mixed lattice with a greater number of available permutations of halide ion organization contributes to a lowering of free energy. Such more complex recrystallization processes tend to be confused with Ostwald ripening in commercial processes because they too tend to be favored by higher temperature (net enthalpic) and the presence of complexing solvents.

Transitions in crystal habit and the resulting effects on growth rates have also been considered [50]. In this work, the thiosulfate levels for optimum chemical sensitization (see Sec. VI.A) were also found to be in the order octahedra $<$ tetradecahedra $<$ cubic, suggesting that the crystal substrate morphology changes the nucleation energy of sensitizer specks. That is, too high a concentration of the sensitizer may result in an excess number of specks, which then compete for the available photoelectrons. The halide compositional depth profile in micrometer-sized silver halide crystals has been correlated with the graded bandgap as a function of distance from the surface [51]. In this way the photographic properties of models such as halide-converted silver chloride emulsion crystals could be rationalized in terms of the probable distribution of photoelectrons and photoholes.

An example of a material using monosize cubic silver halide emulsions is the ILFORD ILFOCHROME master micrographic film. This is a silver-dye-bleach material in which the final image is made up of azo dyes. The image dyes are incorporated during the manufacturing process and are bleached out in proportion to the developed silver in the image during processing. A full range of materials covering all direct positive applications can be provided, and a combination of black-and-white and color technology is used in the same material.

The main benefits of ILFORD ILFOCHROME master micrographic film, due to its extremely fine grain and the use of silver-dye-bleach technology, are its high resolution and acutance. Its contrast and high color saturation are optimized for microfilming or reproducing copies from color reflection originals such as maps, drawings, schematics, catalogs, and prints. It is manufactured on 0.10 mm and 0.18 mm polyester base to provide the necessary flexibility and dimensional stability. The schematic layer structure and its function during the processing sequence is illustrated in Fig. 6. The effective speed is rated at about 1 ASA because of the incorporated dyes in the material; unlike conventional color materials, these dyes do not have to be generated during processing. Exposure times of 0.5–1 s are typical in order to store an A4 size original to a standard 25 times reduction. During exposure, these dyes absorb part of the light. However, this contributes to the acutance and storage stability characteristics of the material. The processing (P5 process) consists of three baths for development (a black-and-white development step, bleach, and fix). Because the final image is made up of azo dyes, the micrographic film offers extremely high archival stability and resistance to light fading.

VI. EMULSION TECHNOLOGY

In photographic emulsion technology, many recent improvements have been brought about by applying the knowledge of how to make emulsion microcrystals of a uniform size ("monosize" emulsions) and of thin tabular shape ("twinned" or "T-grain" emulsions) and of differing halide composition between central and surface regions of the crystals ("core-shell" or "double-structure" emulsions). Such approaches are being actively explored

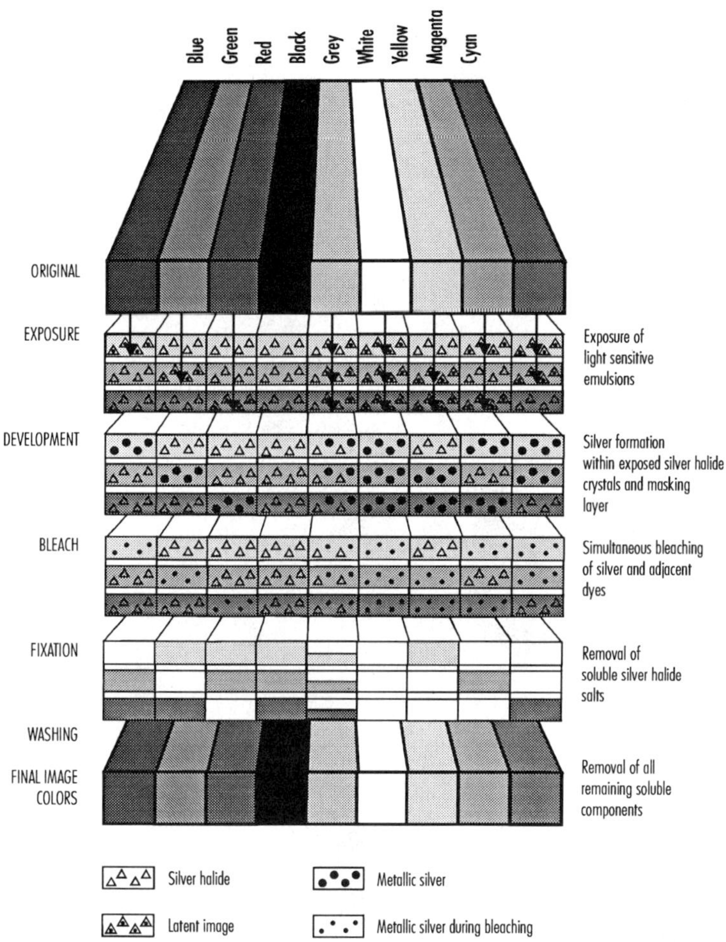

FIG. 6 The schematic layer structure of ILFOCHROME micrographic film and the processes of forming a color image by the silver dye bleach process. (Diagram by S. Garvey.)

by many commercial manufacturers. The technology is based on a knowledge of crystal shape or habit; some common crystal forms adopted by the silver halides are illustrated in Fig. 7.

Some other key features of crystal control are also discussed in this section.

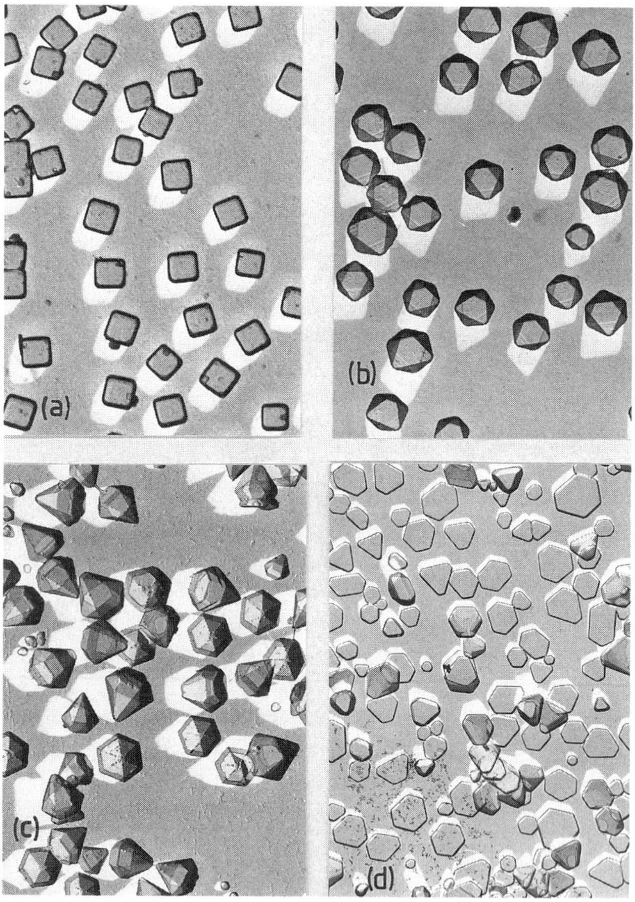

FIG. 7 Electron micrographs showing some common crystal shapes or habits of silver halides. (a) AgBr cubes; (b) AgBr octahedra; (c) AgI bipyramids; (d) AgIBr twinned tabular octahedra. (Electron micrographs by T. Groves and T. Goodman.)

A. Size, Size Spread, Halide Composition, and Crystal Habit

Even in monosize silver halide dispersions, not all the crystals will have the same photosensitivity. Some of the factors responsible for this additional sensitivity spread have been identified [52].

Crystal habit influences the surface structure and reactivity of silver halides. For example, the reactivity of AgBr dodecahedra have been explained in this way [53]. The surface characteristics of (220) dodecahedra of AgBr are intermediate between those of (200) cubes and (111) octahedra. Cubic surfaces comprise alternating silver and bromide ions, octahedral faces (theoretically) would be all bromide ions and dodecahedral surfaces would consist of alternating rows of all silver ions and all bromide ions. The number of nearest neighbors for an ion would be 3 in an octahedral face, 4 in a dodecahedral face, and 5 in a cube face. The order of surface reactivity would be expected to be in this order, as is observed for recrystallization processes, because reactions occur first at sites where the fewest bonds have to be broken. Other studies have shown that the AgBr crystal habit influences the competitive adsorption of bromide ion and an anionic carbocyanine dye [54]. The influence of crystal habit on the growth of AgI dispersions has also been investigated [55].

It has been suggested that solvation effects during the growth of AgBr crystals have a determining influence on the resulting crystal habit [56]. This follows from the influence of Pb^{2+}, I^-, Cd^{2+}, urea, trimethylammonium bromide, and various kinds of gelatin on the crystal habit of AgBr. Changes in habit were ascribed to changes in solvation; the more icelike the water, the more octahedral the AgBr becomes. That structuring of the water that produces the (100) habit also enhanced the monomeric form of the dye methylene blue—that is, the dye acted as an indicator for a particular water structure. The solvation degree was affected by strongly adsorbing dipoles like ammonia.

Epitaxial or composite AgCl/AgI photographic emulsions have been described [57,58]. AgCl crystallites were found to grow epitaxially on basal faces of the AgI. The latent image centers were directed toward the AgCl parts, so they were developable in a monochrome developer. The whole composite crystals were, however, developable in a more powerful chromogenic (color) developer.

Note that the term "photographic grain" has been used with three rather different meanings in silver halide photography. It can refer to the impressions of graininess in a grainy image, to a particle of developed silver that forms part of the image, or to the undeveloped silver halide microcrystals of the original film. Electron microscopy [59] has been employed to illustrate each of these three uses.

B. Preparation of Silver Halide Emulsions

When silver halide crystals are precipitated from aqueous solution, the relative concentrations of the excess silver or halide ions determine which crystal forms are stable. For AgBr, cubic crystals with external (100) faces are produced unless bromide ions are present in much higher concentration than silver ions, in which case octahedral crystals result. By controlling the respective ionic concentrations, intermediate crystal forms may be prepared, such as cubes or octahedra with truncated corners, termed tetradecahedra.

In recent years, more sophisticated control techniques have been employed in emulsion-making to maintain uniform stirring, constant temperature, and ion excess concentrations. As a result, all manufacturers are now able to routinely produce emulsions with very uniform grains of regular size and shape. A method of obtaining a desired wider spread is discussed by Wey [60].

Conventional emulsions were grown by Ostwald ripening methods, that is, by competitive growth at elevated temperatures in the presence of ammonia or organic silver halide solvents. The grains of commercial conventional emulsions therefore had quite complex shapes. For example, in camera films, flattened plates of hexagonal or triangular outline coexist with smaller numbers of needles, tetrahedra, and trapezoidal forms. Most of these shapes occur as a result of the marked effect of twinning on crystal growth. In the conditions under which these emulsions are made, silver halide solvents such as excess bromide ion, potassium thiocyanate, or ammonia may be present at elevated temperatures (50–80°C). The growth needs of all crystals can then not be satisfied, and growth becomes competitive. Those with emergent twin planes, particularly those with planes parallel to each other, become predominant as they possess a self-propagating surface geometry that promotes the addition of fresh material and leads to the production of thinner tabular crystals. Studies of the influence of excess Br^- ion and NH_3 concentration on the shape of AgIBr crystals in gelatin solutions have been reported [61].

Electron microscopy has been the chief method used to investigate the crystal habit of the silver halides. For example, hexagonal bipyramidal AgI particles have been observed in excess I^- (without gelatin) [62]. Studies have also been made on AgBr sols [63]. These sols were relatively dilute by photographic emulsion standards, and they were produced by the method of forming monodisperse AgBr sols by dilution of complex solutions of AgBr in excess Br^-.

Subsequently studied were the effects of added surface-active agents, thiourea, tetraazaindenes, sodium dodecyl sulfate, and gelatin [64]. The relative importance of retarded growth due to adsorption on different faces and of twinning was discussed. Adsorption can also shift the isoionic or isoelectric points of silver halide sols by affecting the uptake of potential-determining ions such as Ag^+ [65].

Understanding of recent photographic materials with increased sensitivity and efficiency has benefited from developments in colloid science and analytical techniques. Further examples can be found in Refs. 66–68.

There are seven different kinds of crystal forms now known for photographic silver halide [69]. They are the cube, octahedron, and rhombic dodecahedron and four new forms—the icositetrahedra, trioctahedra, tetrahexahedra, and hexaoctahedra. The latter can be prepared by additional growth onto core cubic or octahedral emulsion grains in the presence of various organic compounds that inhibit the growth at particular faces.

Reber [70] found that some solvents provided an alternative system to pure water for some emulsion precipitations. In particular, he studied concentrated AgX complex solutions in aqueous *N*-methylformamide and their use in the preparation of photographic emulsions. The usual order of aqueous solubilities for the silver halides, namely AgCl > AgBr > AgI, was inverted in *N*-methylformamide. The emulsions were obtained by mixing aqueous gelatin with an NMF complex of silver halide.

Variations of temperature and halide composition affect the position of the AgX energy levels with respect to sensitizing dyes, etc. [71]. Additions of iodide to AgBr changed the valence band primarily. Small additions of AgCl had little effect, but for large chloride additions the position of the conduction band had changed more than the valence band.

Information on the halide compositional distribution of monosize AgX emulsion microcrystal systems has been gained using elemental mapping techniques [72]. These techniques were based on high resolution imaging secondary ion mass spectrometry (SIMS) [73].

C. Defects, Twinning, Halide Boundaries, and Dislocations

In conventional emulsions, especially with the coarser grain emulsions, internal defects exist [31], and these lead to multiple latent images and prevalent recombination of photoholes and photoelectrons. "Hybridized" grains (epitaxial and double-structured crystals) have been advocated for greater emulsion efficiency [33,74]. With such device-grains, different parts of the grains perform different functions, such as light absorption or latent image formation.

Defect-mediated photochemical processes such as electron transfer and trapping phenomena in silver halides have been investigated using electron paramagnetic resonance (EPR) techniques [75]. This yielded kinetic data for various transition metal electron traps such as rhodium, palladium, and iridium. It was shown that aquation of the metal ion species could to lead to an appreciable increase in the lifetime of the complex species.

Instrumental measurements have been made of the effects of doping silver chlorobromide with trace levels of trivalent rhodium ion on such parameters as the interstitial silver ion concentration [76]. The reactions were considered to be

$$\underset{\text{lattice ion}}{Ag_L^+} \rightarrow Rh^{3+} + \underset{\text{silver ion vacancy}}{2V_{Ag}^-}$$

$$Rh^{3+} + \underset{\text{photoelectron}}{e^-} \rightarrow Rh^{2+}$$

$$V_{Ag}^- + \underset{\text{photohole}}{h^+} \rightarrow Ag_L^+ \cdot h^+$$

Curves of dielectric loss, the photovoltage decay, and the relative photographic sensitivity showed that the dielectric loss reached a maximum as the photographic sensitivity declined. According to this view, the effect of the rhodium complex in increasing photographic contrast depends on electron trapping and the charge balance from interstitial silver ions. "Deep" photoelectron traps, those sites possessing greatest energetic attraction, will be filled up first on exposure. Figure 8 shows the effects of the rhodium addition level during crystal growth for these properties of a silver chlorobromide emulsion.

D. Core-Shell Systems

The term "core-shell emulsion" has been used in two senses: those systems with internal sensitization and those with different halide compositions in the interior and exterior portions of the grains.

The core-shell emulsions originally suggested by Moisar [45] were of great scientific interest for investigating chemical sensitization. The position of the latent image turned out to be the position at which the chemical sensitization had been carried out. For example, if the chemical sensitizers were added at the end of a precipitation of a cubic monodisperse emulsion, the latent image centers were formed on the surface after exposure, as normal. However, if the precipitation was interrupted and gold thiosulfate was added with heating before the emulsion growth was completed, the latent image could be revealed only with an internal (solvent) developer; that is, the centers had been formed in the interior of the grain.

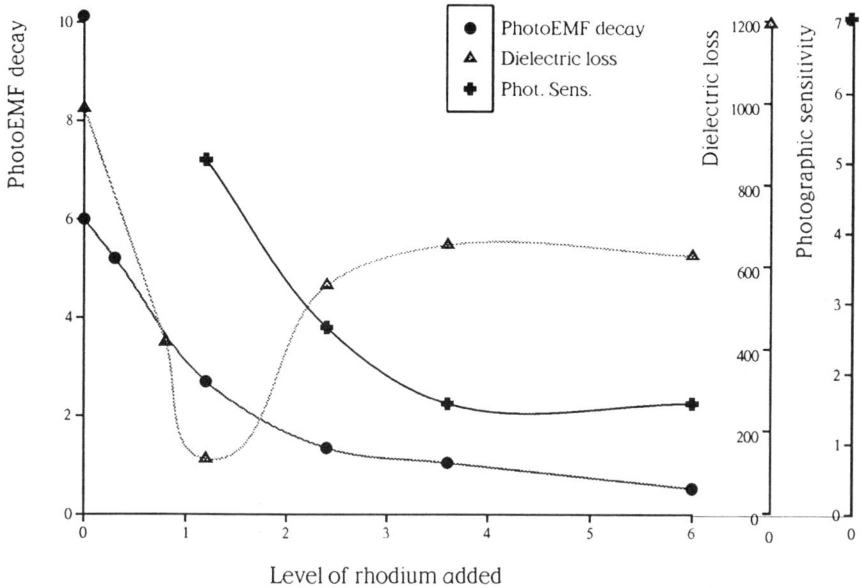

FIG. 8 Effect of rhodium addition during crystal growth on dielectric loss, photo-emf decay, and photographic sensitivity of a silver chlorobromide emulsion. (After Gahler and Roewer [76].)

This competition between the core and shell portions led to relatively efficient positive-image working emulsions for graphics arts materials, for example. A study of competitive electron trapping in core-shell emulsions indicated that on exposure electron trapping occurred at surface and internal trapping sites, but in the presence of internal sites the surface-trapped electrons were lost to room air [77].

With iodide-rich cores and AgBr shells, the emulsions may be found to have increased speed. This was eventually attributed to higher light absorbance and to a charge separation—namely a separation of photoholes and photoelectrons during exposure so that photoholes moved toward the interior while photoelectrons tended to remain near the surface and so could be trapped at sensitizer specks.

Such photographic AgX emulsions containing grains of differing internal and external compositions, sometimes termed double-structure grains, have been the focus of much research and commercial interest. For example, the characteristics of double-structure grains as photographic AgX emulsions have been studied using octahedral grains with various compositional structures [78]. Internal iodide was said to trap photoholes.

Further characterization has been achieved by dielectric loss studies of electrically double-structured grains. Dielectric loss curves for various double-structure grains with specified iodide contents in core or shell showed that uniform AgBr and AgIBr cubes had different maximum frequency peaks [79]. For the core-shell systems, the iodide in the core had a peak intermediate between those of the pure halides. The iodide in the outer shell had a peak corresponding to that of uniform iodobromide.

A greater speed/granularity ratio has also been claimed for a Fuji high speed negative film (Fuji color AX 8514) that features core-shell emulsions [80].

VII. EMULSION SENSITIZATION

A. Chemical Sensitization

The function of chemical sensitization is to improve the efficiency of generation of photocarriers and their separation to increase the yield of developable centers after the exposure of the material. Thus in the normal negative system, a light exposure produces photoholes and photoelectrons. In the "primitive," unsensitized grains, simple recombination of these photocarriers reduces considerably the probability of a grain becoming developable. Chemical decoration of the surface of the emulsion microcrystals with silver or mixed silver/gold sulfide or reduction centers (silver atom clusters) is invariably used to favor electron trapping at the sensitivity specks, consume photoholes by bleaching reactions, or prevent the regression or decay of latent image centers by stabilizing the centers after exposure.

Common sensitizers used are sodium thiosulfate, providing sulfur; sodium tetrachloroaurate or sodium dithiocyanatoaurate, supplying gold; and/or reducing stannous chloride, producing subdevelopable silver clusters of a few atoms only. The effectiveness of the chemical sensitizers has been compared by Spencer [41] in the form of curves of the number of grains developed versus the logarithm of the number of photons received per grain (Fig. 9).

Moisar [81] carried out a classic kinetic and sensitometric investigation of sulfur ripening. The kinetics of Ag_2S formation were studied with monodisperse AgBr emulsions with either cubic or octahedral shapes. The dispersity of image centers was identified as being responsible for the difference in speed. The chemical sensitization of internal and surface-sensitized cubic grains was also studied [82].

A study of the sensitivity and chemical sensitization of AgBr grain faces of different crystallographic indices showed that unsensitized cubo-octahedral faces formed latent image specks preferentially on cubic faces [83]

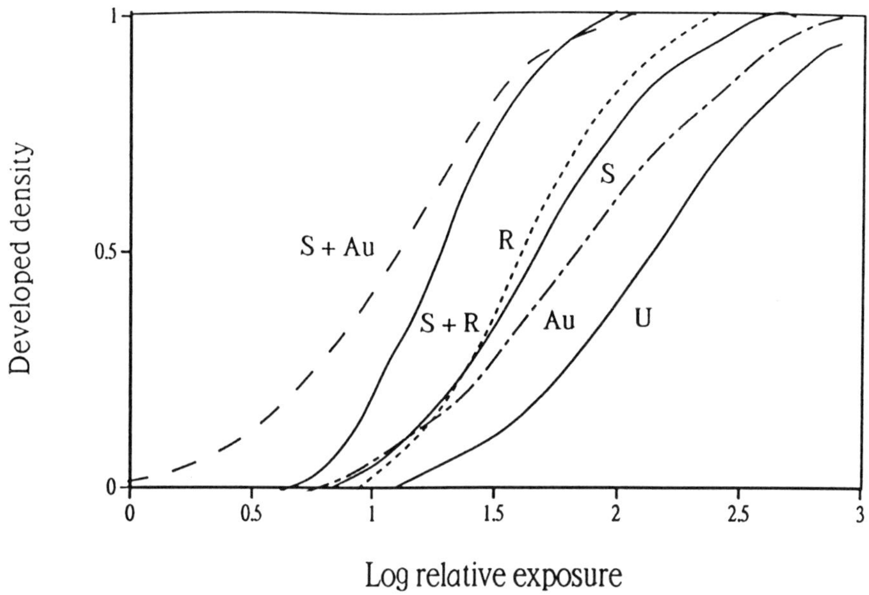

FIG. 9 Plots of number of grains developed versus logarithm of the number of photons received per grain to show the effects of chemical sensitizers. (After Spencer [41].)

and that thiosulfate addition led to a shift in the direction of the sensitization to octahedral faces. Reduction sensitization apparently increased the number of specks. Use of model monosize emulsion systems has led to a proposed mechanism for solvent-induced fog [84].

Reversal systems are often revealing in that they indicate the balance of photocarrier trapping between the interior and surface of emulsion grains. The emulsion characteristics influencing the photographic response of photohole trapping by reduction sensitization centers in such systems have been investigated using internal reduction and external gold sensitizations [85]. A comparison of four emulsions with different crystallographic habits and regularity was made for emulsions internally reduction sensitized and treated with surface gold for reversal systems.

Table 1 shows the main functions of chemical sensitizers and their supposed mode of action. Chemical sensitizers and "finish modifiers," agents added to stabilize or improve the efficacy of the chemical or spectral sensitizers, have been reviewed elsewhere [11].

Ideal reciprocity performance means to a photographer that the product

TABLE 1 Functions of Chemical Sensitizers

Sensitizer	Product and function
(a) S sensitization	
Sodium thiosulfate	Ag_2S (+ Ag?) Surface trapping reduces LIRF, stabilizes $Ag\cdots Ag_3$. Tetraazaindenes stop digestion process, prevent storage fog.
(b) (S + Au) sensitization	
Sodium tetrachloroaurate	AgAuS, etc. Stronger-electron-trapping reduces HIRF, stabilizes Ag_n; smaller latent image; developable Ag_4 / Ag_2Au / Au_3 centers.
(c) (S + R) sensitization	
Stannous chloride, etc.	Ag_2(?) Photohole trapping reduces LIRF; (R + Au)-sensn = Fog

of the necessary exposure intensity and exposure time is a constant, so that, for example, the total exposure of the material does not need to be increased for flash or extended time exposures. For practical photographic materials this is not usually the case, and LIRF and HIRF (low intensity and high intensity reciprocity failure, respectively) are the measures of this nonideality. As Table 1 shows, to some extent this behavior depends on the sensitization process applied by the manufacturer.

The understanding of sulfur digestion mechanisms owes a great deal to the work of Cash [86], who clarified the chemical mechanism of sulfur sensitization and the nucleation and growth of sulfide specks on AgBr. Further work [87] on the effect of gold salts on the chemical mechanisms of sulfur sensitization of AgX photoemulsions and the stabilization of adsorbed sulfide at the crystal surface produced evidence for the chelation of adsorbed sulfide by auric ions.

The effects of stabilizers on chemical sensitizers of AgX emulsions have been rationalized [88]. Sulfur digestion is stopped by hydroxytetraazaindene stabilizers but only retarded by 1-phenylmercaptotetrazole or bromide ion. It was suggested that sulfur digestion is the result of a surface redistribution of silver sulfide to form sensitivity nuclei, whereas gold sensitization is the result of a direct chemical reaction that cannot be similarly retarded. The effect of crystal habit was considered: doubly twinned crystals have grooves for effectively concentrating Ag_2S.

A recent approach to sensitization emphasized the solid-state properties and photographic activity of crystalline Ag_2S and $AgAu_2S$ specks at the surface of AgX [89]. The photographic behavior during chemical sensitization was explained in terms of the solid-state properties of crystalline cubic Ag_2S. The crystallographic basis for the epitaxy of silver sulfide sensitizer specks and the silver halide lattice of emulsion microcrystals has been established by high resolution electron microscopy [90].

The energies of activation and kinetics of fog and sensitivity formation during chemical sensitization have been determined photographically by Locker [91] for monodisperse AgIBr cubic emulsions. The redox stability of latent image specks of R-, S-, or (S + Au)-sensitized octahedral AgBr emulsions have been shown to be important to the stability of the latent image [92].

A survey of the experimental evidence—including the increases found in both internal and surface sensitivity and the depression of low intensity reciprocity failure—indicated that reduction sensitization specks behave as photohole traps [93]. At high levels of sensitization, however, a decrease in internal sensitivity was found. It again appears reasonable that this behavior is due to the formation of larger silver specks, which are then more likely to trap electrons than photoholes. Whether silver specks are photohole traps or photoelectron traps was therefore concluded to be a matter of charge and size.

Studies of the sensitizing ability of the stabilizing agents hydroxytetraazaindenes revealed that a commonly used stabilizer, 4-hydroxy-6-methyl tetraazaindene, increases the speed of cubic AgBr crystals [94]. This was attributed to photohole trapping.

Attempts have been made to extend hydrogen sensitization, known for years as a clean hypersensitizing method for coated emulsion layers [95], to liquid monodisperse AgBr emulsions, but with limited success [96]. Excessive fog was often apparent.

B. Spectral Sensitization

A full review of chemical and spectral sensitization and of latent image formation in silver halide photography was given by James [14]. The use of polymethinecyanine dyes in photography, including holography and infrared photography, has been summarized by Ficken [97].

Monodisperse cubic AgCl and AgBr emulsions have been used in a study of the orientation of individual J-aggregates (see introduction) on cubic AgCl and AgBr emulsion grains [98]. Large J-aggregates were grown on the faces of AgCl and AgBr gelatin-grown cubes. The orientation of the polarized J-band and the direction of axial symmetry were measured relative to the (100) direction by low temperature luminescence microscopy. Chemisorption was not necessary for preferential orientation, and it was likely that benzoxazole dyes were nucleated along surface steps of the same direction.

New aspects of the electron transfer mechanism for spectral sensitization and supersensitization were highlighted in a study featuring picosecond spectroscopy [99]. A study of the J-aggregate sizes of the dyes adsorbed to emulsion crystals was made in this important mechanistic report.

In normal emulsion grains, inefficiencies in spectral sensitization may arise through the following mechanism. As the coverage of dye on a surface is increased, more light is absorbed by the dye aggregates, thus generating more photoelectrons. The accompanying photoholes generated within the dye can travel through each large aggregate and hence can bleach small subimage silver centers. A method that can partition the dye aggregates into smaller adsorbed units may well be more effective.

Examples of improvements in spectral sensitization have been chosen from the products ILFORD ILFOCHROME Classic clear film (CC F7) and the translucent equivalent (CT F7). These are silver dye bleach materials as discussed in Sec. V.B. The new products contain photographic emulsion crystals of a new type termed SMART grains. Images of such a grain are shown in Fig. 10. These photographs were taken at the University of Basle by one of the more promising recent instrumental techniques for such surfaces—atomic force microscopy. The sample is scanned with an extremely fine probe attached to a cantilever arrangement. As the material of the probe comes under the influence of the surface relief, a fluctuating induced electrical signal is recorded. The method was applied to study the surface properties of SMART emulsion crystals. These are based on an optimum arrangement of spectral sensitizer molecules along their surface structures. The structures, for example, terraces of atomic dimensions, are formed deliberately. Such emulsions are therefore designated SMART (surface modification by atomic resolution terraces).

The sensitivity of the new products is much increased (by a factor of between 8 and 15) compared with the comparable conventional materials. The methods have been described by Steiger [100] and by Steiger and Brugger [101] with alternate steps of halide or thiocyanate additions to a set pAg and of dye additions to effect the changes. They have proposed that as the surface of the crystals becomes covered by terraces and steps on which the J-aggregated dyes have nucleated, photoelectron–photohole recombination is decreased, leading to higher photographic speed and reduced low intensity reciprocity failure.

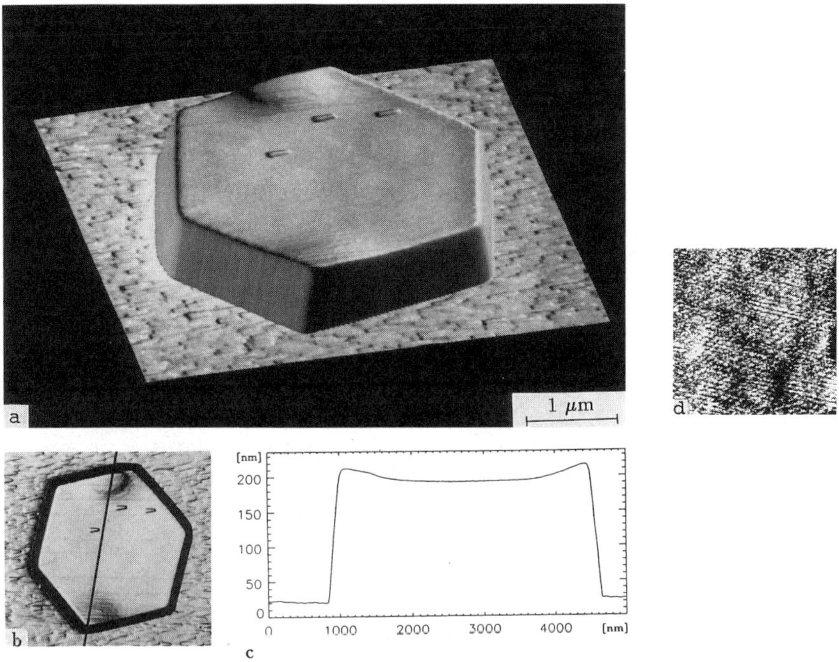

FIG. 10 Photographic emulsion crystals of a new type, termed SMART grains, imaged by atomic force microscopy (a). (c) shows the profile of the crystal measured along the path indicated by the black line in (b). (d) is a square area of image of side 20 nm showing the extreme surface detail resolvable by this technique. (Courtesy of University of Basle; reproduced from Haefke et al. [144].)

VIII. EMULSION CHARACTERIZATION

A. Surface Chemistry and Physics

Tan [68] has given a general introduction to the silver halides in photography, including surface chemistry and surface physics aspects. Various sulfur-containing crystal growth restrainers or accelerators adsorbed onto AgBr grains have been studied by various surface analytical techniques including X-ray photoelectron spectroscopy [102]. The effects were associated with the coordination of the ligands with the surface Ag^+ ions. A study of the structure, surface chemistry, and electronic properties of quantum semiconductor crystallites has considered the behavior of nanometer-sized crystallites of CdSe [103]. Reference was made to earlier work on AgI.

B. Instrumental and Analytical Methods

Mcthods of microanalysis of AgX photographic materials were reviewed in 1979 [67]. Electron microscopy, X-ray diffraction, and electron beam diffraction have been used to investigate the mechanism of the halide conversion process of colloidal AgCl microcrystals by Br^- ions [104]. A method for analysis of the halide distribution in tabular photographic AgX microcrystals by backscattered-electron imaging have been used to reveal the microstructure of these crystals [105].

Several papers describing the application of recent instrumental techniques were included in the Proceedings of an international symposium on the characterization of silver halides held in Kyoto in 1989, with abstracts published in Ref. 106. Takeguchi demonstrated a particularly fine method of determining the internal structure of composite iodobromide crystals. This employed ultramicrotome sectioning and electron microscopic examination combined with electron dispersive analysis to determine the iodide content of the concentric compositional shells revealed.

The complex color assembly in a current motion picture color film (EK5384 positive color film) has been studied by analytical techniques at several structural levels—the material macrostructure, the silver halide emulsion microcrystals and other colloidal particles, and the small particles and clusters of the latent image, Ag_2S, etc. [107].

A summary of some of the instrumental methods recently employed for photographic emulsions is given in Table 2.

TABLE 2 Summary of Instrumental Methods Employed for Photographic Emulsions

Technique	Acronym	Feature studied
Electron microscopy	EM	Size, size spread, surface features
(Transmission, scanning)	TEM, SEM	
Energy-dispersive X-ray analysis	TEM/EDX, etc.	Halide composition spread of individual crystals
X-ray photoelectron spectroscopy	XPS	Surface composition; electronic structure
Secondary ion mass spectroscopy	SIMS	Elemental maps; surface composition
X-ray diffraction	XRD	Halide composition; spread of bulk sample
Dielectric loss spectroscopy	DL	Ionic properties
Microwave photoconductivity	PC	Electronic properties, especially photoelectron lifetime

C. Complementary Instrumental Studies

Integrated studies have been made of silver halide surfaces by microcalorimetry, XPS, imaging SIMS, diffuse reflectance spectroscopy, and atomic force microscopy [108]. Measurements by microcalorimetry and diffuse reflectance spectroscopy revealed the nucleation and growth stages during J-aggregation of a cyanine dye on cubic AgBr emulsion crystals treated with iodide. For the studied cyanine dye, the integral molar adsorption enthalpy on the iodide-doped AgBr surface was higher than that on the pure AgBr surface. At low iodide concentration, only dye monomers and dimers were formed. The relationship between differential adsorption enthalpy data and dye coverage revealed the influence of dye–dye interactions on the adsorption enthalpy.

The presence of I^- on AgBr influenced the nucleation of J-aggregates. XPS and X-AES showed that the silver ion binding energy at concentrations of less than 7 at % was significantly larger than that of surface AgI or of surface AgBrI. It was supposed that the chemical state of this surface iodide was influenced by surface states located near corners and edges of the cubic AgBr grains as revealed by imaging SIMS experiments. Atomic force microscopy on (100) single crystals of AgBr helped the mechanistic understanding of the J-aggregation.

Application of ion scattering spectrometry and secondary ion mass spectrometry on AgX materials has provided interesting new information on compositional and distributional analysis of AgX materials [73,109]. In another study, various crystal models made in the Ilford Laboratories were characterized by scanning ion microprobe analysis at the University of Chicago [72]. The model monosize systems chosen illustrated the variety of halide compositional profiles possible with current precipitation techniques for photographic emulsions. The high spatial resolution of the University of Chicago scanning ion microprobe allowed chemical information to be visualized with submicrometer detail. Table 3 gives a summary of the experimental conditions used for the precipitation and illustrates the variety of composite grains that can be produced by the appropriate monosize growth precipitation sequences. For three of the emulsions a, c, and d described in Table 3, schematic designs and the Br^- SIMS scans are shown in Fig. 11.

IX. DESIGN OF TABULAR EMULSIONS

A. Twinning and General Properties

The most recent color films and conventionally processed black-and-white films use technological approaches based on applications of monosize technology, primarily in the area of tabular crystals (thin tabular crystals also

TABLE 3 Characteristics of Emulsion Models Studied by SIMS—Emulsion Type and Features

(a)	Thin tabular AgIBr emulsion with AgCl growths. Further growth of 2 mol % AgCl after addition of green sensitizing dye. These types were shown by Maskasky [58].
(b)	Dumb-bell AgI/AgBr.[a] Composite crystals formed during epitaxial growth on bipyramidal crystals.
(c)	TM tabular Crystals[a] with high-I cores recrystallized from AgI, aspect ratio 5 : 1.
(d)	Cubic monosize AgBr crystals with iodide-rich shell; intermediate 10% layer after tetradecahedral core (gives an enhanced direct positive effect; see Kriebel [175.]).
(e)	Thin tabular AgIBr crystals with iodide-rich inner shell[b].
(f)	Thin, tabular AgIBr crystals with iodide-rich outer shell[b].

[a]Described more fully in Sec. IX.B.
[b]Described in Sec. IX.D.

known as T-grains). Emulsions of thin, flat silver halide crystals can be coated as thinner layers than previous types, owing to the preferred alignment of the major faces of the crystals parallel to the base. This alignment is partly due to the flow properties of the aqueous emulsion that prevail under the high shear forces during coating, and partly due to the forces exerted on the crystals during drying. The optical cross section is thereby increased. Furthermore, the high surface/volume ratio has enabled a higher useful coverage of spectral sensitizer [4,110]. These factors enable an increase in speed of a dyed emulsion at a given graininess. The characteristics of such an emulsion is therefore a large difference between sensitized speed and (primitive) unsensitized speed.

These crystals can also have excellent developability and lead to low silver coverage (good covering power). They also have a particular advantage for X-ray films. Double-coated X-ray films are normally placed between radiographic screens containing phosphors to generate more light (or allow a lower exposing dose of radiation to be administered to the patient). Such dyed tabular grains reduce the crossover of light exposure through the film base to the other side, and hence the sharpness of the image is improved.

A possible mechanistic problem with such large grains has been discussed by Lohmann [4]. That is, is the drift range of photoelectrons large enough to allow the efficient collection of photoproducts to be collected efficiently by such tabular grains? Some calculations gave this parameter as approximately 1.0 μm, whereas many tabular grains in such products exceed several micrometers in diameter.

The preparation and advantages of tabular grains have been discussed by various authors [110–117]. The structure, growth mechanisms, and formation of multiple twinned crystals in photographic AgBr and AgCl emulsions have been reviewed [118]. Reference 110 gives comprehensive information concerning tabular grains and information on sensitized high aspect ratio AgX emulsions and photographic elements. The history of development of Kodak T-grain emulsions in color films over the last 10 years has been traced [119]. The main advantages of T-grains were delineated: increased minus blue to spectral sensitivity, improved speed/granularity ratio, reduced background radiation sensitivity at equal speed/granularity ratio, and increased sharpness through improved optical characteristics.

The use of AgX crystals in new types of photographic emulsions has also been reviewed [5]. An account has been given of the development of high sensitivity motion picture quality color negative films using these components from Fuji and Eastman Kodak [120,121].

A much earlier paper discussed the crystal growth and reactivity sites of twinned tabular grains [122]. This publication showed large tabular crystals of AgBr and discussed the possible influence of the increased number (5%) of interstitial silver ions beneath the surface of the octahedral faces to restore electrical neutrality.

A study of the particular nucleation and growth kinetics of twinned crystals has shown that twins may be preferentially formed when (111) faces are subject to high supersaturation [123,124]. Considerations on the growth mechanism and size distribution of AgBr tabular grains led to an interesting assignment of (110) character to the rounded reentrant side faces [124].

The concentration of Ag_i in tabular AgBr grains has been shown to be considerably smaller than in the crystals of untwinned emulsions [125]. The twin planes may provide stable sites for Ag_i and therefore internal image sites. Transmission electron microscopy of twinned AgBr crystals has shown that silver particles can form preferentially on twin planes [126].

A description has been given of novel AgBr tabular grain edge growths and their use in determining the separation between parallel twin planes in the host grain [127]. Two techniques used in further experiments have enhanced the understanding of AgX tabular grain growth [128]. One was low temperature luminescence to study the growth history of individual AgBr tabular grains that had been banded with small amounts of iodide. The other was a characterization of individual AgBr or AgIBr tabular grains as to whether they have odd or even numbers of parallel twin planes—indicated by the relative orientation of truncated pyramidal growths grown on major faces.

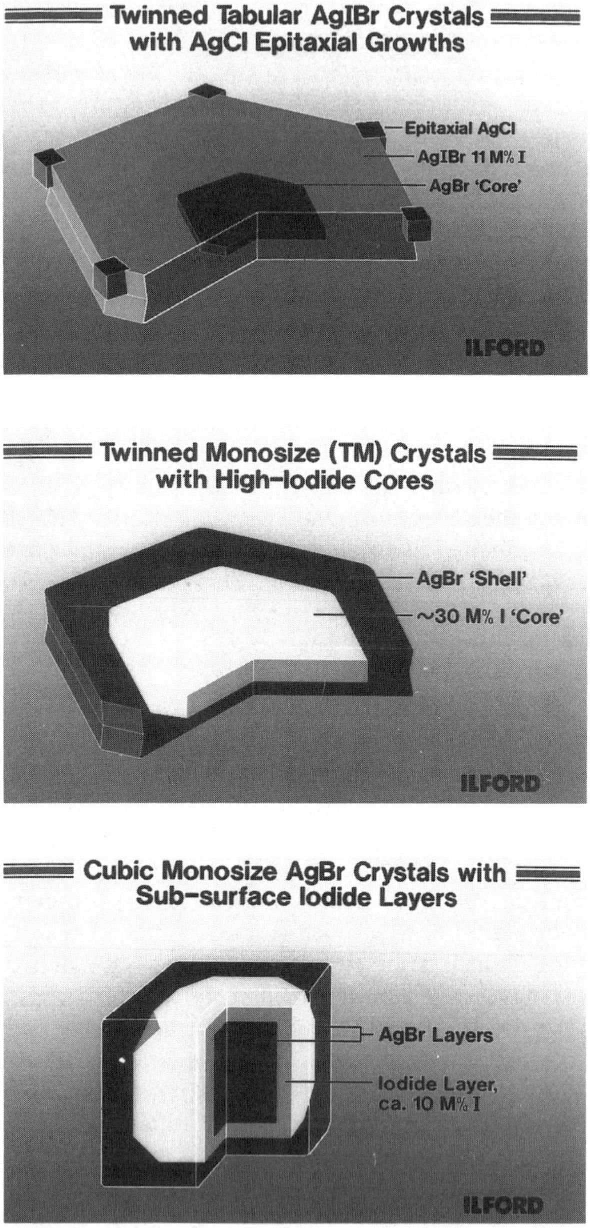

FIG. 11 Schematic diagrams of model crystal structures (left) and scanning SIMS images of actual examples (center and right). Center: ISI (i.e., total ion) images. Right: top, Cl^- image; middle, I^- image; bottom, Br^- image. (After Ref. 72.)

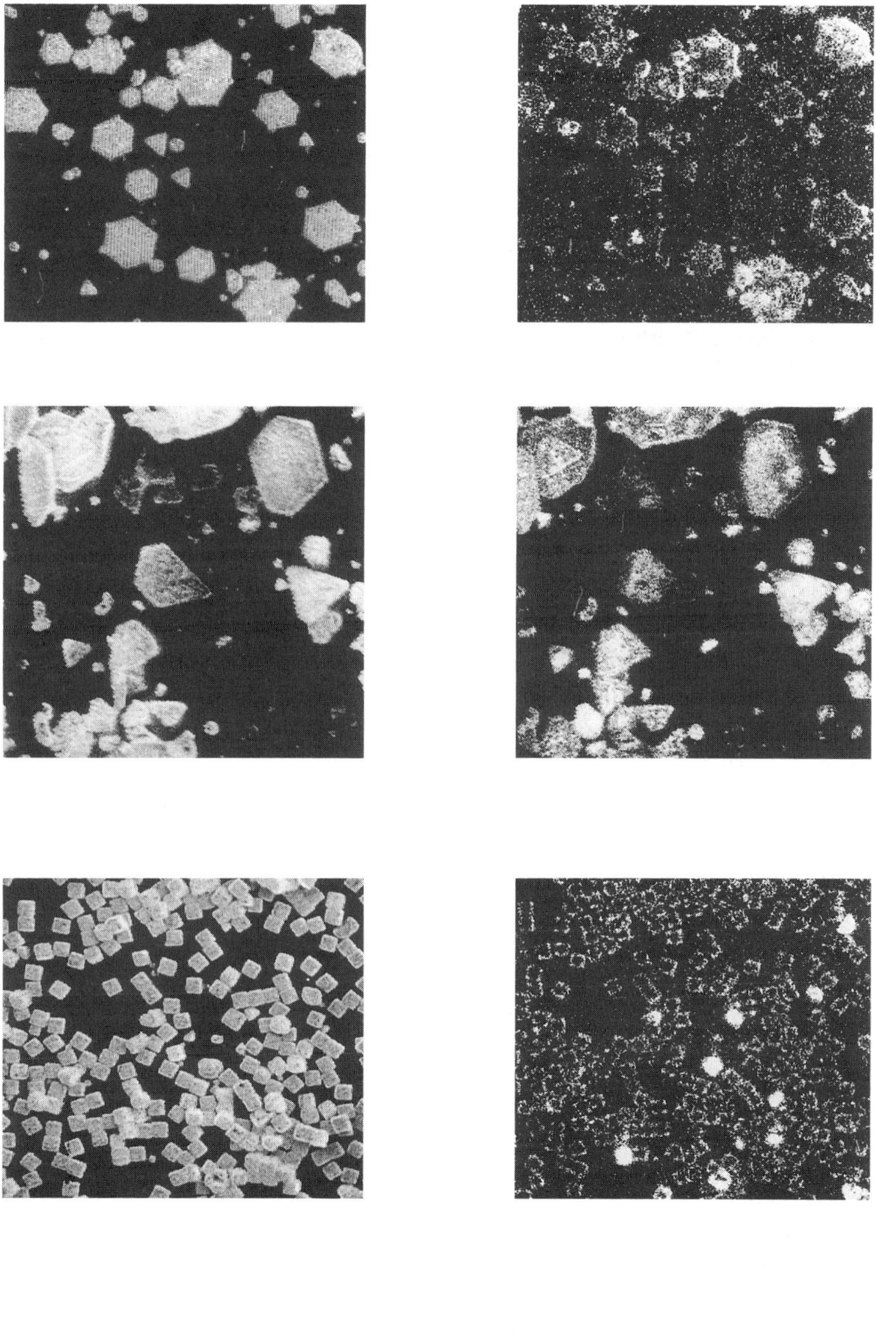

B. Epitaxial Methods

The composition of light-sensitive systems, particularly epitaxial AgX microcrystals, was reviewed by Breslav and Peisakhov [47]. The processes of making photographic emulsions with epitaxial and tabular AgX microcrystals were considered. So too were the specific features of photographic processes occurring in such systems and their advantages over conventional AgX emulsions. In considering the work by Maskasky on well-defined epitaxial growths of AgCl on AgI "host" crystals (described in Ref. 58), they emphasized the following. Although a speed corresponding to the whole crystal volume was claimed, a low granularity corresponding to the smaller AgCl part only was observed with selective processing. This behavior could not be observed with crystals having an AgI-rich interior and an AgCl shell but rather indicated to some extent that the silver chloride and silver iodide were behaving as separate materials.

In the second part of their review, Breslav and Peisakhov [129] considered composite light-sensitive systems spectrally sensitized by semiconductors thought to be in electrical contact with the silver halide grains.

Lewis [113] described how *twinned monosize emulsions* (TM emulsions) could be prepared by using "seed" twinned nuclei prepared by a conventional double-jetting process and further growth by balanced double-jetting. Later methods of preparing twinned monosize emulsions formed by the recrystallization of high iodide or pure AgI seed crystals have been disclosed by Maternaghan in patents [111,112,130]. The first of these showed how epitaxial processes could be used with an Ostwald ripening stage for the preparation of AgX emulsions of the twinned type. Even more monodisperse twinned crystals of silver iodobromide, iodochlorobromide, or iodochloride could be produced by revised methods, eliminating the necessity for the Ostwald ripening. The objectives were to gain control of twinning, size distribution, halide uniformity, halide compositional profile, and the efficient use of silver.

Examination of twinned monosize silver iodobromide microcrystals by X-ray diffraction (XRD) revealed large changes in the diffraction patterns as AgBr was added to hexagonal β-AgI seed crystals [131]. The final twinned crystals have two solid solution phases, the one of higher iodide concentration forming the core. The twinned monosize crystals were derived from microcrystals of hexagonal β-AgI. Aqueous solutions of NaBr and $AgNO_3$ were added to the silver iodide emulsion at controlled flow rates with vigorous stirring while maintaining a constant silver ion concentration in the vessel. During this procedure the β-AgI crystals recrystallized into a cubic silver iodobromide (solid solution) phase. The addition of AgBr was continued until the desired iodide content was reached (typically 5–10 mol% based on total halide).

Samples of the microcrystals were extracted during the growth of a final emulsion of 10 mol% iodide with the aid of an enzyme to yield a dry powder free of gelatin. These were analyzed using a Siemens D500 high resolution X-ray diffractometer operating in transmission geometry with Cu Kα radiation and a position-sensitive detector. The range of diffraction angle (2θ) was scanned from 75 to 83° to give the best resolution and ensure good separation of reflections from β-AgI and cubic AgBrI. The results were correlated with replica electron micrographs taken from a similar series.

The diffraction pattern of AgI showed reflections from the β phase [i.e., (302) and (006) at 75.9° and (205) at 79.8°], where the crystals existed in the form of truncated hexagonal pyramids. At 95% AgI, little change could be seen in the diffraction pattern. The electron micrograph (Fig. 12a) revealed some epitaxial growth of AgBr on the basal (001) faces of the silver iodide grains. It was believed that the defects introduced into the AgBr lattice at this stage caused the crystal twinning. At 61% AgI (Fig. 12b), a new peak emerged at 79.7°, corresponding to the (422) reflection of a cubic AgBrI phase of 29% iodide. Some βAgI was still present. The electron micrograph shows that the original growths on the basal faces had grown much larger and most of the βAgI had dissolved, the AgI becoming incorporated in the growing twin crystals. Some of the twins were still held together by the AgI (dumb-bells) while others had broken apart, revealing the "raspberry-like" AgI.

At 45% AgI, the reflection from AgBrI was much more intense and still corresponded to around 30 mol % AgI. A very small proportion of β-AgI still remained. The electron micrograph (Fig. 12c) reveals many dumb-bells, and the AgBrI twin crystals are much larger in size. At 25% AgI (Fig. 12d), the reflection from β-AgI has gone and two (422) reflections are visible from cubic AgBrI corresponding to phases of 28 and 14% AgI contents. The electron micrograph shows no evidence of β-AgI.

Further addition of AgBr increased the size of the crystals, burying the phases of high iodide content. The two phases (21 and 6 mol % iodide) are still clearly evident at 10 mol % iodide overall (Fig. 12e). The electron micrograph shows the increase in size of the twinned crystals and the approach to a monosize population.

Mapping of the halide distributions by electron microscopic techniques confirmed that the high iodide phase formed the core of these emulsion grains and the low iodide phase the outside shell. This diffraction pattern is a distinctive feature of core-shell emulsion grains.

The mechanisms involved in recrystallization and twin formation now had to be considered. Twinned crystals could be formed, but how exactly was the AgI seed favoring their selection? One outstanding advantage of using preformed AgI instead of iodide salt additions was that twinned crystals could be formed at relatively low pAg. This enabled a uniform population of twins to be produced prior to the monosize growth step.

Two explanations for twin formation were originally considered. The first was that the AgI crystals were "growth sites" for the subsequent growth of twins. This seemed unlikely, as face-centered-cubic lattice AgBr should not be able to grow with a hexagonal lattice AgI core because of the radically different crystal structures. The second explanation was that they could be functioning as an "iodide reservoir," merely supplying iodide ions at a controlled rate. This, however, could not explain why various pro-

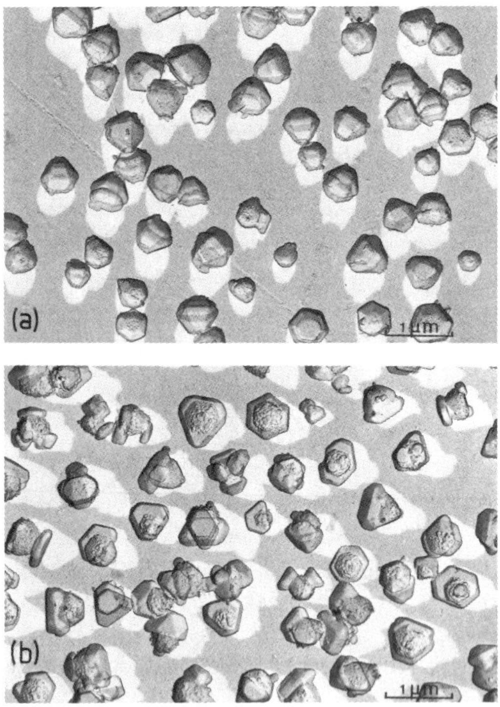

TM emulsion
95 mol % I

TM emulsion
61 mol % I

FIG. 12 The epitaxial growth of AgBr on AgI seed crystals is illustrated for successively lower I contents: (a) 95 mol %; (b) 61 mol %; (c) 45 mol %; (d) 25 mol %; (e) 10 mol %. The "dumb-bell" crystals typical of an intermediate stage of recrystallization are exemplified. (Courtesy of Siemens plc; reproduced from Ref. 131.)

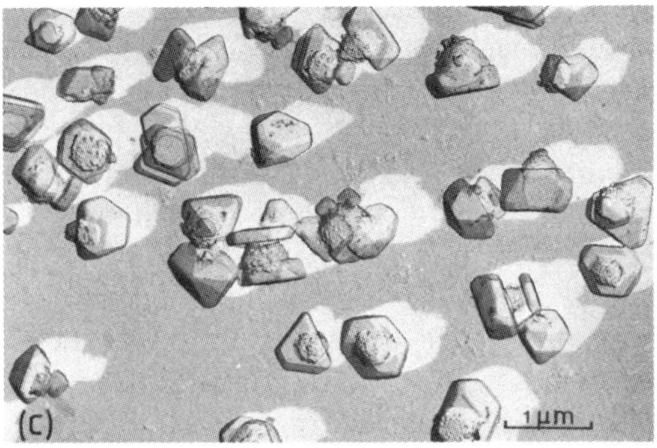

TM emulsion
45 mol % I

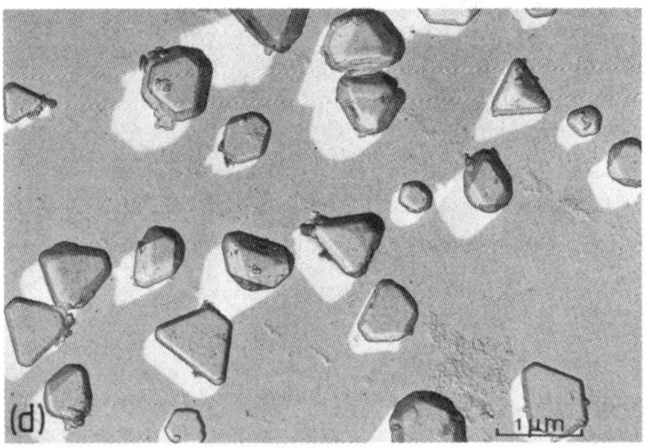

TM emulsion
25 mol % I

TM emulsion
10 mol % I

grammed iodide salt addition profiles failed to produce corresponding results.

In fact, both mechanisms were partially operative. Electron micrographs of the early stages of a modified scheme with larger AgI seed crystals showed that the silver iodobromide twinned crystals were formed by direct nucleation of epitaxial layers on the basal faces of AgI crystals, with corresponding dissolution of the AgI. The light and dark parallel striations on the sides of the crystals correspond to the alternating "ridge-and-groove" structure expected for multiple parallel twin planes in such a crystal. In this position, a minimum value for the number of twin planes can be estimated by simply counting the lines.

"Epitaxis" can be defined as the growth of a crystalline layer under the structuring influence of the substrate or "parent" material. The function of the ripening step could then be seen as dissolution of the small, nontwinned crystals that had nucleated independently in the bulk aqueous gelatin phase and transfer of material to the high iodide twinned crystals. The final stages of twin formation are marked by recrystallization of residual AgI.

Thus a crystallographic conversion takes place during recrystallization, from hexagonal lattice β-AgI via epitaxial growth to cubic lattice AgIBr or AgICl. AgI crystals of the hexagonal lattice type have the shape of hexagonal pyramids or bipyramids. The basal faces of the truncated pyramids comprise external lattice planes of index (0001) or (000_1). Only a few mole percent of AgBr could be accommodated. Silver halide crystals containing less than about 40 mol % iodide have the face-centered-cubic lattice structure and may have the forms of octahedra [external faces comprising (111) lattice planes]. Twinned crystals normally have the shapes of triangular or hexagonal plates.

Both hexagonal and face-centered-cubic forms can be regarded as being built up from successive stacking of close-packed, homoionic lattice planes (0001) and (111) for AgI and AgBr, respectively. The lattice forms differ chiefly in the relative alignment or sequence of these ion layers. It could be calculated that the difference in interionic spacing between these layers is only about 14% (16% between AgI and AgCl). Epitaxial growth is therefore possible between AgI and AgBr. This had been reported for conversion of AgBr by addition of KI solution [132].

Electron micrographs can be used to further illustrate the main features of the recrystallization. The following processes can be recognized.

1. Epitaxial junctions are formed as growth of the new silver iodobromide phase occurs on the basal (truncated) faces of the silver iodide crystals.
2. Iodide ion incorporation in the new crystals takes place, which favors both external octahedral faces and stacking defects, particularly twinning.

3. Formation of twin planes results from the high, "saturated" iodide content of the growing phase; twinning is also probably encouraged by the discontinuity in interlayer forces experienced at the epitaxial junction.
4. Dissolution of AgI crystals occurs, to maintain the equilibrium solubility. The iodide is supplied to the twins either by diffusion through the solution or by anionic diffusion through the epitaxial boundary. Complete dissolution of the AgI crystals occurs but is preceded by release of the tabular twinned crystals.
5. Renucleation occurs simultaneously with epitaxial growth unless AgBr precipitation is made relatively slow. The secondary population can be removed by addition of an Ostwald ripening solvent such as ammonia, sodium thiocyanate, or excess ammonium bromide.

Preparation sequences have been devised that add solutions of silver nitrate and sodium or ammonium bromide or chloride or mixtures of the two, at relatively low rates to the AgI "seed" emulsion so that no independent nucleation occurs, and therefore no ripening step is necessary. This reduces diffusion of iodide from the "inner" cores of the tabular crystals as growth of further AgBr shells takes place.

The "dumb-bell" crystals typical of an intermediate stage of recrystallization are exemplified in Fig. 12. The tabular crystals are formed during precipitation of AgBr on the basal faces of the AgI crystal "sandwiched" between the two. Crystals with more complicated symmetry are often formed.

STEM/EDX electron microscopy analyses of individual crystals in the electron microscope [133] confirmed the probable "core-shell" structure for twinned monosize emulsion crystals. The specimen containing 5 mol % of AgI derived from the recrystallization of 0.6 μm AgI crystals shows increasing iodide content toward the center, reaching a peak of ca. 8 mol % AgI. A sample taken earlier at the 10 mol % iodide level, however, shows the AgI content rising from about 4 mol % at the edge to 20 mol % in the central core. The crystals were tabular but of low aspect ratio in this case, so that allowing for the deposition of bromide-rich shells the validity of a "core" of ca. 25–30 mol % AgI seems directly confirmed.

The methods described allowed the control of the average crystal size, size distribution, and crystal shape of the twinned monosize emulsions to be achieved as follows. The average crystal size of the resulting emulsion depended directly on the size of the original AgI seed size. The size distribution could be altered by using a different size distribution of the AgI "seed," by blending two seeds of differing size together or, of course, by blending two twinned monosize emulsions of differing sizes together. The shape or habit of the resulting crystals could be made cubic or octahedral as required by adjusting the conditions during the final (monosize) growth step. The composition of twin types in the final emulsions were typically >70% tabular twins and about 30% single twins, so the selection of the favorable twinned tabular crystal type was high.

The aspect ratio could be further increased by continued growth at high pAg. The required overall AgI content could be achieved by simply fixing the quantity of AgBr precipitated compared with the molar quantity of initial AgI seed. The appropriate iodide distributional profile through the crystals was provided by the AgI recrystallization, which gave a high AgI "core," or by additional AgI-containing shells, which could be produced by appropriate additions of the soluble iodide salt in the halide stream during the later stages of growth. The main advantage was to provide a series of emulsions of high sensitivity in which crystal size and iodide content could be independently varied.

The photographic properties are summarized in Table 4. Figure 13 illustrates how an iodide-rich core prevents recombination of silver and halide produced by exposure to light. The electron-trapping level of a chemical sensitization product is illustrated for a conventional grain with a particular sensitizing dye, where the levels of the highest occupied (HOMO) and lowest unoccupied (LUMO) orbitals are important in light excitation and subsequent photoelectron transfer (sensitization) processes for the dye. The lower diagram indicates the efficiency improvement by partitioning photoholes into the progressively iodide-enriched interior of the grain. The trapping of photoelectrons at sensitizer specks may thus be less impeded.

TABLE 4 Photographic Properties of Twinned Monosize Emulsions

High speed/crystal size	Twinning, high-I core
High contrast	Narrow size distribution
(Ag usage) covering power	Excellent due to high aspect ratio
Sharpness (scattering)	Reduced scattering due to monodispersity
Extended blue sensitivity	High-I core
Speed-to-crystal size ratio maintained to high size	High-I core reduces recombination. (Fig. 13)
High internal sensivity (unsensitized)	Internal defects
Good direct-positive behavior	Internal defects, surface electron acceptors
Development characteristics	Can be modified by subsurface or surface I or CI inclusions; blending of components

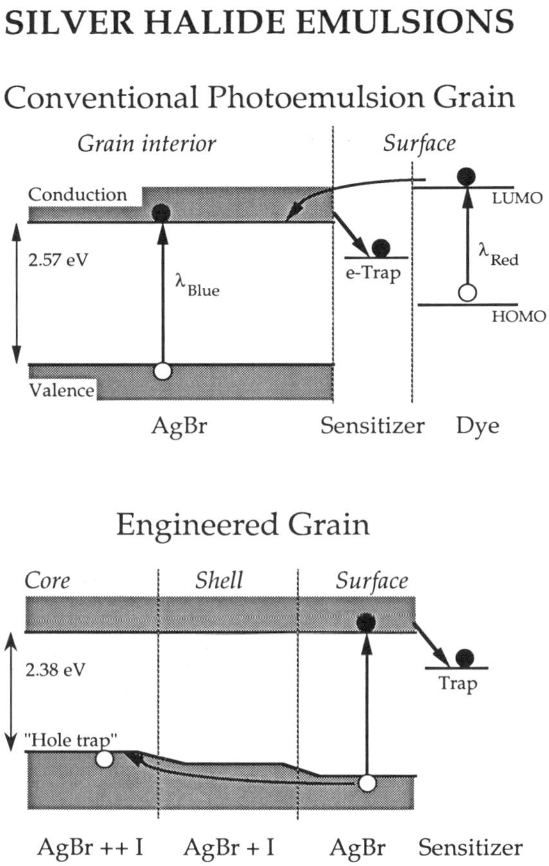

FIG. 13 Schematic diagram illustrating the effect of an iodide-rich core, etc., in preventing recombination of photoholes and photoelectrons. The upper diagram represents a conventionally sensitized emulsion. The lower diagram represents an emulsion with iodide-rich cores. Valence and conduction electron bands are depicted. HOMO and LUMO are, respectively, the levels of the highest occupied and lowest unoccupied molecular orbitals of the sensitizing dye in the conventional material. (Courtesy of R. Levi-Setti and J. M. Chabala.)

C. Applications

Twinned monosize emulsions based on the recrystallization of AgI have been used since 1978 by ILFORD AG for higher speed silver dye-bleach materials for small portrait applications (e.g., passport photography). The extension to monochrome camera films is described in a later section. The interpretation of the crystal growth and the X-ray diffraction data has been confirmed [134]. The equilibrium phase diagram for AgIBr mixed crystal formation from AgI and AgBr by double jet precipitation was found to be applicable only if the pAg maintained during growth was high, which conferred high solubility. The mechanism of recrystallization via two-phase microcrystals was thus confirmed.

The principle of using the silver iodide seed crystals as "templates" to form the required size and shape of final core-shell twinned crystals has been extended using single pyramidal crystals of AgI [135] and tabular AgI crystals [136]. The repercussions of the starting shape of the AgI crystals is illustrated schematically in Fig. 14.

Irving [137,138] has recently described the production of hollow grains.

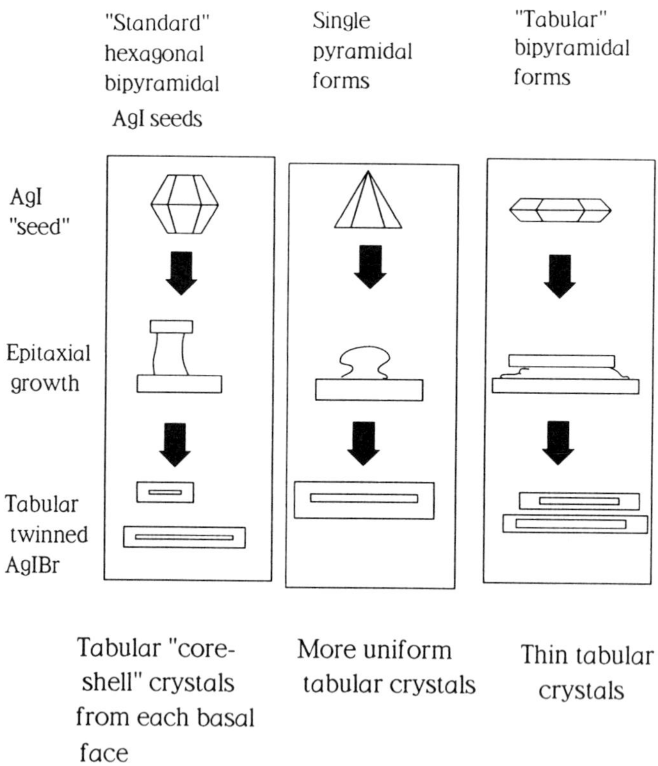

FIG. 14 The principle of using different shapes of silver iodide seed crystals as "templates" to form the required sizes and shapes of derived twinned crystals. (After Refs. 111, 135, 136.)

First AgI seed crystals with a novel crystal form were prepared. They were composed of at least 90 mol % AgI and had a crystal geometry consisting of four hexagonal bipyramids joined at their base to form a common tetrahedron. Using a template recrystallization essentially similar to that described above, the hollow crystals were then formed. The growth scheme is schematically illustrated in Fig. 15 to show how this is achieved.

D. Core-Shell Tabular Crystals

More published information on core-shell tabular crystals is becoming available, but due to the high commercial interest much information comes from patent disclosures. It can be stated that tabular crystals are advan-

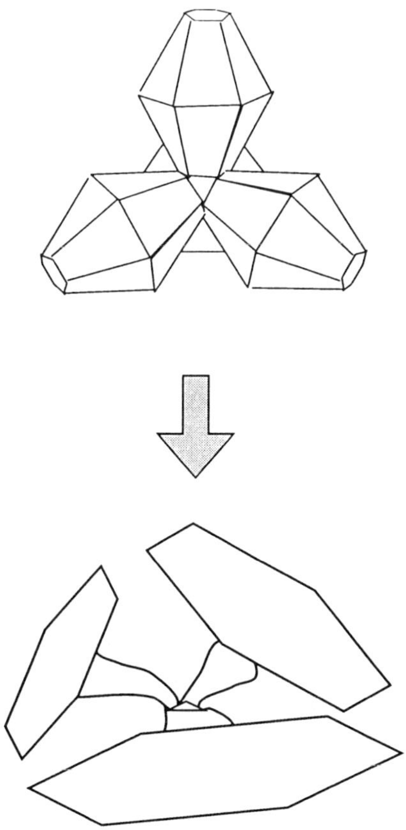

FIG. 15 Use of novel AgI seed crystals to prepare "hollow" silver halide emulsion grains. As each set of epitaxial tabular crystals enlarges, a hollow cavity is formed. (After Irving [137,138].)

tageous for the same reasons that led them to being invariably employed in previous high speed conventional emulsions—for example, their good response to chemical and spectral sensitization treatments, high maximum potential speed, and high efficiency of silver utilization.

The core-shell character also give improved photographic properties. In this context, the physical, ionic, and electronic properties of the phase boundaries within the crystals are recognized as being important, and some reasons for this have been suggested [139]. The concept of heterojunctions, more usually used in connection with electronic devices, has been applied to the internal interfaces of such composite emulsion grains.

It has been suggested that there are three important aspects of the preparation and properties of tabular core-shell grains: uniformity of crystal habit, monodispersity, and the compositional difference between core and shell [140]. The speed is related to the iodide content: increasing iodide content causes the crystal habit or shape to change progressively to octahedral and then tabular forms. The effects of iodide inclusion in photographic core-shell emulsions are summarized elsewhere [19,78,141].

The difficulty in achieving a narrow crystal size distribution in such emulsions was reported by one laboratory to increase with increasing grain size, rather than with the growth rate as previously believed [142]. A method partly using Ostwald ripening as the growth method was adopted as a result.

E. Nucleation and Growth at High Halide Concentrations (T-Grains)

Many publications and patents on the thin twinned tabular crystals termed "T-grains" have concentrated on the crystal growth processes and modifications to the crystal shape by various additives or conditions. In fact, such tabular grains have been known for many years as objects of scientific curiosity. The advance in recent years has probably been in chemical sensitization. Grains of such large surface area have had problems of latent image dispersity due to many chemical sensitization centers being formed at different parts of each large grain; these sites may then compete for the available photoelectrons, and none may then reach the critical silver cluster size sufficient for developability. Much of the recent progress has come from concentrating the potential sites of latent image formation at only a few places such as grain corners or grain edges, or from deliberately creating defects at which latent image formation would be particularly favored.

Sowinski and Wightman [143] described how tabular grains could be employed to minimize the sensitivity of photographic films to environmental radiation. The contribution of natural background radiation to aging of photographic film should be dependent on the volume of the silver halide grains, whereas the spectrally sensitized photographic speed should depend on their surface area. This was probed using two emulsion types: a normal conventional emulsion and a tabular emulsion of high aspect ratio. The sensitivities to a test radiation and photographic speeds were determined. The sensitivity to natural radiation was less dependent on the aspect ratio than the photographic speed, which was much higher for the tabular crystal type. In this way the problem of fogging by background radiation for high speed films could be reduced.

Atomic force microscopy offers exciting possibilities for resolving the details of silver halide surface microstructure. Haefke et al. [144] have reported on the appearance of tabular emulsion microcrystals and single crystals with differing external lattice faces. Keyes et al. [145] have imaged tabular grains of AgBr treated in situ with I^- or Br^- solutions by means of a solution cell. Silver iodobromide growths formed rims near the edges of the tabular grains on exposure to the KI, and steps averaging 15 nm could be resolved.

The three-dimensional I^- ion distribution in tabular Ag(Br,I) microcrystals has been investigated by scanning Auger microscopy and energy-dispersive spectroscopy methods [146]. At high pAg, the I^- ions added early on were transferred to the surface and became enriched at the surface. The structure of the AgIBr crystals was not of the sandwich type.

The growth kinetics of the iodide-enriched shells of twinned tabular crystals have been investigated by Chabala et al. [147], using complementary data from XRD, SIMS, and EDX. Various iodobromide shells were precipitated onto a pure AgBr "seed" emulsion under conditions of high bromide ion excess, so that the crystal edges grew faster than the major faces and the crystals grew more tabular (i.e., the growth was more "two-dimensional" than "three-dimensional"). It might have been expected that the iodide deposition would correspond to the pattern of overall growth of material at edges and faces. In fact, this was often not the case, partly owing to recrystallization and crystal habit effects.

The X-ray diffraction data showed peaks at the expected primary phases but also phases showing evidence of iodide conversion. Diffractograms from two emulsions, grown with core/shell/shell mole percent iodide concentration bands of 0/8/16 and 0/16/8 are given in Fig. 16. Emulsion A (0/8/16) shows three distinguishable peaks at effectively pure AgBr, 8% I, and 14% I; the crystals show evidence of recrystallization, conversion, and aspect ratio changes. Emulsion B (0/16/8) displays a core composition of almost pure AgBr again, but with a broad peak centered on 11% I (and with perhaps a minor peak at ca. 18% I), indicating that iodide conversion of the inner shell, as well as intracrystal transfer, has occurred.

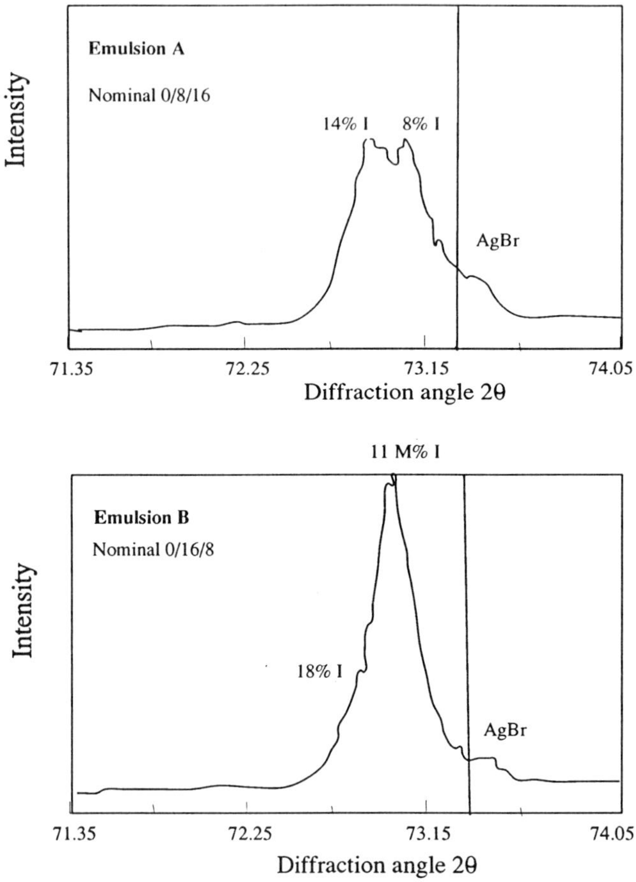

FIG. 16 X-ray diffractograms from two emulsions grown with core/shell/shell mol % concentration bands of 0/8/16 and 0/16/8. Emulsion A (0/8/16) shows three distinct peaks corresponding to ca.100, 8, and 14% I. (After Chabala et al. [147].)

The SIMS images confirm the existence of an AgBr-rich core regardless of how the iodide was introduced into the precipitation. Figure 17 shows how the iodide distribution maps from the two emulsions compare. Triangular and hexagonal cores devoid of iodide ion are apparent as well as two distinct banding patterns. EDX compositional profiles across individual crystals of the two emulsion types showed three distinct compositional regions for each, corresponding to the differing iodide contents of the core and shells.

This example is one of many recent studies that illustrate how the subtle details of the construction of photoemulsion crystals, which is important for their photographic function, can be revealed by such a mutually reinforcing combination of instrumental methods. This has allowed specific photographic behavior to be associated with defined emulsion crystal microstructure and has thus aided the design of more complex and efficient imaging systems.

F. Recent Developments

Some recent patents and publications are mentioned here to indicate current trends in silver halide imaging and in monodisperse silver halide emulsions in particular.

Aida [148] describes a "sandwich" kind of emulsion with internal high I core. Such a "stratified structure" with additional I-containing layers on the major surface faces was claimed to promote developability and greater dye adsorption. The outer layers were formed by growth at a lower pAg or by conversion to AgI. A lower level of gold salts to sensitize the crystals was specified. Other stratified or layered structures have been described (see, e.g., Ref. 149).

Blending of emulsions with low and high aspect ratio with stated size differences is described by Newmiller [150]. The examples cover the achievement of a better speed/granularity ratio by appropriate blending. The optimal blend ratios and mean sizes of the component emulsions for image quality are discussed. Dobson and Wharton [151] discuss the blending requirements for optimal speed/granularity ratio when the high aspect ratio emulsions are less than 8 : 1 and are twinned monosize emulsions with a high iodide core.

In methods proposed to increase the monodispersity of T-grain emulsions, Zola and Bryant [152] discuss the modification of current T-grain methods to obtain more monodisperse twinned emulsions (again with high aspect ratio). The principle in terms of known classical nucleation theory (see Ref. 14) is to shorten the nucleation period by various means.

In a comparison of semiconductors, Steward [153] examined some AgX core-shell systems and the dislocations arising from the mismatch at the internal junctions. Edge dislocations were much more visible with high iodide core crystals than with the corresponding high bromide core equivalents. The existence of dislocations had a significant effect on the photographic response. She concluded that the existence of high concentration differentials did not necessarily give dislocations but that the location of iodide was an important factor in determining the dislocation content and thereby influenced the photographic response.

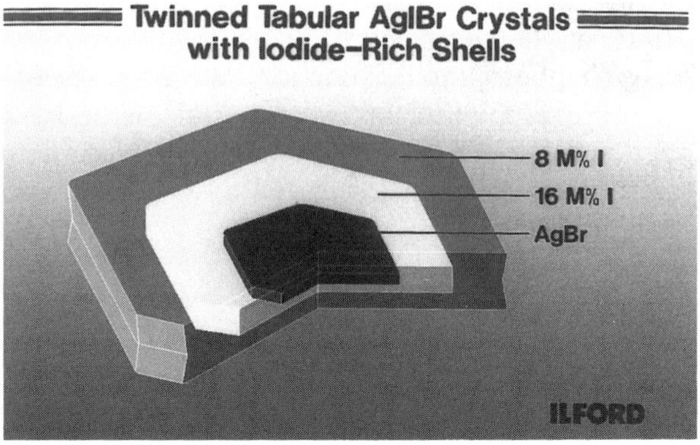

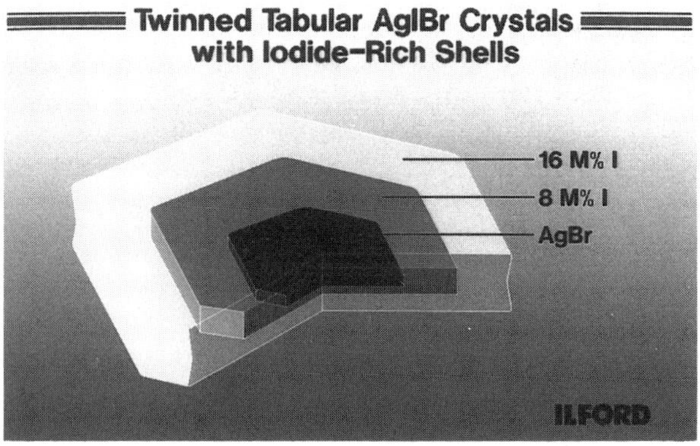

FIG. 17 Schematic diagrams of model crystal structures (left) and scanning SIMS images of actual examples (center and right). Center: Br^- images; right: I^- images.

Technical aspects and features of the new Fuji Velvia color reversal film were disclosed by Ishimaru et al. [154]. This paper claims that location of development centers on grains with dislocations provides greater edge developability and therefore better reversal characteristics.

Harvey et al. [155] describe a core-shell AgX emulsion with subdevelopable Ag centers at the interlayer boundary. The subdevelopable centers are small clusters of Ag atoms that are claimed to increase sensitivity without a corresponding increase in fogging. A particularly favorable boundary is given as that between internal AgIBr and a outer layer of AgBr (Fig. 18). The subdevelopable Ag atoms are formed by controlled reduction. This

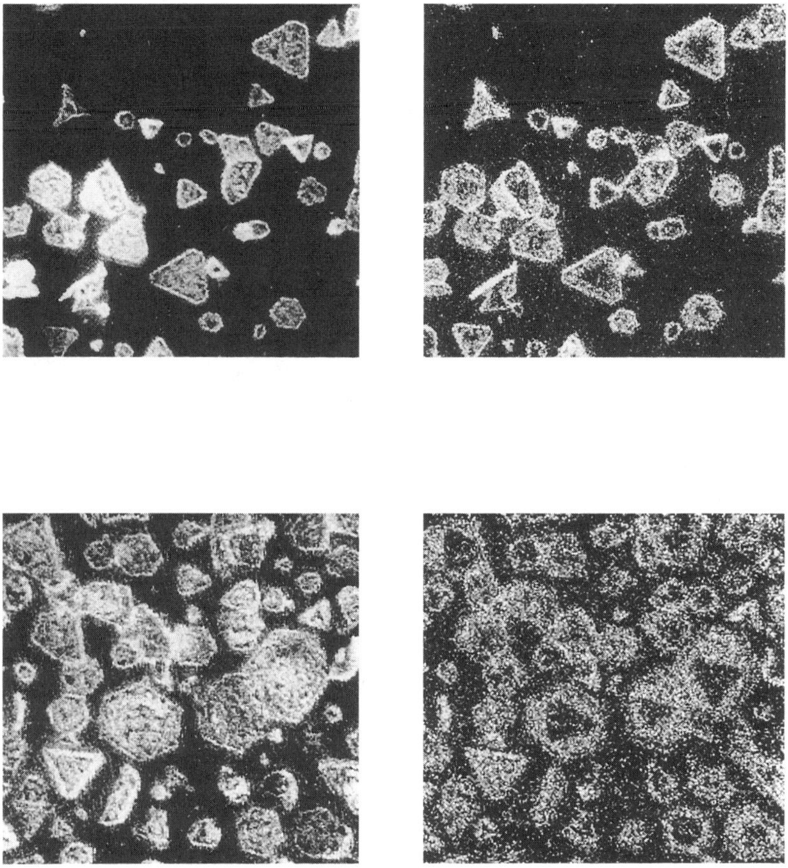

may entail reducing the pAg below 7 or raising the pH to 7–10 at 50°C or by adding a reducing agent such as stannous chloride or glutaraldehyde before growing the outermost shell by Ostwald ripening or by double- or single-jet growth. A typical tabular grain for treatment has an inner IBr core, an intermediate AgBr shell enclosing the subdevelopable atoms, and a surface shell. Alternatively, an AgIBr shell may be formed on an AgBr core, followed by reduction treatment to form clusters of Ag atoms, and then enclosed in an outer AgBr shell. The emulsions are claimed to show a marked increase in speed and contrast as well as improved low intensity reciprocity failure. The effects were seen in both black-and-white and chromogenic processing.

Takada [156] also exemplified reduction before completion of the outer

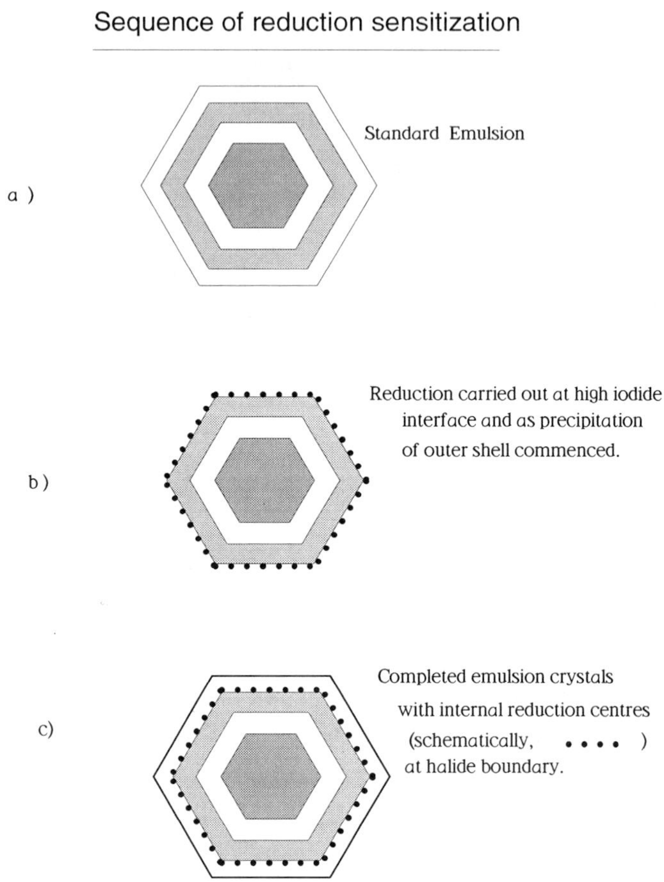

FIG. 18 Schematic diagram illustrating reduction of a core-shell AgX emulsion to form subdevelopable Ag centers (small clusters of Ag atoms) at the interlayer boundary. (After Ref. 155.)

shell, which should contain greater than 5 mol % iodide. He disclosed that subsurface silver centers give greater dye-sensitized speed or allow a higher dye level to be used. The surface layer is formed in the presence of a sensitizing dye or antifoggant or with a solvent such as thiocyanate. Yamada [157] describes the reduction of AgX emulsions of approximately 3 mol % iodide. The combination of stabilizing or antifoggant thiosulfonic acids, tetraazaindene, KSCN, and surface iodide is disclosed. Reduction was carried out either before or after the outer layer was formed.

Urabe [158] discloses emulsions containing more than half tabular crystals of aspect ratio greater than 2 : 1, in which the major faces are of (111) character and in which more than 30% of the tabular grains have dislocations in the major faces. The dislocations are formed by a series of processes such as preparation of tabular crystals, epitaxial growth by addition of material derived from numerous fine AgCl or AgClBr grains on the major surfaces, conversion of a large number of the fine crystals on the tabular grains by a physical ripening or recrystallization step, and creation of dislocations at the major faces of the crystals by growth of a further silver halide shell by precipitation. The final emulsion was claimed to have superior sensitometric properties and resistance to stress.

Saito [159] describes an apparatus for producing highly uniform mixing of one solution with two or more reactant solutions. This apparatus consists of a porous member sunk beneath the surface of the liquid in the reaction vessel. The porous member may be a membrane or an assembly of fine tubes through which the reactants are pumped. The chemical reactions during precipitation or sensitization of monodisperse crystals were said to be more controlled and more reproducible with such a device.

Piggin and Bishop [160] disclosed a process for preparation of tabular grain silver iodobromide emulsions in which silver iodobromide laminae had been formed on the major faces. The iodide content of these laminae was higher than those of the original host grain, and indeed also higher than the average of the final composite grain. The silver iodobromide layer was formed by a recrystallization of an initially formed silver iodide layer on the host grains.

A related process for forming such "sandwich" grains is described by Lin et al. [161], who claim its advantages are high sensitivity and much reduced stress sensitivity. These authors remark that the properties of high iodide layers up to 40 mol % iodide are truly remarkable and have only become apparent. An interesting feature of these crystals is the high surface coverage of the outer iodide layer, which leads to a characteristic alteration to the usual shape of twinned hexagonal crystals.

Brust and Leipold [162] claim an improved method of forming monodispersed tabular grains. This method refers to ripening a population of small twinned crystals at a slightly elevated temperature and then carrying out most of the growth at a slightly reduced temperature compared with the ripening temperature and at pBr 1–3.5. Yagi and Heki [163] have described a growth method for monodispersed silver halide tabular grains, starting from spherical seed crystals. Tsaur and Kam-Ng [164,165] have described a process and additives that are aimed at producing tabular grains that are more uniform in size than the usual T-grains prepared at low pBr (high halide excess). In this process, a population of seed nuclei containing parallel twin planes are formed in the presence of, for example, a polyoxyethylene/polyoxypropylene copolymer. A portion of the crystals is removed by a ripening stage, and then the remaining tabular crystals are grown in the normal way. These two disclosures cover the use of a polyoxyalkylene block copolymer containing three terminal hydrophilic oxyalkylene groups linked through a hydrophobic oxyalkylene block linking unit (at least 4% of the molecular weight of the copolymer, or the inverse structure).

Antioniades et al. [166] describe a process for making silver iodobromide grains with a size coefficient of variation of less than 25%. The process starts with small, monodisperse untwinned octahedra that are transferred into a second vessel, where conditions are changed to form a population of predominantly tabular crystals containing parallel twin planes. The process responsible is disclosed as twinning induced by high bromide ion concentrations. Jagannathan [167] also discusses the formation of twinned crystals.

Breslav and Kolesnikov [168] recently attempted a rationalization of the alternative approaches to the structures and surface properties of monodisperse twinned emulsion microcrystals. It was remarked that these forms contain chemically built heterojunctions—composite materials with boundary zones between different structures or compositions and different electronic band structures. Photoelectrons and photoholes tend to migrate into iodide-deficient and iodide-rich areas, respectively. Ripening using silver halide solvents such as thiosulfate and thiocyanate led to complete core-shell restructuring and homogenization. The iodide redistribution and dye adsorption were found to influence the progress of the subsequent photographic development.

The growth of AgCl-rich tabular crystals has become the focus of some effort, especially from Eastman Kodak recently, presumably to improve processing characteristics [169].

G. ILFORD 400 and 100 DELTA Materials

Those advances in color film technology that have derived from improved efficiency of emulsion crystal growth have found their way into monochrome applications such as X-ray and rapid-processed medical films. Some of the improvements in image quality in color films, though, have derived from the incorporation of organic compounds that release development-inhibiting agents (DIR compounds) during chromogenic processing to limit the growth of the dye clouds formed and hence the perceived granularity of the final image. This is not easy to achieve with a monochrome film and developer combination.

The design of ILFORD 100 and 400 DELTA monochrome camera films illustrates how effective improvements in granularity (an objective measurement of the perceived graininess) of the image can be achieved by the application of controlled crystal growth materials. The structure of the emulsions is radically different from that of the preceding conventional crystals, and the crystal size and size spread can be selected using the "controlled crystal growth" methods discussed above. In this way, the crystal structure can be carefully controlled; also, the way the emulsion is grown uses new methods and equipment. The structure of the film assembly has also been revised to suit the particular characteristics of the sensitive emulsion.

Each crystal in the 100 DELTA emulsion, for example, is composed of three main layers. The innermost layer or core of the crystal has a higher silver iodide content, which has higher light absorption than the other layers and controls the speed of the film. The next layer and the outer shell are both development control layers. The inner of these two layers, which has a moderate silver iodide content, controls the granularity and general image quality. The outer layer, with a lower silver iodide content, controls the development rate and ensures that the film develops rapidly and is reproducible in manufacture.

The schematic diagrams of Fig. 19 show the shape and structure of the DELTA crystals compared with conventional crystals. Two electron micrographs are shown in Fig. 20 that compare a typical DELTA emulsion with a conventional emulsion. Unlike most conventional films, DELTA products have a single emulsion layer, which has been kept as thin as possible. This ensures better sharpness and sparkle and a faster development rate and makes the film easier to manufacture. The extremely high sharpness of 100 DELTA, for example, is due to both the advanced crystal technology and the physical thickness of the single emulsion layer, which has been kept to a minimum.

X. FUTURE OF SILVER HALIDE PHOTOGRAPHY

A. Development of Competing Electronic Systems

Kriss [170] has described the challenge and the promise of electronic imaging. The chief challenges were seen as the difficulty of electronic systems in competing effectively in the area of high quality hardcopy and the continuing price differential in favor of silver halide photography.

In an interesting comparison of color negative film technology with that of an electronic still camera, Ikenoue and Tabei [171] concluded that silver halide color negative films have distinct advantages over electronic systems in exposure latitude, in the number of pixels (picture-capturing elements), and in the efficiency of capturing color information and aliasing. The eco-

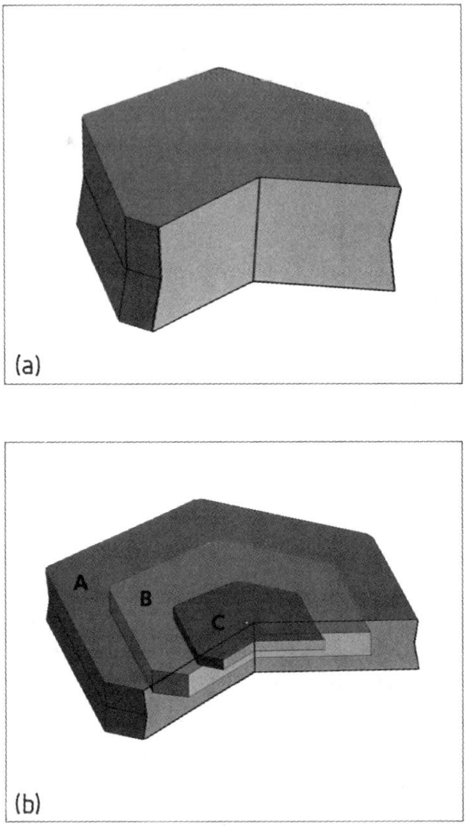

FIG. 19 Schematic diagrams showing the differences in shape and structure of (a) conventional crystals and (b) DELTA crystals.

nomic production of charge-coupled-device arrays with a competitive high pixel density was still difficult. They also cited the large improvements yet possible in light absorption and latent image formation for the silver halide system before its theoretical limits are reached.

Harvey [172] has described Eastman Kodak's view of the convergence of silver halide and electronics systems, including the release of the Kodak PhotoCD system using electronic media for storage and replay of photographic images initially taken on conventional silver halide media.

B. Potential for Improvement in AgX Systems

Eachus and Olm [75] have stated that recent developments make the eventual displacement of silver halide photography by electronic systems "cer-

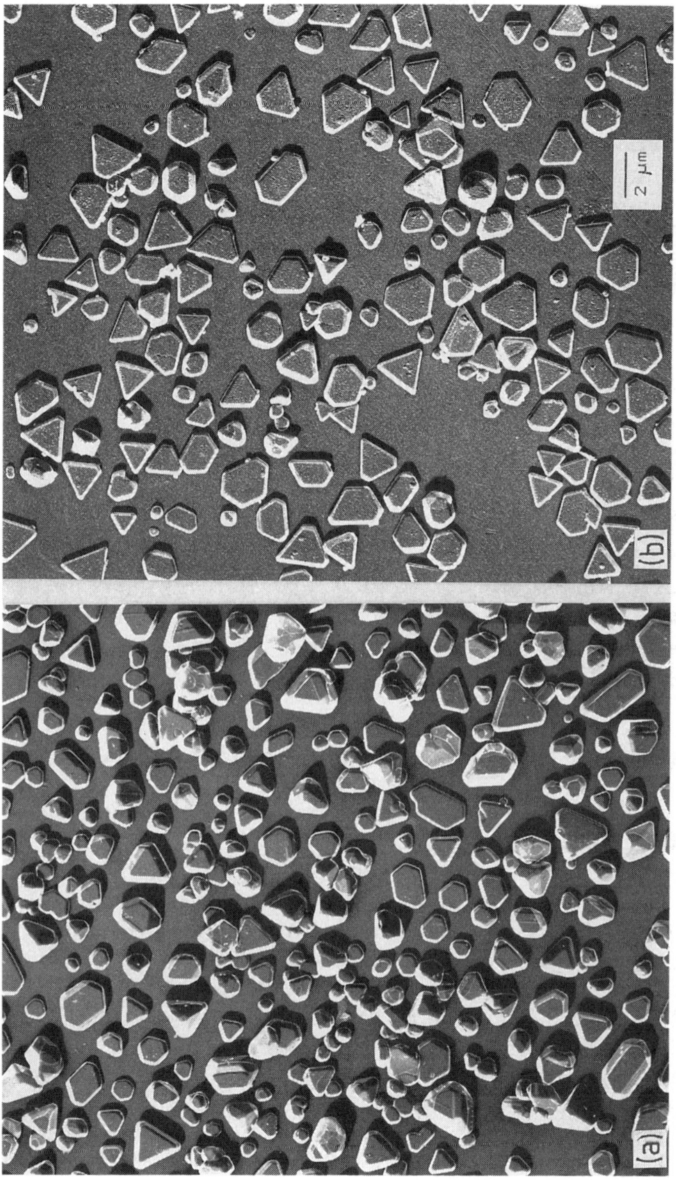

FIG. 20 Electron micrographs comparing (a) a conventional emulsion with (b) a typical DELTA emulsion. (Electron micrographs by K. N. Harvey and T. Groves.)

tain." However, Tani [173] has discussed the future prospects for silver halide photography and stated

> Future progress in imaging technology for taking pictures will solely depend upon innovations in AgX photography. Analyses on the processes for image formation have revealed the existence of much scope for future innovation in AgX photography by conventional and unconventional ways. The extensive study of AgX will contribute, not just to progress but also to progress of the underlying science and technologies.

According to Whiteley et al. [174], in an article entitled "Synergism: photography into the 21st century," silver halide imaging has a synergistic relationship with electronic photography. It remains to see how the consumer will use the additional pictorial imaging capability available.

ACKNOWLEDGMENTS

I am grateful to ILFORD Ltd and International Paper for permission to publish this review. Thanks are also due to many colleagues in the ILFORD laboratories and collaborating external groups. In many cases their contributions are acknowledged individually in the text and references.

REFERENCES

1. P. Krause, *World Silver Survey* 1992, The Silver Institute, Washington D.C., p.1.
2. T.H. James (Ed.), *Theory of the Photographic Process*, 4th ed., Macmillan, New York, 1977.
3. N. Kawanishi, H.K. Christenson, and B.W. Ninham, *J. Phys. Chem. 94*: 4611 (1990).
4. J. Lohmann, *Progress in Basic Principles of Imaging Systems* (F. Granzer and E. Mosiar, Eds.), Vieweg, Braunschweig, 1987, p. 4.
5. A. Sasai, *J. Soc. Photogr. Sci. Technol. Jpn. 47*: 1 (1983).
6. T.J. Maternaghan, in *Encyclopaedia of Materials Science and Engineering*, (M.B. Bever, Ed.), Pergamon, New York, 1986, p. 3512.
7. P. Krause, *Mod. Photogr.* April 1984, p. 72.
8. E.H. Land, *Photogr. J. 114*:1 (1974).
9. J. Bailey and D.N. Rogers, *Chem Br. 21*:251, (1985).
10. Anon., Research disclosure *12*:993 (1989).
11. H.E. Roberts, *J. Imag. Sci. 34*:149 (1990).
12. B. Deryagin and M. Levi, *Coating Photographic Emulsions*, Focal Press, 1974.
13. D.P. Roberts and N.L. Smith, *Radiographic Imaging*, Churchill Livingstone, London, 1988, p. 84.
14. T.H. James, *Adv. Photochem. 13*:329 (1986).

15. D.J. Locker, in *Kirk-Othmer Encyclopedia of Chemical Technology*, (M. Grayson and D. Eckroth, Eds.), 3rd ed., Wiley-Interscience, New York, 1982, Vol. 17, p. 611.
16. D.M. Sturmer and A.P. Marchetti, in *Imaging Processes*, (J.M. Sturge et al., Eds.), Neblette's 8th ed., Van Nostrand Reinhold, New York, 1989, Vol. 5, p. 71.
17. J.H. Coote, *Monochrome Darkroom Practice*, Focal Press, London, 1982, p. 61.
18. K. Radcliffe, *J. Photogr. Sci. 24*:198 (1976).
19. G.C. Farnell and R.B. Flint, *J. Photogr. Sci. 29*:236 (1981).
20. M.R.V. Sahyun, *Chemtech 7*:419 (1992).
21. R.K. Hailstone and J.F. Hamilton, *J. Imaging Sci. 31*:229 (1987).
22. R.K. Hailstone, M. Liebert, and M. Levy, *J. Imaging Sci. 33*:165 (1989).
23. R.K. Hailstone, M. Liebert, and M. Levy, *J. Imaging Sci. 31*:185 (1987).
24. C. McCullough, Ph.D. Thesis, Univ. London, 1990.
25. J.M. Chabala, R. Levi-Setti, and T.J. Maternaghan, *Proc. East/West Conf.*, Maui, Hawaii, October 1992, p. C-41.
26. J.F. Hamilton, *Adv. Phys. 37*:359 (1988).
27. M.R.V. Sahyun, *J. Chem. Educ. 51*:72 (1974).
28. M. Kawasaki, *J. Soc. Photogr. Sci. Technol. Jpn. 61*:515 (1986).
29. H.E. Roberts, *J. Imaging Sci. 29*:175 (1985).
30. J.F. Hamilton, *Photogr. Sci. Eng. 18*:493 (1974).
31. T. Tani, *J. Imaging Sci. 29*:93 (1985).
32. G.C. Farnell, *J. Photogr. Sci. 17*:116 (1969).
33. T.H. James, *J. Imaging Sci. 29*:45 (1985).
34. G.C. Farnell and R.L. Jenkins, *J. Photogr. Sci. 39*:229 (1991).
35. A. Hoffman and V. Weiss, *J. Photogr. Sci. 39*:220 (1991).
36. T. Tani, *Phys. Today 9*:36 (1989).
37. R.C. Baetzold, *Photogr. Sci. Eng. 19*:11 (1975).
38. G.R. Bird, R.C. Jones, and A.E. Ames, *Appl. Opt. 8*:2389 (1969).
39. C. Sutton, *New Sci. 12*:40 (1986).
40. L.M. Slifkin, *MRS Bull. 5*:36 (1989).
41. H.E. Spencer, *J. Photogr. Sci. 24*:34 (1976).
42. E.F.W. Thurston, *J. Photogr. Sci. 38*:34 (1990).
43. E. Klein and E. Moisar, *Ber. Bunsenges. Phys. Chem. 67*:349 (1963).
44. C.R. Berry and D.C. Skillman, *Photogr. Sci. Eng. 6*:159 (1962).
45. E. Moisar, *J. Soc. Photogr. Sci. Technol. Jpn. 54*:273 (1991).
46. J.D. Lewis, *J. Photogr. Sci. 27*:24 (1979).
47. Y.A. Breslav and V.V. Peisakhov, *J. Sci. Appl. Photogr. Cinema (Zh. Nauch.) 30*:249 (1987).
48. C.R. Berry and D.C. Skillman, *J. Photogr. Sci. 16*:137 (1968).
49. T.G. Bogg, M.J. Harding, and D.N. Skinner, *J. Photogr. Sci. 24*:81 (1976).
50. J. Rodgers, *J. Photogr. Sci. 27*:47 (1979).
51. T.M. Kelly and M.G. Mason, *J. Appl. Phys. 49*:4721 (1976).
52. G.C. Farnell, *J. Photogr. Sci. 24*:169 (1976).
53. C.R. Berry, *Photogr. Sci. Eng. 19*:171 (1975).
54. W.L. Gardner, D.P. Wrathall, and A.H. Herz, *J. Imaging Sci. 21*:325 (1977).

55. R.L. Daubendiek, *ICPS Conf. Proc.*, Rochester, 1978, p. 140.
56. F.H. Claes and W. Peelaers, *Photogr. Sci. Eng. 12*:207 (1968).
57. J.E. Maskasky (to Eastman Kodak), U.S. Patent 4,094,684 (1977).
58. J.E. Maskasky, *Photogr. Sci. Eng. 25*:96 (1981).
59. J.F. Hamilton, *Appl. Opt. 11*:13 (1972).
60. J.S. Wey, *J. Imaging Sci. 34*:202 (1990).
61. W. Marcocki and A. Zaleski, *Photogr. Sci. Eng. 17*:289 (1973).
62. R.H. Ottewill and R.F. Woodbridge, *J. Photogr. Sci. 16*:581 (1961).
63. R.H. Ottewill and R.F. Woodbridge, *J. Photogr. Sci. 13*:98 (1965).
64. R.H. Ottewill and R.F. Woodbridge, *J. Photogr. Sci. 13*:104 (1965).
65. A.H. Herz and J.O. Helling, *J. Colloid Sci. 16*:199 (1961).
66. T. Sugimoto, *Adv. Colloid Interface Sci. 28*:65 (1987).
67. K. Kurosaki, *J. Soc. Photogr. Sci. Technol. Jpn. 42*:91 (1979).
68. Y.T. Tan, *MRS Bull. 5*:13 (1989).
69. J.E. Maskasky, *J. Imaging Sci. 30*:247 (1986).
70. J.F. Reber, *Progress in Basic Principles of Imaging Systems* (F. Granzer and E. Moisar, Eds.), Vieweg, Braunschweig, 1987, p. 98.
71. C.R. Berry, *Photogr. Sci. Eng. 19*:93 (1975).
72. T.J. Maternaghan, C.J. Falder, R. Levi-Setti, and J.M. Chabala, *J. Imaging Sci. 34*:58 (1990).
73. B.K. Furman, G.H. Morrison, V.I. Saunders, and Y.T. Tan, *Photogr. Sci. Eng. 25*:121 (1981).
74. T. Tani, *J. Soc. Photogr. Sci. Technol. Jpn. 43*:335 (1980).
75. R. Eachus and M.T. Olm, *J. Soc. Photogr. Sci. Technol. Jpn. 54*:294 (1991).
76. S. Gahler and G. Roewer, *J. Info. Rec. Mater. 18*:294 (1990).
77. S.S. Collier and F.J. Evans, *J. Imaging Sci. 29*:229 (1985).
78. S. Bando, Y. Shibahara, and S. Ishimaru, *J. Imaging Sci. 29*:193 (1985).
79. S. Yamashita, N. Ohshima, and S. Takada, East-West Conf. Proc., Hawaii, November 1988, p. C-115.
80. S. Yamaryo, S. Ishimaru, and K. Takemura, *SMPTE J.*, September 1985, p. 735.
81. E. Moisar, *J. Photogr. Sci. 16*:102 (1968).
82. D.M. Sturmer and L.N. Blackburn, *Photogr. Sci. Eng. 19*:352 (1975).
83. D.C. Birch, G.C. Farnell, and R.B. Flint, *J. Photogr. Sci. 23*:249 (1975).
84. A.N.R. Shephard, R. King, and D.C. Schuman, *J. Imaging Sci. 33*:230 (1989).
85. W. Vanassche, H. Borginon, H. Pattyn, W. Peelaers, and E. Moisar, *J. Photogr. Sci. 22*:121 (1974).
86. D.J. Cash, *J. Photogr. Sci. 29*:140 (1981).
87. D.J. Cash, *Photogr. Sci. Eng. 27*:156 (1983).
88. D. Corbin, A. Gingello, G. Macintyre, and B.H. Carroll, *Photogr. Sci. Eng. 24*:45 (1980).
89. M.K. van Doorselaer, *J. Photogr. Sci. 35*:42 (1987).
90. T. Shiozawa and T. Kobayashi, *Phys. Stat. Sol. 110*:375 (1988).
91. D.J. Locker, *J. Imaging Sci. 20*:242 (1976).

92. T. Matsubara and M. Levy, *J. Imaging Sci. 32*:178 (1988).
93. K. Kuge, *J. Soc. Photogr. Sci. Technol. Jpn. 52*:377 (1989).
94. T. Tani, *J. Soc. Photogr. Sci. Technol. Jpn. 21*:37 (1977).
95. A.G. Smith and G. Fitzgibbons, *J. Imaging Technol. 16*:139 (1990).
96. E. Moisar and F. Grantzer, *J. Imaging Sci. 29*:233 (1985).
97. G.E. Ficken, *Chem. Ind. 10*:672 (1989).
98. J.E. Maskasky, *J. Imaging Sci. 35*:29 (1991).
99. T. Tani, *J. Imaging Sci. 34*:143 (1990).
100. R. Steiger (to ILFORD AG), U.S. Patent 5,077,190 (1991).
101. R. Steiger and P.A. Brugger (to ILFORD AG), U.S. Patent 5,141,845 (1992).
102. T. Tani, *J. Imaging Sci. 32*:227 (1988).
103. L. Brus, *J. Soc. Photogr. Sci. Technol. Jpn. 53*:329 (1990).
104. T. Sugimoto and K. Miyake, *J. Colloid Interface Sci. 140*:335 (1990).
105. X.L. Gao, R. Gijbels, B. Niys, W. Jacob, and Y. Gilliams, *J. Imaging Sci. 33*:87 (1989).
106. T. Tani, *J. Soc. Photogr. Sci. Technol. Jpn. 53*:123 (1990).
107. V. Oleshko, M. Alfimov, A. Galashin, and K. Leonov, *J. Imaging Sci. 36*: 273 (1992).
108. R. Steiger, *SPSE Conf. Proc.* Rochester, N.Y., 1990, p. 34.
109. B.J. Garrison and N. Winograd, *Chemtech 1*:25 (1993).
110. Anon., *Res. Disclosure 1*:20 (1983).
111. T.J. Maternaghan (to ILFORD Ltd), Br. Patent 1 570 581 (1981).
112. T.J. Maternaghan (to ILFORD Ltd, Br. Patent 1 520 976 (1976).
113. J.D. Lewis (to ILFORD Ltd), Br. Patent 1 469 480 (1974).
114. M.E. Sprackling, *J. Photogr. Sci. 32*:21 (1984).
115. M.E. Sprackling, *J. Photogr. Sci. 32*:96 (1984).
116. J.C. Solberg, R.H. Piggin, H.S. Wilgus, and F.J. Evans (to Eastman Kodak), U.S. Patent 4,433,048 (1980).
117. J.E. Maskasky, U.S. Patent 3,241,647 (1981).
118. H.J. Metz, *J. Soc. Photogr. Sci. Technol. Jpn. 53*:124 (1990).
119. J.T. Kofron and R.E. Booms, *J. Soc. Photogr. Sci. Technol. Jpn. 49*:499 (1986).
120. I. Itoh, S. Takada, and S. Ikenoue, *Sci Publ. Fuji Photo Film Co. 30*:149 (1988).
121. S.J. Powell and F.R. Reinking, *SMTE J. 9*:870 (1986).
122. F.H. Claes, *Photogr. Korr 101*:139 (1965).
123. T. Sugimoto, *Photogr. Sci. Eng. 28*:137 (1984).
124. T. Sugimoto, *J. Imaging Sci. 33*:203 (1989).
125. K. Ohseki, S. Urabe, and T. Tani, *J. Imaging Sci. 34*:136 (1990).
126. J.F. Hamilton, *Phil. Mag. 16*:1 (1967).
127. J.E. Maskasky, *J. Imaging Sci. 31*:93 (1987).
128. J.E. Maskasky, *J. Imaging Sci. 31*:15 (1987).
129. Y.A. Breslav and V.V. Peisakhov, *J. Sci. Appl. Photogr. Cinema* (*Zh. Nauch*) *30*:375 (1987).
130. T.J. Maternaghan (to ILFORD Ltd), Br. Patent 1 596 602 (1978).
131. C.W. Mowforth and T.J. Maternaghan, Siemens Anal. Appl. Note No. 314, 1988.

132. E. Klein, E. Moisar, and K. Munch, *Photogr. Korr 102*:59 (1966).
133. M.A. King, M.H. Loretto, T.J. Maternaghan, and F.J. Berry, *Progress in Basic Principles of Imaging Systems*, (F. Granzer and E. Moisar, Eds.,), Vieweg, Braunschweig, 1987, p. 73.
134. W. Schmidt and K. Hoffman, *Progress in Basic Principles of Imaging Systems*, (F. Granzer and E. Moisar, Eds.), Vieweg, Braunschweig, 1987, p. 161.
135. C.W. Mowforth, J.B. Bullock, T.J. Maternaghan, and K.N. Harvey (to ILFORD Ltd), Eur. Patent 0 359 507 (1990).
136. C.W. Mowforth, J.B. Bullock, and T.J. Maternaghan (to ILFORD Ltd), Eur. Patent 0 359 506 (1990).
137. M.E. Irving (to Eastman Kodak), Eur. Patent 478 670 (1990).
138. M.E. Irving, *Proc. 4th East/West Conf.*, Maui, Hawaii, 1991, p.6.
139. F. Grantzer, *J. Imaging Sci. 33*:207 (1989).
140. Z. Hou, S.-Y. Zhuang, C. Tao, and X.-M. Ren, *J. Imaging Sci. 34*:155 (1990).
141. Z. Si-Yong, *J. Imaging Sci. 30*:16 (1986).
142. K. Heki and T. Yagi, *Progress in Basic Principles of Imaging Systems*, (F. Granzer and E. Moisar, Eds.), Vieweg, Braunschweig, 1987, p. 161.
143. A.F. Sowinski and P.J. Wightman, *J. Imaging Sci. 31*:162 (1987).
144. H. Haefke, U.D. Schwarz, E. Meyer, G. Gerth, R. Steiger, and H.J. Guentherodt, *Proc. 4th East/West Conf.* Maui, Hawaii, 1992, p. 17.
145. M.P. Keyes, E.C. Phillips, and W.L. Gladfelter, *J. Imaging Sci. Technol. 36*:268 (1992).
146. Z. Chi and S. Wang, *Photogr. Sci. Photochem.* (*Chin.*). *4*:36 (1989).
147. J.M. Chabala, R. Levi-Setti, K.N. Harvey, T.J. Maternaghan, A. Tyrer, M. Loretto, and F. Berry, *SPSE Conf. Proc.*, Rochester N.Y., 1990, p. 40.
148. S. Aida (to Fuji), Eur. Patent 4 962 015 (1989).
149. S. Takada, N. Ohshima, and S. Ishimaru (to Fuji), Eur. Patent 0 147 854 (1984).
150. R.J. Newmiller (to Eastman Kodak), U.S. Patent 4,685,964 (1990).
151. A. Dobson and M.S. Wharton (to ILFORD Ltd.), Eur. Patent 0 477 772 (1991).
152. P.J. Zola and R.A. Bryant (to Eastman Kodak), Eur. Patent 0 362 699 (1990).
153. A. Steward, Ph.D. Thesis, Univ. Birmingham, England, 1990.
154. S. Ishimaru, H. Ikeda, K. Miyazaki, S. Hirano, and J. Tamano, *Fuji Film R&D Publ. 37*:1 (1992).
155. K.N. Harvey, T.J. Maternaghan, D.J. Edwards, D.J. Turner, and C.W. Mowforth (to ILFORD Ltd), Br. Patent 2 222 094A (1990).
156. S. Takada (to Fuji), Eur. Patent 3 699 424 (1990).
157. S. Yamada, Eur. Patent 404 142 (1989).
158. S. Urabe (to Fuji), Eur. Patent 0 460 656 (1991).
159. M. Saito (to Fuji), U.S. Patent 0,493,625 (1990).
160. R.H. Piggin and C.J. Bishop, Eur. Patent 408 214 (1990).

161. M.-J. Lin, P.J. Zola, and R.H. Piggin, Eur. Patent 048 213 (1990).
162. T.R. Brust and P.A. Leipold, Eur. Patent 506 947 (1991).
163. T.R. Yagi and K. Heki, *Proc. 4th East/West Conf.*, Maui, Hawaii, 1992.
164. A.K. Tsaur and M. Kam-Ng, U.S. Patent 5,147,772 (1991).
165. A.K. Tsaur and M. Kam-Ng, U.S. Patent 5,147,773 (1991).
166. M. Antioniades, R. Jagannathan, D.E. Fenton, and J.L. Hall (to Eastman Kodak), Eur. Patent 0 507 701 (1991).
167. R. Jagannathan, *J. Imaging Sci. 35*:104 (1991).
168. Y.A. Breslav and L.V. Kolesnikov, *J. Photogr. Sci. 39*:2 (1991).
169. J.E. Maskasky (to Eastman Kodak), U.S. Patent 5,061,617 (1990).
170. M.A. Kriss, *J. Soc. Photogr. Sci. Technol. Jpn. 50*:357 (1987).
171. S. Ikenoue and M. Tabei, *J. Imaging Sci. 34*:187 (1990).
172. J.W. Harvey, *Proc. 4th East/West Conf.*, Maui, Hawaii, 1992.
173. T. Tani, *J. Soc. Photogr. Sci. Technol. Jpn. 53*:87 (1990).
174. T.E. Whiteley, R.A. Landholm, and A.M. Kelly, *J. Soc. Photogr. Sci. Technol. Jpn. 53*:95 (1990).
175. A. Kriebel, *Proc. 4th East/West Conf.*, Maui, Hawaii, 1992.

12

Magnetic Recording Materials

DAVID H. DAVIES Ampex Recording Media Corporation, Redwood City, California

I. Introduction—The Impact of Dispersion on Tape Properties 443
A. High frequency losses due to aggregates 444
B. Increased orientation requirements 444
C. Durability and shed minimization 445
D. Particle size distribution control 446
E. Effect of dispersion on signal-to-noise ratio and defect levels 446
F. The increasing difficulty of dispersion 446
G. Reactions at the head/tape interface 447
II. Mechanical Dispersion Methods 449
A. Milling and double planetary mixing (DPM) 449
B. Kneading 450
C. Newer concepts 452
III. Dispersant Materials and Surfactants 452
A. The model surfactant for magnetic dispersion 453
B. The role of the surfactant in magnetic media 454
C. Recommendations for dispersant selection 454
IV. Conclusions 455
References 455

I. INTRODUCTION—THE IMPACT OF DISPERSION ON TAPE PROPERTIES

The emergence of magnetic tape as a format for data, video, and audio recording as well as for the distribution of precorded information of all three varieties is a remarkable recent phenomenon. Magnetic tape consists of magnetic particles dispersed in a flexible polymeric binder coated on a polymeric substrate. The business of producing such tape now exceeds $14 billion annually; no other technical factor in its growth has been more important than the increasing ability to disperse finer and finer magnetic particles. The criticality of dispersion was recognized early and is due to a number of factors.

A. High Frequency Losses Due to Aggregates

Undispersed magnetic material can act as a single magnetic aggregate composed of several particles. Since the density of recording depends on a discrete identifiable bit (or domain edge), the presence of aggregates is equivalent to the use of larger particles, and the ability to record at higher frequencies is diminished. The minimum recordable wavelength, or data bit, is closely related to the average particle size. Modern formats, especially those based on high density digital codes, depend heavily on high frequency recording, and this has increased the degree of importance attached to obtaining uniformly dispersed small magnetic particles.

B. Increased Orientation Requirements

Directly related to the degree of dispersion is the ability to effectively orient the tape in a magnetic field, that is, have the domains line up unidirectionally. The degree of orientation is a measure of the magnetic anisotropy of the individual magnetic domains. Since, as noted above, the domain and the individual particles are closely related, aggregates and undispersed clumps are not readily orientable. Hence the orientation ratio (OR) and the squareness of the hysteresis loop (SQ, see Fig. 1) are dramatically affected. The effects of both of these factors result in a lower "output" tape, that is, a tape that produces lower responses to magnetic read heads for a given write signal. A good discussion of those factors is given in Refs. 1 and 2.

This problem is exacerbated in newer tape formats where the need for higher density recording forces the use of small individual particles. In turn this requires particles of greater magnetic density (σ_s), particles having higher remnant magnetization (M_r) and much higher coercivity (H_c). As a consequence, the level of the external magnetic field strength needed to achieve magnetic orientation during manufacture is greatly increased (approximately as the cube of the coercivity). This greatly stresses the available technology of magnets. For example, whereas it was common as recently as 10 years ago to use 1 kOe field strength (80 kA/m) for orientation, it is now necessary to use 10–15 kOe, which requires massive cooled electromagnets. Hence a much greater burden is placed on the dispersion process to achieve uniform well separated particles.

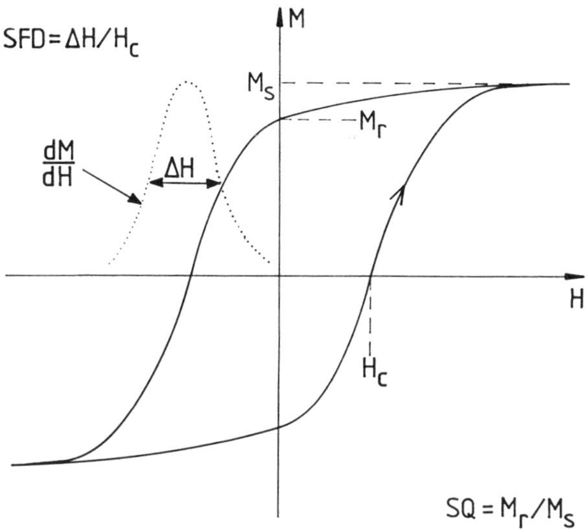

FIG. 1 Hysteresis loop in the plot of magnetization M versus applied magnetic field H for a typical recording medium. ΔH is the half-height width of the differentiated hysteresis loop; H_c is the coercivity; SQ is the magnetic squareness; SFD is the switching field distribution.

C. Durability and Shed Minimization

As particles have become smaller and data transfer rates have increased, the head-to-tape tribology has become greatly magnified as an issue [3]. Modern formats use helical scan systems that result in head/tape relative speeds of 2–3 m/s. Bear in mind that even at this speed the tape must be in intimate physical contact with the head to avoid spacing losses [4]; hence the presence of undispersed clumps of materials in coatings that are often below 1 μm thickness causes surface asperities. These asperities are worn down by the head and build up as debris often called "shed;" Ref. 5 provides an excellent recent review of this problem. This shed is a significant problem as it contaminates the heads, transport guides, and other parts of the tape. In addition, the degree of wear resistance of the tape to repeated head and guide passes is affected by the degree of wetting of the individual particles by the binder polymer. The binder–particle contact must be intimate to achieve the degree of cohesiveness needed for durability. The binders (typically urethane and vinyl materials) are often inhibited from particle wetting by air layers or by the relative adsorption at the surface of the particles by other ingredients such as lubricants [6]. It is necessary to achieve binder–particle wetting as a first step in dispersion and to achieve it to a high degree of surface saturation if a durable tape is to be achieved.

D. Particle Size Distribution Control

Since the individual particles are essentially related to the individual recorded bits or data elements, it is clear that a high degree of particle size control is needed. Very small or broken particles can result in highly magnetic defects that cause a problem called "print-through"; this is the magnetic imprinting of tape layers in a wound spool. A critical factor caused by overgrinding is the magnetic squareness (SQ $= M_r/M_s$) loss in the hysteresis loop. The other parameter affected by particle size distribution is the switching field distribution (SFD, see Fig. 1), which is a key parameter in achieving uniform erasure and magnetic writing properties.

E. Effect of Dispersion on Signal-to-Noise Ratio and Defect Levels

The critical ratio of signal to noise (SNR) is a key parameter in all electronic information channels. Poor dispersion impacts this in several ways [7]. Poor dispersion results in a rough surface, which in turn induces high frequency noise and also, more critically, reduces head–tape contact, thereby minimizing signal levels [4]. Also, undispersed aggregates are poorly recorded and oriented; this results in poor recorded bit edge definition, which increases data error. Finally, very large aggregates are physically and mechanically distinct from the bulk of the tape and hence can be discrete defects that cause "dropouts" (analog loss of signal) or "burst errors" (digital data error).

F. The Increasing Difficulty of Dispersion

The factors above describe some of the issues that relate to the increasing criticality of dispersion quality on tape performance as recording density has increased in recent years. Equally relevant is the increasing difficulty of achieving dispersion; this is due to four factors.

1. The smaller and smaller size of the magnetic particle
2. The changes in the materials of the particles themselves
3. The increasing coercivity of the particles
4. The growing realization that the head–tape interface during high speed helical scan results in chemical reactions that are deleterious and are directly related to the traditional dispersants often used

The first three factors can be illustrated by Table 1 which lists some changes in the parameters of magnetic particles in recent years. The change in average dimensions and the change to a very acicular (rodlike) shape have been dramatic. The specific surface area (SSA), a measure of average surface area and hence particle size, has increased markedly. Even small increases in SSA have a major impact on ease of dispersion. Of equal importance is the change in the particle itself. Doped and undoped oxides have magnetic coercivities of ca. 400–900 Oe. However, the much smaller metal particles (MPs) now used, which are iron coated with a surface protectant layer of various ceramic oxides, can reach 1700 Oe. Generally the degree of difficulty of dispersion can be thought to parallel the difficulties in orientation, which increase by a factor of 10 for each tripling in coercivity. Finally, a more recent change is the development of more exotic particles such as barium ferrite. These materials are characterized by a stacking platelet structure that is very highly ordered. Needless to say, the degree of difficulty in dispersion is extreme, to the point where it has hindered the commercial exploitation of these materials. A good discussion of recent developments in particulate magnetic coatings is given in Ref. 8. Some further information on magnetic pigments is given in Chapter 3, Sec. VIII.E, of this volume.

G. Reactions at the Head/Tape Interface

Last, and very important, is the issue of chemical reactions at the head–tape interface. The very high surface temperatures at the interface caused by frictional heating result in the development of pyrolysis residue, often forming a binding matrix for ultrafine metal particles. This effect, often called head filming, is sometimes also attributed to "frictional polymers" [9,10]. This causes gradual loss of contact and lower and lower output at the tape–head interface. Eventually it must be removed by abrasive cleaning agents. It is common to include these (e.g., Al_2O_3) in the tape itself, but this causes wear of the unfilmed head.

The relevance here is that the classical surfactant materials (lecithin, phosphates, etc.) used in magnetic tapes and described below, are clearly major contributors to the growth of these filming agents. The relatively reactive nature of these materials, their tendency to seek surfaces, which makes them good surface-active agents in the first place, and their migration tendencies all tend to make them culpable.

Hence a major trend in the formulation of magnetic tape is to eliminate, or minimize, the use of classical wetting agents to the extent possible. Unfortunately, this requirement is paralleled by the increasing need to achieve good dispersion for better high frequency recording and by the increasing

TABLE 1 The Magnetic Particle[a]

Approx. year of intro.	Particle nature	Shape	Typical SSA (m^2/g)	Typical length (μm)	Typical width (Å)	Typical coercivity (Oe)	Application (major)
1940s	Ferric oxide	Moderate acicular	35	0.4	400	350	Audio
1960s	Codoped ferric oxide	Moderate acicular	35	0.35	400	900	VHS video
1960s	Chromium oxide	Acicular	45	0.5	300	600	Data, audio
1980s	Metal particle MP-1	Long acicular	60	0.3	200	1550	D2 video
1990	MP-2	Short acicular	61	0.2	150	1650	8 mm video, DAT
1992	MP-3	Ultrashort acicular	65	0.15	100	1750	Hi 8 mm video
1990s	Barium ferrite	Platelet	N/A	0.05	<100	range	High-density floppy

[a]Parameters now vary widely as particles are customized to applications; values given are for consumer commercial particles.

difficulty caused by the use of smaller and smaller particles of greater and greater magnetic energy density.

The resolution of this paradox has been to use mechanical, high shear, processes as the primary method to achieve initial particle dispersion in place of the chemical methods in vogue in recent years. The prewetting is then followed by milling. The next section describes mechanical methods from conventional milling through to the high shear preprocessing methods recently introduced.

II. MECHANICAL DISPERSION METHODS

As explained above, the need in magnetic tape dispersion is to create as great a degree of particle separation as possible while permitting the polymer binder to wet the surface of the particles. The constraints are (1) to avoid the need for low molecular weight reactive dispersants and surfactants and (2) to minimize particle breakage. Practical considerations such as mill time, media cost, contamination, filterability, and safety all enter into the balance as well. This section progresses from conventional mechanical milling to newly emerging methods and finally to radical new concepts.

A. Milling and Double Planetary Mixing (DPM)

Bead or sand milling has become the standard method of operation for magnetic pigment dispersions in recent years although many vendors still retain classical ceramic ball mills (employing ceramic or steel balls); these are used for oxide formulations developed several years ago, for example, in professional audiotape production. In sand milling, the milling medium (which today is rarely sand but most commonly fine, ca. 30 mesh, steel or ceramics) is circulated in a vertical or horizontal cylinder with the pigment-containing mix by means of a disk impeller. Most of the necessary ingredients (abrasives, pigments, solvents, binders, dispersants, lubricants, and conductive agents) are present from the beginning, but in a modified formulation that will be subsequently “let down.” The initial premix often is prepared in a double planetary mixer. The milling passes (commonly 4–10) are often performed in sequence in a continuous bank of mills. The degree of dispersion is monitored through drawdowns (hard coatings), with magnetic characteristics such as squareness loops used as criteria. The magnetic properties are an excellent monitor for particle dispersion. In the typical sand mill the wetting and separation action takes place by the action of disk rotation, which brings mix and medium together. As the disk rotates, the centrifugal forces vary so that high shearing forces are brought into play between layers. These forces result in good separation without the occasional crushing effects that occur in ball mills. Since magnetic pigments are very rapidly harmed by even small percentages of crushed particles (as discussed above), this has become the dominant mode of dispersion in this industry. Factors relevant to the sand milling of magnetic pigments include the nature of the magnetic particle (oxide, metal, etc.), the required rheological properties of the mix, the mix-to-media ratio, the temperature rise, the degree of media contamination allowed, the coating method to be used subsequently (gravity, roll coating, knife coating, etc.), and, of course, the number of passes. Careful letdown is critical to avoid shock and to achieve the final rheological properties. For a given medium, horizontal milling is generally considered more aggressive than vertical, and steel media are considered more aggressive than ceramic. Also used are glass media and a variety of exotic ceramics. The final step of the letdown is filtration, with a range of filter sizes utilized. Most magnetic pigment mixes can be held (uncatalyzed) for several hours under agitation and with temperature control. After catalysis (usually with an isocyanate–solvent blend) they must be used within a few hours. Filtering prior to final coating is common, and a final "polish" milling pass is often used if mix has been sitting unused for a while.

Recent developments include newer media such as zirconia/yttria that are denser and more wear-resistant. This results in shorter mill times and less contamination from the grind medium in the finished mix.

In the last few years a significant amount of attention has been paid to the premix DPM step, and innovations such as Inoue's Tri Mix (Inoue Mfg., Japan) have been introduced. This uses a triple-arm planetary format with stainless blades and a close (ca. 1 mm) tolerance between blades and wall. Pressurized and nonpressurized units are available, as is temperature control. Most magnetic tape formulations are designed to anticipate some heating during the DPM process. Generally this premix step is done at ca. 60% solids and viscosities exceeding 1000 Pa.s (1 M cps). The objective as always is to adequately wet the particle with binder prior to extensive milling. Such a requirement has led to the next process discussed—kneading, also called high shear, high solids preprocessing.

B. Kneading

Since the introduction of pigments with small particles and high surface area (SSA > 45), it has become common to prepare the premix with high pressure and/or continuous kneader units. Only in this way is it possible to achieve the particle–binder contact while avoiding extensive use of classical surfactants. In these dispersantless magnetic pigment formulations it has been shown that premix kneading at shear rates exceeding 1×10^7 Pa can improve final coating gloss by up to 20%, reduce surface roughness (Kosaka R_Z by ca. 10 nm), and increase remnant magnetization by 100–400 Oe. The advantage of kneading over DPM for premixing is approximately a doubling in remnant magnetization when SSA increases from 50 to 70 m^2/g. Below 30 m^2/g it is probably not worthwhile to use a premix kneader. Squareness ratio (see above) can be improved, for example, from 0.81 to 0.85, other factors being equal.

Two kneading methods are in use for magnetic tape mix preparation: continuous and batch high pressure systems. The former have excellent productivity but are harder to control. The latter are highly tolerant of variability and tend to be more effective. Both are followed by conventional sand milling. Typical vendors for kneaders used in magnetic mix preparation are Inoue Mfg., Japan; Kurimoto Ltd., Japan; and APV Chemical Machinery, Saginaw, Michigan. A typical initial kneading composition for a two-shaft continuous extrusion kneader (e.g., Kurimoto T-4) is as follows [11]:

	Parts by weight
Iron oxide particle (SSA 35 m^2/g)	100
Binder solvents (20% vinyl acetone/chloride, 40% MEK, 40% butyl acetate)	30
Vinyl polymer (acetate/chloride)	10
Carbon black	1
Butyl acetate (BA)	2.5

Additional solvents are added during kneading; the mix is then re-kneaded with a blend of polyurethanes in MEK and BA. Finally (and order of additional is critical), fatty acid lubricants such as mysteric and oleic acid are added with other abrasives and solvents. The final mix is then sand-milled in conventional fashion.

Some key elements of the use of kneaders include the provision and control of an inert gas atmosphere if reactive metal particle pigments are used, the selection of the kneader shaft sequence (i.e., the sequence of paddles in mixing, reversing, and forwarding sections), the cooling through the bore and walls, the clearance between paddles and walls, the surface hardness of the kneader materials, the sequence of addition, and the viscosity of the mix.

C. Newer Concepts

As dispersantless magnetic media formulations have become more common and the degree of dispersion more demanding, an extensive effort has developed to achieve more effective milling. In particular, several developments are under way to achieve the good dispersion properties of sand milling, especially the minimization of particle breakage, while preventing a weakness commonly called "by-pass" or "short pass." This problem allows partially aggregated material to pass through the mill unseparated, and it is very much an issue of concern for magnetic tape pigment dispersions.

One such effort is the continuous annular stirred bead mill [12,13] pioneered by Yokohama University. This unit features a rhombic pattern on the mill surface. The rotation of the rotor transmits momentum to the beads that are contained in the annular gap between the rotor cylinder and the inner surface of the mill vessel. A remarkable result is a pluglike flow with very sharp residence time distribution. This produces a very precise, sharp, particle size distribution that is very effective for magnetic particle dispersion. The plug-type flow and uniform speed result in minimal dead space in the mill, thus minimizing "short-pass" problems. The mill is aggressive in its grinding action, much like another related mill, the PEG mill. However, experience to date indicates that there is little particle breakage when it is used to disperse ultrafine (SSA 65 m^2/g) metal particle magnetic pigments.

III. DISPERSANT MATERIALS AND SURFACTANTS[1]

Despite attempts to eliminate the use of low molecular weight reactive species, such as those used in classical surface science applications, the utility of dispersant materials and surfactants in dispersions for magnetic tapes is such that they are still widely used. This is especially so in VHS/Beta consumer applications where the frictional properties at the head–tape interface are less stressful than in professional applications.

Typical materials used in commercial practice are phosphate esters [15], which are produced by several companies and are often included under the GAFAC trade name. Also very common are lecithin [16] and quaternary ammonium compounds [17]. Fatty acids and esters are often incorporated, and they also serve as lubricants. A recent development is the creation of surfactants that will copolymerize with the binder matrix in magnetic dispersions (polyester/polyamine copolymer, Hypermer-MTI, ICI America, Inc.).

---

[1]This section incorporates material from Chagnon and Donadio [14].

A. The Model Surfactant for Magnetic Dispersion

In order to prepare stable dispersions of ultrafine magnetic pigments in organic solvents, a model surfactant was first proposed by Papell in the early 1960s (see Ref. 14) and later revised by Kaiser [18] and Rosensweig [19] in their attempts to prepare stable magnetic colloids. Although the loading densities of the colloids prepared by Kaiser and Rosensweig were about an order of magnitude lower than pigment loading in the dispersions commonly used in magnetic recording media today, modifications of their proposed model have been useful tools in selecting a dispersant (or more properly stated, an idealized molecular structure) for use in magnetic dispersions.

In general, a molecule of the type $R''-R'-R-YH$ where YH is a polar functional group that bonds by means of covalent linkage, chemisorption, physical adsorption, or ionic interaction to the surface of the magnetic particle, has been observed to yield dispersions with the best magnetic properties, highest degree of dispersion, optimum milling behavior, and best performance on tape [20]. YH is typically $-COOH, -NH_2, -SO_3H, -PO_3H, -M^+Cl^-$, or SH, can be mono- di-, or trisubstituted on the terminal carbon atom as in

$$\underset{(a)}{R-CH_2-COOH} \qquad \underset{(b)}{R-CH(COOH)_2} \qquad \underset{(c)}{R-C(COOH)_3}$$

and can be a single or mixed species, such as

$$\underset{(d)}{R-CH(COOH)_2} \quad \text{or} \quad \underset{(e)}{R-\underset{\displaystyle NH_2}{\underset{|}{CH}}(COOH)}$$

If YH is a cation or anion, it is most effective when its charge is of the same magnitude but of opposite sign to the surface charge on the particle being dispersed.

R is an aliphatic chain (C_4–C_{20}), aromatic ring, or cyclic aliphatic group. Addition of polar functional groups along the chain (such as primary hydroxyls) can in some cases result in stronger interactions between YH and the particle by acting as additional YH adsorption sites or by increasing the ionization of the YH group. If R is aliphatic, the length of the R chain is useful in changing the magnitude of the charge on YH.

R' is a linking group that causes a change in electron density, separating the polar $R-YH$ portion of the molecule from the nonpolar tail.

R'' is a hydrophobic tail that has a solubility parameter similar to that of the binder–solvent matrix and is of sufficient chain length to separate the magnetic particles so that no particle–particle magnetic field interference occurs.

B. The Role of the Surfactant in Magnetic Media

The primary role of a surfactant for magnetic media is not unlike that of surfactants used in other particulate dispersions, that is, to disperse particles uniformly in an organic liquid medium. Because of the nature of the end product, however, several other factors must also be considered. A typical formulation for manufacture of magnetic tape is

	Parts by weight
Magnetic oxide	100
Surfactant	3
Polymer binder(s)	25
Solvent	200
Abrasive	1
Carbon black	3
Lubricant	1

Based on this formulation, the conditions that the surfactant must satisfy, in addition to uniform dispersion, include

1. Optimization of magnetic properties
2. Reduction in milling time required to achieve dispersion
3. Compatibility with the binder system in order to avoid coatings with poor surface quality
4. Effectiveness at low concentrations in order to avoid softening or plasticization of the binder molecule and to maximize the amount of magnetic material in the coating
5. Minimization of reactivity at the magnetic head and avoidance of durability loss due to migration from the coating

C. Recommendations for Dispersant Selection

The selection of a surfactant for magnetic recording media requires a good understanding of the dispersion quality that one wants to achieve, process limitations, and end product usage. Proper selection of the surfactant is necessary to optimize both processing time and tape performance. Selection of a good surfactant can be based on the following considerations.

1. Optimization of the polar YH functional head group of the surfactant. This can be achieved experimentally or by choosing a surfactant with a charge of similar magnitude and opposite sign to that of the surface of the particle to be dispersed.
2. Choosing an R″ functionality whose solubility parameter is similar to that of the solvent–binder matrix. δ, the solubility parameter of R″, can be calculated from Hildebrand units or heats of vaporization ΔH_v.

$$\delta = \left[\frac{D(\Delta H_v - RT)}{M}\right] \quad (1)$$

where D is the density and M the molecular weight of R″H, R is the gas constant, and T the absolute temperature [21].
3. Choosing a surfactant molecule that has an R chain that minimizes spacing losses between particles. In general, if R exceeds 20 carbon atoms in length, the magnetic signal can be decreased sufficiently to result in a loss of tape output.
4. Selecting a material that can copolymerize into the binder matrix to prevent migration and interaction at the head–tape interface.

IV. CONCLUSIONS

The areas of active research and development in magnetic tape dispersion can be identified as

1. New dispersant materials that are inert in head/tape interface interaction and do not migrate to the heads.
2. New dispersant materials that do not cause durability issues. Possibly this can be achieved by reacting the dispersant into the binder matrix as a copolymer.
3. High shear premill processing to get prewetting of the particle by the binder and to predisperse the mix.
4. Novel milling methods that use unique mill designs (annular, peg, etc.) and novel mill media made from advanced ceramics, such as zirconia and yttria.

It is anticipated that the rising difficulties of dispersion will be more than adequately matched by the creativity of the materials scientists and process engineers.

REFERENCES

1. D. Jiles, *Introduction to Magnetism and Magnetic Materials*, Chapman and Hall, London, 1991.
2. R.H. Perry and A.A. Nishimura, in *Kirk-Othmer Encyclopedia of Chemical Technology*, Wiley, New York, 1981, 3rd ed., pp. 732–753.

3. B. Bhushan, *Tribology of Magnetic Storage Devices*, Springer-Verlag, New York, 1990.
4. R.L. Wallace, *Bell System Tech J. 30*:1145 (1973).
5. H. Osaki, E. Oyanagi, H. Aonuma, T. Kanou, and J. Kurihara, *IEEE Trans. Magn. 28*:76 (1992).
6. S. Dasgupta, *Prog. Org. Coatings, 19*:123 (1991).
7. M.D. Clarke, P.R. Bissel, R.W. Chantrell, and R. Carlson, *J. Magn. Magn. Mat. 95*:17 (1991).
8. M.P. Sharrock, *IEEE Trans. Magn. 25*:4374 (1989).
9. J.J. Lauer and P.M. Blanc, *Polymer Prepr. ACS 29*:275 (1988).
10. C.M. Stahle and T.D. Lee, *Adv. Inf. Storage Syst. 4*:79 (1992).
11. C. Mizumo et al., Fuji Photo Film Co., U.S. Patent 4,946,615 (1990).
12. M. Kamiwano et al., Inoue Mfg. Co., U.S. Patent 4,919,347 (1990).
13. M. Kamiwano, M. Kaminoyama, K. Arai, Y. Inoue, and K. Masi, Paper presented at the Topical Conference on Mixing, AIChE 1992 Annual Meeting, Miami, November 1992.
14. M.S. Chagnon and R. Donadio, in *Surfactants in Emerging Technologies* (M.J. Rosen, Ed.) (Surfactant Sci. Ser., Vol. 26), Marcel Dekker, New York, 1987, p. 67.
15. S. Hendrix and F.P. Lootz, Agfa Gavaert, Br. Patent 1 080 614 (1967).
16. J.P. Talley, Ampex Corp., U.S. Patent 3,144,352 (1964).
17. S.A. Weber, Ampex Corp., U.S. Patent 3,647,539 (1972).
18. R. Kaiser, U.S. Patent 3,784,471 (1974).
19. R. Rosensweig, U.S. Patent 3,917,538 (1975).
20. W. Bottonberg and M.S. Chagnon, U.S. Patent 4,315,827 (1982).
21. J.H. Hildebrand and R.L. Scott, *The Solubility of Nonelectrolytes*, 3rd ed., New York, 1949; *Regular Solutions*, Prentice-Hall, Englewood Cliffs, N.J., 1962.

13

Dispersions as Pharmaceutical Dosage Forms

DUNCAN Q. M. CRAIG Centre for Materials Science,
The School of Pharmacy, University of London, London, England

I.	Principles of Drug Delivery	457
II.	Controlled Release and Drug Targeting	459
	A. The biological fate of colloidal particles	462
	B. The use of antibodies in drug targeting	469
III.	Examples of Colloidal Systems Used in Controlled and Targeted Drug Delivery	471
	A. Microspheres and microcapsules	471
	B. Nanoparticles	473
	C. Liposomes	475
	D. Nonionic surfactant vesicles (niosomes)	483
	E. Biological microcapsules	484
	F. Magnetic particles	487
IV.	Conclusions	488
	References	489

I. PRINCIPLES OF DRUG DELIVERY

The biological activity of a drug substance will be primarily dependent on the nature of the interaction between that drug and the target tissue. However, in order to exert that effect, the drug must be present at the target site in quantities sufficient to elicit the desired response; hence it is essential that the drug be delivered to the appropriate site within the body at the appropriate rate. This can be controlled by the way in which the active substance is prepared (formulated) prior to administration as a medicine. For example, the same quantity of drug may have a different effect when given as an oral solution than when given as a solid tablet due to the differences in the rate at which the drug is absorbed from the gut in these two instances. The preparation of medicines from pure drugs is therefore a wide and important field within the pharmaceutical sciences, incorporating elements of engineering and materials science, physical chemistry, polymer science, solution chemistry, and biochemistry, among others.

All dosage forms contain substances (excipients) that while not biologically active themselves, impart properties to the medicine that improve the product manufacturing process and, more important, alter the manner in which the drug interacts with the body. For example, disintegrants are added to tablet formulations to ensure that the dosage forms break up in the body, allowing rapid dissolution and absorption of the drug. In the case of liquid dosage forms, solvents are obvious examples of excipients, while surface-active agents, buffers, and viscosity-modifying agents are also frequently used.

A large number of dosage forms involve the dispersion of a drug or particles containing drug in a liquid medium; hence it is essential for the formulator to have a knowledge of the science and technology of dispersions. Indeed, the use of disperse systems in the pharmaceutical sciences is so widespread that a thorough review of every aspect would be impractical. Instead, this chapter will give a general discussion of the use of disperse systems as drug carriers, particularly with a view to controlled drug release and drug targeting.

When a drug is introduced into the body, it must cross a series of lipid membranes in order to reach the site of action. These membranes are composed largely of phospholipids and cholesterol, with a number of incorporated proteins. The mechanism of transport through biological barriers is still a matter of debate, but in general it is believed that most (but by no means all) drugs cross these membranes via passive diffusion. While drugs can be given by a number of routes of administration, the oral route is by far the most frequently used, as it is convenient for the patient and allows the satisfactory absorption of a wide range of medicaments. As the gastrointestinal (GI) tract is responsible for the absorption of nutrients from food, this organ has a number of structural features that can be exploited for the absorption of drugs. For example, the microvilli of the small intestine provide a large surface area over which absorption may occur, while additional mechanisms involving emulsification via bile salts are thought to be involved in the absorption of some lipid-soluble drugs. However, the GI tract also metabolizes ingested material; hence drugs administered by this route must be stable to the changes in pH and the presence of enzymes encountered in the gut. Many drugs, particularly proteins, cannot be given orally, as the quantity of intact drug absorbed across the gut membrane is too small and inconsistent to be of benefit to the patient; hence these drugs must be given by an alternative route. Furthermore, blood leaving the gut tissue is carried directly to the liver via the hepatic portal vein, where many drugs are extensively metabolized before entering the venous circulation.

As a result of these and other difficulties, many drugs must be given via alternative routes. Examples include parenteral administration (generally synonymous with any route involving injection), topical administration, inhalation, rectal administration, and, for a few drugs, intranasal administration. Parenteral administration is particularly important, as it bypasses many of the problems associated with oral drug delivery. The distribution of injected drugs will depend on the route of administration. Drugs injected into the venous circulation are directed first to the heart and then to the lungs, where the blood flows through the alveolar capillary bed before returning to the heart in an oxygenated state, after which the drug enters the arterial system. Injection directly into the arterial system results in the drug being directed to the capillaries of whichever organ is in the path of the circulation. It should be mentioned here that although intraarterial administration is often cited as a method of directing drugs to specific organs, there are a number of difficulties associated with this route. Specifically, the depth of the arteries within the body, the high pressure of the blood in these vessels, and the considerable risks involved in administering the injection make intravenous, intramuscular, or subcutaneous administration preferable in the vast majority of cases. A summary diagram of the circulation is given in Fig. 1, showing the general routes involved and the relationship between the arterial, capillary, and venous systems. Once the drug is in the capillary beds, its diffusion into the tissues is possible, leading to the biological response.

II. CONTROLLED RELEASE AND DRUG TARGETING

While the majority of drugs can be given via conventional dosage forms (e.g., tablets, capsules, injections of solutions), there are a number of occasions when these approaches are inappropriate. In particular, conventional dosage forms often involve repeated administration in order to maintain the drug concentration in the blood at a sufficiently high level for a therapeutic effect. This frequent administration may result in inconvenience for the patient, leading to poor compliance with the prescribed dose regimen. Furthermore, repeated administration may result in considerable fluctuations in blood levels between doses, sometimes causing either toxic side effects or an inadequate response to the drug when blood levels are at

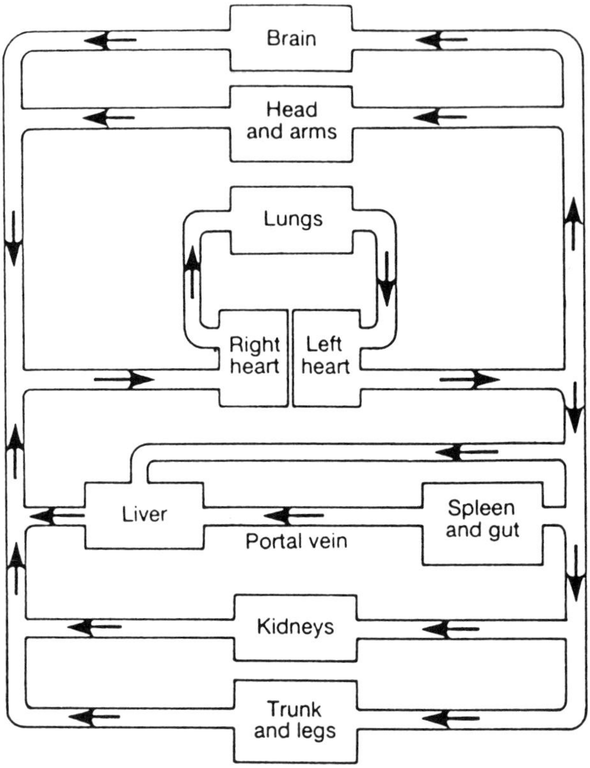

FIG. 1a Summary of principal circulatory pathways. (From Ref. 1.)

a maximum or minimum, respectively. One method of overcoming these problems has been the development of controlled release systems, whereby the drug is released from the dosage form at a constant (or near-constant) rate over a prolonged period of time. These dosage forms may be given by the oral route in devices such as polymeric matrices from which the drug diffuses slowly or from spheronized pellets coated with an insoluble polymeric film. Alternatively, controlled-release systems may be given by injection, for example as aqueous particulate systems for intravenous administration or as oily suspensions for intramuscular use. Controlled-release preparations have also been designed for administration via the nasal, vaginal, and topical routes as well as via the lung.

A related approach is the development of systems for drug targeting. Although it is desirable for drugs to diffuse into the tissue of interest, a significant proportion will be distributed to other regions of the body,

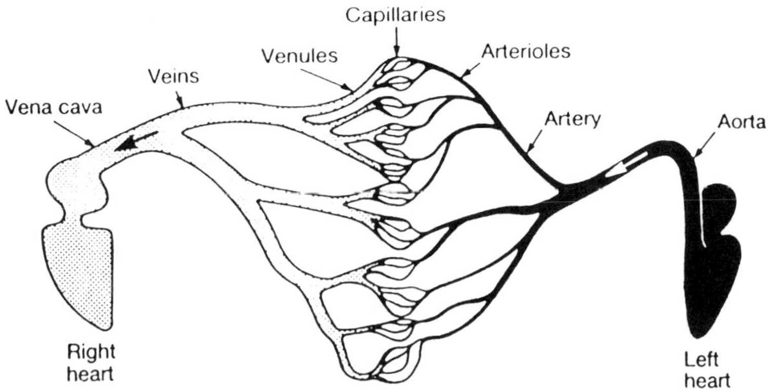

FIG. 1b Diagram showing the relationship between the blood vessels of the circulation. (From Ref. 1.)

thereby raising two difficulties. First, the drug may act on organs other than the target tissue, increasing the probability of adverse effects. Second, the concentration of the drug at the site of action may be low owing to unfavorable distribution characteristics or poor absorption. It is therefore preferable to target the drug to the tissue of interest rather than to allow free circulation in the bloodstream. Furthermore, it may be necessary in some cases to target not merely the organ but specific cells or even intracellular sites within that organ.

A particularly pertinent example of the necessity for drug targeting is provided by the treatment of solid tumors. The size of an organ and the number of cells it contains are governed by a number of mechanisms. However, for reasons that are not yet fully understood, these control mechanisms may become deficient, causing an abnormal proliferation of cells. The resulting tumor may be treated by surgery, radiation therapy, or drug therapy, with other techniques such as the application of electromagnetic energy and gene therapy also being explored. Drug therapy attempts to exploit the differences between cancerous and healthy cells, thus allowing selective toxicity toward tumors. However, due to the intrinsic similarity between these cells it is often only possible to exploit the rapid rate of cell division found in tumors rather than to design drugs that recognize specific biochemical features that are not found in healthy tissue. These drugs, known as cytotoxic agents, kill cells that are in a state of rapid division. Unfortunately, they are also toxic to rapidly dividing healthy cells, particularly those in the intestine, the hair follicles, the bone marrow, the testes, and the fetus. Furthermore, a particular problem with solid tumors is that the drug must be present in sufficiently high concentrations to penetrate the tumor. Tumors are generally not recognized until they are several centimeters in diameter, by which time they will have formed a dense core surrounded by a more permeable outer layer. Penetration of the drug into the core may be highly impeded, rendering drug treatment of solid tumors difficult [2]. One method of overcoming these difficulties has been to inject the drug directly into the tumor itself, although problems may arise due to rapid diffusion of the drug away from the tumor site into the general circulation [3]. An alternative approach is therefore to develop methods of specific targeting that will maximize the concentration of the drug in the proximity of the tumor.

Both targeting and controlled release of drugs can be achieved by appropriate use of carrier systems, either to deliver the drug to a specific area of the body or to retard the release of the drug. Disperse systems have been used extensively for both of these purposes, particularly for targeting. These dosage forms are loosely termed colloidal drug delivery systems and are composed of aqueous suspensions of particles containing drugs, usually administered by injection. The requirements for colloidal systems used for targeting have been discussed by Davis and Illum [4] and can be summarized as follows. The drug should selectively accumulate at the required site of action, with residual concentrations in other tissues or in the circulation being very small. The carrier particles must be capable of carrying a suitable quantity of the drug, for if they do not, the dose required may be too large for practical administration. The drug should also be released at the appropriate rate at the site of action, although this is often difficult to measure in practice and is usually assessed either in vitro or indirectly in vivo, for example by measuring blood levels of the drug. The colloid should be stable in vitro and in transit to the required site in vivo. The system should show minimal toxicity and immunogenicity and must also be biodegradable to prevent accumulation of the carrier. Finally, the system must be relatively simple to produce on a large scale. A further requirement that could arguably also be included is that the materials used in the colloid should preferably already have been recognized as nontoxic by licensing authorities such as the Committee for the Safety of Medicines in the United Kingdom and the Food and Drug Administration in the United States. If a new carrier material is used, extensive (and hence expensive) toxicity tests will be required for that excipient in addition to those required for the drug.

A. The Biological Fate of Colloidal Particles

The colloidal systems described in this chapter are administered primarily by injection, as particles are generally not absorbed intact from the GI tract (although the validity of this statement will be discussed later). The fate of colloidal particles in the bloodstream depends on the size and surface characteristics of the particle and the route of administration [4–6]. The effect of particle size on the distribution after intravenous administration is summarized in Fig. 2.

Particles greater than approximately 7 μm in diameter will accumulate in the lung, as this organ contains the first capillary network that the particles encounter. The small diameter of the lung capillaries results in physical entrapment of these larger particles. At smaller particle sizes, physical entrapment is unlikely, and the principal mechanism of their removal from the circulation is via the immune system. The processes associated with this removal represent an interesting fusion of biology with colloid chemistry and will be discussed in some detail.

The human immune system comprises a highly complex and interactive set of processes designed to protect the body from potentially harmful foreign matter. The cells and tissues responsible for the physical removal of particles from the circulation were formerly known as the reticuloendothelial system (RES). The term was originally introduced by Aschoff [8] after studies showed that certain dyes were taken up from the bloodstream into phagocytic mononuclear cells distributed in a number of tissues within the body. Phagocytosis is a process whereby foreign particles are taken into a cell by inclusion within a vacuole, as shown in Fig. 3. Once incorporated, the foreign matter is taken to the lysosomes within the cell, these being organelles that contains a number of degradative enzymes. The particle is then broken down and the digested material is excreted from the cell. The term "reticuloendothelial system" originated from the belief that the cells involved were endothelial and were associated with fibers composed of reticulin. It is now known that neither of these criteria is valid, the cells involved being tissue-bound white blood cells (macrophages), and the term "mononuclear phagocyte system" (MPS) has therefore been suggested as an alternative [9]. Although this description is considerably more accurate, the phrase reticuloendothelial system is still widely used.

The principal tissues associated with the MPS are the liver, the bone marrow, and the spleen, of which the liver is usually the most important. For example, Bradfield [10] has shown that after sheep red blood cells were intravenously injected into mice, 90% were taken up by the liver, 5% in the spleen, and a smaller percentage in the bone marrow. Within the liver, the Kupffer cells are primarily responsible for particle uptake by phagocytosis, although hepatocytes and endothelial cells also play a role [11]. Although macrophages are also located in the lungs, they tend to be associated with the alveolar membrane at the air–membrane interface rather than in the alveolar capillaries; hence this does not tend to be an important site for circulating particle uptake.

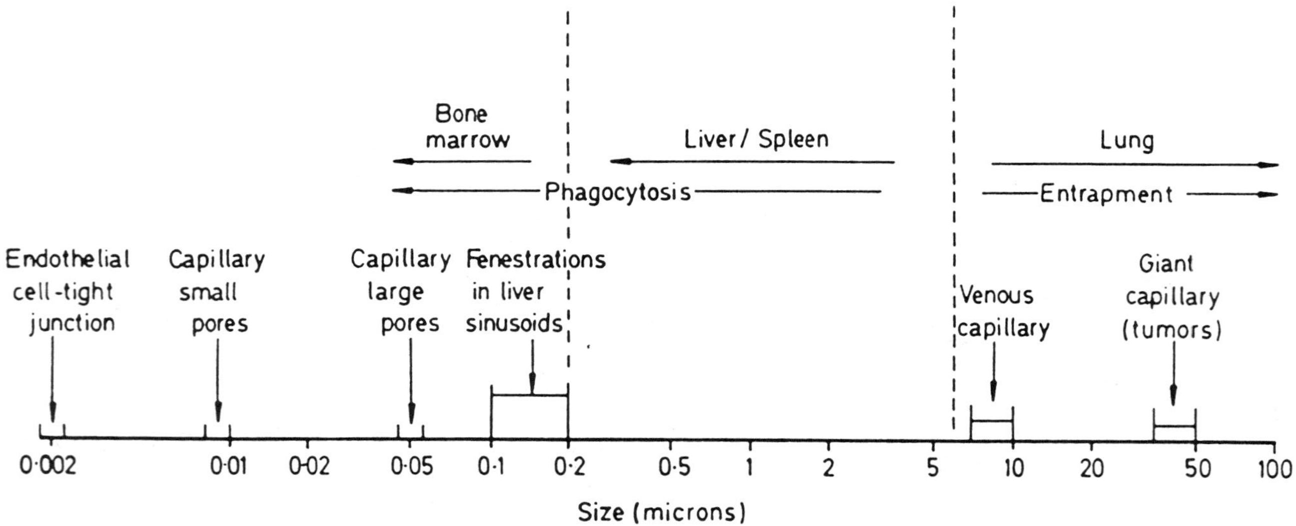

FIG. 2 The effects of particle size on the biological fate of injected particles. (From Ref. 7.)

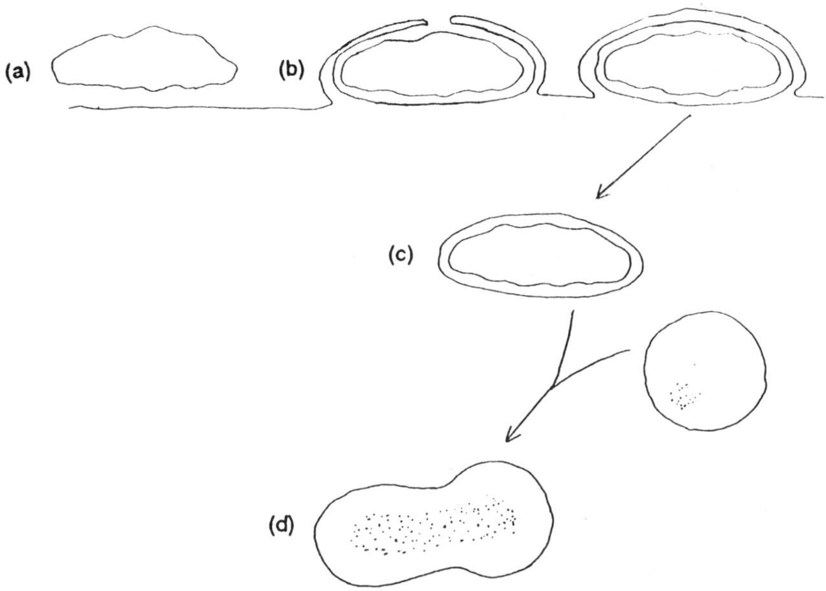

FIG. 3 Diagramatic representation of the phagocytosis process. (a) Particle in contact with phagocyte surface; (b) phagocytic membrane extends to surround particle; (c) particle enters cytoplasm of phagocyte; (d) fusion with primary lysosome.

The MPS presents a major barrier to the use of colloidal drug delivery systems, as any injected colloid will rapidly be removed from the circulation and digested. For example, a study by Illum and Davis [12] has indicated that up to 90% of a particulate dose may be removed from the circulation within 1 min of injection. This can be utilized in the treatment of diseases whereby cells of the MPS system are to be targeted. For example, leprosy, tuberculosis, and leptospirosis all involve infection of macrophage cells, so it is theoretically possible to target these cells by incorporating drugs into colloidal particles that will be taken up by the diseased macrophages. It has also been suggested that immunomodulators, which stimulate the macrophages into an active state, may be incorporated into colloids in order to cause an antitumor effect, as by having a more active macrophage population the resistance to the spread of tumor cells could be increased [13].

A further example of targeting to the MPS is given by the treatment of leishmaniasis, a widespread protozoal disease involving extensive infection of the MPS. Alving et al. [14] demonstrated that delivery of the drug meglumine to infected hamsters in a colloidal system resulted in a marked improvement in response compared to administration of the drug alone.

This can be attributed to targeting of the drug to the MPS as a result of the normal colloidal uptake mechanism, giving higher local concentrations of meglumine than result from simple intravenous injection of the drug alone.

While the treatment of these and other diseases via colloidal systems is a very interesting approach, the MPS remains a serious barrier to the targeting to any tissues other than those associated with the immune system itself. One way of bypassing the MPS is to suppress the immune system; then injected colloids will remain in the circulation for prolonged periods of time, although the suppression of the immune system is in itself a dangerous course of action due to the lowering of patient resistance to infection [15]. However, interest has been generated by the observation that when the MPS is partially suppressed or the colloid characteristics altered appropriately, small particles may be taken up by circulating white blood cells and delivered to the lungs [16]. This presents possibilities for targeting drugs to the lungs for the treatment of cancers and other local conditions.

The answer to the avoidance of the MPS ultimately lies in understanding the mechanisms involved in the uptake process. This topic has been studied extensively, and some exciting breakthroughs have been achieved in using this knowledge in the design of more effective colloidal carrier systems. A brief summary of the principal mechanisms will therefore be given here, although a number of articles and reviews are available for further information [17–20].

When a colloidal particle is injected into the bloodstream, components of the blood will absorb onto the particle surface, a process known as opsonization. The substances that are responsible for this process (opsonins) are macromolecules and include immunoglobulin G, complement c_3b, and fibronectin [21]. The mechanisms involved are complex, but in general flexible molecules such as nonglobular proteins and polysaccharides are believed to adhere to the particle by multipoint attachment, leading to a large number of comparatively weak bonds per molecule. Desorption is therefore difficult, as this would involve simultaneous detachment from all the contact points. More rigid molecules such as globular proteins are believed to undergo structural rearrangement upon adsorption onto the circulating particle, possibly by breaking van der Waals bonds between hydrophobic moieties within the molecule. This bond breaking increases the flexibility of the protein as a whole, thereby also increasing the entropy of the system and rendering the adsorption process effectively irreversible.

After opsonization, the particles are recognized by the MPS, and phagocytosis into macrophages occurs. It is generally believed that although it is difficult (but not impossible) to prevent uptake of the opsonized particles, it is possible to prevent the opsonization process itself by altering the surface characteristics of the particles. The recognition process may involve the size, surface charge, and chemical nature of the colloid [22,23]. Particles within the size range 300 nm to 2 μm tend to localize in the Kupffer cells of the liver, while particles of diameter 100–300 nm tend to localize in the hepatocytes [24]. Illum et al. [6] have suggested that negatively charged particles may be preferentially taken up in the liver, while positively charged particles show an initial accumulation in the lung. However, they point out that the charge on the particle prior to injection will not necessarily be the same as the charge on the circulating particle once opsonization has taken place; hence caution is required when interpreting these results. Indeed, as it has been suggested that all colloids become negatively charged once injected, there is some confusion over the role of initial particle charge in the opsonization process [25]. In general, particles with hydrophobic surfaces tend to be taken up at a greater rate than particles with hydrophilic surfaces [26]; hence one of the effects of the opsonization process may be to render the particles more hydrophobic and therefore more likely to be taken up by the MPS.

This information suggests that if the surface characteristics of the particles are altered, the distribution will likewise be affected. This has indeed been found to be the case. For example, particles treated with gelatin become sensitive to opsonization by fibronectin [27]. Studies on red blood cells showed that treatment with glutaraldehyde leads to opsonization by immunoglobulin G, whereas treatment with *N*-ethylmaleimide leads to opsonization by complement c_3b [28]. A number of other studies have shown the effect of surface characteristics on the opsonization and uptake process [29–33]. An interesting further development is the use of surface-active agents to coat colloidal particles, as not only can this render the particles more hydrophilic and hence less likely to be recognized by the MPS, but it may also reduce particle–macrophage adhesion via steric stabilization. Illum et al. [32] showed that coating polystyrene microspheres with surfactants (poloxamers, poloxamine, and xanthan) could inhibit in vitro uptake of particles by mouse peritoneal macrophages. In vivo studies [33] confirmed that tissue distribution of microspheres could be radically altered, depending on the surfactant used. This therefore presents the possibility of controlling the fate of colloidal particles by the use of appropriate coating techniques.

A further problem associated with drug targeting is that even if the MPS is avoided, the particles must then be able to leave the circulation in order to enter the target organ. Clearly, particles of the sizes used for drug delivery cannot simply diffuse through capillary membranes in the same manner as small, individual molecules can. Instead, the particles must escape the circulation, a process known as extravasation. The possible sites for this process are summarized in Fig. 4. Capillaries can be considered to fall into

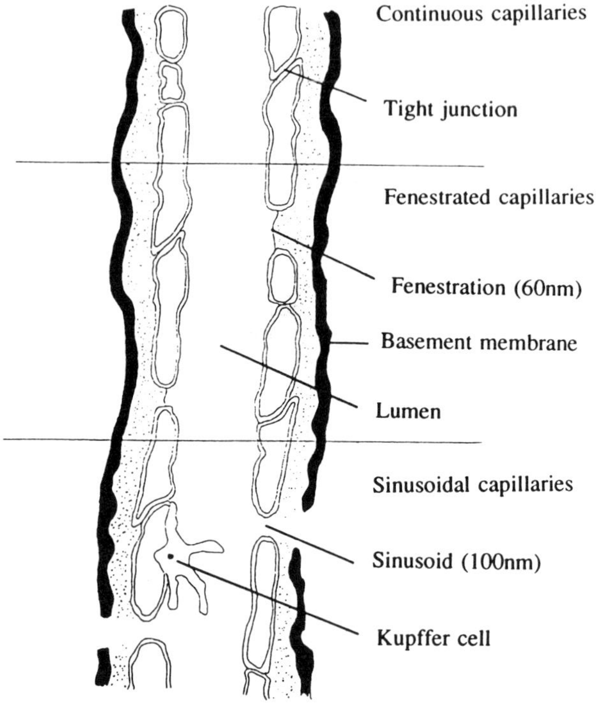

FIG. 4 Diagramatic representation of the lining of a blood vessel. (Adapted from Ref. 4.)

three categories [34]. First, continuous capillaries, which consist of tightly packed cells, underneath which is a continuous basement membrane. There is therefore no opportunity for extravasation from these beds unless they have been ruptured by mechanical trauma. Second, fenestrated capillaries, which have continuous basement membranes; the capillary cells themselves, however, have small discontinuities (fenestrae) of approximately 50–60 nm. Particles may therefore be able to pass through these holes but are unlikely to pass through the membrane itself. Third, sinusoidal or discontinuous capillaries. These, which are found in the liver, spleen, and bone marrow, contain gaps of approximately 100 nm that pass through both the capillary cells and the basement membrane. Extravasation is therefore possible in these regions, although the above considerations raise the question as to whether there is any point in avoiding the MPS if the only sites for extravasation are the same as those to which the MPS takes the particles in any case.

However, two further points should be stated with respect to the above argument. First, if the particles are to be retained by physical entrapment in the capillary network of an organ followed by slow release, extravasation is not a problem, as the drug will diffuse across the capillaries while the particles themselves will remain trapped in the blood system. Second, the above description refers to healthy tissue. Considerably less is known concerning the structure of capillaries located within tumors [35–37]. The size and integrity of capillaries appear to be highly variable in these instances, while the high interstitial pressure and density of solid tumors may counteract the diffusion of drugs into the tissues.

B. The Use of Antibodies in Drug Targeting

Although a review of immunology is well outside the scope of this discussion, the use of antibodies and antigens in drug targeting is so widespread that a brief description of the principles involved is necessary. As mentioned in the previous section, the body possesses a number of mechanisms to counteract the harmful effects of foreign matter such as bacteria and viruses. These mechanisms can be broadly classified into two categories, nonspecific and specific paths. The nonspecific path involves mechanical mechanisms such as coughing, the presence of secretions that help destroy bacteria such as enzymes in tears, cells capable of destroying foreign matter such as the macrophages described earlier, and a group of macromolecules that circulate in the bloodstream such as interferon and the constituents of the complement system. The second set of mechanisms, the specific path, are concerned with the specific recognition of foreign molecules known as antigens. This is achieved via the humoral system, which consists of circulating proteins known as immunoglobulins or antibodies, which attach to the antigens, and the cell-mediated system, which involves the activity of cells (T lymphocytes) that are capable of destroying antigenic material.

An antigen can be defined as a substance that is capable of reacting specifically with antibodies or T lymphocytes or alternatively as a substance that can stimulate the formation of antibodies and the activation of the cell-mediated system. The area of the antigen molecule that specifically reacts with the antibodies is known as the antigenic determinant. For example, human albumin has at least six different determinants, which means that six antibodies of different specificities may be produced after injection into a different species. The basic antibody unit consists of two heavy chains and two light chains, as shown in Fig. 5.

The Fab portion of the molecule binds specifically to the antigen, while the Fc portion serves other functions such as determining the distribution of the antibody around the body and the interaction with the complement

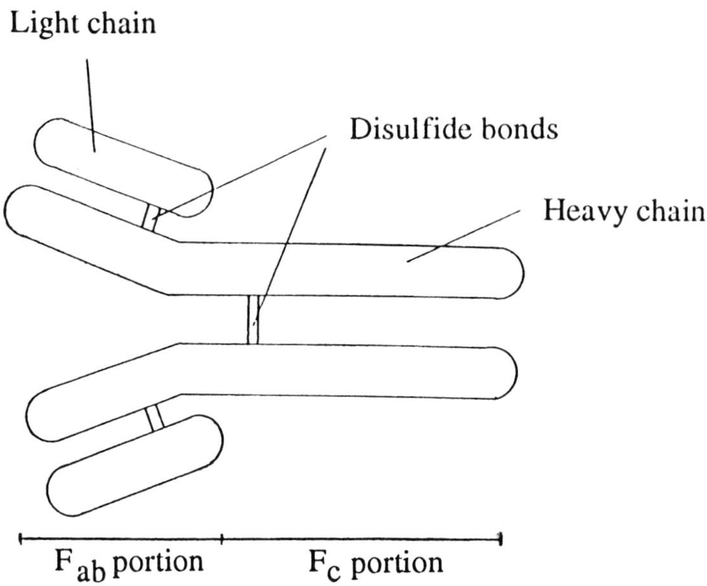

FIG. 5 Basic structure of an antibody.

system. There are a number of different types of antibodies, which have a range of binding sites and hence a range of functions. The antibodies (particularly IgM) can act as binding agents, causing agglutination of the antibody–antigen complexes, or alternatively may act as adjuvants for other mechanisms of the immune system, for example by activating the complement system once attached to the antigen.

This knowledge is of great relevance to drug targeting, particularly for the treatment of cancer, as when cells become malignant, new surface antigens may appear. In untreated patients, the tumor may reach such proportions that the normal immune system is unable to cope despite recognizing the cancer cells. However, the presence of these antigens provides a means of differentiating between healthy and cancerous tissue, as it is possible to produce antibodies specific to those tumor cells. The antibodies can then be bound to drugs or drug carrier systems, usually by inducing a direct reaction between the drug or carrier and the antibody or by incorporating a spacer molecule between the two species. This approach has been used extensively for drug targeting and will be mentioned at various points in the subsequent discussion.

III. EXAMPLES OF COLLOIDAL SYSTEMS USED IN CONTROLLED AND TARGETED DRUG DELIVERY

A. Microspheres and Microcapsules

"Microspheres" and "microcapsules" are general terms for colloidal drug delivery systems in which a drug is incorporated into a polymeric sphere in the micrometer size range. Strictly speaking, the difference between the two categories lies in the manner in which the drug is dispersed within the system. In the case of microspheres, the drug is present as a dispersion throughout the particle, whereas in microcapsules the drug is present as a single mass surrounded by a polymeric barrier. In practice, however, the two descriptions are often used interchangeably, and for simplicity the term "microsphere" will be used here.

A number of methods for preparing microspheres have been suggested, including emulsion polymerization and heating a solution of the drug and carrier in the presence of cross-linking agents [38]. The carriers are usually biodegradable in order to avoid accumulation in the body. Examples of materials that have been used include gelatin, ethylcellulose, albumin, polylactides, and lactide/glycolide copolymers.

A large number of studies have been conducted in which the microspheres have been used to release drugs at a controlled rate or to target particular organs or both; hence only a brief discussion of selected examples will be given here. Several examples of the use of microspheres for controlled release have been given. For example, Ratcliffe et al. [39] used albumin microspheres to deliver steroids to the knee joint for the treatment of arthritis. When the colloid is administered via intraarterial injection, the particles are taken up by the macrophages in the synovial cavity and transported into the synovium, thus eliminating the need for painful injections directly into the knee joint. The steroids are released slowly in order to give prolonged relief, and the spheres degrade to albumin monomers that are found naturally in the body. Similarly, Nakono et al. [40] used polylactic acid and polyethylene carbonate microspheres to release local anesthetics over a prolonged period of time by subcutaneous injection.

A further area in which microspheres have been used extensively has been the delivery of proteins. It is known that a number of tumors are dependent on the presence of hormones, which are themselves peptides or proteins; hence it is possible to treat certain cancers by reducing blood levels of a particular hormone. Hormone levels are controlled by a complex set of feedback mechanisms whereby one will control the release of another. An example of this is luteinizing hormone-releasing hormone (LHRH), which acts indirectly to suppress progesterone levels in women and testosterone levels in men; tumors that are dependent on the presence of these two substances can be treated by administering a drug that acts in the same manner as LHRH (an agonist). Sanders et al. [41] described the use of lactide/glycolide copolymer microspheres containing the LHRH agonist nafarelin, whereby the spheres allow sustained release of the drug over a prolonged period of time.

The treatment of tumors using microsphere systems provides a good example of the diversity of approaches that have been adopted. Targeting to lung tumors by incorporating drugs in large (>10 μm) albumin microspheres has been reported [42]. Intravenous administration of particles this size leads to selective uptake in the lung, although problems were reported as there appeared to be a rapid initial release of the drug in the lungs (burst effect), which could result in damage to healthy tissue. Several studies have examined intraarterial drug delivery for targeting to tumors [43]. Injection of starch microspheres containing adriamycin into the hepatic artery has been used to treat tumors in the liver [44]. Measurement of the venous concentration of the drug (i.e., the concentration of drug escaping the liver) indicated that the majority of the drug was being retained in the liver tissue, thus allowing organ-specific delivery of potentially highly toxic drugs.

An interesting variation on the treatment of cancers has been the development of chemoembolization techniques. These involve cutting off the blood supply to a tumor by blocking the capillaries leading to that tissue. This approach was investigated by Arfors et al. [45], who studied the occlusion (blockage) of blood vessels within intestinal tissue in the pig. Lindell et al. [46] examined the effects of occluding blood flow to the liver and demonstrated that blood flow to liver tumors appeared to be reduced to a greater extent than was found for healthy liver tissue. This therefore presents the possibility of using microspheres to cause tissue necrosis by permanent occlusion or else to deliver drugs by using spheres of an appropriate size range [47–49]. Chemoembolization has also been used in the treatment of lung cancer. Ionizing radiation can be used in the treatment of this condition, although this approach may also cause extensive damage to healthy tissue. However, the susceptibility of healthy tissue to ionizing radiation decreases when that tissue is starved of oxygen (a condition known as ischemia). Rothman et al. [50] demonstrated that it is possible to cause temporary ischemia in healthy lung tissue by injecting biodegradable starch microspheres that are not taken up in the tumor. In this way, the healthy tissue can be selectively protected from the ionizing radiation, thus reducing the adverse effects of the treatment.

There are a number of further uses for microspheres, including their use as adjuvants for vaccines. These drugs stimulate the immune system in order to afford protection against certain disease states. Encapsulation of vaccines within microspheres may increase the immunogenic effect, thereby rendering them more effective [51,52]. Furthermore, microspheres can be used for the protection and controlled release of high molecular weight soluble proteins, thus removing the necessity of multiple injections for substances that are normally rapidly metabolized on intravenous injection [53].

There are, however, problems associated with the use of microspheres. For example, a number of studies have reported that the microspheres are themselves immunogenic, causing an immune response that could endanger the patient. The mechanisms involved appear to be complex; a study by Artusson et al. [54] using polyacrylic starch microspheres showed that there was no immune response to the spheres alone. However, after incorporating the protein bovine serum albumin, an immune response was noted for both the protein and the spheres; hence the presence of the drug was sensitizing the body to the spheres.

Other effects associated with the physical properties of the spheres can also cause problems. The emulsion polymerization process must be carefully controlled in order to achieve spherical, monodisperse particles. Bissery et al. [55] reported that variations in droplet size and particle–particle interactions during the solvent evaporation stage of polylactide-poly(β-hydroxybutyrate) microsphere manufacture can lead to the formation of irregular clumps of particles, which caused mice injected with them to die of asphyxiation, presumably due to blockage of the respiratory capillaries.

Despite these difficulties, microspheres are still attracting considerable interest as a means of delivering a number of drugs. One aspect of this approach that has caused considerable discussion has been the use of starch microspheres to deliver drugs, particularly proteins, via the nasal mucosa [56]. The nasal mucosa is comparatively permeable, and it is well established that a number of drugs can cross this membrane, including peptides such as vasopressin. However, a problem has been to find a means of overcoming the rapid removal of foreign particles from the nose via the mucociliary system. This problem may be overcome by using bioadhesive microspheres that form gel-like structures when in contact with mucus. For example, albumin, starch (Spherex), and DEAE-dextran (DEAE-Sephadex) microspheres were shown to prolong clearance times from the nasal mucosa [57]. It has also been suggested that incorporating agents that enhance drug penetration through the mucosa may be of use [58,59], although the safety of some penetration enhancers is a matter of debate.

B. Nanoparticles

Nanoparticles are colloidal drug delivery vehicles having a diameter between 200 and 500 nm. They are therefore related to microspheres, the difference lying in the smaller size. These systems were originally described by Speiser [60,61] and were designed to be biodegradable, nontoxic, and relatively stable on storage. A variety of materials have been used for their manufacture, including natural polymers such as proteins and celluloses (e.g., bovine serum albumin, ethylcellulose, casein, and gelatin) as well as synthetic polymers such as polymethyl methacrylate (PMMA).

A number of manufacturing methods have been described in the literature. For example, nanoparticles are manufactured by a coacervation method, whereby use is made of the phase behavior of ternary systems containing the polymer in question. In the case of gelatin nanoparticles, a solution of gelatin in water is prepared and the active ingredient added. The system is then desolvated by adding a solvent-competing solute such as sodium sulfate. By controlling the process such that desolvation occurs just outside the coacervation region, it is possible to manufacture colloid-sized particles rather than larger aggregates [62]. A surfactant may also be included to solubilize the drug, and the resulting nanoparticles can be freeze-dried and reconstituted when required. Alternatively, in situ polymerization of alkylcyanoacrylate monomers can be achieved by adding the monomer to aqueous solutions of surfactant and shaking vigorously under controlled pH conditions or by exposure to radiation, leading to the generation of nanoparticles [63,64].

Drug entrapment within nanoparticles can be achieved by adding the drug to the polymerization medium or to the suspension of formed nanoparticles, whereby the drug is incorporated by adsorption and absorption. For example, Kreuter [65] studied benzocaine-PMMA systems and reported a nanoparticle/water partition coefficient of 400, indicating a very favorable uptake of this drug. The distribution of nanoparticles shows a pattern similar to that reported for other colloidal systems. Investigations into the distribution of ^{14}C-labeled PMMA nanoparticles showed approximately 60% to be taken up by the liver, with the remainder being located in the bone marrow, spleen, and lungs [66].

A number of applications for nanoparticles have been described. They have been used as diagnostic agents, carrying radioisotopes to various organs to allow imaging. One of the most common preparations for imaging the liver is technetium-99m–sulfur colloid that has a particle size of 1 μm [67]. As an alternative, gelatin nanoparticles containing ^{99m}Tc have been described [68], although the poor specificity of liver uptake was disappointing. This again relates to the question of which factors determine the distribution of particles around the body. Work has also been conducted on treating the surface of gelatin and albumin nanoparticles with fluorescein isothiocyanate. Oppenheim [68] suggested that some tumor lines appear to preferentially take up such treated nanoparticles, thus allowing the possibility of drug targeting. Similarly, Marty et al. [69] incorporated nitroxynil into nanoparticles for the treatment of liver parasites (flukes), again taking advantage of the tendency of nanoparticles to accumulate in the liver.

Perhaps the most exciting application of these particulate systems lies in the possibility of their being transported across the gut wall intact after oral administration. Studies have indicated that latex particles of diameter 200 nm may cross through the GI tract across gut-associated lymphoid tissue known as Payers patches [70]. Jani et al. [71–73] recently lent support to this hypothesis by demonstrating that polystyrene latex particles of sizes from 50 nm to 3 μm could indeed be taken up across the Payers patches, with the smaller particles showing the greatest uptake. The implications for this work in terms of transporting proteins through the gut are considerable.

C. Liposomes

Liposomes are phospholipid vesicles containing aqueous compartments between one or more concentric lipid bilayers. These structures were first discovered by Bangham et al. [74], who found that when an aqueous medium was added to a dry phospholipid film, spherical bilayered structures were formed. These vesicles are known as multilamellar liposomes (MLVs), and a schematic representation of the structure is given in Fig. 6. These vesicles rapidly generated considerable interest as potential artificial membranes because of their chemical and physical resemblance to naturally occurring membranes. A number of related applications have been de-

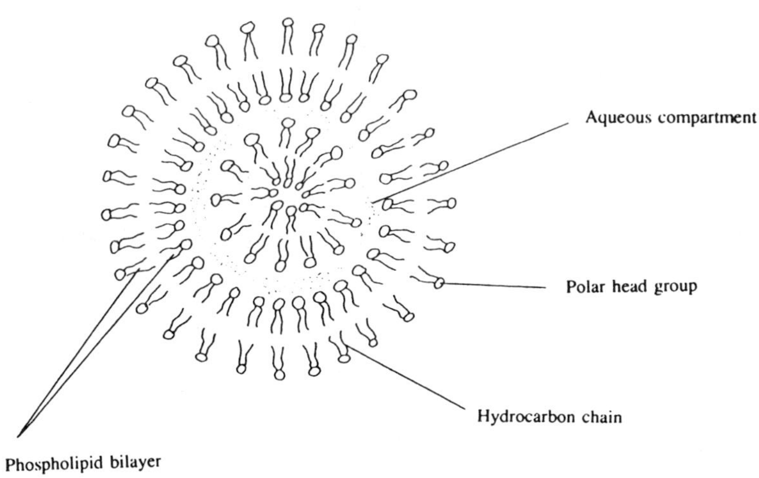

FIG. 6 The structure of a multilamellar liposome.

scribed [75], including the examination of transport mechanisms across membranes, lipid–protein interactions, and the function of membrane enzymes.

Liposomes also attracted interest as potential drug delivery vehicles as it soon became apparent that liposomally entrapped drugs were not distributed and metabolized by cells and tissues in the same manner as the free drugs. Other pertinent factors include the comparative ease of production (at least at a bench level), the reduced risk of toxicity compared to other vehicles because phospholipids occur naturally, and the possibility of entrapping both aqueous and nonaqueous drugs within the liposome. It is now established that water-soluble drugs are retained within the aqueous compartments, whereas lipid-soluble drugs become associated with the lipid bilayer [76,77]. Furthermore, entrapping a drug in a liposome may protect it from a hostile environment such as one that contains degradative enzymes. As will be discussed, the possibility of using these vesicles to target drugs and to release drugs at a controlled rate has proved particularly interesting. The number of publications and patents relating to liposomal products is far too great to attempt to fully cover the field here; therefore this discussion will outline the general principles of liposome behavior and will highlight some of the more important advances made in this field.

The natural phospholipids commonly in use as liposomal materials include egg phosphatidylcholine, phosphatidylserine, phosphatidylglycerol, and sphingomyelin. Synthetic phospholipids include dipalmitoylphosphatidylcholine (DPPC), distearoylphosphatidylcholine (DSPC), and dimyristoylphosphatidylcholine (DMPC). In general, synthetic phospholipids are preferred due to the instability of many natural phospholipids and the difficulties in obtaining samples of standard quality.

Since the introduction of MLVs, a number of further liposome types have been described. These include small unilamellar vesicles (SUVs), which may be produced by sonication of MLVs [78], by injecting an ethanolic solution of phospholipid into the aqueous phase [79], by dialyzing a detergent-containing phospholipid solution [80,81], or by adjusting the pH of the aqueous medium containing the liposomes [82]. MLVs tend to be in the size range 100 nm to 10 μm, while SUVs are 20–100 nm in diameter. Large unilamellar vesicles (LUVs) can also be produced by the injection method using suitable solvents [83], these vesicles being 70–190 nm in diameter.

There has been interest in developing methods to increase the drug entrapment efficiency. Szoka and Papahadjopoulos [84] produced liposomes known as reverse-phase evaporation vesicles (REVs), whereby the aqueous drug solution is emulsified in the presence of phospholipids dissolved in diethyl ether. By evaporating the organic solvent under high vacuum, large unilamellar vesicles are formed that may have an entrapment efficiency of

35–65%. Schneider [85] described a method whereby a water-in-oil-in-water (w/o/w) emulsion is formed, the organic phase containing the phospholipid and the inner aqueous compartment containing the drug. Schneider reported an entrapment efficiency of up to 80% using this method.

The mechanism by which liposomes form from phospholipids is not yet fully understood, although Lasic [86] suggested models involving hydration of the solid phospholipid followed by swelling and "pinching off" of the liposome. In comparison, the physical chemistry of the formed liposome is reasonably well characterized. At low temperatures, the lipids are arranged in a close-packed form (known as the gel state, as these systems are effectively liquid crystalline), which at a higher temperature become considerably less ordered over a narrow temperature range to form a fluid state. The bilayer becomes thinner, and the cross-sectional area per molecule increases owing to the greater degree of rotational motion. If cholesterol is incorporated within the liposome, the phase transition becomes "smaller," with an abolition of the phase transition at approximately 50% cholesterol. Papahadjopoulous et al. [87] suggested that the cholesterol renders the membrane more fluid below the phase transition temperature but more ordered above this temperature, and thus the difference in structure above and below the temperature is reduced. Whatever the explanation, it is well established that the release rate of most drugs increases from a low level at low temperatures to a high level in the immediate proximity of the phase transition. On further heating the system, the release rate decreases again [88]. A knowledge of phase transition phenomena is therefore necessary in order to manipulate the release characteristics of the liposomes as well as to understand the behavior of biological membranes. In particular, it has been suggested that at the phase transition temperature, the bilayer contains regions of both fluid and gel-state lipid [88]. The existence of these separate regions may play a role in natural phenomena such as cell–cell fusion [89] as well as explaining why drug release is rapid in this temperature region.

With certain notable exceptions, liposomes have been studied principally for administration via the parenteral route. As with most colloidal systems, injection results in rapid clearance from the blood via the MPS, with the majority of radiolabeled liposomes being found in the liver and smaller percentages being found in the spleen and bone marrow [90,91]. Factors such as the charge on the liposome [92,93], dose size [94], and vesicle size [95] have all been reported to influence the deposition behavior of liposomes. A number of studies have also described the attachment of antibodies to liposomes (immunoliposomes) for more specific targeting such as to lung endothelial cells [96]. Furthermore, Crommelin et al. [97] have been investigating the "target cell dragging" concept, whereby the drug-loaded immunoliposome attaches to a circulating target cell, thereby rendering the complex susceptible to MPS recognition. The complex is phagocytosed and the drug is released inside the macrophage, thereby exerting a highly localized effect on the target cell. While still in an early stage of development, this approach could be of considerable use in increasing the specificity and hence reducing the side effects of potent drugs.

The mechanism of uptake by the macrophages is in itself of interest. It was generally assumed that liposomes were taken up by the macrophages via phagocytosis, as depicted in Fig. 3. Indeed, this mechanism has been shown to occur as there is a marked decrease in liposome uptake in the presence of cytochalasin B, which blocks the endocytosis process [98]. However, this inhibition was considerably less marked when the liposomes used were above the phase transition temperature of the constituent phospholipids, that is, when the liposomes were in a "fluid" state. This has been ascribed to membrane fusion, whereby the lipid of the liposomes fuses with the phospholipid in the cell membrane, thus releasing the contents of the liposomes directly into the cytoplasm. As this mechanism avoids the necessity of liposome breakdown via the lysosomes, there is some interest in using this fusion process as a means of increasing the effectiveness of drugs that might otherwise be rapidly degraded within the cell. Furthermore, incubating biological cells with liposomes may result in the transfer of proteins from the cell to the vesicle bilayer, thus allowing the extraction of proteins that would be damaged by most other extraction techniques [99]. However, it is now thought that spontaneous fusion of liposomes with cell membranes is a comparatively rare event [100].

Other proposed mechanisms include adsorption of the liposomes onto the cell surface, thus releasing the drug in close proximity to the cell [101]. Furthermore, lipid exchange is believed to take place, with phospholipids from the liposome exchanging with those in the serum or cell membrane. This may be a source of artifacts when using radiolabeled liposomes, as the results may imply that the liposomes are located in the region where lipid exchange has taken place, which may well not be the case.

As with all colloidal drug delivery systems, the problem often lies in finding methods of circumventing the MPS. A number of approaches have been described, including swamping the MPS by coadministering carbon particles or empty liposomes [102–104]. This approach has led to a reduction in liver uptake of the liposomes, although not to a sufficient extent as to be of practical significance. Papahadjopoulos and Gabizon [105] showed that the inclusion of certain glycolipids in the liposome formulation prolonged circulation times. For example, surface-modified liposomes containing vasopressin may remain in the circulation for up to 4 days, thus presenting the possibility of effective sustained release preparations [106]. From these studies, a number of methods have been developed for producing surface-modified liposomes using polyethylene glycol lipid derivatives [107]. It is thought that the recognition of the liposomes is prevented by steric stabilization caused by the polyethylene glycol chains, thus altering the nature of the interaction between the liposomes and the circulating opsonins. This approach has led to the introduction of Stealth liposomes, marketed by Liposome Technology Inc., which involve steric stabilization using polyethylene glycol distearyl phosphatidylethanolamine [108]. Surface modification has also been studied as a means of targeting liposomes, for example, by coating with antibodies or carbohydrate derivatives specific to a certain cell type [109,110].

There have been a very large number of recent studies involving the use of liposomal drug delivery systems, and a comprehensive review of these areas is given by Gregoriadis [111], so only a brief outline of the main areas will be given here. Entrapment of anticancer drugs within liposomes may lead to a reduction in side effects. In particular, doxorubicin has been extensively studied as it has a broad spectrum of activity against solid tumors and leukemias. However, the use of the drug is limited due to the side effect of serious and irreversible cardiotoxicity; hence any method of reducing this effect would have considerable clinical application. Since the initial work on liposomally encapsulated doxorubicin by Forssen and Tokes [112], a number of studies have shown the cardiotoxicity to be markedly reduced when the drug is administered in a liposomal carrier [113]. In addition to reduced side effects, liposomes have also been shown to enhance antitumor activity [114]. Juliano and Layton [75] suggested that the observed effects may be due to enhanced time spent by the drugs in the circulation, longer residence times in the target tissues, protection of the drugs from metabolic inactivation, and alterations in the tissue distribution of the drug. In the latter, enhanced levels are found in the organs associated with the MPS and lower levels are found in other tissues such as the heart or kidney, which is important for drugs that show cardiotoxicity or nephrotoxicity. Furthermore, there may be additional mechanisms by which liposomes facilitate the passage of drugs across biological membranes. This is an important aspect of their use, as many tumor cell populations may develop resistance to anticancer drugs by reducing uptake through the cell membrane [115]. Studies by Papahadjopoulos et al. [116] showed that cells resistant to actinomycin-D become considerably more sensitive to the drug when the latter was administered as a liposomal suspension. They suggested that this may be due to enhanced uptake of the liposomally bound drug by the cancer cells.

Liposomes can also be used to deliver antimicrobial compounds. In particular, the drug amphotericin B is used for patients with widespread (disseminated) fungal infections involving the MPS. This is a particular problem for patients whose immune system is compromised. However, use of the drug is limited by acute and chronic side effects, the latter involving nephrotoxicity and central nervous system effects. Liposomal entrapment of the drug was found to result in a remarkable increase in efficacy and a decrease in side effects, possibly due to the liposomes being taken up by the fungal cells in preference to mammalian cells [117]. This work has led to the production of a commercial product, Ambisome (Vestar Ltd.), which is currently attracting considerable attention and may well prove to be the "flagship" product for pharmaceutical liposome preparations.

Other liposomally bound antibiotics that have been investigated include ampicillin for the treatment of *Listeria monocytogenes* infections in the liver and spleen [118]. A further area of interest has been the possibility of activating macrophages so as to render them tumoricidal using a range of substances such as lymphokines, muramyl dipeptides, and interferon [119,120]. This approach is in itself of great interest, as the macrophages, being natural to the body, are capable of differentiating between healthy and cancerous tissue and could therefore theoretically be ideal agents with which to treat tumors. Fidler et al. [121] showed that liposomally delivered muramyl dipeptides may activate macrophages in the lung in vivo, causing a marked reduction of established metastases in mouse lung. It has been suggested that the liposomes are taken up by circulating white blood cells, which then migrate to the lung, whereupon they mature into macrophages and are thus available for the eradication of tumor cells [122]. Studies have also been conducted on liposomally entrapped immunomodulators for treatment of liver tumors [123].

Liposomal systems have also been studied in the treatment of lysosomal storage diseases. Lysosomes are cell constituents that are responsible for breaking down potentially toxic materials found in the cell. Storage diseases result from enzyme deficiencies that reduce the ability of the lysosomes to decompose toxic materials, hence damaging the cells [124]. One method of treating such conditions is to replace the missing enzymes within the cell. Methods have been developed for entrapping enzymes within liposomes [125], and a number of studies have suggested that correction of the condition is possible using this approach [126,127], although the use of liposomes is limited to the treatment of tissues to which the vesicles can gain access, hence precluding their use in many cases. However, conditions such as Gaucher's disease and Pompe's disease, which affect cells associated with the MPS, may well be treated in this way. As with microspheres, there is also considerable interest in using liposomal systems as adjuvants for vaccines to increase their potency [128,129].

A further exciting new application of liposomes is in enzyme therapy; enzymes such as the tumoricidal protein L-asparaginase can be encapsulated to reduce toxicity and to protect the enzyme from circulating antibodies [130] as well as to allow controlled release [131]. Furthermore, intracellular delivery of enzymes such as superoxide dismutase has been described [132]. Radiation therapy and certain chemical agents such as paraquat may lead to the generation of O_2-, which is toxic to the host organism. There is some evidence that pretreatment with superoxide dismutase gives some protection against the damage caused by this radical [133], although the protection is limited due to the inability of the enzyme to enter host cells [134]. There is evidence that liposomal entrapment of this enzyme may result in enhanced protection, although there is some dispute regarding the extent and mechanism of this process [135].

There have been a number of studies on oral administration of liposomes, particularly with a view to delivering proteins. Patel and Ryman [136] and Dapergolas and Gregoriadis [137] reported decreases in blood sugar levels after oral administration of insulin-loaded liposomes, thus suggesting an oral treatment for the treatment of diabetes. However, it is not clear whether this is due to enhanced transport through the gut or to protection of the protein from enzymatic degradation in the gastrointestinal tract. Factor VIII, a protein required for the treatment of hemophilia, has also been given orally in liposomal form, although there is some disagreement regarding the effectiveness of this treatment [138,139]. The usefulness of orally delivered liposomes is still a matter of discussion, and work is still ongoing to assess the validity of this approach.

There has been considerable success in using liposomes for delivery to the nose and lungs [140,141]. In particular, delivery to the lungs may be used either to deliver phospholipids to the alveolar surface or to prolong the duration of action of drugs used to treat respiratory conditions [142]. Again, the advantage of liposomal drug delivery lies in the possibility of using carrier materials that are themselves endogenous to the lung, in terms of both the cellular membranes of the alveoli and lung surfactant, which is a mixture of phospholipids (largely DMPC), cholesterol, and various nonserum proteins. This extracellular material is responsible for maintaining the structure of the alveoli by altering the surface energy at the alveolar–air interface, thus rendering expansion of the surface during inspiration easier. The mechanisms for the clearance and reutilization of lung surfactant are likely to have a strong influence on the biological fate of inhaled liposomes. Although recent studies have indicated that soybean phosphatidylcholine liposomes are well tolerated [143], long-term administration of liposomes may saturate the normal clearance mechanisms. There is some evidence that the macrophage population increases in order to remove the excess phospholipid, although more work is required to verify this [144].

The earliest application of liposomes to the lung was as a means of replacing the phospholipids found in lung surfactant for the treatment of respiratory distress syndrome [145], in which the lungs lack the levels of surfactant required to function properly. Subsequently, gamma scintigraphy has been widely used in the study of liposome deposition in the lung, showing that the liposomes remain intact over a period of 24 h [146,147]. There is therefore considerable potential for using liposomes as sustained-release vesicles for the treatment of conditions such as asthma [148], carcinomas [149], and bacterial and viral infections [150].

Delivery of liposomes to the skin (topical administration) has also been studied, with increased effectiveness of encapsulated triamcinolone compared to a traditional ointment formulation [151]. Liposomes are now used in a number of cosmetic preparations, the manufacturers claiming that the liposomes penetrate the largely impermeable stratum corneum intact to release the moisturizing agents in the underlying epidermis. Liposomes can also be used to deliver drugs to the eye for controlled release of those drugs [152].

There are also a number of practical considerations that need to be addressed in producing liposomal products. These include the need to produce large numbers of identical liposomes that must have an appropriate size distribution, quality control of phospholipids obtained from natural sources, and the chemical and physical stability of the liposome suspension, including the need to avoid flocculated systems, which would be extremely dangerous if injected due to the possibility of embolism caused by a liposome aggregate.

The physical stability of liposomes appears to be a function of the lipid and drug composition as well as the degree of curvature of the liposome membrane [153]. For storage purposes, the liposomal suspensions can be freeze-dried or lyophilized from organic solvents. Addition of cholesterol to the bilayer or cryoprotectants such as lactose have been reported to be effective in reducing damage to the liposomes during freeze-drying [154]. Further problems lie in the need for the liposomes to be sterile prior to injection. As heat sterilization would damage the bilayer structure, filtration is the only viable option. However, this is limited to liposomes of diameters less than approximately 0.5 μm, as larger liposomes are damaged when passed through filters that are sufficiently fine to remove bacterial cells. Other problems include the in vivo stability, particularly with respect to membrane proteins and serum lipoproteins, which have been reported to damage the lipid bilayer unless a protecting agent such as cholesterol is included [154,155]. Furthermore, although it is known that the lipid bilayer is a relatively flexible structure, studies have indicated that hydrodynamic shear can cause drug leakage and a reduction in liposome size [76]; hence it is not certain whether the stresses exerted on the liposome during circulation will have an effect.

D. Nonionic Surfactant Vesicles (Niosomes)

Although the use of phospholipids as vesicular delivery systems is now widespread, these molecules are by no means the only materials that can form such vesicles. A number of alternatives to phospholipids have been studied, including unsaturated fatty acids (ufasomes [156]), saturated fatty acids [157], amphipathic peptides [158], and lipids that occur naturally in the stratum corneum of the skin [159]. In particular, nonionic surfactant vesicles (NSVs) have been investigated as alternatives to liposomes. A number of suitable molecules have been identified, including alkyl polyoxyethylene ethers, alkyl polyglyceryl ethers, and steroidal lariat ethers. A more comprehensive list of vesicle-forming surfactants is given by Florence and Baillie [160]. NSVs have also been referred to as niosomes, although this term has become a registered trademark of L'Oreal; hence its use is now restricted.

There are several parallels between NSVs and liposomes, in terms of both their physical chemistry and biological activity. The physicochemical properties of NSVs and liposomes are undoubtedly similar. NSVs are manufactured in a similar manner to liposomes [161], with the surfactant molecules forming bilayers on hydration, which in turn form vesicles. Furthermore, cholesterol can be incorporated into the bilayers and may affect the physicochemical properties such as stability and the rate of drug efflux [162]. It is less clear whether the biological fate of NSVs is also similar to that of liposomes, as the processes involved are not fully understood in either case. The membrane interactions reported for liposomes may not necessarily be seen for synthetic NSVs, although nonionic surfactants have been shown to interact with low density lipoproteins in vitro [163,164]; hence it is possible that in vivo interactions also occur. Florence and Baillie [160] have argued that as the surface charge of colloidal particles may determine membrane activity, the lack of charges on NSV surfaces may be of relevance in determining the extent of the membrane interaction. However, the significance of the initial surface charge on the biological fate of colloidal particles is by no means clear, as opsonization by plasma proteins will take place shortly after injection; hence the charge of the particles at the site of action is unknown. It should also be borne in mind that the significance of membrane interactions between liposomes and biological tissues is still controversial, as some workers do not consider membrane fusion to be of particular relevance to drug release.

As with virtually all colloidal systems, NSVs are rapidly taken up by the MPS, so investigations into their uses have concentrated on the treatment of diseases involving the MPS and the prolongation of the presence of drugs in the circulation. In particular, the use of NSVs containing sodium stibogluconate for the treatment of leishmaniasis has been studied [165–167]. Although significant reductions in the level of parasites found in the liver were reported, the high specificity of the NSVs for this organ was found to be a disadvantage, as parasites living in the spleen and bone marrow were unaffected by the treatment [168].

Investigations have also been carried out using the anticancer drugs methotrexate [169,170] and doxorubicin [171,172]; their incorporation into NSVs resulted in a reduction in the rate of their metabolism and excretion. Although these studies were not conducted in order to target the drugs to a specific organ, the results are of interest as they suggest a means of maintaining therapeutic concentrations of these drugs over a prolonged period of time. Furthermore, as the cardiotoxic effects of doxorubicin are largely bassociated with the presence of the metabolite doxorubicin 7-deoxybaglycone, for example, protection of these drugs from metabolism may in turn protect the patient from some of the side effects of the drugs.

E. Biological Microcapsules

An alternative method of preparing sustained-release or targeted drug delivery systems has been to use biological cells as drug carriers. Among other cell types, the use of resealed red blood cells (erythrocytes) has been extensively investigated. This method was suggested simultaneously by Ihler et al. [173] and Zimmermann [174].

The erythrocytes are made permeable (lysed) by one of a number of methods. The simplest procedure is to place them in a solution of a lower osmotic pressure than the cell contents, so that water enters the erythrocytes. The cell increases in size to approximately 1.6 times the original volume before the induced stress causes pores to appear [175]. These pores are 200–500 Å in size and may be open long enough to allow intra- and extracellular liquid exchange, especially at low temperatures. Increasing the ionic strength and incubating at 37 °C restores the osmotic properties of the cells and causes the pores to close [176]. A drug present during the lysis procedure will be incorporated within the cells, thus allowing a relatively simple method of encapsulation.

A problem with this method is the poor level of entrapment, being in the region of 5–10% of the exogenous compound. A variety of alternative methods have been proposed for increasing the entrapment ratio, including the use of dialysis membranes and electrical hemolysis. The latter method involves passing fields of ca. 10^4 V/cm across suspensions of cells, leading to the development of stable pores in the membrane, which are then resealed by incubation at 37°C [177]. The advantages of this method are that it is possible to control the size of the pore and the pores are stable enough to allow complete equilibration between the intra- and extracellular media [178]. Isotonic methods have also been described whereby small molecules such as ethylene glycol and ammonium chloride are included in the incubation medium. These molecules can cross the cell membrane relatively freely, causing ingress of water due to the increased osmotic pressure within the erythrocytes [179].

An intriguing development of this technology has been the method of cell fusion, whereby the molecule of interest is entrapped within either a biological or synthetic (e.g., liposomal) cell. The drug-loaded vesicle is then fused with other culture cells to give a drug delivery vehicle with properties of both original materials [180,181]. Fusion can be induced by fusinogenic agents such as polyethylene glycol or by the presence of the Sendai virus. Zimmermann et al. [182] also suggested the use of electric fields to induce fusion.

The normal half-life of erythrocytes in the circulation is approximately 27 days before the cells are phagocytosed by the spleen. However, modification of the erythrocyte surface, either deliberately or induced by the lysing process, may dramatically alter both the half-life of the cells and the site of their metabolism. For example, coating the cells with the antibody IgG leads to increased liver uptake [183,184]. Similarly, treatment of glutaraldehyde, which cross-links molecules on the surface of the erythrocytes, may lead to uptake by the liver or spleen, depending on whether the treatment is heavy or light, respectively [185]. Cooper [184] suggested that this is due to the spleen being capable of recognizing relatively subtle changes in the surface structure of the erythrocytes. More dramatic changes activate the MPS, which will then tend to take the recognized cells to the liver. Other methods of modifying the surface of erythrocytes include the enzymatic alteration of cell surface carbohydrates, an example being the treatment with sialidase to remove sialic acid from the external glycoproteins and glycolipids of the erythrocytes. This in turn leads to a reduction in the circulating half-life and sequestration of the erythrocytes in the liver and spleen [186,187]. Clearly, therefore, it is possible to alter the uptake distribution of these dosage forms in a number of ways.

There are several advantages to using resealed erythrocytes compared to other vesicular delivery systems. Erythrocytes can be taken from the patient, filled with the drug in question, and reinjected. Consequently, they can be considered to be nonimmunogenic and biodegradable, the former being dependent on there being negligible change to the surface characteristics of the cells. Furthermore, either the resealed erythrocytes can be used for targeting to different regions of the MPS (ie., the liver or spleen), or use can be made of the extended circulation time of unmodified erythrocytes. The entrapped drug is also shielded from the normal immune response. For example, studies have shown that animals that have been sensitized to a protein do not show any reaction to that protein in an encapsulated form [188].

The disadvantages associated with the technique are largely associated with the alterations in surface structure, which may affect both the distribution profile in the body and the kinetics of drug release from the vesicles. Zimmermann [189] noted that there are differences in the uptake distribution of erythrocytes prepared using the various methods, probably due to differences in the surface damage caused during the manufacturing processes. In particular, hemoglobin may be lost during hypotonic lysis, which may itself cause changes in the surface characteristics. Small changes in the ionic strength of the surrounding medium have been reported to cause dramatic changes in the loss of membrane proteins [190]. Furthermore, the drug itself may affect the membrane structure, so care must be taken to ensure that the vesicles reach the appropriate site and release the drug at the desired rate. In comparison to synthetic vesicles, resealed erythrocytes have the advantages of immunocompatibility and controllable circulation time but have the disadvantages of there being little flexibility regarding either the size of the vesicles or the permeability of the membrane to various drugs.

A number of specific uses for resealed erythrocytes have been described. For example, studies have shown their potential as enzyme carriers for lysosomal storage diseases. Phagocytosis of the erythrocytes by macrophages results in the direct delivery of the entrapped enzymes to the lysosomes. This approach has been studied with particular reference to Gaucher's disease, as the MPS is the main storage site for the unmetabolized material, glucocerebroside [191]. In addition to high molecular weight molecules such as enzymes, the delivery of smaller molecules has also been studied. For example, desferrioxamine is an iron chelator that can be used in cases of iron supplement overdose or for long-term transfusion patients. However, it is relatively ineffective at removing iron stored in the MPS and is also an expensive drug. Targeting to the MPS in resealed erythrocytes therefore has advantages in terms of both therapeutic efficacy and reductions in the dose, and hence cost, required to treat the condition [192]. Other uses include slow release of anticancer drugs such as methotrexate [193] and the use of resealed erythrocytes as "processing centers," as the circulating cells contain enzymes that metabolize unwanted substances. For example, Zimmermann et al. [194] encapsulated urease for the treatment of kidney failure, which requires degradation of serum urea. Similarly, certain leukemias require the presence of asparagine. Updike et al. [195] entrapped an enzyme to metabolize asparagine in order to treat this condition, although the remission induced by this therapy was found to be short-lived.

While resealed erythrocytes have been discussed here in some detail, there are a number of other biological carriers that have been described, including yeast cells [196] and unfertilized fish eggs [197]. Furthermore, there has been considerable interest in the use of replication-deficient retroviruses as carriers for genetic material [198]. This approach has potential for a number of conditions and is likely to be of greater importance in the future.

F. Magnetic Particles

There has also been interest in the use of magnetic particles as an aid to targeting specific cells, particularly in the treatment of cancers. Most such systems involve the incorporation of Fe_3O_4 (magnetite) or other ferrite materials into natural or synthetic polymer particles. A number of preparation methods have been described, including emulsion polymerization of acrylate monomers in the presence of finely divided magnetite [199] and the manufacture of a microsuspension of magnetite, monomers, and initiators in water [200]. Magnetic particles containing natural polymers such as alginates can be prepared by adding the polymer and magnetite to water and pouring the suspension into a water-immiscible liquid. The resultant particles can then be cross-linked, if required, and antibodies can be attached to the surface.

One of the uses of magnetic particles has been the removal of cancerous neuroblastoma cells from bone marrow [201]. Samples of bone marrow were incubated with mouse monoclonal antibodies that are specific for the cancer cells, and magnetic particles carrying anti-mouse antibodies were added. The cancer cells became coated with particles. Application of a magnetic field then resulted in the separation of the cancer cells and excess magnetic particles from the healthy cells. This raises the possibility of removing the bone marrow from the patient, “cleansing” it using the magnetic particles, and reinfusing it back into the patient.

Widder et al. [202] used magnetic albumin spheres to target tumors in vivo. By applying a magnetic field to the area of the body in question, the spheres can be held in that position to concentrate the drug at the desired site of action. This has the dual advantages of avoiding the MPS because the particles are not transported to the liver and phagocytosed at the same rate as freely circulating particles, and reducing the side effects of the drug because the levels of drug in the general circulation are lowered. Widder et al. demonstrated these advantages by giving an intraarterial infusion of radiolabeled magnetic albumin microspheres to rats and placing a magnetic field across the tails. Twenty-four hours after the removal of the magnetic field, 50% of the activity was found to be present at the target site, this percentage being the same as that found 30 min after removal of the field. In a further study, Widder et al. [203] demonstrated that rats inoculated with the Yoshida rat sarcoma (which was again located in the tail) showed a 75% remission rate for animals treated as above with doxorubicin-loaded microspheres, as opposed to an 80–100% mortality rate in control groups (untreated, intravenous doxorubicin in solution, and drug-loaded magnetic spheres without the application of the magnet). Furthermore, a single dose of 0.5 mg/kg given in the spheres showed more favorable results than repeated doses of 5 mg/kg doxorubicin given intravenously. While the therapy is more likely to be successful in an organ such as a tail to which a magnetic field can be easily and selectively applied, the results are nevertheless of considerable interest.

IV. CONCLUSIONS

A broad outline has been given of the ways in which disperse systems can be used as drug delivery vehicles, particularly for controlled release and targeted systems. The review has necessarily been confined to providing an idea of the approaches used rather than a comprehensive discussion of every aspect of these applications. However, the general principles behind the use of colloidal drug delivery systems are largely the same for all the approaches described here, in that the drug is incorporated into a biocompatible carrier particle, which then either directs that drug to a specific tissue or allows its slow release over a period of time. These studies also serve to illustrate that the activity of a drug is not solely a function of the chemical structure but is highly dependent on the manner in which the drug is introduced into the body.

The most important question is that of where this line of research is likely to lead in the future. The arguments for the development of colloidal drug delivery systems include the considerable need to develop treatments for conditions affecting the MPS such as leishmaniasis and malaria, as well as the need for circulating slow-release systems. Furthermore, as major advances have been made in developing means of avoiding MPS uptake, it is likely that in the future, far greater control of the biological distribution of these particles will be possible. Finally, the need to find means of formulating proteins is as great as ever, particularly in terms of enabling proteins to cross the gut wall. The possibility of allowing nonparenteral administration of drugs such as insulin by their incorporation into colloidal particles is of considerable importance and is an area where there is scope for considerably more research.

There are perhaps three factors acting against the use of the colloidal systems described here. First, the advent of a number of new approaches has to some extent focused attention away from dosage form design to the reexamination of the therapeutic agent itself, particularly in terms of using proteins as drugs. This shift in emphasis is hardly surprising when one considers that the idea of successfully treating diseases such as genetic disorders is now a distinct possibility rather than a distant goal, thanks to the recent advances made in biotechnology. Second, there has perhaps been some disappointment in the effectiveness of colloidal drug delivery systems, largely because of the problems involving the MPS and extravasation as well as the difficulties in penetrating solid tumors, irrespective of the specificity of the carrier. Third, there has been comparatively little research conducted into the technology involved in producing these systems on a large scale; hence the feasibility of these dosage forms in terms of ease and cost of production is not certain in many cases. However, these problems will be solved as the understanding of the basic processes involved increases; it is unlikely that they will preclude the use of colloidal drug delivery systems in the future.

In conclusion, therefore, colloidal drug delivery systems provide numerous opportunities for developing more effective therapies for a range of conditions. Now that the limitations of these drug delivery systems are more fully realized, the opportunity arises to work within those limits and concentrate efforts on the development of more effective and safer medicines, as well as to examine ways of overcoming the difficulties associated with these systems.

ACKNOWLEDGMENT

Thanks are due to Dr. Kevin Taylor for helpful comments during the preparation of this manuscript.

REFERENCES

1. J.B. Bray, P.A. Cragg, A.D.C. Macknight, R.G. Mills, and D.W. Taylor, *Lecture Notes in Human Physiology*, 2nd ed., Blackwell Scientific, Oxford, 1989, pp. 373–374.
2. V.A. Levin, C.S. Patlak, and H.D. Landahl, *J. Pharmacokin. Biopharm. 8*: 257 (1980).
3. E.A. Swabb, J. Wei, and M. Gullino, *Cancer Res. 34*:2814: (1978).

4. S.S. Davis and L. Illum, in *Site-Specific Drug Delivery* (E. Tomlinson and S.S. Davis, Eds.), Wiley, New York, 1986, pp. 93–110.
5. L. Illum, in *Novel Drug Delivery and Its Therapeutic Applications* (L.F. Prescott and W.S. Nimmon, Eds.), Wiley, New York, 1989, pp. 273–279.
6. L. Illum, S.S. Davis, C.G. Wilson, N.W. Thomas, M. Frier, and J.G. Hardy, *Int. J. Pharm. 12*:135 (1982).
7. L. Illum and S.S. Davis, in *Drug Targeting* (P. Buri and A. Gumma, Eds.), Elsevier, New York, 1985, p. 66.
8. L. Aschoff, *Lectures on Pathology*, Paul Hoeber, New York, 1924, pp. 1–33.
9. R. Van Furth, *Mononuclear Phagocytes*, Blackwell Scientific Publications, Oxford, 1970.
10. J.W.B. Bradfield, *Br. J. Exp. Pathol. 61*:617 (1980).
11. D.P. Ptananing-van Dalen, A. Brouver, and D.L. Knook, *Gastroenterology 81*:1036 (1981).
12. L. Illum and S.S. Davis, *FEBS Lett. 167*:79 (1984).
13. I.J. Fidler, Z. Barnes, W.E. Fogler, R. Kirsh, P. Bugelski, and G. Poste, *Cancer Res. 42*:496 (1982).
14. C.R. Alving, E.A. Steck, W.L. Hanson, P.S. Loizeaux, W.L. Chapman, and V.B. Waits, *Life Sci. 22*:1021 (1978).
15. G. Poste and R. Kirsh, *Biotechnology 1*:869 (1983).
16. I.J. Fidler, *Science 208*:1469 (1980).
17. G.J. Fleer and J. Lyklema, in *Adsorption from Solution at the Solid–Liquid Interface* (G.D. Parfitt and C.H. Rochester, Eds.), Academic, New York, 1983, pp. 153–220.
18. F.T. Hessellink, In *Adsorption from Solution at the Solid–Liquid Interface* (G.D. Parfitt and C.H. Rochester, Eds.), Academic, New York, 1983, pp. 377–412.
19. W. Norde, in *Microspheres and Drug Therapy* (S.S. Davis, L. Illum, J.G. McVie, and E. Tomlinson, Eds.), Elsevier, New York, 1984, pp. 39–59.
20. S.M. Moghimi and H.M. Patel, *Biochim. Biophys. Acta 984*:379 (1989).
21. S.C. Silverstein, J. Michl, and S.S.J. Sung, in *Transport of Macromolecules by Cellular Systems* (S.C. Silverstein, Ed.), Dahlem Konferenzen, Berlin, 1978, pp. 245–264.
22. M.J. Hsu and R.L. Juliano, *Biochim. Biophys. Acta 720*:411 (1982).
23. J.W.B. Bradfield, in *Microspheres and Drug Therapy* (S.S. Davis, L. Illum, J.G. McVie, and E. Tomlinson, Eds.), Elsevier, New York, 1984, pp. 25–37.
24. E. Tomlinson, *Int. J. Pharm. Technol. Prod. Manuf. 4*:49 (1983).
25. K.J. Wilkins, A.E. Senyei, and D.F. Ranney, *Adv. Pharmacol. Chemother. 16*:213 (1979).
26. C.F. Van Oss, C.F. Gillam, and A.W. Neuman, *Phagocytic Engulfment and Cell Adhesiveness*, Marcel Dekker, New York, 1975.
27. P.M. Cararelli, F.A. Blumenstock, T.M. Saba, and F.J. Rourkes, *J. Leuk. Biol. 36*:477 (1984).
28. G. Kaplan, *Scand. J. Immunol. 6*:797 (1977).

29. D.J. Wilkins and O.P.A. Myers, *Br. J. Exp. Pathol. 47*:568 (1966).
30. I. Sjoholm, B. Ekman, A. Koberr, I. Tjungstedt-Pahlman, B. Seiving, and T. Sjodin, *Mol. Pharm. 16*:767 (1979).
31. H. Beukers, F.A. Deierkauf, C.P. Blom, M. Deierkauf, C.C. Scheffers, and J.C. Riemersma, *Chem.-Biol. Interac. 33*:91 (1980).
32. L. Illum, I.M. Hunneyball, and S.S. Davis, *Int. J. Pharm. 29*:53 (1986).
33. L. Illum and S.S. Davis, *J. Parent. Sci. Technol. 36*:242 (1982).
34. M. Bundgaard, *Ann. Rev. Physiol. 42*:325 (1980).
35. J.N. Weinstein and L.D. Leserman, *Pharm. Ther. 24*:207 (1984).
36. G. Poste, in *Receptor Mediated Targeting of Drugs* (G. Gregoriadis, G. Poste, J. Senior, and A. Trouet, Eds.), Plenum, New York, 1984, pp. 427–474.
37. R. Jain, *Proc. Int. Symp. Controlled Release Bioact. Mat. 19*:130 (1992).
38. W.E. Longo, H. Iwata, T.A. Lindheimer, and E.P. Goldberg, *J. Pharm. Sci. 71*:1323 (1982).
39. J.H. Ratcliffe, I.M. Hunneyball, A. Smith, C.G. Wilson, and S.S. Davis, *J. Pharm. Pharmacol. 36*:431 (1984).
40. M. Nakano, N. Wakiyama, T. Kojima, K. Juni, R. Iwaoka, S. Inoue, and Y. Yoshida, in *Microspheres and Drug Therapy* (S.S. Davis, L. Illum, J.G. McVie, and E. Tomlinson, Eds.), Elsevier, New York, 1984, pp. 327–335.
41. L.M. Sanders, J.S. Kent, G.I. McRae, B.H. Vickery, T.R. Tice, and D.H. Lewis, *J. Pharm. Sci. 73*:1294 (1984).
42. S.S. Davis, I.M. Hunneyball, L. Ullum, J.H. Ratcliffe, A. Smith, and C.G. Wilson, *Drugs Exp. Clin. Res. 9*:633 (1985).
43. N. Willmott, H.M.H. Kamel, J. Cummings, J.F.B. Stuart, and A.T. Florence, in *Microspheres and Drug Therapy* (S. S. Davis, L. Illum, J.G. McVie, and E. Tomlinson, Eds.), Elsevier, New York, 1984, pp. 205–215.
44. H. Teder, B. Nilsson, K. Johnson, C. Hellekant, K. Aspergren, and K.F. Aronsen, in *Anthracyclines and Cancer Therapy* (H.H. Hansen, Ed.), CRC Press, Boca Raton, Fla., pp. 166–176.
45. K.-E. Arfors, J.O. Forsaberg, B. Larsson, D.H. Lewis, B. Rosengren, and S. Odman, *Nature 262*:500 (1976).
46. B. Lindell, K.F. Aronsen, U. Rothman, and H.O. Sjogren, *Res. Exp. Med. 171*:63 (1977).
47. A.E. Young, *Br. Med. J. 283*:1144 (1981).
48. V.P. Chuang and S. Wallace, *J. Am. Med. Assoc. 20*:1151 (1981).
49. S. Dakhil, W. Emsinger, K. Cho, J. Niederbuber, K. Doan, and R. Wheeler, *Cancer 50*:631 (1982).
50. U. Rothman, K.-E. Arfors, K.F. Aronsen, B. Lindell, and G. Nylander, *Microvasc. Res. 11*:421 (1976).
51. J.H. Elridge, C.J. Hammond, J.A. Mealbroek, J.K. Stuas, R.M. Gilley, and T.R. Tice, *J. Controlled Release 11*:205 (1990).
52. H.T. Wang, E. Schmitt, D.R. Flanagan, and R.J. Lindhardt, *J. Controlled Release 17*:23 (1991).
53. S. Cohen, T. Yoshioka, M. Lucarelli, L. Hwang, and R. Langer, *Pharm. Res. 8*:713 (1991).

54. P. Artusson, I.L. Martensson, and I. Sjoholm, *J. Pharm. Sci. 75*:697 (1986).
55. M.-C. Bissery, F. Valeriote, and C. Thies, in *Microspheres and Drug Therapy* (S.S. Davis, L. Illum, J.G. McVie, and E. Tomlinson, Eds.), Elsevier, New York, 1984, pp. 217–228.
56. Y.W. Chien, K.S.E. Su, and S.-F. Chang, *Nasal Systemic Drug Delivery*, Marcel Dekker, New York, 1989, pp. 89–198.
57. L. Illum, H. Jorgensen, H. Bisgaard, O. Krogsgaard, and N. Rossing, *Int. J. Pharm. 39*:189 (1987).
58. A.A. Hussain, R. Kimura, and C.H. Huang, *J. Pharm. Sci. 70*:466 (1981).
59. A. Pontiroli, M. Alberetto, and G. Pozza, *Br. Med. J. 290*:1390 (1985).
60. P. Speiser, *Acta Pharm. Suecica 10*:381 (1973).
61. P. Speiser, *Progr. Colloid Polym. Sci. 59*:48 (1976).
62. J.J. Marty and R.C. Oppenheim, *Aust. J. Pharm. Sci. 6*:65 (1977).
63. P. Couvreur, M. Roland, and P. Speiser, U.S. Patent 4,329,332 (1982).
64. P. Couvreur, V. Lenaerts, D. Leyh, P. Guiot, and M. Roland, in *Microspheres and Drug Therapy* (S.S. Davis, L. Illum, J.G. McVie, and E. Tomlinson, Eds.), Elsevier, New York, 1984, pp. 103–107.
65. J. Kreuter, in *Topics in Pharmaceutical Sciences* (D.D. Breimer and P. Speiser, Eds.), Elsevier, New York, 1983, pp. 359–370.
66. J. Kreuter, U. Tauber, and V. Illi, *J. Pharm. Sci. 68*:1443 (1979).
67. J.G. McAfee and G. Subramanian, in *Clinical Scintillation Imaging*, 2nd ed. (L.M. Freeman and P.M. Johnson, Eds.), Grune and Stratton, New York, 1975, pp. 13–114.
68. R.C. Oppenheim, in *Drug Delivery Systems* (R.L. Juliano, Ed.), Oxford Univ. Press, New York, 1980, pp. 177–188.
69. J.J. Marty, R.C. Oppenheim, and P. Speiser, *Pharm. Acta Helv. 53*:17 (1978).
70. A.R. Thompson, J.M. Payne, B.F. Sansom, R.J. Garner, and B.J. Miles, *Nature 188*:586 (1960).
71. P. Jani, A.T. Florence, G.W. Halbert, and J. Langridge, *J. Pharm. Pharmacol. 41*:809 (1989).
72. P. Jani, A.T. Florence, G.W. Halbert, and J. Langridge, *J. Pharm. Pharmacol. 42*:821 (1990).
73. P. Jani, A.T. Florence, and D.E. McCarthy, *Int. J. Pharm. 84*:245–252 (1992).
74. A.D. Bangham, M.M. Standish, and J.C. Watkins, *J. Mol. Biol. 13*:238 (1965).
75. R.L. Juliano and D. Layton, in *Drug Delivery Systems* (R.L. Juliano, Ed.), Oxford Univ. Press, New York, 1980, pp. 189–236.
76. D. Stamp and R.L. Juliano, *Can. J. Physiol. Pharmacol. 57*:535 (1979).
77. R.L. Juliano and D. Stamp, *Biochim. Biophys. Acta 586*:137 (1978).
78. C.H. Huang and J.P. Charlton, *J. Biol. Chem. 246*:2555 (1971).
79. S. Batzri and E.D. Korn, *Biochim. Biophys. Acta 298*:1015 (1973).
80. Y. Kagawa and E. Racker, *J. Biol. Chem. 246*:5477 (1971).
81. M.H.W. Milsmann, R.A. Schwendener, and H.G. Weder, *Biochim, Biophys. Acta 512*:147 (1978).
82. H. Hauser and N. Gains, *Proc. Natl. Acad. Sci. USA 79*:1683 (1982).

83. D. Deamer and A.D. Bangham, *Biochim, Biophys. Acta 443*:629 (1976).
84. F. Szoka, Jr., and D. Papahadjopoulos, *Proc. Natl. Acad. Sci. USA 75*: 4194 (1978).
85. M. Schneider, in *Drug Targeting* (P. Buri and A. Gumma, Eds.), Elsevier, New York, 1985, pp. 119–134.
86. D.D. Lasic, *Period. Biol. 85*:147 (1983).
87. D. Papahadjopoulos, M. Cowden, and H.K. Kimelberg, *Biochim, Biophys. Acta 330*:8 (1973).
88. D. Papahadjopoulos, K. Jacobson, S. Nir, and T. Issacs, *Biochim, Biophys. Acta 311*:330 (1973).
89. G. Poste and D. Papahadjopoulos, *Ann. N.Y. Acad. Sci. 308*:164 (1978).
90. G. Gregoriadis and B.E. Ryman, *Eur. J. Biochem. 24*:485 (1972).
91. R.L. Juliano and D. Stamp, *Biochem. Pharmacol. 27*:21 (1978).
92. M.M. Jonah, E.A. Cerny, and Y.E. Rahman, *Biochim, Biophys. Acta 401*: 336 (1975).
93. W.E. Magee, M.L. Talcott, S.X. Straub, and C.Y. Vriend, *Biochim, Biophys. Acta 451*:610 (1976).
94. R.M. Abra and C.A. Hunt, *Biochim, Biophys. Acta 666*:493 (1981).
95. G. Poste, R. Kirsh, and T. Koestler, in *Liposome Technology* Vol. 3 (G. Gregoriadis, Ed.), CRC Press, Boca Raton, Fla., 1984, pp. 1–28.
96. B.J. Hughes, S. Kennel, R. Lee, and L. Huang, *Cancer Res. 49*:6214 (1989).
97. D.J.A. Crommelin, G. Storm, U.K. Nassander, and P.A. M. Peeters, *Proc. Int. Symp. Controlled Release Bioact. Mat. 19*:20 (1992).
98. G. Poste and D. Papahadjopoulos, *Proc. Natl. Acad. Sci. USA 79*:1603 (1976).
99. A. Darszon, C.A. Vandenberg, M.H. Ellisman, and M. Montal, *J. Cell Biol. 81*:446 (1979).
100. L.B. Margolis, in *Liposomes as Drug Carriers* (G. Gregoriadis, Ed.), Wiley, New York, 1988, pp. 75–92.
101. R.E. Pagano and L. Huang, *J. Cell Biol. 67*:49 (1975).
102. R.M. Abra, M.E. Bosworth, and C.A. Hunt, *Res. Commun. Chem. Pathol. Pharmacol. 29*:349 (1980).
103. Y.J. Kao and R.L. Juliano, *Biochim, Biophys. Acta 677*:453 (1981).
104. R.L. Souhami, H.M. Patel, and B.E. Ryman, *Biochim, Biophys. Acta 839*:1 (1981).
105. D. Papahadjopoulos and A. Gabizon, *Ann. N.Y. Acad. Sci. 507*:64 (1987).
106. M.C. Woodle and D. Papahadjopoulos, *Methods Enzymol. 17*:193 (1989).
107. D. Papahadjopoulos, T.M. Allen, A. Gabizon, E. Mayhew, K. Matthay, S.K. Huang, K.-D. Lee, M.C. Woodle, D.D. Lasic, C. Redermann, and F.J. Martin, *Proc. Natl. Acad. Sci. USA 88*:11460 (1991).
108. M.C. Woodle, M. Newman, L. Collins, C. Redermann, and F. Martin, *Proc. Int. Symp. Controlled Release Bioact. Mat. 17*:77 (1990).
109. G. Gregoriadis and D. Neerunjun, *Res. Commun. Chem. Path. Pharmacol. 10*:351 (1975).
110. G. Gregoriadis and D. Neerunjun, *Biochem. Biophys. Res. Commun. 65*: 537 (1975).
111. G. Gregoriadis (Ed.), *Liposomes as Drug Carriers*, Wiley, New York, 1988.

112. E.A. Forssen and F.A. Tokes, *Biochem. Biophys. Res. Commun. 91*:1295 (1979).
113. A.A. Gabizon and Y. Barenholz, in *Liposomes as Drug Carriers* (G. Gregoriadis, Ed.), Wiley, New York, 1988, pp. 365–379.
114. A. Rahman, A. Ganjei, and J.R. Neefe, *Cancer Chemother. Pharmacol. 16*: 28 (1986).
115. C.J. Goldenberg, *Cancer Res. 35*:1087 (1975).
116. D. Papahadjopoulos, G. Poste, W.J. Vail, and J. Biedler, *Cancer Res. 30*: 2988 (1976).
117. G. Lopez-Berestein, R. Mehta, R.L. Hopfer, K. Mills, L. Kain, K. Mehta, K. Fainstein, M. Luna, E.M. Hersh, and R.L. Juliano, *J. Infect. Dis. 147*: 939 (1983).
118. I.A.J.M. Bakker-Woudenberg, A.F. Lokerse, F.H. Roerdink, D. Regts, and M.F. Michel, *J. Infect. Dis. 151*:917 (1985).
119. L. Adinolfi, P.F. Bonventre, M.A. van der Pas, and D.A. Eppstein, *Infect. Immun. 48*:409 (1985).
120. I.J. Fidler, *Cancer Res. 45*:4714 (1985).
121. I.J. Fidler, S. Sone, W.E. Fogler, and Z.L. Barnes, *Proc. Natl. Acad. Sci. USA 78*:1680 (1981).
122. M.E. Key, J.E. Talmadge, W.E. Fogler, C. Bucana, and I.J. Fidler, *J. Natl. Cancer Inst. 69*:1189 (1982).
123. T. Daemen, A. Veninga, F.H. Roerdink, and G.L. Scherphof, *Cancer Res. 46*:4330 (1986).
124. R. Hirshhorn and G. Weissman, *Prog. Med. Genet. 1*:49 (1977).
125. G. Weissman, H. Karchak, M. Finkelstein, J. Smolen, and S. Hoffstein, *Ann. N.Y. Acad. Sci. 308*:235 (1978).
126. G.P. Reynolds, J.H. Baker, and R.H. Reynolds, *Nature 275*:754 (1978).
127. G. Gregoriadis, *Ann. N.Y. Acad. Sci. 308*:333 (1978).
128. A.C. Allison and G. Gregoriadis, *Nature 252*:252 (1974).
129. T.S. Vlasov, V.S. Salov, V.P. Torchillin, and V.R. Berdichevsky, *J. Epidemiol. Microbiol. Immunobiol. 8*:12 (1982).
130. G. Gregoriadis and A.C. Allison, *FEBS Lett. 45*:71 (1974).
131. A.W. Segal, G. Gregoriadis, and C.D.V. Black, *Clin. Sci. Mol. Med. 49*:99 (1975).
132. A.K. Tanswell and B.A. Freeman, *J. Appl. Physiol. 63*:347 (1987).
133. D.N. Granger, D.A. Parks, and M. Hollworth, *Paediatr. Surg. Int. 1*:15 (1986).
134. B.A. Freeman and J.D. Crapo, *Lab. Invest. 47*:412 (1982).
135. P.C. Panus and B.A. Freeman, in *Liposomes as Drug Carriers* (G. Gregoriadis, Ed.), Wiley, New York, 1988, pp. 473–482.
136. H.M. Patel and B.E. Ryman, *Biochem. Soc. Trans. 5*:1739 (1977).
137. G. Dapergolas and G. Gregoriadis, *Lancet, ii*:824 (1976).
138. H.C. Hemker, W.T. Hermens, A.D. Muler, and R.F.A. Zwaal, *Lancet, i*:70 (1980).
139. C.J. Kirby and G. Gregoriadis, *J. Microencap. 1*:33 (1984).

140. M.J. Oyarzun, J.A. Clements, and A. Baritumo, *Am. Rev. Resp. Dis. 121*: 709 (1980).
141. R.L. Juliano, D. Stamp, and N. McCullough, *Ann. N.Y. Acad. Sci. 308*:411 (1978).
142. K.M.G. Taylor and J.M. Newton, *Thorax 47*:257 (1992).
143. D.A. Thomas, M.A. Myers, B. Wichert, H. Schreier, and R.J. Gonzalez-Rothi, *Chest 99*:1268 (1991).
144. W.J. Martin and J.E. Standing, *J. Pharmacol. Exp. Ther. 244*:774 (1988).
145. H. Ivey, S. Roth, and J. Kattwinkel, *Paediatr. Res. 11*:573 (1977).
146. S.J. Farr, I.W. Kellaway, D.R. Parry-Jones, and S.G. Woolfrey, *Int. J. Pharm. 26*:303 (1985).
147. S.A. Barker, K.M.G. Taylor, and M.D. Short, *Proc. 18th Int. Symp. Controlled Release Bioact. Mat. 18*:289 (1991).
148. K.M.G. Taylor, G. Taylor, I.W. Kellaway, and J. Stevens, *Pharm. Res. 6*: 633 (1989).
149. S.G. Woolfrey, G. Taylor, I.W. Kellaway, and A. Smith, *Proc. Int. Symp. Controlled Release Bioact. Mat. 12*:274 (1985).
150. R.L. Debs, R.M. Straubinger, E.N. Brunette, J.M. Lin, and E.J. Lin, *Am. Rev. Respir. Dis. 135*:731 (1987).
151. M. Mezei and V. Gulasekharam, *Life Sci. 26*:1473 (1980).
152. V.H.L. Lee, P.T. Urrea, R.E. Smith, and D.J. Schanzlin, *Surv. Ophthalmol. 29*:335 (1985).
153. J.T. Mason and C. Huang, *Ann. N.Y. Acad. Sci. 308*:29 (1978).
154. D.J.A. Crommelin and E.M.G. Van Bommel, *Pharm. Res. 1*:159 (1984).
155. T. Hernandez-Caselles, J. Villalain, and J.C. Gomez-Fernandez, *J. Pharm. Pharmacol. 42*:397 (1990).
156. J.M. Gebicki and M. Hicks, *Chem. Phys. Lipids 16*:142 (1976).
157. W.R. Hargreaves and D.W. Deamer, *Biochemistry 17*:3759 (1978).
158. R. Neumann and H. Ringsdorf, *J. Am. Chem. Soc. 108*:487 (1986).
159. P.W. Wertz, W. Abraham, L. Landmann, and D.T. Downing, *J. Invest. Dermatol. 87*:582 (1986).
160. A.T. Florence and A.J. Baillie, in *Novel Drug Delivery and Its Therapeutic Application* (L.F. Prescott and W.S. Nimmo, Eds.), Wiley, New York, 1989, pp. 281–296.
161. A.J. Baillie, A.T. Florence, L.R. Hume, G. Muirhead, and A. Rogerson, *J. Pharm. Pharmacol. 37*:237 (1985).
162. A. Rogerson, J. Cummings, N. Wilmott, and A.T. Florence, *J. Microencaps. 4*:321 (1987).
163. I.G. Tucker and A.T. Florence, *J. Pharm. Pharmacol. 35*:705 (1983).
164. A. Helenius and K. Simons, *Biochemistry 10*:2542 (1971).
165. S. Stafford, A.J. Baillie, and A.T. Florence, *J. Pharm. Pharmacol. 40*:26P (1988).
166. A.J. Baillie, G.H. Coombs, T.F. Dolan, and J. Laurie, *J. Pharm. Pharmacol. 38*:502 (1986).

167. C.A. Hunter, T.F. Dolan, G.H. Coombs, and A.J. Baillie, *J. Pharm. Pharmacol. 40*:161 (1988).
168. K.C. Carter, A.J. Baillie, J. Alexander, and T.F. Dolan, *J. Pharm. Pharmacol. 40*:370 (1988).
169. M.N. Azmin, A.T. Florence, R.M. Hanjani-Vila, J.F.B. Stuart, G. Vanlerberghe, and J.S. Whittaker, *J. Pharm. Pharmacol. 37*:237 (1985).
170. M.N. Azmin, A.T. Florence, R.M. Hanjani-Vila, J.F.B. Stuart, G. Vanlerberghe, and J.S. Whittaker, *J. Microencaps. 3*:95 (1986).
171. C. Cable, J. Cassidy, S.B. Kaye, and A.T. Florence, *J. Pharm. Pharmacol. 40*:31P (1988).
172. D.J. Kerr, A. Rogerson, G.J. Morrison, A.T. Florence, and S.B. Kaye, *Br. J. Cancer 58*:432 (1988).
173. G.M. Ihler, R.H. Glew, and F.W. Schnure, *Proc. Natl. Acad. Sci. USA 70*: 2663 (1973).
174. U. Zimmermann, *Jahresbericht der Kerforschungsanlage Julich GmbH*, Nuclear Research Centre, Julich, 1973, pp. 55–58.
175. P. Seeman, *J. Cell Biol. 32*:55 (1967).
176. H. Bodeman and H. Passow, *J. Membrane Biol. 8*:1 (1972).
177. U. Zimmermann, G. Pilwat, C. Holzapfel, and K. Rosenheck, *J. Membrane Biol. 30*:135 (1976).
178. K. Kinosita, Jr., and T.Y. Tsong, *Science 268*:438 (1977).
179. S. Jausel-Husken and B.J. Deuticke, *J. Membrane Biol. 63*:61 (1981).
180. K. Gersonde and C. Nicholau, *Naturwissenchaften 66*:567 (1979).
181. K. Gersonde and C. Nicholau, *Blut 39*:1 (1979).
182. U. Zimmermann and P. Scheurich, *Planta 151*:26 (1981).
183. J.H. Jandl and M.E. Kaplan, *J. Clin. Invest. 41*:1145 (1960).
184. R.A. Cooper, in *Hematology*, 2nd ed. (W.J. Willimas, E. Beutler, A.J. Erslev, and R.W. Rundles, Eds.), McGraw-Hill, New York, 1977, pp. 216–229.
185. J. DeLoach, S. Peters, O. Pinkard, R. Glew, and G. Ihler, *Biochim. Biophys. Acta 496*:507 (1977).
186. D. Aminoff, W.F.V. Bruegge, W.C. Bell, K. Sarpolis, and R. Williams, *Proc. Natl. Acad. Sci. USA 74*:1521 (1977).
187. J.M. Janick, R. Schauer, K.H. Andres, and H. von Deuring, *Cell Tissue Res. 186*:209 (1978).
188. M.B. Fiddler, L.D.S. Hudson, and R.J. Desnick, *Biochem. J. 168*:141 (1977).
189. U. Zimmermann, in *Targeted Drugs* (E.P. Goldberg, Ed.), Wiley, New York, 1982, pp. 153–200.
190. C.D. Mitchell and D.J. Hanahan, *Biochemistry 5*:51 (1966).
191. E. Beutler, G.L. Dale, E. Guinto, and W. Kuhl, *Proc. Natl. Acad. Sci. USA 74*:4620 (1977).
192. R. Green, J. Miller, and W.H. Crosby, *Blood 50*:78 (1977).
193. D.A. Tyrell and B.E. Ryman, *Biochem. Soc. Trans. 4*:677 (1976).
194. U. Zimmermann, F. Riemann, and G. Pilwat, *Biochim. Biophys. Acta 436*: 460 (1976).

195. S.J. Updike, R.T. Wakamiya, and E.N. Lightfoot, Jr., *Science 193*:681 (1976).
196. J.L. Shank, U.S. Patent 4,001,480 (1977).
197. I.Y. Moroz and B.P. Luzkin, *J. Ichthyol.* (*English Transl.*) *16*:1964 (1976).
198. G. Wilson, in *Site-Specific Drug Delivery* (E. Tomlinson and S.S. Davis, Eds.), Wiley, New York, 1986, pp. 149–164.
199. A. Rembaum, R.C.K. Yen, and R.S. Molday, *J. Macromol. Sci., Chem. A13*:603 (1979).
200. J.-C. Daniel, J.-L. Schuppiser, and M. Tricot, U.S. Patent 4,358,388 (1982).
201. J.G. Treleaven, F.M. Gibson, J. Ugelstad, A. Rembaum, T. Philip, G.D. Caine, and J.T. Kemshead, *Lancet i*:70 (1984).
202. K.J. Widder, A.E. Senyei, and D.G. Scarpelli, *Proc. Soc. Exp. Biol. Med. 158*:141 (1978).
203. K.J. Widder, P.A. Marino, R.M. Morris, and A.E. Senyei, in *Targeted Drugs* (E.P. Goldberg, Ed.), Wiley, New York, 1982, pp. 201–230.

14

Milk as a Food Dispersion

DAVID S. HORNE Hannah Research Institute, Ayr, Scotland

I.	Introduction	500
II.	Milk Composition	501
	A. Individual milk proteins	502
III.	Casein Micelles	505
	A. Micellar dissociation	506
	B. Micellar size and size distribution	507
	C. Variations in micellar composition with size	509
IV.	Routes to Micellar Destabilization and Aggregation	512
	A. Acid coagulation	513
	B. Heat-induced coagulation	513
	C. Ethanol stability	515
	D. Rennet coagulation	517
V.	Micellar Interactions	518
	A. Failure of the DLVO theory	518
	B. Steric stabilization	519
	C. Ethanol stability and steric stabilization	519
	D. Chymosin treatment and micellar surface structure	524
VI.	Micellar Structure	526
VII.	Casein Micelles and Fractals	529
VIII.	Concluding Remarks	534
	References	534

I. INTRODUCTION

All mammals secrete milk for the sustenance and nourishment of their newborn young. As foodstuffs, these complex biological fluids are a blend of all essential dietary constituents—lipids, proteins, carbohydrates, and minerals—in proportions generally specific to the nutrient requirements of each species. Most mammals consume milk only for the first months of their life. Humans, however, have recognized the nutritional properties of the milk of certain species and domesticated them from the earliest days of history to provide themselves with a constant, ready supply of this foodstuff. In a further exercise of our ingenuity, we have sought to preserve this milk and its constituents in a variety of ways. The results of this are seen in such foodstuffs as butter, cheese, and fermented products that date from Biblical times and in the more modern products such as ultrahigh-temperature treated and evaporated milks, milk powders, ice cream, or even cream liqueurs. The majority of these are manufactured from bovine (cow's) milk, but the milks of sheep, goat, and buffalo are frequently used in certain regions of the world for the entire range of products or, more commonly in the Western world, for specific varieties such as Roquefort cheese from ewe's milk.

Dairy foods thus span the entire range from powdered solids through viscoelastic gels to emulsions and liquid dispersions. This chapter, however, is concerned only with fluid milk, the liquid milk as secreted by the cow. The major constituent of cow's milk is water; the remainder consists largely of lipids, proteins, and carbohydrates synthesized within the mammary gland. Also present are mineral components and some water-soluble and lipid-soluble materials transferred directly from blood plasma, specific blood proteins, traces of enzymes, and intermediates of milk component synthesis.

Many readers will consider milk to be a low concentration emulsion of lipids stabilized by protein in an aqueous phase. This is true of freshly secreted milk, in which the large fat droplets of 1–12 μm diameter are coated with fat globule membrane protein. It is also true of homogenized milk, which is the most common form of liquid milk sold in the United States and is produced by passing milk through a small orifice under high pressure. This process breaks up the fat globules, increases their total surface area, and brings into play other surface-active milk proteins as stabilizing entities. But fresh milk, as well as being the emulsion described, is also a dispersion of aggregated protein/mineral complexes of colloidal dimensions in that same aqueous phase. These colloidal particles are the casein micelles, and their properties form the substance of this chapter on milk as a food dispersion. The fat portion can be largely removed from the milk, either by centrifugation in the laboratory or by a cream separator on a commercial scale, to leave the product we know as skim milk. Unless otherwise stated, all studies referred to in this chapter involve skim milk or casein micelles drawn from it. The various sections deal with the composition of milk; the casein and serum proteins; the size, composition, and structure of the casein micelles; the stability and integrity of those micelles; the interactions involved; and how our understanding of these has been furthered by study of the various techniques and methods employed to destabilize the micellar system.

II. MILK COMPOSITION

It is important to emphasize at the outset that, as a natural product, milks can often have highly variable compositions. Within even bovine milk, substantial variations arise from climatic or seasonal differences, from breed of cow, diet, stage of lactation, age, and health. All of this has to be borne in mind when conducting milk-based experiments or comparing results between laboratories. Milk researchers do not have the luxury of a constant substrate for their studies. Table 1, from Muir [1], outlines the gross composition of a typical mid-lactation bovine milk. The milk solids content averages 12.8%. Of these solids, milk sugar, lactose, is the major component, and fat and protein form the bulk of the remainder.

Breed has a distinct effect on the gross composition of milk. The milks of the Channel Islands breeds, Guernsey and Jersey, have an unusually high content of the major components, fat in particular. Even within a breed, however, a wide variation in composition can be observed between individual cows in a herd. These differences are partly the result of physiological and environmental factors and partly genetic in origin.

TABLE 1 The Average Composition of Bulk Milk

Constituent	Concentration (g/L)	Proportion solids (%)
Fat	37.0	28.9
Protein:	34.0	26.6
casein	27.6 }	
whey protein	6.4 }	
Non-protein nitrogen	1.9	1.5
Lactose	48.0	37.5
Ash	7.0	5.6
Total solids	127.9	100.0

Source: Ref. 1.

Profound changes in milk composition also occur as lactation progresses. Colostrum, the milk secreted at the beginning of lactation, is rich in fat and protein but low in lactose. This milk is also particularly rich in immunoglobulins, proteins associated with conferring disease resistance, but their presence persists for only a few days. As lactation progresses, there is a rapid reduction in fat and in protein content, to minimum values within 6–8 weeks of lactation. Lactose content also increases to a maximum 2–4 weeks into lactation. Thereafter, for the remainder of lactation, these changes are reversed. There are also comparable changes in the major minerals of milk, calcium and phosphate, during lactation.

Dietary effects on milk composition are complex. Diet affects mainly the fat content and its composition, which can be influenced by both the type and amount of fat included in the diet as well as by the nonlipid components of that diet. The protein content of milk and its composition are, however, largely resistant to dietary manipulation. On the other hand, genetic variants of these proteins are found, some of which result in higher total protein levels being present in the milk. Possibilities therefore exist of increasing overall protein content by selective breeding programs.

A. Individual Milk Proteins

The proteins in milk fall into two distinct types, whey proteins and caseins. The physicochemical and technological properties of milk are largely determined by the behavior of the caseins. Almost all of these caseins in freshly secreted milk are contained in the aggregated colloidal particles known as the casein micelles, the particles of our dispersion. Figure 1 gives the typical protein distribution in a midseason, midlactation skim milk. As can be seen here, the caseins constitute over 80% of the total protein. Originally defined as the phosphoproteins precipitated from milk at pH 4.6, it is now recognized that this family of proteins is made up of four distinct primary members. These are designated α_{s1}-, α_{s2}-, β-, and κ-caseins and exist in milk in the approximate weight ratio 4 : 1 : 4 : 1 [2]. The primary structures of all four have been completely elucidated [3]. They differ from one another in sequence and in their content of phosphoseryl residues.

All the caseins are phosphorylated at serine, or rarely threonine, residues to varying extents. α_{s1}-Casein contains eight such esterified residues, but occasionally an additional serine residue is also phosphorylated. α_{s2}-Casein carries between 10 and 13 phosphates per molecule, whereas β-casein invariably contains five. κ-Casein contrasts with the other caseins in that it has only one serine phosphate ester group, but it is also the only casein post-translationally glycosylated, and this to varying extents in different molecules in the same milk. In general the phosphate side groups are concen-

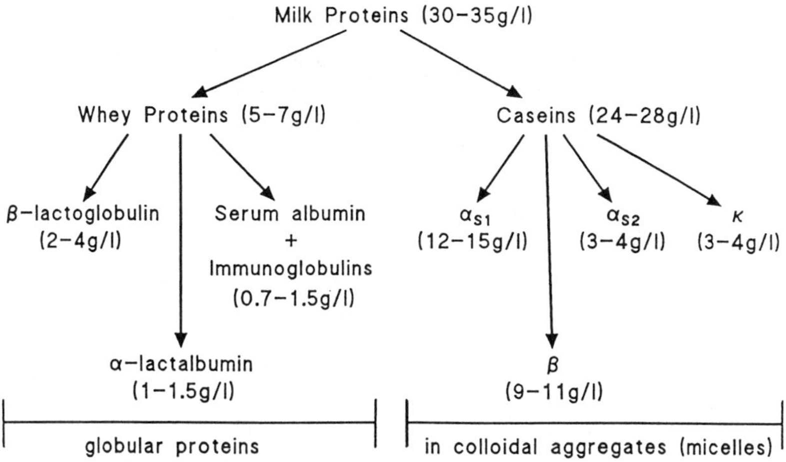

FIG. 1 Distribution of protein classes and individual proteins in skimmed cow's milk.

trated in clusters toward the N-terminal end of the molecules, and, as the remainder of the amino acid chain is largely hydrophobic, this gives the molecules, especially β-casein, amphiphilic properties. This allows the caseins to act as highly effective surface-active agents, making them excellent stabilizers of foams and emulsions, and is responsible for their widespread use in that role in the food industry.

Because of their high content of phosphoseryl residues, the α_{s1}- and β-caseins bind polyvalent cations strongly, most relevantly Ca^{2+}, which leads to charge neutralization, aggregation, and eventually precipitation. Many of the technologically important properties of milk—for example, heat stability, alcohol stability, rennet coagulability, and the strength and syneresis properties of rennet gels—are strongly influenced by the calcium content of a milk. Also, in the manufacture of certain dairy products, for example sterilized concentrated milk, processed cheeses, and cream liqueurs, it is necessary to reduce the ionic calcium level by adding chelating agents, usually ortho- or polyphosphates or citrate, to prevent gelation.

Calcium binding by the caseins is influenced by temperature, pH, and ionic strength. At low concentrations of calcium, the Ca^{2+} ions are probably bound exclusively to phosphoseryl residues, but at higher concentrations binding to aspartyl or glutamyl residues may also occur. The precipitation of α_{s1}- and β-caseins by Ca^{2+} is the result of a reduction of the overall charge on binding of the cation to the protein; at neutral pH α_{s1}-casein has a net charge of −22 units and β-casein a charge of −12 units. As charge is reduced, the repulsion between the molecules is decreased, and aggregation, effectively isoelectric precipitation, begins to take place. In the case of α_{s1}-casein, octamers are formed initially, but these aggregate and precipitate at higher Ca^{2+} levels [4]. Since the aggregation of β-casein depends strongly on hydrophobic interactions and hence is temperature-dependent, the precipitation of β-casein by Ca^{2+} is more strongly dependent on temperature than is that of α_{s1}-casein, so much so that β-casein remains soluble in the presence of Ca^{2+} at temperatures less than 20°C.

κ-Casein, which contains only one phosphoseryl residue, does not bind Ca^{2+} strongly and hence is not precipitated by calcium. Furthermore, κ-casein will associate with α_s- and/or β-casein and stabilize them against precipitation by Ca^{2+}, forming instead stable colloidal particles whose size depends on the proportion of κ-casein and the concentration of Ca^{2+}. This stabilizing ability of κ-casein is lost if the molecules of κ-casein are cleaved by the enzyme chymosin, the active constituent of the rennet preparation used in cheesemaking.

The caseins also differ from the whey proteins in their lack of organized secondary structure. Thus, unlike the whey proteins, they are relatively unaffected by heat treatment, suffering no recognizable denaturation either in the sense of a conformational change or as a rearrangement of disulfide bridges. Indeed the caseins are almost devoid of sulfhydryl groups; only α_{s2}- and κ-casein possess such residues. α_{s2}-Casein contains two 1/2-cystine residues per molecule, which exist naturally as intramolecular disulfide bonds. κ-Casein, which also contains two 1/2-cystine residues per molecule, normally exists as disulfide-linked polymers. In raw milk, κ-casein does not appear to form disulfide-linked polymers with α_{s2}-casein or the whey proteins, but following heat treatment of milk, the globular serum proteins denature, and β-lactoglobulin, in particular, forms a disulfide-linked complex with κ-casein [5] with major effects on many technologically important properties of milk.

The heterogeneity of milk proteins is further complicated by their exhibiting genetic polymorphism. This genetically controlled substitution generally involves only one or two amino acids but occasionally deletions of considerable lengths of polypeptide chain. It was first recognized in the whey protein β-lactoglobulin but has since been demonstrated in all of the principal milk proteins. Recent studies have shown that genetic polymorphism has technological significance, especially in relation to cheese manufacture. Since the frequency of occurrence of these genetic variants is species- and breed-dependent, this raises the possibility of selective breeding to yield milks of desired properties, fashionably termed "designer milk." Before such drastic steps are taken, however, a greater understanding of the effects of genetic polymorphism on the processing properties of milk must first be secured.

κ-Casein is the only glycosylated member of the casein family. Despite considerable research, information on the number and type of oligosaccharide groups is inconsistent and incomplete. The level of glycosylation is believed not to influence the micelle-stabilizing ability of κ-casein but does influence its susceptibility to hydrolytic reaction with the enzyme chymosin, susceptibility decreasing with increasing carbohydrate content.

In contrast to the caseins, the whey proteins (Figure 1) are typical globular proteins. The major partners are β-lactoglobulin and α-lactalbumin. In their native state, their many cysteinyl residues result in the formation of intramolecular disulfide bonds that stabilize the protein structures. They lack phosphate groups and do not bind large numbers of calcium ions. Their degree of aggregation or association is minimal in their native state; β-lactoglobulin exists as a dimeric molecule. At temperatures in excess of 70°C, the whey proteins denature, a process accompanied by the extensive rupture and randomization of the stabilizing disulfide bonds. When this takes place in heated milk, the aforementioned interaction between β-lactoglobulin and κ-casein can also occur.

III. CASEIN MICELLES

The description of the individual casein proteins has been extensive because these molecules form the building blocks of the dispersed colloidal particles of skim milk, the casein micelles. In normal milk, 95% of the caseins exist in these aggregated particles, whose molecular weights average approximately 10^8 daltons and whose mean hydrodynamic diameters are about 200 nm. Their size and light-scattering power are the source of milk's white appearance. The micelles consist of α_s- and β-caseins stabilized by κ-casein, in association with Ca^{2+} and calcium phosphate. Each individual micelle may contain many thousands of casein molecules, but there is also a size distribution, and small micelles are most numerous. Assembly of the micelles occurs largely in the Golgi vesicles of the secretory cells of the mammary gland. The micelles are thus excreted as intact entities, though a degree of postsecretory aggregation of micelles has been detected [6].

There are two aspects to the stability of casein micelles as a colloidal dispersion. Although the bulk of this chapter is concerned with aggregation processes that follow destabilization, micellar environmental conditions can be altered in such a way as to induce disintegration of the micelles or preferential dissociation of some of the individual caseins. The factors that govern the processes of aggregation and dissociation are obviously different: the first phenomenon almost certainly depends on the behavior of the surface layers of the micelle and how they interact, whereas the second depends more on the internal structure of the particles and how their components are held together.

The internal structure of the casein micelle is not firmly established. Several models have been proposed, but none of them is wholly satisfactory. The particles are basically spherical but show a rough surface [7]. The simplest model envisages the structure as a core of α_s- and β-caseins and calcium phosphate surrounded by a coat of κ-casein [8]. An alternative model represents the casein micelle as an aggregate of "submicelles" that are themselves limited aggregates of casein [9], thus giving the raspberry-like appearance observed in electron micrographs. Neither of these models explains the role of calcium phosphate. A simple extension of the submicelle model suggests that the subunits are linked by calcium phosphate [7], but more recent studies suggest that the calcium phosphate and the α_s and β-caseins are linked by the participation of their phosphoseryl residues in the structure of the micellar calcium phosphate [10]. This latter model is more appealing in that it ascribes a detailed functional role to the calcium phosphate; however, some doubts as to the necessity for such a role are raised by studies of the dissociation behavior of casein micelles following acidification of milk [11]. A common factor in most of the models is that the micellar κ-casein appears to be close to, or on, the surface of the casein micelles. Further evidence for this and its consequences for micellar stability are discussed in greater detail later in this chapter.

A. Micellar Dissociation

Dissociation of the micellar casein can be achieved in several ways. Although in freshly secreted milk there is little soluble casein and nearly all of the casein is incorporated into the micelles, when milk is stored at low temperatures, 4–6°C, as much as 40% of the β-casein dissociates from the micelles [12]. Little of the α_s- or κ-casein passes into solution, and it has been assumed that these two components provide the structural framework of the micelle. This low temperature dissociation of β-casein appears to be reversible, insofar as only slight changes in behavior are found when the milk is rewarmed.

Dissociation of casein also occurs when the pH of milk is decreased from its natural value of about 6.7 [11], the results of which are temperature-dependent. Again β-casein is found to be the largest single component in the liberated casein. At 30°C, less than 10% dissociation occurs when milk pH is reduced to 5.0. This contrasts with the behavior at 4°C, where a maximum of 60% of the casein is dissociated before isoelectric precipitation takes place at pH values around 5.0. All caseins are liberated at all temperatures. The fraction of κ-casein found in the serum, though increasing with temperature, is virtually independent of pH, while the proportion of α_s- and β-casein appear to compensate for each other as the pH is lowered, particularly at 4°C, with the fraction of β-casein released increasing when that of α_s-casein decreases.

Three processes can be identified that affect the dissociation of caseins from micelles when pH is lowered. These are (1) the hydrophobic effect [13], where dissociation is greatest at low temperatures and decreases as temperature increases but should have no obvious pH dependence; (2) the dissociation of calcium and phosphate from the micelles, which causes a loss of casein that increases as the pH decreases; and (3) the isoelectric precipitation of the caseins as the pH falls to a critical value. The last of these is both pH- and temperature-dependent. As for the second, studies of the release of calcium phosphate show similar release of mineral at both 4 and 30°C, but the patterns and extent of casein dissociation are distinctly different, indicating perhaps more than one association or form of micellar linkage in the casein micelle.

Even when milk pH and temperature are maintained constant, dissociation of casein from the micelle can be induced by dialyzing milk against buffers that are depleted in calcium or phosphate [10]. When buffers nearly saturated with respect to micellar calcium phosphate were employed, no significant casein dissociation was observed until the free ionic calcium level was reduced below 1 mM, less than half the level found normally in milk. The ease of release of the caseins was then $\kappa > \beta > \alpha_{s1} > \alpha_{s2}$. When buffers high in free calcium (6 mM) were used, both micellar calcium and phosphate in the milks were decreased over the dialysis period (up to 110 h), but no significant casein dissociation took place until more than 30% of the phosphate had been released. The release of the caseins followed a pattern similar to that described above with serum levels, $\kappa \approx \beta > \alpha_{s1} > \alpha_{s2}$. These results suggest that colloidal phosphate is of considerable importance in maintaining micellar integrity but that short-term stability can be preserved by maintaining a high free Ca^{2+} level in the buffer.

B. Micellar Size and Size Distribution

The casein micelles in milk have a distribution of sizes. Though the proportions of the individual caseins in milk are fairly constant, micelles of different sizes have been shown to be of varying composition, with a higher proportion of κ-casein being found in the smaller micelles [14–17]. Since it is possible that micelles of different compositions may behave differently during milk processing, considerable effort has been expended in attempting to define micellar size distributions and the factors controlling them.

The studies [14–17] reported above determined micellar sizes by electron microscopy, detecting some micelles up to 680 nm in diameter. Casein micelles are highly hydrated species, however, and early electron microscopic measurements were thought to be subject to artifactual errors introduced during sample preparation. Measurements of average micelle size using dynamic light scattering initially gave values considerably in excess of the electron microscopic values and appeared to confirm original doubts [18]. A subsequent study [19] of the micelles from the milks of two cows by both electron microscopy and a combination of light-scattering techniques suggested that the root of the problem lay in the width of the micellar size distribution function. The early electron microscopic results had been weighted toward the smaller micelles because insufficient numbers of micelles had been counted to adequately describe the large size tail of the distribution. The light-scattering results were intensity-weighted. The weighting factor comprised a product of size raised to the sixth power and a particle-scattering factor whose inverse dependence on particle size had the effect of reducing the size exponent in the final weighting factor from 6 to 4 or 3 over the size range encompassed by the micellar size distribution. This yields a light-scattering estimate of average size biased toward the larger particles. Only by recognizing these factors was the discrepancy between the average sizes measured by the two techniques resolved.

Further measurements showed that the average sizes obtained from different milks were a function of milk composition. Variations in average micellar size by up to a factor of 2 were found not only between individual cow milks but also in milks from the same animal during the course of its lactation [20]. When the survey was broadened to include bulked milk samples from local creameries, marked seasonal variations associated with feed patterns were observed. In these milks, highly significant correlations were demonstrated between average micellar size and composition, positive with respect to colloidal calcium phosphate content and negative with respect to casein-bound calcium content [21]. Of the soluble milk components tested, only citrate content was found to be positively correlated with average micelle size.

The majority of these studies of the compositional factors influencing micelle size were made using a turbidimetric technique developed by Holt et al. [22,23]. This relates average micelle size derived from an assumed functional form for the size distribution to the gradient of the double logarithmic plot of diluted micellar turbidity and wavelength.

Direct measurement of the casein micelle size distribution is, of course, one of the possible outcomes of electron microscopic studies, but to achieve reasonable statistical confidence, a large number of micelles have to be included in the sample—an extremely tedious and time-consuming chore that is not free of artifactual error. Hence this technique has not been widely applied to micelle distribution studies. In principle, it should be possible to extract size distribution parameters from dynamic light-scattering measurements on diluted micellar suspensions where single scattering prevails. The correlation function decay curve from such polydisperse systems is multiexponential as a result of the intensity-weighted contribution from each individual size class [24]. In a monodisperse system, a single exponential would be obtained, and from its relaxation time a diffusion coefficient and hence particle size would be calculated. Deconvoluting the correlation function for a polydisperse system to yield the size distribution weighting is fraught with difficulty, and much effort has been expended in deriving robust analytical routines to solve this ill-conditioned inversion problem.

Two of these approaches have been applied to dynamic light-scattering measurements on casein micelle suspensions. In the first of these attempts, Horne [25] assumed a two-parameter functional form for the size distribution curve, including variation of the average diffusion coefficient with scattering angle in his measurements, to derive values for the two descriptive parameters of the distribution function. In the second attempt, Horne and Dalgleish [26] applied the "exponential sampling" procedure developed by Ostrowsky et al. [27]. Previous applications of direct inversion techniques had been restricted to particles small compared to the wavelength of light. This simplified the derivation of the weight fraction size distribution from the relaxation time distribution that the analysis of the correlation function actually yields. Casein micelles, however, range in size up to 50% or more of the light wavelength, and consequently the derived line-width distribution is angle-dependent, though the correction factors are calculable from the known form factor or $P(\theta)$ functions. Horne and Dalgleish [26] exploited this to establish greater confidence in the uniqueness of their calculated fits by imposing the following condition: When their derived relaxation time functions are corrected for form factor weighting, the "best-fit" solutions for the weight fraction distributions should be independent of scattering angle. This approach foundered when applied to a casein micelle suspension containing the full range of sizes found in milk, because the particle form factors oscillate with scattering angle for the larger micelles, a situation with which the exponential sampling analysis procedure is unable to cope. Removal of the largest micelles by centrifugation or filtration gave a micellar suspension to which the proposed strategy could be applied with complete success. Further developments in the inversion of dynamic light scattering data have come about, but their application to determining casein micelle size distributions is still awaited.

C. Variations in Micellar Composition with Size

Permeation chromatography on gel or controlled pore glass (CPG) substrates has also been used to demonstrate the width of the casein micelle size distribution. The technique has been more widely used, however, to examine the composition of the fractionated micelles as a function of their size, with only qualitative verification of micellar size decreasing with elution volume being demonstrated. Thus Ekstrand and Larsson-Raznikiewicz [28,29] noted that fractionation according to size took place by demonstrating a monotonic decrease in the ratio of the turbidities of the fractions measured at 320 and 750 nm. McGann et al. [14] demonstrated the size separation capability of CPG columns by using electron microscopy to measure average sizes after pooling their fractions into three groups in order of their elution. Only Griffin and Anderson [30] have attempted to derive a micellar size distribution by fractionating on CPG, measuring the size of the micelles in each collected fraction by dynamic light scattering. Though they suggested that their profiles gave a size distribution and expressed this as a cumulative weight fraction, what they in fact measured was the casein micelle size distribution convoluted with the column spreading function. This latter function is extremely complicated, changing across the profile because the limited size of the pores in the glass beads denies access to the largest micelles, thereby diminishing the effects of column fractionation in a stepwise fashion. This results in a bimodal chromatographic profile for what is confidently expected to be a monomodal size distribution function. Though useful for assessing qualitative differences, particularly when the results are plotted in a cumulative fashion, chromatographic techniques have yet to yield quantitative data on micellar size distributions.

All of these chromatographic studies are subject to the further criticism that during the fractionation process, unless suitable eluent buffers are chosen with care, the micelles can be subjected to experimental conditions that can permit partial dissociation, thus perhaps distorting the size distribution function. Griffin and Anderson [30] avoided this problem by fixing the micelles by cross-linking the proteins with glutaraldehyde. Such an option was not open to those interested only in micellar composition and its variation with size because the cross-linking reaction makes subsequent analysis for individual caseins almost impossible.

Differential high speed centrifugation avoids the need for an eluting and possibly dissociating buffer. Though this too has been criticized as perhaps compressing the micelles and thereby altering their voluminosity, the technique has been successfully used by various authors to study variation in micellar composition with size [12,16,31]. In the most recent of these studies, Dalgleish et al. [31] separated the micelles into eight fractions and measured the average size in each resuspended fraction using dynamic light scattering. After precipitation of the micellar casein, analysis of the proteins was performed using fast protein liquid chromatography (FPLC).

The results for the Friesian milk sample tested showed that the content of κ-casein increased as the micellar size decreased, from 5.7% in the largest micelles to 12.9% in the smallest. This trend confirmed earlier observations, and the range of values was very similar to that found by Donnelly et al. [15] for CPG-fractionated micelles. However, the centrifugation results differed on the contents of α_s- and β-caseins. In the centrifuged fractions, the increase in content of κ-casein was almost exactly balanced by a decrease in the micellar content of β-casein, and the proportions of both α_{s1}- and α_{s2}-caseins were independent of micellar size.

The changes in casein composition of the centrifuged fractions were not accompanied by large changes in their contents of calcium and phosphate [31]. The phosphate content appeared to decrease slightly as the micellar size decreased, but the calcium content remained essentially constant. This resulted in an increase in the ratio of calcium to inorganic phosphate from 1.90 for the largest micelles to 2.00 for the smallest. As the micelles decrease in size, their levels of casein phosphate decrease as the β-casein (five phosphoserines) is replaced by κ-casein (one phosphoserine). This may suggest a lesser requirement for calcium phosphate in the smaller micelles and consequently a lower phosphate content.

Converting these protein compositions into a distribution pattern between micellar surface and interior demanded the adoption of a value for the micellar hydration. This allowed a volume for an individual protein molecule and hence a surface layer thickness to be defined. Accepting as the most likely value 3 g of H_2O per gram of protein [32], the plots of Fig. 2 showing the fractional contents of the individual caseins as a function of

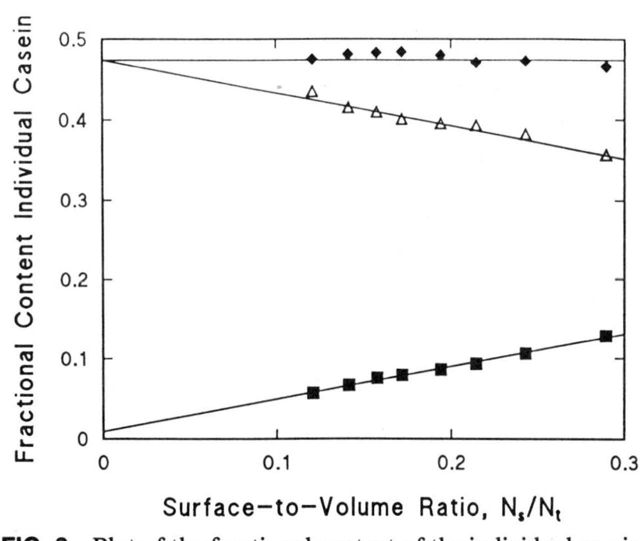

FIG. 2 Plot of the fractional content of the individual caseins in micelles of different sizes, as defined by their surface-to-volume ratio. (♦) Sum of α_{s1}- and α_{s2}-caseins; (△) β-casein; (■) κ-casein.

surface/volume ratio were computed. It was clear from these calculations that virtually all of the κ-casein was located on the surface of the micelle. The calculated content of κ-casein in the interior of the micelle ranged from 0.6 to 1.2% when hydration was varied between 1 and 5 g/g. Conversely, about 47% of the interior protein was β-casein. Since α_s content was independent of size, the interior also contained approximately 48% α_s-casein and was thus composed of almost equimolar amounts of β- and α_s-caseins.

Calculations of the proportions of the different caseins on the surface, particularly of β-casein, depended rather more on the value chosen for protein hydration. The preferred value of 3 g/g gave a calculated κ-casein content of 42% and a β-casein content of 7%. With the α_s-casein content remaining at the 48% demanded by its constancy with size, it is clear that the micellar surface is predominantly composed of α_s- and κ-caseins, with only small amounts of β-casein. These observations that some proportion of all of the individual caseins may be found on the micellar surface are consistent with earlier studies of protein hydrolysis using immobilized enzymes where peptides originating from all the caseins were detected [33,34].

IV. ROUTES TO MICELLAR DESTABILIZATION AND AGGREGATION

Casein micelles can be induced to aggregate by several treatments, some of which are exploited in the manufacture of a range of dairy products. These treatments include acidification of the milk, heating, addition of ethanol or other organic solvents, limited proteolysis by enzyme action, and addition of calcium. All of these treatments can be carried out singly or in combination. Each process modifies the micelle structure in such a fashion as to render the micelle unstable, some more extensively than others.

In addition to the adventitious uses in some dairy manufacturing processes, micellar destabilization can also create serious problems during the production or storage of some dairy products, such as coagulation during the sterilization of concentrated milks or age-gelation of sterilized concentrates, especially of milks subjected to UHT treatment (ultrahigh temperature of 140°C for 5–10 s). In view of their great importance to the dairy industry, considerable research effort has been and is still being expended in studying these destabilization processes. Some are so disruptive of micellar integrity that it is doubtful whether the process relates to the aggregation of the original dispersed colloidal particles. Others perhaps affect only the surface structure of the micelles, and it is with these that we are largely concerned here. For completeness, however, a brief résumé of the more important destabilization processes and their benefits in processing follows.

A. Acid Coagulation

By definition, the caseins are those milk proteins that are precipitated by acidification of the milk to their isoelectric points, approximately pH 4.6. This is the basis of the initial stages of the production of sodium caseinate. If the milk temperature is maintained below 6°C, the caseins remain more or less soluble. As the temperature at precipitation is raised, a progressively coarser precipitate is obtained. The disintegration of the micelle and the solubilization of the micellar calcium phosphate, with accompanying release of the caseins into the serum phase as the pH of the milk is lowered toward 4.6, have already been discussed. The subsequent precipitation of the caseins at pH 4.6 cannot therefore be strictly regarded as a micellar aggregation process, though a great deal of the structural integrity of the colloidal particle can be retained under certain circumstances.

Precipitate formation following acidification occurs when the mixture is continuously stirred. If the milk is left quiescent, then a gel may be formed. This is exploited in the formation of yogurt and cottage cheese. In yogurts, acidification is achieved by inoculating the milk with a culture of specific bacteria that produce lactic acid as they grow over a period of hours at around 40°C. For cottage cheese production, starter bacteria are also employed, but the recipe additionally generally requires the inclusion of low levels of the enzyme chymosin and a lower reaction temperature of 25–30°C. The gelation mechanism and the stability of acid gels have been reviewed by Roefs [35] and by Fox and Mulvihill [36].

B. Heat-Induced Coagulation

Heating of milk to 120–140°C in a sealed system, as required for in-can sterilization or production of UHT products, can also affect micellar stability, but the physicochemical effects are not fully understood. Two reactions have been identified. First, at 80–100°C, the whey proteins denature and bind to the casein micelles, principally to the κ-casein component through disulfide bridges. Second, perhaps as a consequence of the first reaction, the micelles aggregate to form a gel or precipitate depending on the reaction conditions. These reactions can occur during processing and assume importance in concentrated milks where precipitation takes place at lower temperatures.

Heat stability of milk is generally defined as the time required to induce visible coagulation when the milk is heated at 140°C. If the pH of the milk is adjusted over the range 6.0–7.5 prior to heating, a heat stability versus pH profile characteristic of the original milk can be obtained. These curves fall into two categories as exemplified in Fig. 3—Type A, for which the plot has a pronounced minimum, and Type B, for which there is a mono-

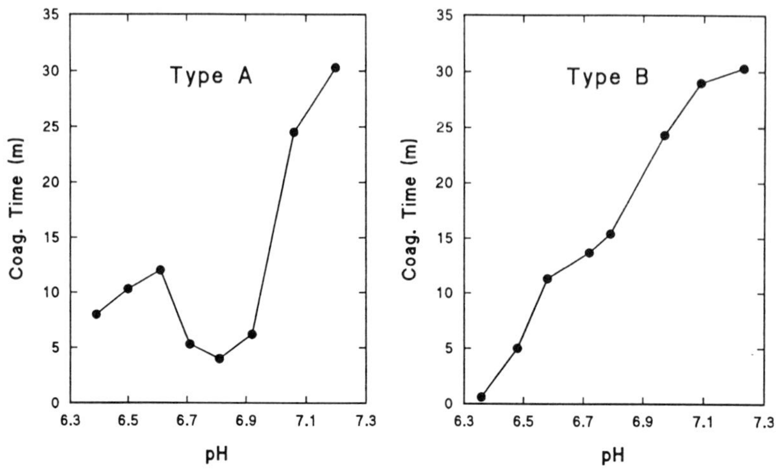

FIG. 3 Typical heat stability profiles; plots of time to achieve coagulation of milk at 140°C as a function of adjusted milk pH. These illustrate the two types of curves generally encountered.

tonic increase in coagulation time as pH is increased. It is possible in suitable circumstances to induce a milk showing Type A characteristics to become Type B, and vice versa. The ways in which this can be done and some suggested mechanistic reasons for their action have been reviewed by Horne and Muir [37]. These authors suggest that the destabilization reaction occurring within the minimum in a Type A situation is controlled by the formation of the β-lactoglobulin/κ-casein complex referred to above. Outwith the minimum, the most important single milk component controlling heat stability is the organic compound urea, which is present at low levels in all milks. Evidence for this came in the first instance in studies by Holt et al. [38], who demonstrated a very highly significant correlation between the coagulation time of farm and creamery milk at its natural pH and the milk urea content. Subsequent studies in Ireland [39] confirmed this observation: 89% of the seasonal variance in heat coagulation time at natural pH was accounted for by variations in urea level. There is clear evidence that the urea level in milk is controlled by feeding practice [40]. In two separate series of experiments, the direct link between diet, blood urea level, milk urea concentration, and heat coagulation time was clearly established [41]. Urea level is subject to dietary manipulation and control, and through this the heat stability behavior of milk at its natural pH is also adjustable. The mode of action of the urea is still unclear, but a plausible mechanism is given in the review by Horne and Muir [37].

It should also be noted that Singh and Fox [42] propose an alternative explanation for the origins of the minimum in the heat stability versus pH profile. They suggest that the β-lactoglobulin/κ-casein complex dissociates from the micelle at pH values above the profile maximum, thus leaving a κ-casein-depleted micelle core with increased calcium sensitivity and reduced heat stability.

C. Ethanol Stability

Stability tests based on the addition of alcohol to milk are of long standing. In the late nineteenth century, they, or simple boiling tests, provided an objective assessment of the keeping quality of milk for retail sale. If a precipitate formed when an equal volume of 70% alcohol solution was added or when milk was boiled, then that batch of milk was rejected. As late as the 1930s such tests were accepted as indicators that milk was going "off."

Even as the necessity for such tests declined with rapidly improving bacteriological quality in milk, stability defects associated with season, diet, and stage of lactation came to be recognized. Stability tests based on alcohol addition proved to be unreliable indicators of these defects and, in particular, of the suitability of milk for processing into evaporated or condensed products. It was not until the early 1980s that an extensive study of the susceptibility of milk to alcohol-induced coagulation was carried out and a mechanistic basis was proposed for the reaction. Its importance in this chapter on milk as a colloidal dispersion is clear, because of all the methods of inducing micellar aggregation, it and chymosin-induced aggregation are perhaps the least disruptive to the casein micelle.

These studies of the ethanol stability test have been summarized recently by Horne [43], and a comparison of the response of milk in alcohol and heat stability tests was the subject of an earlier review [37]. Though initially carried out as an academic exercise seeking to further understanding of casein micelle interactions and stability, the research on ethanol stability found rapid exploitation in the production of cream liqueurs, particularly in the important area of extending their shelf life. This aspect has been reviewed by Banks and Muir [44].

In the modern version of the alcohol test, ethanol stability is defined as the strength (% v/v) of the ethanol solution required to induce instantaneous precipitation when two parts of the ethanol solution are rapidly mixed with one part of milk. Again, as with heat stability, adjustment of the milk pH above and below the natural value changes the ethanol stability in a manner characteristic of that milk. These ethanol stability versus pH profiles always have a sigmoidal shape (Fig. 4) that can be characterized by

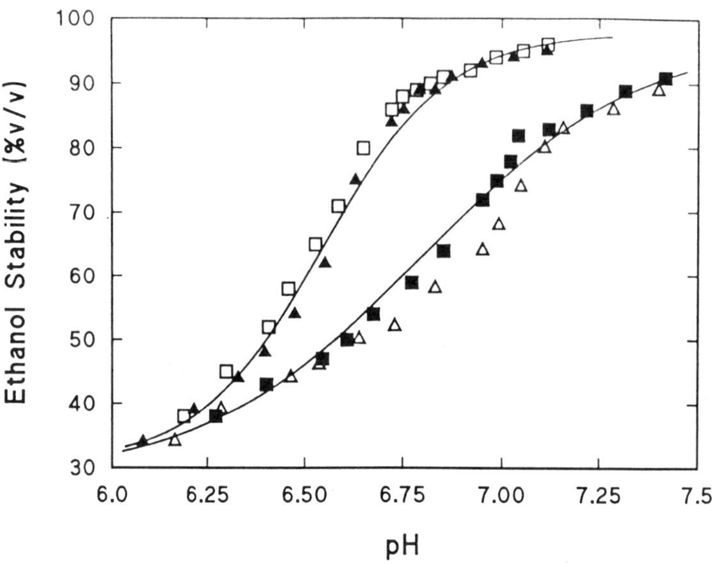

FIG. 4 Typical ethanol stability versus pH profiles. (▲) Milk A; (■) milk B. The figure also illustrates the importance of milk serum components in controlling ethanol stability. These components can be interchanged by dialysis of one milk against another, giving stability profiles denoted by the open symbols: (△) milk A after dialysis against milk B; (□) milk B after dialysis against milk A. The solid lines are "best-fit" profiles for the experimental data for milks A and B, calculated using Eq. (1).

a simple mathematical form defined by four parameters: S_{min} and S_{max}, the asymptotic minimum and maximum ethanol stabilities at low and high pH, respectively; p*K*, denoting the transition or inflection point of the profile along the pH axis; and *b*, the gradient of the logarithmic transform of the profile function, this function being given by the equation

$$\frac{S_{max} - S}{S - S_{min}} = 10^{(\mathrm{pH} - \mathrm{p}K)/b} \tag{1}$$

Studies relating the profile parameters to milk composition demonstrated that the level of ionic calcium plays a dominant role in controlling ethanol stability, particularly in dictating the value of the profile p*K* parameter [45]. The involvement of ionic calcium, which is known to precipitate the individual caseins by an isoelectric precipitation mechanism, binding to the molecules and neutralizing their negative charge [4,46], points to a strong electrostatic involvement in the micellar interaction potential. Further evidence of this is furnished by the observations that (1) the effects of chemical modification of milk protein on ethanol stability can be explained as due to changes in protein charge [47] and (2) other alcohols and similar solvents are as effective as ethanol as milk protein precipitants when compared on their ability to reduce the dielectric constant of the system to the critical level [48]. As we will see later, however, the view that ethanol-induced coagulation is the result of the removal of an electrostatic repulsion component from the overall free energy equation is somewhat simplistic.

D. Rennet Coagulation

The most technologically significant destabilization of casein micelles is that caused by limited proteolysis of κ-casein. This phenomenon is exploited as the first stage in the manufacture of most cheese varieties, some 10^7 tons of cheese being produced annually worldwide. Most proteinases are capable of enzymatically coagulating milk, but κ-casein is uniquely sensitive to acid proteinases, especially chymosins. Chymosin, the active enzyme of calf rennet, is highly specific in its action, splitting the κ-casein molecule into what are termed its *para*-κ-casein and macropeptide moieties by breaking a single peptide bond, the Phe—Met bond between residues 105 (phenylalanine) and 106 (methionine). Because such specific destruction of κ-casein by chymosin leads to destabilization of the micelle, we are led to two conclusions: (1) that micellar stability depends on κ-casein and (2) that the κ-casein must be accessible to the enzyme. Such conclusions were among the earliest arguments in favor of a structural model of the casein micelle where κ-casein forms the surface layer.

Following cleavage of the Phe–Met bond, the macropeptide diffuses into the aqueous serum phase of the milk. This peptide, known also as the glycomacropeptide, since about two-thirds of the peptides released contain carbohydrate, is highly polar owing to its amino acid profile and sugar complement. Studies of the electrophoretic mobility of casein micelles and their renneted or fully proteolyzed counterparts [49,50] have demonstrated that its removal causes a considerable decrease in the zeta potential of the micelle from about −20 to about −10 mV. Again such a decrease can occur only if the κ-casein is in a suitable location to dominate micellar charge, that is, on the micellar surface.

Aggregation does not start immediately but only some time after the enzyme has been added to the milk. Indeed, depending on solution factors, principally pH, a considerable fraction of the κ-casein has to be cleaved by the enzyme before the micelles will aggregate and gel [51]. At natural milk pH, this fraction can be up to 90%. It is therefore common in studies of the rennet coagulation of milk to consider the process as a consecutive series of reactions, aggregation and gel firming following proteolysis. Despite much research effort, the mechanistic features of the reaction are not fully understood. Summaries of the influence of compositional and environmental factors on the various stages of the reaction are available in recent reviews [52–54].

V. MICELLAR INTERACTIONS

A. Failure of the DLVO Theory

Central to any discussion of milk as a colloidal dispersion is the nature of the micellar interaction potential. In the classical Deryaguin-Landau-Verwey-Overbeek (DLVO) theory of colloidal stability, dispersion stability is ensured when electrostatic repulsion contributions dominate over van der Waals attraction in the interaction free energy equation. Diminution or loss of the repulsive component produces a negative free energy, allowing the particles to flocculate.

Attempts have been made to describe the stability of the casein micelle system in terms of the DLVO theory. Some experimental observations on micellar stability seem at least qualitatively to fit the requirements of the theory. At natural milk pH, the caseins carry negative charge [50,55], and initially it was thought that this charge, if on the micellar surface, would provide sufficient electrostatic repulsion to satisfy stability requirements. Micellar zeta potential values are decreased by about a factor of 2 during chymosin proteolysis, the first stage of rennet coagulation [49,50]. A lower dielectric constant decreases electrostatic repulsion and leads to instability. This would fit with observations on alcohol destabilization [48]. A lower micellar surface charge and hence a lower surface potential would lead to less stable particles. This could be invoked to explain the role of pH, calcium, and chemical modification in altering ethanol stability [45,47].

A closer examination of the DLVO predictions suggests that this theory fails to fully account for micellar stability. First, though Green and Crutchfield [50] predicted stability for native casein micelles and instability for chymosin-treated micelles on the basis of their zeta potential measurements, Payens [55] pointed out that the energy barrier for the stable system was located at such a small intersurface distance (ca. 0.5 nm) as to be physically meaningless, lying well within the orbit of surface roughness and short-range forces. Second, as Payens [55] also indicated, the casein micelle system does not respond to electrolyte addition in the manner predicted by the Schulze-Hardy rule, a corollary of the DLVO theory. This latter failure begs the question of the appropriateness of the DLVO theory as a descriptor of casein micelle stability, for the prerequisites of the theory are not fully met. The casein micelles are neither smooth nor uniformly conductive.

They owe their charge in part to specific binding of divalent ions, a charge-regulating mechanism whose secondary response to ionic strength changes is beyond the scope of the DLVO theory. This allows the casein micelle system to tolerate divalent ion concentrations greater than the critical level of 1 mM predicted by the Schulze-Hardy rule [55]. Further, at high ionic strengths, instead of coagulating, the micelles tend to dissociate [56]. Electrostatic charge is important in the control of casein micelle stability—there are too many indicators of this—but its contribution cannot be quantified in terms of the DLVO theory.

B. Steric Stabilization

The failure of the DLVO theory can be considered as evidence that micellar stability is ensured by steric stabilization [57]. In this mechanism, the surfaces of the colloidal particles are covered with flexible macromolecules extending into the suspending medium ("hairy surface"). Their presence prevents close approach of the particles, the interpenetration or compression of the extended loops leading to a loss of conformational entropy which contributes to the free energy of repulsion. Theories for the calculation of steric repulsion now exist [58,59], but too many imponderables remain for them to be successfully applied to casein micelles.

The previous observations regarding the surface location of κ-casein suggest that the glycomacropeptide portion is a prime candidate for the role of sterically stabilizing macromolecule [32], since its removal by chymosin has the expected result of destabilizing the micelle. Moreover, during chymosin treatment in a highly diluted system, Walstra et al. [60] observed that the hydrodynamic diameter of casein micelles decreases by about 10 nm before increases due to the aggregation reaction become apparent—consistent with the removal of a macropeptide sterically stabilizing layer. Independent confirmation of the depilatory action of chymosin has since been provided by a number of laboratories [61–63]. All of these studies were carried out in dilute solution to achieve the single scattering conditions required for application of the dynamic light-scattering techniques used. Employing a single-mode, single-optical-fiber system, Horne and Davidson [64] demonstrated that micellar size decreases by about the same amount (14 nm) before gelation ensues in a renneted undiluted milk. Other evidence in support of this steric stabilization model comes from proton NMR studies, which reveal considerable flexibility and high levels of mobility in the C-terminal of the κ-casein polypeptide chain [65,66].

C. Ethanol Stability and Steric Stabilization

The effects of ethanol on casein micelle size provide yet more evidence in favor of a protruding, sterically stabilizing component. When dilute solution dynamic light scattering was used to measure casein micelle size as a function of ethanol concentration [61], the size was first observed to decrease with increasing alcohol concentration, pass through a minimum value, and then increase rapidly and strongly at still higher alcohol concentrations (Fig. 5 inset). This latter increase depended on both time and micelle concentration, indicating an aggregation reaction.

It was recognized that the decrease in radius could be the result either of a uniform global shrinkage of the micelle or of a conformational change in the outer stabilizing layer. Support for the occurrence of a purely surface phenomenon came from an experiment similar to that carried out by Walstra et al. [60], in which the micelles were treated with chymosin (Fig. 5). In the absence of ethanol, chymosin treatment produced a decrease in radius of about 5 nm, similar to Walstra's original observations. In the presence of 16% ethanol (v/v), with no chymosin present, the micellar radius was reduced by approximately 8 nm. On adding chymosin to this latter mixture, no further change in micellar radius was recorded before aggregation induced by chymosin action ensued. The reaction was still exhibiting a lag phase or clotting time dependent on enzyme concentration, indicating that although the polypeptide "hairs" had been collapsed by the ethanol and the steric component removed from the interaction equation, reduction of some other repulsive component (presumably electrostatic) through enzyme action was still required to bring about aggregation.

The explanation proposed for the change in micellar radius on exposure to ethanol was the collapse of the stabilizing protein hairs onto the micellar surface. Desorption of the sterically stabilizing layer was ruled out. By varying the concentration of the casein micelles in the diluted suspension, the rapid increase in size at high ethanol concentrations was confirmed as due to aggregation. Reducing the micellar concentration inhibited the aggregation phase and showed that the minimum size was indeed a true minimum, defining a micellar core below which size could decrease no further. The difference between this micellar core size and the average radius measured in the absence of ethanol was taken as the thickness of the steric stabilizing barrier. An investigation [67] of the effects of environmental conditions on the layer thickness demonstrated that increasing the level of Ca^{2+} in the diluting buffer decreased the thickness but that this decrease arose mainly from within the micelle, as if the inner reaches of the stabilizing layer were being incorporated into the micellar core. Decreasing buffer pH had a similar effect. The other major result of this study, plotted in Fig. 6, found a positive correlation between the critical ethanol concentration required to induce micellar aggregation and the barrier layer thickness—thicker barriers required more ethanol to induce complete collapse. This correlation held whether changes in barrier thickness were brought about

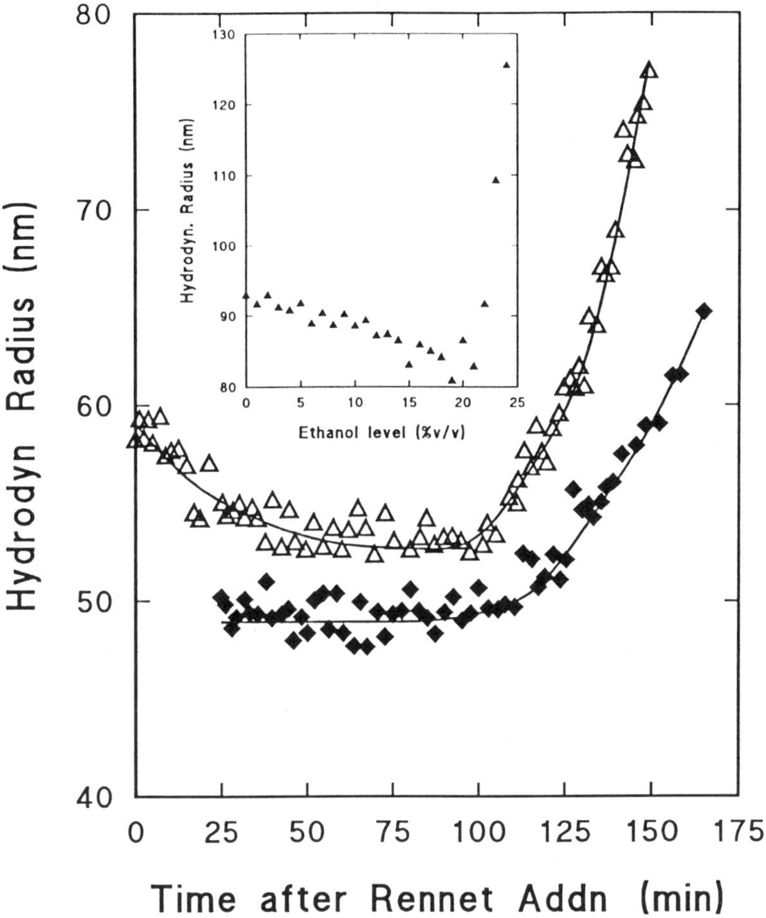

FIG. 5 Changes in micellar hydrodynamic radius of diluted (×400) milk after the addition of rennet solution. (△) No ethanol present; (◆) with 16% v/v ethanol in the buffer solution and rennet solution diluted 10. An additional arbitrary 25 min has been added to the reaction time for the ethanol mixture to displace this curve and display the data more clearly. The inset shows the hydrodynamic radius of casein micelles (mean of four consecutive measurements) as a function of added ethanol concentration.

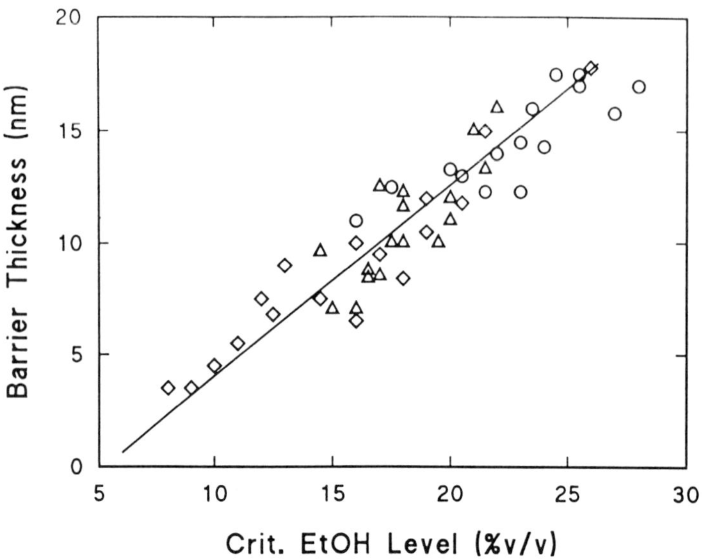

FIG. 6 Barrier layer thickness plotted against critical EtOH level required to induce aggregation of casein micelles from milk (diluted ×400). Barrier layer thickness is defined as the difference between micelle radius in alcohol-free buffer and minimum radius measured in the presence of alcohol in the same base buffer system. Observed variations were obtained by adjusting (△) pH, (◇) ionic calcium level, and (○) ionic strength of buffer.

by varying the pH, ionic strength, or calcium ion concentration of the dilution buffer system and when barrier thickness was found to increase with micellar size.

Qualitatively the observed changes in critical ethanol level could be explained by a mechanism involving electrostatic repulsion between charged polyelectrolyte "hairs" [67]. Take, for example, the case where increase in the Ca^{2+} level decreases the critical ethanol concentration. Here the effect of calcium ion concentration could be envisaged as resulting from specific calcium ion binding reducing the net negative charge of the protein, thereby decreasing interhair repulsion and making the hairs more susceptible to collapse—effects similar to those observed previously in calcium-induced aggregation of individual caseins [4,46]. Similarly, the increase in barrier strength with increasing pH (its increasing resistance to collapse) was consistent with the increase in charge brought about by the greater ionization of the casein phosphate groups, which have one of their p*K* values around pH 6.5.

One further effect of having charged, sterically stabilizing hairs is to push the electrostatic repulsion between micelles out to distances much further from the core surface than would be the case if the hairs were uncharged. Holt and Dalgleish [62] attempted to calculate the electrostatic potential on the basis of a theoretical model incorporating a partially draining outer layer. Calculated layer thickness was greater than the decrease in micelle size observed after the addition of rennet by about a factor of 2. Attempts to model the measured decreases in electrophoretic mobility on rennet treatment were less successful, however, in that the measured variation with calcium ion concentration could be reproduced by a surface location of less than 10% of the available κ-casein of the micelle [62].

Such observations argue that a model of the micellar surface more complicated than individual charged hairs should be considered. Moreover, though the simple charged hair model accommodates the behavior seen when ionic species within the buffer are varied, it does not address the changes in barrier thickness that also result from these variations. Horne and Davidson [67] therefore suggested that the steric stabilizing layer would be better viewed as a weak gel. In reality, with casein micelles, there is no actual separate adsorbed layer to be treated as a gel, and this refinement is tantamount to describing the casein micelle as a microgel particle formed by extensive cross-linking of the constituent protein molecules. This microgel model provides a physically more meaningful picture in which the protein hairs are linked not only to the core surface but also to each other. The effects of decreased charge, described previously as permitting hair collapse, are incorporated into this model as promoting increased cross-linking and incorporation of the surface layers into the more rigid structure that is the micelle core. In other words, this model allows for increases in core radii through the formation of further bonds in the region of the original core surface, bonds that are unlikely to differ in chemical or physical nature from those already existing in the micelle interior. Similarly, the effects of increased charge, with increasing pH, for example, would be explicable as the promotion of greater repulsion between the hairs leading to a reduction in cross-linking, the protein hairs now free to extend and give rise to the increased hydrodynamic radii seen in alcohol-free buffers [67].

This microgel model also predicts that in situations where high levels of cross-linking are present, the stabilizing sheath should be less responsive to the action of ethanol (i.e., more difficult to collapse), and vice versa [67]. When the stabilizing layer is thin and the system is consequently less stable, the presence of the cross-links makes the hairs less readily collapsible and less susceptible to the effects of added ethanol, compared to the converse situation where the system is sterically stable but has few interhair cross-links so that paradoxically the hairs collapse more readily in the presence of ethanol.

The magnitude of the initial gradient of the plot of hydrodynamic radius versus ethanol concentration was taken as an indicator of the susceptibility of the steric stabilizing sheath to alcohol-induced collapse. The magnitude of this gradient increased linearly with buffer pH. As predicted by the microgel model, the steric stabilizing barrier was becoming softer due to the decreasing number of cross-links diminishing the gel strength as pH was increased. The increase in softness was not accompanied by a decrease in micellar stability, however, because as the barrier became softer it also became thicker, and that total thickness was the controlling factor that regulated the overall stability of the micellar system. Greater concentrations of ethanol were required to fully collapse the thicker barrier and allow aggregation to proceed. A similar analysis explained the effects of changing the concentration of ionic calcium.

The susceptibility of the steric stabilizing sheath to collapse decreased as the calcium concentration was increased because the sheath became more rigid as the increased binding of calcium neutralized greater amounts of charge, reduced localized interhair repulsion, and promoted the formation of cross-links between the hairs. The barrier became thinner as the inner layers became more rigid and were taken into the micelle core. Though it became progressively more difficult to induce conformational collapse as the outer gel layer thinned with increasing calcium level, the overall stability of the micellar system declined, because again the overriding parameter controlling that stability was the barrier thickness alone.

D. Chymosin Treatment and Micellar Surface Structure

Chymosin treatment has been mentioned already in the context of its usefulness in elucidating details of the nature of the intermicellar potential. Chymosin breaks down the κ-casein of the micelle and by this means causes the loss of both charge and steric stabilization, allowing the micelles to aggregate. However, this aggregation does not begin until some time after the enzyme has been added to milk—not, in fact, until a considerable fraction of the milk has been broken down [51]. Dalgleish and Holt [68] developed a simple geometrical model to explain why this delay should occur. The essence of this model relates to the need to form gaps in the steric stabilizing layer of the micelle, "hot spots" through the interaction of which micelles can form aggregates. The model envisages the micelles as being initially completely covered by a hydrated surface layer of κ-casein, but as the proteolysis of κ-casein proceeds following random enzymatic attack, the possibility of creating gaps also increases, allowing access to the inner core. Interaction between micelles is postulated to occur only when areas of the core can touch. The known thickness of the stabilizing layer and the micellar diameter allow calculation of the area of this gap. Assuming the breakdown of κ-casein to be a random process, Dalgleish and Holt [68] calculated the probability that gaps of the required size were produced. These calculations generally agreed with experimental observations, but refinements are required to take account of changing solution conditions such as pH and temperature.

More recently, de Kruif and coworkers [69] developed a theory to explain the behavior of the milk viscosity during the initial stages of rennet treatment. Their measurements reproduce the initial drop in viscosity followed by a rapid rise first observed by Scott Blair and Oosthuizen [70]. Their theory suggests that part of the drop in viscosity is attributable to the decrease in micellar volume fraction accompanying the cleavage of the κ-casein hairs. This loss of the steric stabilization component allows the van der Waals attraction between the cores to become operative, and it is this attraction that brings about the initial increase in viscosity. The depth of the attractive potential and therefore the viscosity depend on the degree of proteolysis of the κ-casein. The latter was separately measured and used to estimate a well depth that in turn allowed the viscosity of the milk during renneting to be calculated and compared with the experimentally observed behavior.

This latter approach is interesting because it attempts to characterize the rennet-treated casein micelle as an adhesive hard sphere and to apply to it the considerable body of colloid theory evolved for this type of particle. It is likely that in the near future we will see further steps in this direction. Such studies of the effects of rennet or ethanol are concerned, however, not with the interactions of casein micelles as they exist in milk but with how these micelles interact after they have been modified in some way. No matter how slight the modification, some fraction of the potential energy function is removed by the reaction system.

Few studies have attempted to verify the suggestion that it is legitimate to consider the native casein micelle as a hard sphere. They are certainly not solid in the mechanical sense because it is possible to concentrate milk by ultrafiltration to a level where the hairs must be compressed. Horne [71] has recently presented some preliminary viscosity data on concentrated milks that suggest that these systems behave as weakly flocculated suspensions. The flow curves (stress as a function of shear rate) for these milks show yield values of a few hundred millipascals, which show a power-law dependence on milk concentration with an exponent value of ca. 4.0, behavior similar to that observed by Buscall et al. [72] in their studies of flocculated suspensions. The yield values were also interpreted in terms of the elastic floc model of Hunter [73] as the energy required to disrupt the flocs and restore the micelles to individuality. By further assuming a lattice structure for the concentrated milks, Horne [71] was able to demonstrate that the depth of the attractive well increased exponentially as the surface–surface separation decreased. The concentrated milks were, however, extremely weak gels, and the attractive well depths probed only a few tenths of kT.

Our discussion of casein micelle interactions began with the tacit assumption that the micelles could be treated as hard-sphere particles. At the concentration levels extant in secreted milk, this appears a reasonable approximation. The viscosity of milk shows no appreciable difference from Newtonian behavior and a value perhaps one-third greater than that of water. Only when the system in concentrated do we begin to probe the limitations of that assumption and the strengths of the micellar interactions. In spite of much research on micellar destabilization reactions, such knowledge is still lacking today.

VI. MICELLAR STRUCTURE

The discussions on the nature of the micellar interaction potential have highlighted some of the models proposed to describe the structure of the surface layers of the micelle, but little has been said of the inner core of the particle. Most of the surface structure models were developed to describe one aspect of micellar behavior measured in a particular experiment. Thus the geometrical model of Dalgleish and Holt [68] and its requirements for clearing surface "hot spots" explain the reluctance of the micelle to aggregate until most of the κ-casein hairs have been cleaved. Only with almost complete baldness are sufficiently large gaps created on enough micelles for the reaction to proceed to complete polymerization. Similarly, Horne and Davidson [67] developed their microgel model to describe the collapse of the steric stabilizing layer on exposure to ethanol. A unification of these themes in a single model is still awaited.

Such diversity of opinion does not exist when micellar internal structure is considered. Though the actual model has evolved over many years, the almost universally accepted view is that the casein micelle is composed of subunits or submicelles. It is my contention, however, that much of the evidence for this model structure is equivocal and is explicable in other ways.

Evidence for a submicellar structure came originally from electron microscopy, where the picture showed the casein micelle as being roughly spherical but with a raspberry-like appearance [74–77]. More recently, neutron and X-ray scattering data, as well as NMR relaxation measurements, have been interpreted in terms of this subunit model [78–83]. All of these techniques highlight the inhomogeneity of the micelle interior, indicating that there are regions of particularly high density and scattering power interspersed with areas that have an especially low density. From such measurements it is possible to estimate a characteristic size for these subunits of the order of 10–15 nm. What these scattering and spectroscopic techniques are unable to demonstrate, however, is that these subunits are capable of a discrete, separate existence or that their formation is a necessary step in the synthesis and assembly of the casein micelle in the mammary gland.

When casein micelles are broken down by any of a variety of disrupting agents, aggregated protein particles again with diameters ca. 10 nm are observed [75,77,84,85], and these have been described as submicelles. Since similarly sized aggregates can be readily produced by mixing appropriate proportions of the individual caseins under the same ionic strength and pH regime as is found in milk [9], it has been suggested that these polymeric aggregates are the submicelles that then interact via calcium phosphate linkages [86] to form casein micelles.

Rollema and Brinkhuis [87] compared the NMR spectra of intact casein micelles, casein association polymers, and products produced by micellar disintegration via calcium depletion. Whereas the dissociation products and association polymers showed similar spectra, indicating considerable flexibility and mobility, they differed markedly from the micellar spectrum, which epitomized an inflexible, rigid structure. Such a micellar spectrum could not be generated from a structure composed of the flexible subunits as separate entities held together by calcium phosphate clusters.

Further criticism of the Slattery model can be expressed as follows. To drive the association equilibrium of the individual caseins toward generating aggregates of subunit size requires relatively large concentrations of the proteins. This presumably must occur in the absence of phosphate to avoid the further aggregation that forms the second stage of micellar assembly. However, inorganic phosphate is involved in the enzymatic synthesis of lactose, and this synthesis takes place in the Golgi apparatus of the mammary gland concurrently with casein protein emergence and micellar assembly [88]. The caseins therefore emerge into a bath of calcium and inorganic phosphate. Horne [89] demonstrated that at these levels of calcium and phosphate, the inorganic salts coprecipitate with the protein. In the absence of phosphate, no precipitate is formed, but with phosphate included, precipitation occurs at protein concentrations far below those needed to produce aggregates of submicellar dimensions via the association equilibrium route. In these circumstances the Slattery-Schmidt stepwise model for micellar assembly does not seem viable.

Horne et al. [90] presented a computer simulation of micellar assembly that reproduces many features and properties of the casein micelle. Based on a polyfunctional polymerization approach, this model treats α_s-casein as a trifunctional monomer, β-casein as a bifunctional chain propagator, and κ-casein as a monofunctional chain terminator.

The structures grown by the simulation procedure are very open, highly ramified, and porous. The deep fjords they develop would allow penetration of small reagents to the very heart of the micelle, providing the accessibility seen by Garnier and Ribadeau Dumas [91]. These interstices would trap water hydrodynamically, contributing to the high voluminosity observed for casein micelles [32]. Equally, areas of high local density were observed in these clusters, around perhaps a particularly reactive grouping of monomers. This argues against the necessity of postulating the existence of discrete submicellar entities when the requisite heterogeneous structure develops as a consequence of the random nature of the polycondensation process.

Treating κ-casein as a chain terminator meant that this protein was found primarily on the cluster perimeter, though the evolving cluster was allowed to circumvent blocked sites if it so chose, leading to some κ-casein being buried in the cluster interior. Varying the proportions of the individual caseins within the pool of monomers demonstrated that high levels of κ-casein gave rise to small clusters (micelles): all growth sites were rapidly closed off. Reducing the number of κ-casein units allowed further growth, qualitatively matching the observed proportionality between κ-casein content and surface-to-volume ratio found in casein micelles [15,31]. Moreover, because saturation of functionalities on a given monomer also stopped chain extension from that site, β- and α_s-caseins also contributed to the outer "closed" surface of the computer-simulated micelle cluster. All these components were therefore allowed some surface location on the micelle, consistent with the accessibility to insoluble cross-linked papain observed by Ashoor et al. [33] and with the surface composition data of Dalgleish et al. [31].

Perhaps more relevant to the discussion of micellar stability and its origins, the clusters created in this simulation of polyfunctional assembly evolved with the dangling surface loops and tails necessary for steric stabilization. In the simulated structures, these are rigid and immobile. In a micelle so constructed, however, their mobility, coupled with the emergence of further reactive sites along their length (as might occur if electrostatic barriers were neutralized), would permit the further cross-linking and growth of the micellar core envisaged earlier in this chapter, where gel-like properties were assigned to the surface layers.

VII. CASEIN MICELLES AND FRACTALS

Aggregation processes play an important role in many areas of science and technology. They are particularly important in a discussion of the colloidal properties of casein micelles because such reactions lead to the gels and curds that are so important in the manufacture of the dairy products yogurt and cheese. The results of computer simulation studies using different models for the aggregation reaction stimulated interest in defining the structure of colloidal aggregates in terms of the concept of the fractal dimension [92]. Qualitatively, fractal objects show dilation symmetry. They are self-similar or scale-invariant in the sense that if the degree of resolution of our observation is increased, a magnified portion appears identical to the entire object viewed under lower resolution. Quantitatively, aggregate structures or clusters can be characterized by their fractal dimension D, a measure of the cluster packing density such that

$$N(r) = (r/r_0)^D \tag{2}$$

where $N(r)$ is the number of seed particles of radius r_0 contained inside a sphere of radius r within the aggregate or cluster. For a growth process we can take r to be some characteristic measure of cluster size (e.g., its radius of gyration) and interpret Eq. (1) as a law relating cluster mass to size at various stages of growth. The exponent D, the fractal dimension, describes the distribution of matter in the volume the cluster occupies. For a compact object (an arrangement of particles in a d-dimensional Euclidean space), one should have $D = d$. In general, for a fractal object, $D < d$ and is typically not an integer. D being less than d, the larger the fractal object, the less dense the distribution of matter in it.

Since the fractal nature of colloidal aggregates was first experimentally verified [93], there have been many attempts to build realistic theoretical models able to reproduce these structures. There is as yet no general theory for irreversible growth, no method for predicting the fractal dimension of a growing aggregate. Instead, several computer simulation models have been developed to describe the growth situation with the particles or clusters moving within prescribed constraints.

The pioneering model was due to Witten and Sander [94]. In this diffusion-limited particle-cluster model, particles are added one after another to a single growing cluster. The aggregation process starts with an immobile seed particle. Then a second particle is released at some distant, randomly chosen boundary point to follow a random walk toward the central area. When it reaches a nearest-neighbor position with respect to the first, it sticks rigidly and irreversibly to form a dimer. The computerized simulation continues with the release of a third particle at the boundary, and so on until several thousand, perhaps several million, particles have been released, the cluster growing immovably all the time. No rearrangement of the cluster is allowed. When such a simulation is carried out in three dimensions, a value of 2.5 is obtained for the fractal dimension of the diffusion-limited particle-cluster aggregate.

The second basic simulation model, the cluster-cluster aggregation model, was introduced independently by Meakin [95] and by Kolb et al. [96]. This model starts with an assembly of identical particles randomly distributed in a box. Each of the particles is free to move randomly in space. When two particles collide, they irreversibly stick together to form a rigid dimer, which is also allowed to diffuse randomly in the box while retaining its orientation. This dimer can, on contact, stick to another dimer or to a single particle, and so on. After each collision, the two colliding clusters form a new, rigid, larger cluster. The process can be continued until a single aggregate remains alone in the box.

A typical product of the cluster-cluster aggregation is much more open, much more ramified than a cluster resulting from the particle-cluster simulation. This comes about because of the manner in which such a cluster is generated. A pair of colliding clusters cannot enter into one another because of their rigid sticking on first particle–particle contact, whereas in a particle-cluster simulation there is always the possibility of deeper penetration of a single particle into the growing cluster. This open structure is reflected in the lower value obtained for the fractal dimension in the cluster-cluster model. In simulations carried out in three dimensions, a value of about 1.7 is obtained for D. This is very close to the experimental value obtained by Weitz et al. [97] in their studies of the morphology of aggregates produced from colloidal particles under a diffusion-limited kinetic regime.

Many modifications of this cluster-cluster model have been considered—varying the type of trajectory, varying or not varying the mobility of the cluster in accordance with its increasing mass—and it has been found that the most important stage in cluster-cluster aggregation is the sticking process. In the original model, a cluster stuck on first contact of one of its constituent particles with one of the particles of its collision partner. The sticking probability here is unity. A variant of this model reduces the sticking probability to some small fraction, so that on average only one collision in perhaps a thousand is successful and permits aggregation. In effect, the system becomes reaction-limited. The clusters grown by this model are more compact because their collisions sample more packing configurations. This is reflected in an increase in the fractal dimension to a value of approximately 2.0 for aggregates grown according to the reaction-limited cluster-cluster aggregation model.

Different growth regimes can therefore give rise to structures with different fractal dimensions. There is a link between the dynamics of the aggregation process and the structure of the resulting cluster. If the cluster is then incorporated into or is a gel, are the physical, viscoelastic properties of that gel a reflection of the dynamics of the original aggregation reaction? Such considerations brought about the measurement of the fractal dimension of the products of various reactions where casein micelles were aggregated and coagulated.

The most direct method of determining the fractal dimension of an aggregate is to apply its definition and quantitatively analyze digitized electron micrographs. This has not so far been applied to casein micelle aggregates. It is also possible to measure the fractal dimension by exploiting the scattering properties of the aggregate using either X-ray or neutron scattering or, if the cluster is large enough, light scattering. Since the relationship between scattered intensity and fractal dimension is with wave vector rather than scattering angle, Horne [98] demonstrated that the fractal dimension of an aggregate is accessible through wavelength dependence of the turbidity of the aggregating suspension. Weitz et al. [97] also showed that for diffusion-limited aggregation, the log-log plot of hydrodynamic radius and reaction time provides another route to D, the reciprocal of the gradient of this straight line being the fractal dimension.

Horne [98] applied the turbidity technique to aggregates produced in the ethanol-induced coagulation of casein micelles, showing them to be fractal and, for the particular micelle concentration chosen, to have a fractal dimension of 2.3. This value was also found to be invariant across the ethanol concentration range 26–34% v/v despite initial rate studies showing that the aggregation passed from reaction-limited to diffusion-limited kinetics at an ethanol level of 29% v/v. This constant value of 2.3 is somewhat higher than the value of 1.7 obtained in simulations of diffusion-limited cluster-cluster aggregation or even reaction-limited cluster-cluster aggregation with their fractal dimension of 2.0, but it is lower than the value of 2.5 obtained with the particle-cluster model. In a similar study of the rennet-induced aggregation of casein micelles, the turbidity technique gave a value of 2.4 for the fractal dimension of the aggregates [98].

Other methods of measuring the fractal dimension have been developed for application to fully formed casein gels. These methods derive the fractal dimension from the exponent of the power law scaling relationship between the measured parameter and the volume fraction of the primary particles. In the case of permeability, the relationship is given by the equation [99]

$$B \propto \phi^{2/(D-3)} \qquad (3)$$

The predicted dependence of the elastic properties of the gel on volume fraction is more complicated. Two possible models have been suggested

[100,101]. In the first, the stress-carrying strands of the structure are short and straight. The elastic modulus (G) of the gel is assumed to be proportional only to their elongation, and this leads to the relation

$$G \propto \phi^{2/(3-D)} \tag{4}$$

In the second model, the strands are not straight, and any deformation leads to either straightening or further bending. Here the modulus of the gel is proportional to the bending modulus of the strands, which gives

$$G \propto \phi^{3/(3-D)} \tag{5}$$

A compilation of the values of the fractal dimension for different casein gels by various methods is given in Table 2. In all instances the various measures of D are in reasonable agreement with one another, but all differ quite significantly from the predictions of the model simulations. Various explanations for these discrepancies have been proposed [102]. The majority relate to removal of simplifying assumptions from the original theory, allowing cluster rotation, annealing, or rearrangement, for example. Some, such as cluster restructuring, have been tackled in simulation studies [103–105], but such rearrangements increased the fractal dimension of the fully developed structures only slightly, from 2.0 to 2.1.

Another problem arises because the simulation studies have been carried out with relatively low numbers of particles and mimic effectively infinite dilution conditions. In contrast, experiments, even those for light scattering, are carried out at relatively high number concentrations. Kolb et al. [106] suggested that higher fractal dimensions should be found at higher particle concentrations, arguing that in this regime the developing clusters interpenetrate much more. They did not, however, speculate on the magni-

TABLE 2 Fractal Dimensions of Casein Gels Made by Rennet, Ethanol, and Acid-Induced Coagulation[a]

	Method			
	Rennet	Ethanol	Acid 1	Acid 2
Turbidity	2.4	2.3	–	2.27
Permeability/ϕ	2.23	–	2.39	2.36
Modulus/ϕ	2.17	–	2.24	2.36
Fracture/ϕ	–	–	2.20	2.20
Cluster size/ϕ	–	–	2.35	2.35

[a]Acid 1 gels are made by cold acidification followed by warming, Acid 2 gels by slow acidification following addition of glucono-δ-lactone.

tude of the effect nor on how it would vary with particle concentration. This suggestion led Horne [107] to measure the fractal dimension of aggregates produced in ethanol-induced aggregation as a function of the initial micellar concentration, giving the plot shown in Fig. 7. The reactions were all carried out under diffusion-limiting conditions, every collision leading to aggregation. At the lower micellar concentrations, the fractal dimensions were derived from measurements of the hydrodynamic radius as a function of reaction time because of the inherently greater sensitivity of photon counting procedures. In the region where this technique overlapped with the turbidimetric procedure, the results agreed fairly closely. From the data plotted in Fig. 7, extrapolating to zero concentration gave a value for D of 1.78 ± 0.08 while a value of 2.56 ± 0.04 was obtained for the asymptote at infinite concentration. By increasing the micellar concentration, it is therefore possible to produce a transition from aggregates of fractal dimension 1.7 and therefore possibly the result of a cluster-cluster mechanism to aggregates with $D = 2.5$, which could be the result of particle-cluster kinetics, all within the diffusion-limited regime.

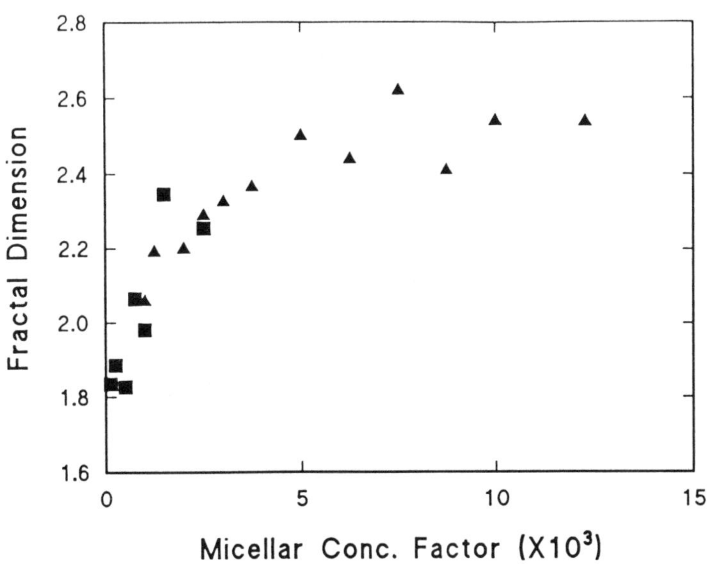

FIG. 7 The concentration dependence of the measured fractal dimension of aggregates produced in ethanol-induced aggregation of casein micelles. The reactions were carried out in the diffusion-limited regime in a buffer containing 30 % v/v ethanol. Measurements made by (■) the turbidimetric technique and (▲) photon correlation spectroscopy.

No such variations in fractal dimension with concentration have been commented upon for data derived from mechanical properties of gels. Upward bends have been observed in plots of log(G) versus log(ϕ), but these have been attributed to changes in the mechanical nature of the gels switching from being governed by one to the other of the structural models rather than to a concentration-dependent change in the actual aggregation mechanism. In view of the controversy raging over the applicability of the structural models, it is expected that some time will elapse before the situation is resolved.

Finally, in this section on casein micelles and fractals, it may be noted that the simulation of micelle growth via a polycondensation mechanism as described in the previous section also produces a fractal cluster. This growth process was carried out on a two-dimensional lattice, and straight lines were obtained in log-log plots of cluster mass versus radius of gyration, the slope being the fractal dimension. The value obtained was, however, dependent on the input functionality distribution, that is, on the proportions of the individual caseins assigned to the originating particle pool. As the number of routes to creating structures multiplies, each apparently with its own particular value for fractal dimension, the concept seems to be devalued and its usefulness lessened. Further theoretical and experimental investigations of the properties of fractal aggregates must be pursued if the fractal concept is to progress beyond being nothing more than a descriptive tool.

VIII. CONCLUDING REMARKS

From the foregoing sections, it should now be apparent that milk presents a colloidal dispersion par excellence. It should also be apparent that the properties of the casein micelle system have been the subject of scrutiny by each and every one of the full panoply of physical techniques available to the modern colloid scientist. The interest is, of course, not merely academic, since the dairy industry is a major contributor to the gross national product and export potential of many countries throughout the world.

REFERENCES

1. D.D. Muir, in *The Technology of Dairy Products* (R. Early, Ed.), Blackie, Glasgow, 1992, p. 24.
2. D.T. Davies and A.J.R. Law, *J. Dairy Res. 48*:83 (1980).
3. W.N. Eigel, J.E. Butler, C.A. Ernstrom, H.M. Farrell, Jr., V.R. Harwalkar, R. Jenness, and R.McL. Whitney, *J. Dairy Sci. 67*:1599 (1984).

4. D.G. Dalgleish, E. Paterson, and D.S. Horne, *Biophys. Chem. 13*:307 (1981).
5. W.H. Sawyer, *J. Dairy Sci. 52*:1347 (1969).
6. B.E. Brooker and C. Holt, *J. Dairy Res. 45*:355 (1978).
7. D.G. Schmidt, in *Developments in Dairy Chemistry*, Vol. 1, Proteins (P.F. Fox, Ed.), Elsevier Applied Science, London, 1982, p. 61.
8. D.F. Waugh and R.W. Noble, *J. Am. Chem. Soc. 87*:2246 (1965).
9. C.W. Slattery and R. Evard, *Biochim. Biophys. Acta 317*:529 (1973).
10. C. Holt, D.T. Davies, and A.J.R. Law, *J. Dairy Res. 53*:557 (1986).
11. D.G. Dalgleish and A.J.R. Law, *J. Dairy Res. 55*:529 (1988).
12. D.T. Davies and A.J.R. Law, *J. Dairy Res. 50*:67 (1983).
13. C. Tanford, *The Hydrophobic Effect: Formation of Micelles and Biological Membranes*, 2nd ed., Wiley, New York, 1980.
14. T.C.A. McGann, W.J. Donnelly, R.D. Kearney, and W. Buchheim, *Biochim. Biophys. Acta 630*:261 (1980).
15. W.J. Donnelly, G.P. McNeill, W. Buchheim, and T.C.A. McGann, *Biochim. Biophys. Acta 789*:163 (1984).
16. C.W. Slattery, *Biochemistry 17*:1100 (1978).
17. R.A. Sullivan, M.M. Fitzpatrick, and E.K. Stanton, *Nature 183*:616 (1959).
18. C. Holt, D.G. Dalgleish, and T.G. Parker, *Biochim. Biophys. Acta 328*:428 (1973).
19. C. Holt, A.W. Kimber, B. Brooker, and J.H. Prentice, *J. Colloid Interface Sci. 65*:555 (1978).
20. C. Holt and L. Baird, *J. Dairy Res. 45*:339 (1978).
21. C. Holt and D.D. Muir, *J. Dairy Res. 45*:347 (1978).
22. C. Holt, T.G. Parker, and D.G. Dalgleish, *Biochim. Biophys. Acta 400*:283 (1975).
23. C. Holt, *Biochim. Biophys. Acta 400*:293 (1975).
24. B.J. Berne and R. Pecora, *Dynamic Light Scattering*, Wiley-Interscience, New York, 1976.
25. D.S. Horne, *J. Colloid Interface Sci. 98*:537 (1984).
26. D.S. Horne and D.G. Dalgleish, *Eur. Biophys. J. 11*:249 (1985).
27. N. Ostrowsky, D. Sornette, P. Parker, and E.R. Pike, *Opt. Acta 28*:1059 (1981).
28. B. Ekstrand and M. Larsson-Raznikiewicz, *Milchwissenschaft 39*:591 (1984).
29. B. Ekstrand and M. Larsson-Raznikiewicz, *Biochim. Biophys. Acta 536*:1 (1978).
30. M.C.A. Griffin and M. Anderson, *Biochim. Biophys. Acta 748*:453 (1983).
31. D.G. Dalgleish, D.S. Horne, and A.J.R. Law, *Biochim. Biophys. Acta 991*: 383 (1989).
32. P. Walstra, *J. Dairy Res. 46*:317 (1979).
33. S.H. Ashoor, R.A. Sair, N.F. Olson, and T. Richardson, *Biochim. Biophys. Acta 229*:423 (1971).
34. B. Chaplin and M.L. Green, *J. Dairy Res. 49*:631 (1982).
35. S.P.F.M. Roefs, Structure of acid casein gels, Ph.D. Thesis, Agricultural Univ., Wageningen, Netherlands, 1986.

36. P.F. Fox and D.M. Mulvihill, in *Food Gels* (P. Harris, Ed.), Elsevier, Applied Science, London, 1989, p. 121.
37. D.S. Horne and D.D. Muir, *J. Dairy Sci. 73*:3613 (1990).
38. C. Holt, D.D. Muir, and A.W.M. Sweetsur, *J. Dairy Res. 45*:183 (1978).
39. P.M. Kelly, P. Hughes, M.K. Keogh, and J.A. Phelan, *Ir. J. Food. Sci. Technol. 3*:62 (1979).
40. M. Journet, R. Verité, and B. Vignon, *Le Lait 55*:212 (1975).
41. W. Banks, J.L. Clapperton, D.D. Muir, A.K. Powell, and A.W.M. Sweetsur, *J. Sci. Food. Agric. 35*:165 (1984).
42. H. Singh and P.F. Fox, *J. Dairy Res. 54*:523 (1987).
43. D.S. Horne, in *Advanced Dairy Chemistry* (P.F. Fox, Ed.), Elsevier, Applied Science, London, 1993. p. 657.
44. W. Banks and D.D. Muir, in *Advances in Food Emulsions and Foams* (E. Dickinson and G. Stainsby, Eds.), Elsevier, Applied Science, London, 1988, p. 257.
45. W.J. Donnelly and D.S. Horne, *J. Dairy Res. 53*:23 (1986).
46. D.S. Horne and D.G. Dalgleish, *Int. J. Biol. Macromol. 2*:154 (1980).
47. D.S. Horne and T.G. Parker, *J. Dairy Res. 49*:449 (1982).
48. D.S. Horne and T.G. Parker, *Int. J. Biol. Macromol. 3*: 399 (1981).
49. D.G. Dalgleish, *J. Dairy Res. 51*:425 (1984).
50. M.L. Green and G. Crutchfield, *J. Dairy Res. 38*:151 (1971).
51. D.G. Dalgleish, *J. Dairy Res. 46*:643 (1979).
52. D.G. Dalgleish, in *Cheese: Chemistry, Physics and Microbiology*, Vol. 1 (P.F. Fox, Ed.), Elsevier Applied Science, London, 1987, p. 63.
53. P.F. Fox, in *Developments in Food Proteins*, Vol. 3 (B.J.F. Hudson, Ed.), Elsevier Applied Science, London, 1984, p. 69.
54. M.L. Green and A.S. Grandison, in *Cheese: Chemistry, Physics and Microbiology*, Vol. 1 (P.F. Fox, Ed.), Elsevier Applied Science, London, 1987, p. 97.
55. T.A.J. Payens, *J. Dairy Res. 46*:291 (1979).
56. Z. Saito, *Neth. Milk Dairy J. 27*:143 (1973).
57. C. Holt, *Proc. Int. Conf. Colloid and Surface Science, Budapest*, Vol. 1, 1975, p. 641.
58. J.M.H.M. Scheutjens and G. Fleer, *Macromolecules 18*:1882 (1985).
59. B. Vincent, *Adv. Colloid Interface Sci. 4*:193 (1974).
60. P. Walstra, V.A. Bloomfield, G.J. Wei, and R. Jenness, *Biochim. Biophys. Acta 669*:258 (1981).
61. D.S. Horne, *Biopolymers 23*:989 (1984).
62. C. Holt and D.G. Dalgleish, *J. Colloid Interface Sci. 114*:513 (1986).
63. M.C.A. Griffin, *J. Colloid Interface Sci. 115*:499 (1987).
64. D.S. Horne and C.M. Davidson, *Int. Dairy J. 3*:61 (1992).
65. M.C.A. Griffin and G.C.K. Roberts, *Biochim. J. 228*:273 (1985).
66. H.S. Rollema, J.A. Brinkhuis, and H.J. Vreeman, *Neth. Milk Dairy J. 42*: 233 (1988).
67. D.S. Horne and C.M. Davidson, *Colloid Polym. Sci. 264*:727 (1986).

68. D.G. Dalgleish and C. Holt, *J. Colloid Interface Sci. 123*:80 (1988).
69. C.G. de Kruif, in *Food Colloids and Polymers* (E. Dickinson, Ed.), Royal Society of Chemistry, Cambridge, U.K., 1993, p. 55.
70. G.W. Scott Blair and J.C. Oosthuizen, *J. Dairy Res. 28*:165 (1961).
71. D.S. Horne, in *Food Colloids and Polymers* (E. Dickinson, Ed.), Royal Society of Chemistry, Cambridge, U.K., 1993, p. 260.
72. R. Buscall, I.J. McGowan, and C.A. Mumme-Young, *Faraday Disc. Chem. Soc. 90*:115 (1991).
73. R.J. Hunter, *Foundations of Colloid Science*, Vol. II, Oxford Univ. Press, Oxford, 1989, p. 992.
74. P.D. Shimmin and R.D. Hill, *J. Dairy Res. 31*:121 (1964).
75. D.G. Schmidt and W. Buchheim, *Milchwissenschaft 25*:596 (1970).
76. D.G. Schmidt, P. Walstra, and W. Buchheim, *Neth. Milk Dairy J. 27*:128 (1973).
77. W. Buchheim and U. Welsch, *Neth. Milk Dairy J. 27*:163 (1973).
78. P.H. Stothart and D.J. Cebula, *J. Mol. Biol. 160*:391 (1982).
79. P.H. Stothart, *J. Mol. Biol. 208*:635 (1989).
80. H. Pessen, T.F. Kumosinski, and H.M. Farrell, Jr., *J. Dairy Res. 56*:443 (1989).
81. T.F. Kumosinski, H. Pessen, H.M. Farrell, Jr., and H. Brumberger, *Arch. Biochem. Biophys. 266*:548 (1988).
82. T.F. Kumosinski, H. Pessen, J.T. Prestrelski, and H.M. Farrell, Jr., *Arch. Biochem. Biophys. 257*:259 (1987).
83. H.M. Farrell, Jr., H. Pessen, and T.F. Kumosinski, *J. Dairy Sci. 72*:562 (1989).
84. R.J. Carroll, H.M. Farrell, Jr., and M.P. Thompson, *J. Dairy Sci. 54*:752 (1971).
85. L. Pepper and H.M. Farrell, Jr., *J. Dairy Sci. 65*:2259 (1982).
86. D.G. Schmidt, *Neth. Milk Dairy J. 34*:42 (1980).
87. H.S. Rollema and J.A. Brinkhuis, *J. Dairy Res. 56*:417 (1989).
88. N.J. Kuhn and A. White, *Biochem. J. 168*:423 (1977).
89. D.S. Horne, *J. Dairy Res. 49*:107 (1982).
90. D.S. Horne, T.G. Parker, and D.G. Dalgleish, in *Food Colloids* (R.D. Bee, P. Richmond, and J. Mingins, Eds.), Royal Society of Chemistry, Cambridge, U.K., 1989, p. 400.
91. J. Garnier and B. Ribadeau Dumas, *J. Dairy Res. 37*:493 (1970).
92. B.B. Mandelbrot, *The Fractal Geometry of Nature*, Freeman, San Francisco, 1982.
93. S. Forrest and T.A. Witten, *J. Phys. A12*:L109 (1979).
94. T.A. Witten and L.M. Sander, *Phys. Rev. Lett. 47*:1400 (1981).
95. P. Meakin, *Phys. Rev. Lett. 51*:1119 (1983).
96. M.K. Kolb, R. Botet, and R. Jullien, *Phys. Rev. Lett. 51*:1123 (1983).
97. D.A. Weitz, M.Y. Lin, and C.J. Sandroff, *Surface Sci. 153*:147 (1985).
98. D.S. Horne, *Faraday Disc. Chem. Soc. 83*:259 (1987).
99. L.G.B. Bremer, T. Van Vliet, and P. Walstra, *J. Chem. Soc. Faraday Trans. I 85*:3359 (1989).

100. L.G.B. Bremer, B.H. Bijsterbosch, R. Schrijvers, T. Van Vliet, and P. Walstra, *Colloids Surf. 51*:159 (1990).
101. L.G.B. Bremer, Fractal aggregation in relation to formation and properties of particle gels, Ph.D. Thesis, Agricultural Univ., Wageningen, Netherlands, 1991.
102. P. Walstra, T. Van Vliet, and L.G.B. Bremer, in *Food Polymers, Gels and Colloids* (E. Dickinson, Ed.), Royal Society of Chemistry, Cambridge, U.K., 1991, p. 369.
103. R. Botet and R. Jullien, *Phys. Rev. Lett. 55*:1943 (1985).
104. P. Meakin, *J. Chem. Phys. 83*:3645 (1985).
105. P. Meakin, *J. Colloid Interface Sci. 112*:187 (1986).
106. M. Kolb, R. Jullien, and R. Botet, in *Scaling Phenomena in Disordered Systems* (R. Pynn and A. Skjeltorp, Eds.), Plenum, New York, 1985, p. 71.
107. D.S. Horne, *J. Dairy Res. 56*:535 (1989).

Index

Acrylic polymer dispersions (organic continuous) in textile printing
  as binders of pigment print, 320–322
  as dispersing agents for pigment, 326
  as printing table adhesives, 329
  as thickeners of disperse dye pastes, 328, 329
  as thickeners of pigment pastes, 322–326
  as thickeners of reactive dye pastes, 327, 328
Addition polymerization, 272–274
Agricultural chemicals, 298, 299, 305
Alkali-activated thickeners (aqueous-based), 283–293
Alkali-activated thickeners (oil-based), 322–328
Alkali-soluble thickeners, 283–288
  chemical aspects of, 283, 284
  in emulsion paints, 285–288
  mechanism of action of, 284–286
  rheological effects of, 285, 286
Alkali-swellable thickeners, 263, 283, 288–290
  advantages of, 289, 290
  chemical constitution of, 283, 284, 288
[Alkali-swellable thickeners]
  in emulsion paints, 289
  mechanism of action of, 289
  particle size of, 289
  shear thinning character of, 289
Amorphous silica and silicate pigments, 256
Antibodies, 469, 470
Aqueous binder-resin dispersions for inks, 275–283
  in aqueous ink formulation, 278–281
  glass transition temperature of, 277, 278
  historical development of, 275, 276
  overlacquers and, 275, 282
  in pigment grind base preparation, 281, 282
Aqueous continuous microcapsule dispersions, 293–306
  aging of, 295
  applications of, 303–306
  density distribution in, 295
  dispersion properties of, 294, 295
  flocculation of, 295
  particle morphology of, 294
  rationale for, 295, 296
  terminology for, 294

Aqueous continuous polymer "emulsions", 275–306
Aqueous continuous rheology modifiers, 283–293
Aqueous printing inks, 185–187, 275–283
  aqueous alkali-soluble polymers for, 277
  aqueous dispersions of polymers for, 277
  binder resins for, 275–283
  color development characteristics of, 281
  composition of grind base for, 281, 282
  dried film properties of, 276–277
  formulation of, 278–281
  for impervious substrates, 185–187
  liquid ink properties of, 276
  pigment grind base preparation for, 281–282
  polymers as grinding vehicles for, 281
  for porous substrates, 185
  properties required of, 185, 187
Artist's colors, 135
Artist's oil paint, 49
Associative alkali-soluble thickeners, 283, 290–293
  chemical aspects of, 290, 291
  in emulsion paints, 292
  micellar structure of, 291–293
  mode of action of, 291, 292
  rheological effects of, 292
  thickening efficiency of, 292
Atomic force microscopy, 376, 406, 409, 425
Automotive paint, 8, 9, 11–12, 19, 21, 59, 63
Azo pigments, 145, 147

Barium sulfate pigments, 256
Beer's law, 116
Binders (adhesives) for paper coatings, 257–260
  glass transition temperature of, 259
  polymer latex types of, 259, 260
[Binders (adhesives) for paper coatings]
  requirements of, 257, 258
  solution types of, 258, 259
Biological microcapsules for drug targeting, 484–487
  drug incorporation in, 484, 485
  uses of, 486
Blue wool lightfastness scale, 225
Bulk polymerization, 273

Cadmium pigments, 132–135
Calcined clay pigments, 255
Calcium carbonate pigments, 252, 253
Carbon black pigments, 178, 195–198, 200, 239–241
  dispersibility of, 198
  for man-made fibers, 239–241
  for newsink, 195–198
  properties of, 197
Carbonless copy paper, 256, 293
α_{s1}-Casein, 502, 503, 506, 507, 511, 512, 528
  chemical character of, 502, 503
α_{s2}-Casein, 502–504, 506, 507, 511, 512, 528
  chemical character of, 502, 503
β-Casein, 502, 503, 506, 507, 511, 512, 528
  chemical character of, 502, 503
κ-Casein, 502–505, 510–514, 517, 519, 523–528
  chemical character of, 502, 503
  chymosin, reaction with, 505, 517–519
  β-lactoglobulin, reaction with, 504, 505, 514
  NMR studies of, 519
  stability of casein micelles and, 504
Casein micelles, 500, 505–534
  alcohol, effect on surface structure of, 519, 520, 524
  calcium ion, effect on surface structure of, 522, 524
  charge stabilization of, 517–519, 522–524

[Casein micelles]
charge/steric stabilization of, 522–523
chymosin, effect on surface structure of, 519, 520, 524, 525
coagulation of
by acid, 513
by chymosin (rennet), 517, 518, 524, 525
by ethanol, 515–517, 519–524, 531
by heating, 513–515
colloidal stability of, 505
destabilization and aggregation of, 512–518
discrete-subunit cluster model of, 506, 526, 527, 528
dissociation of, 506, 507
DLVO theory and stability of, 518, 519
hard sphere model of, 524–526
hydration of, 507, 508, 511, 512, 528
interactions between, 518–526
internal composition of, 511
internal structure of, 506, 526–528
microgel structural model of, 523
particle size analysis of
by centrifugation, 510
by dynamic light scattering (PCS), 508–510
by electron microscopy, 507, 508, 510
by light scattering/turbidimetry, 508
by permeation chromatography, 509, 510
polyfunctional polymerization model of, 528, 534
size of, 505, 507–509, 527–528
size/composition relationship of, 507–512
size distribution of, 507–509
stabilization by κ-casein, 504, 517
steric stabilization of, 519–524, 528
surface charge of, 516–519, 522–524
[Casein micelles]
surface composition of, 506, 511, 512
surface layer, molecular conformation of, 519–522
surface structure of, 519–525
zeta potential of, 517
Casein micelle aggregates
fractal dimension of, 529–534
and micellar concentration in gels, 532–534
from power law scaling relationships, 531, 532
from turbidimetry, 531
theoretical models of growth of, 529, 530
Cement, 335, 336, 340–343, 347
aggregates in, 342
air bubbles in, 339
cracks in, 339, 341–343, 347
crack bridging in, 343
DSP, 341
elasticity modulus of, 342, 343
flaws (*see* cracks in)
fracture surface energy of, 342
macro-defect-free (MDF), 342
microstructure in, 342
mixes for
deaeration of, 339–341
deaggregation of lumps in, 341, 342
dispersing agents for, 339–341, 343
dispersion state, effects on, 341–343
extrudable types of, 340
flocculation of, 341
fluidity of, 341–343
polymer thickened, 341–342
solids content of, 336, 342
thickening agents for, 341
water content, effect on, 341
water soluble polymers in, 341, 342
polymer removal by pyrolysis, 342
porosity of, 341–343
strength of, 339–343
reinforcing agents for, 343

Ceramics, 335, 336, 343–347
aggregates in, 344–347
air bubbles in, 339, 340
alumina, 339
dispersions for
aggregate formation in, 345
deaeration of, 339–341
deaggregation of, 346, 347
destabilization of, 345
dispersing agents for, 340, 344
dispersion state, effect on, 343–347
drying of, 345–347
flocculation of, 343, 345, 347
molding, 343, 344
particle size of, 343
polymer mixes with, 346–347
polymer thickened, 346, 347
pressure casting of, 340
sedimentation in, 344
viscosity of polymer mixes, 346, 347
flaws in, 339, 344–347
fracture surface energy of, 344
microstructure of, 346
porosity of, 343
processing of, 343–344
sintering of, 343
sources of flaws in, 339
strength of, 344–347
zirconia powder processing for, 346
Ceramic tile adhesive, 290
Chip-form pigment dispersions, 184–185
Cholesteric temperature indicators, 352, 353, 369
Chromate and molybdate pigments, 137, 138
Chromium oxide pigments, 135–137, 447, 448
Clay, 336, 340, 344
Clay (kaolin) pigments, 251, 252, 260
crystallographic properties, 251, 252
flocculate structure of dispersions of, 260, 261
for paper coating, 252
[Clay (kaolin) pigments]
rheology of dispersions of, 261
sources of, 252
Coacervates for microencapsulation, 296–300
aminoplasts as, 298
chemical composition of, 298–300
of drugs, 474
incompatible polymers in, 299
interfacial tension and, 297
interpolymer complex formation in, 297
particle size control of, 297
preparation of, 297–299
simple, 299
sol-gel transition in, 297
Colloidal aggregation, computer simulation of, 529, 530
Colloidal drug delivery systems (*see* Drug carrier systems)
Concrete (*see* Cement)
Condensation polymerization, 272
Copper phthalocyanine pigments, 145–161, 163, 165–168, 171, 173, 213
Cosmetics and toiletries, 304, 482

Daniel flow point, 40–41, 202
Decorative paints, 8, 9, 17–21, 59, 64
Diketopyrrolopyrrole (DPP) pigments, 147–150, 154, 156, 157
Dioxazine pigments, 147, 148
Drug administration routes, 458, 459
Drug carrier particles, 462, 489
adsorption of blood components on (*see* opsonization and)
biological fate of, 462–469
capillary size in tissues and, 463, 467–469
effect of particles size on, 463, 467, 474, 475, 477
effect of surface character on, 463, 466, 467, 477
extravasation and, 467–469, 489
macrophages and, 463, 465–467, 471, 478, 480, 486
opsonization and, 466, 467

[Drug carrier particles]
phagocytosis and, 463, 466, 478, 479
reticulo-endothelial system and, 463
biological microcapsules, 484–487
circumvention of MPS by, 466, 467, 478, 479
erythrocytes as, 484, 485
liposomes as, 475–483
magnetic particles as, 487, 488
microspheres as, 467, 471–473
nanoparticles as, 473–475
nonionic surfactant vesicles as, 483–484
opsonization of, 466, 467
requirements of, 462
steric stabilization of, 467, 479
surface character of, 467
surface treatment of, 466, 467, 478, 479
Drug carrier systems, biological hindrance of, 458, 459
by capillaries in tissues, 463, 467–469
in extravasation, 467, 468
by gastro-intestinal tract, 462
by lipid membranes, 458
by mononuclear phagocyte system (MPS), 465, 466
Drug delivery, principles of, 457–459
Drug targeting, 457–497
antibodies, use of in, 469, 470, 477, 479
of antimicrobial compounds, 479, 480
in enzyme therapy, 480, 481
treatment of diseases of the MPS by, 465, 484, 488
treatment of lysosomal storage diseases by, 480, 486
treatment of respiratory conditions by, 481, 482
treatment of tumors by, 461, 462, 465, 466, 470–472, 474, 479, 480, 487, 488
Drugs
controlled release of, 459–460, 471–473, 486, 488
from microspheres, 471, 472
from resealed erythrocytes, 486, 487
treatment of arthritis by, 471
treatment of tumors by, 471, 472, 486
prolonging effectiveness of, 478, 481, 483, 484
Durability of paint films, 91–103
Dynamic light scattering (*see also* Photon correlation spectroscopy, PCS), 172, 173, 507–510, 520

Electronic imaging systems, 376, 433–436
Electro-optic devices, applications of, 349
Electro-optic liquid crystal dispersions, 355–369
light efficiency of, 355
nematic curvilinear aligned phase of, 356
polymer dispersed, 358–365
polymer stabilized, 368
Electro-optic shutters, 359, 361–369
Emulsion paint (*see* Latex paint)
Emulsion polymerization, 274, 471, 473, 487
Extenders, 14–15, 51, 78, 255

Fillers, 214, 318
Flocculants (*see* Polyelectrolyte flocculants)
Flushed pigment dispersions, 190–193
Fresnel's equation, 77

Gloss of paint films, 89, 90, 95, 99, 100

Indigo pigments, 147–149
Industrial paints, 8, 19, 63

Inorganic materials, strength of, 335–348
aggregates, effect on, 338, 346, 347
air bubbles, effect on, 338, 339, 347
bending test for, 338
crack size and, 337–339, 342, 344, 347
deaeration and, 340–342
deaggregation and, 341, 342, 346, 347
dispersing agents and, 339–344, 346, 347
dispersion state and, 339, 336, 339–347
elasticity modulus and, 337
flocculation in dispersion and, 335, 339–341
fracture surface energy and, 337, 338, 342, 344
Griffith equation and, 337–339, 342, 344–347
microstructure and, 335, 336, 342
natural crack size and, 338
natural flaws, effect on, 339, 340, 347
porosity, effect on, 336
tensile test for, 337
theory of, 337–339
Inorganic pigment (colored) dispersions, 105–142
abrasiveness of, 114, 135, 137
absorption of light by, 110, 112, 115, 116, 139
chroma of, 125, 126, 135
color of, 106–112, 123–139
dispersion hardness of, 114
fastness of (*see* lightfastness and weather resistance)
flocculation of, 112, 114, 137, 140
gloss of, 129
lightfastness of, 123, 130, 132, 134–136, 139, 140
opacity of, 118, 123, 125, 130, 134, 135, 137, 139
optical theory of, 115–117
particle shape in, 115, 123–127, 129, 136, 140, 141
[Inorganic pigment (colored) dispersions]
particle size in, 112, 114–117, 123–126, 128, 134–141
particle size analysis of, 114
particle size distribution in, 112, 114, 115, 123, 125
pigment volume concentration of, 115, 116, 137
refractive index of, 115
scattering of light by, 110, 115, 116
tinting strength of, 126, 129, 130, 134, 138, 139
weather resistance of, 123, 130, 132, 134, 136
Inorganic pigments (colored), 14, 15, 105–142
acid value of, 112
aggregation of, 112, 114, 128, 130, 139, 140
bulk density of, 114
chemical composition of, 107, 117, 118, 132–134, 137–141
crystal chemistry of, 110–112
crystal lattice structures of, 107–110, 129, 131–134, 136–138
density of, 112–114, 131, 138
dispersibility of, 114, 121, 129, 130, 132, 135, 139, 140
electrical conductivity of, 112
electron microscopy of, 123, 127
hardness of, 112, 114, 135, 136
history of, 106–107
isomorphism of, 132
in man-made fibers, 239
manufacturing processes for, 119–123, 129–130, 134–138
natural types of, 107, 117, 118
oil absorption by, 112, 113, 129, 131
particle characteristics of, 112–114
pH of, 112
physical properties of, 107, 114–115
in plastics, 210–212, 235, 236
primary particles of, 112
refractive index of, 117, 136, 139–141
resistance properties of, 112, 123, 130, 132, 134–140

[Inorganic pigments (colored)]
sieve residue of, 112–114, 129
specific surface area of, 112, 129, 130, 139
thermal stability of, 128, 130, 132, 134, 135, 139
transparent types of, 139, 140
Interference pigments, 141
Iron blue pigments, 139
Iron oxide pigments, 117–129
black, 127
brown, 117, 128
electron microscopy of, 123, 127
magnetic, 141, 142, 447, 448
manufacture of, 119–123
Laux process, 122, 123
precipitation process, 120, 121
roasting process, 119, 120
micaceous, 126
micronized, 128–129
naturally occurring, 117–118
properties of, 123–129
red, 46, 123–126
thermal stability of, 128
transparent, 125, 126
yellow, 127
Isoindoline pigments, 147, 148, 154
Isoindolinone pigments, 147, 148, 155

Kubelka-Munk theory, 44, 45, 115

Latex paint, 13, 15, 19–21, 57, 63, 285–290, 292, 293
Laundry detergents, 306
Liposomes, 302
Liposomes as drug carriers, 475–483
chemical composition of, 476
in cosmetic preparations, 482
drug entrapment efficiency, 476
drug location in, 476
in enzyme therapy, 480, 481
mechanism of formation, 477
phase transition temperature, 477, 478
physical stability of, 482
size of, 476
[Liposomes as drug carriers]
steric stabilization of, 479
surface modification of, 478, 479
topical application of, 482
in treatment of respiratory conditions, 481, 482
Liquid crystal dispersions
electro-optic, 355–371
birefringence of particles in, 358
configuration of particles in, 356–357
dielectric anisotropy of, 358
light scattering efficiency of, 355
morphology of particles in, 357
nematic curvelinear aligned phase (NCAP), 356, 358
polymer dispersed (PDLC), 356, 358–367
clarity in ON state of, 361–362
colored displays using, 365, 368, 369
contrast of, 365
driving voltage for, 362, 364
dyes for coloration of displays in, 365
electro-optic shutters from, 359, 361–364
electro-optics of, 361–364
haze-free shutters from, 365–367
infrared shutters from, 365
light scattering by, 362, 365, 369
manufacture of, 358–361
particle morphology of, 359, 360
particle size control in, 360, 361, 365
in polymer microspheres, 367, 368
polymer types for, 358–361
refractive indices of components of, 362, 365–367
switchable windows, 358, 369
switching speed of, 362, 364, 365
polymer encapsulated liquid crystals (*see* NCAP)
polymer microspheres, 367, 368
refractive indices of, 355

Liquid crystal displays, 353–355
ferroelectric, 355
image formation in, 354, 355
light efficiency of, 355
resolution of, 355
supertwist, 353, 355
twisted nematic (TN), 353, 354
Liquid crystalline continuous phase, dispersions with, 355, 367
Liquid crystal polymer gels, 368
polymer stabilized cholesteric, 368, 369
Liquid crystals, 349–352
anisotropy of, 350, 351
birefringence of, 351
director in, 351, 353, 354
electro-optic applications of, 351
lyotropic, 350
NCAP dispersions of, 356, 358
orientation director of, 350, 351
polymer dispersed (PDLC), 358–367
polymer stabilized dispersions of, 368, 369
refractive indices of, 351
temperature range of, 351
thermotropic, 350, 351
cholesteric, 349, 351, 353
nematic, 349, 351, 353, 355, 358, 367
smectic, 351
Liquid crystal temperature indicators, 352, 353, 369
Liquid dispersion polyelectrolytes, 309, 310
incorporation into application media, 310, 311
molecular features and effectiveness of, 312–314
particle size of, 309
solid-liquid separation and, 311–319
surfactant activator for, 309
Liquid inks
aqueous (*see* Aqueous printing inks)
bead milling of, 182–184
dispersibility of organic pigments in, 163
[Liquid inks]
disposal of, 187
large particulate species in, 189
optical properties of, 187–189
pigment particle size effect on, 187–189
resistance properties of, 181, 185, 187
solvent based, 180–191
adhesion to substrates, 181, 182
binder resins for, 182
blending of, 181
flexographic printing of, 179–182
formulation of, 180–182, 184
gloss of, 184, 185
gravure printing of, 180
laboratory-scale bead mills for, 182–184
manufacture by bead milling, 182–184
manufacture from pigmented chips, 184–185
nitrocellulose media for, 180, 181
optical properties of, 183–185
particle size of pigment in, 185–187
pigment concentrates for, 181–182
rheological properties of, 182, 185
Lithographic inks (*see* Paste inks)
Lustrous pigments, 140–141

Magnetic pigment dispersions
particles in
aggregation of, 444
coercivity of, 141, 444, 447
magnetic anisotropy of, 444
magnetic density of, 444
remnant magnetization of, 444, 451
shape of, 447
size of, 447
size distribution, control of, 447
surfactants for
lecithin, 447, 452
molecular structure (idealized) of, 453
phosphate ester, 447, 452

[Magnetic pigment dispersions]
polyester/polyamine copolymeric, 452
quaternary ammonium, 452
requirements of, 454
selection criteria for, 454
Magnetic pigments, 141, 446–452
application of, 448
barium ferrite as, 447
chemical constitution of, 141, 448
chromium oxide, 447, 448
difficulty of dispersion of, 446, 447
dispersion of
by annular stirred bead milling, 452
by ball milling, 449
by bead milling (*see* by sand milling)
ingredients for, 449–450
kneading of premixes for, 450, 451
milling media for, 449, 450
premixing for, 449, 450
by sand milling, 449–451
dispersion formulations for, 452, 454
iron oxide as, 141, 447, 448
metal type of, 141, 142, 447, 448, 452
specific surface area of, 447, 450, 451, 452
Magnetic recording materials, 443–456
Magnetic tape, 443–449
aggregation of particles in, effects of, 444–446
cleaning agents for, 447
defect levels in, 446
dispersion, importance of, 444–447
durability of, 445
future requirements of dispersions for, 455
gloss of, 451
magnetic squareness of, 444–446
orientation of magnetic domains in, 444
particle size distribution in, 446
print through in, 446
pyrolysis of at head/tape interface, 447, 449
[Magnetic tape]
shed from, 445
signal to noise ratio of, 446
surface roughness of, 445, 451
surfactants in, 447
switching field distribution of, 446
wear resistance of, 445
Man-made fibers
acrylic, 208, 209, 242
cellulose acetate, 242
dry spun, 242, 243
economic significance of, 208
mass coloration of, 209–210
melt spun, 220, 227, 231, 238–242
pigmented
bloom resistance of, 212
fastness properties of, 209
lightfastness of, 209, 212, 239–242
mechanical properties of, 213, 241
migration resistance of color in, 213, 242
optical microscopy of, 222
rubbing fastness of, 213, 232
transparency of, 213
weather resistance of, 212, 242
wet fastness of, 213
pigment dispersions for
abrasiveness of, 213, 238, 241
carrier resin for, 239, 240
for dry spun fibers, 242
filterability of, 221, 241, 242
heat resistance of, 211, 212, 239
liquid-form, 240, 242
mechanical properties of fiber, effect on, 213, 221, 241
for melt-spun fibers, 220, 238, 241
optical microscopy of, 222, 242
particle size in, 222, 242
pressure rise testing of, 241
processing aids for, 214
requirements of colorants in, 211–213
spinneret blockage by, 241
for wet-spun fibers, 242
polyamide, 208–210, 221, 240

[Man-made fibers]
polyester, 208–210, 221, 222, 240–241
polyolefin, 208–221, 238
regenerated cellulose, 209, 242
special dyes for, 209–211, 241
wet-spun, 242
Marine paints, 8, 20
Metallic paints, 12, 153
Metallic pigments, 140
Microcapsule dispersions (*see* Aqueous continuous)
Microencapsulation
applications of, 303–306, 471–473
by complex coacervation, 296–300
by emulsion methods, 301–302
by interfacial polymerization, 300–301
of liquid crystals, 353, 356
Microspheres, pharmaceutical use of, 303, 304, 467, 471–473
as carriers for vaccines, 472
for chemoembolization, 472
for controlled release of proteins, 471–473
for nasal delivery of drugs, 473
problems with, 473
in treatment of cancer, 472
Mie theory, 116
Milk, 499–538
calcium content and ethanol stability of, 516
chymosin coagulation of (*see* rennet coagulation of)
composition of, 501–505
ethanol stability of, 515–517
genetic polymorphism of proteins in, 504
heat stability of, 513
homogenized, 500
protein types in, 502–505
rennet coagulation of, 504, 517
skim milk, 501
sterilization of, 513
urea content of, 514
whey proteins in, 502, 504, 505, 513
Milk products, 500, 512
cheese, 303, 304, 500, 504, 517, 529
cottage cheese, 513
cream liqueurs, 500, 503, 515
concentrated milk, 503, 512
homogenized milk, 500
UHT products, 512, 513
yogurt, 513
Mineral processing, 314, 315
Molybdate pigments, 137–138
Montmorillonite pigments, 255, 256
Multiple emulsions, 301

Nacreous pigments, 140–141
Nanoparticles as drug carriers
manufacture of, 474
pharmaceutical use of applications of, 474, 475
size of, 473
uses of, 474, 475
Newsink, 179, 195–198
cold-set web offset, 195–198
deinking of, 198
formulation of, 195–196
large particulate species in, 198
lithographic printing performance of, 197
manufacture of, 198
printed image quality of, 198
rub resistance of, 197–198
water based flexographic, 198
Nonionic surfactant vesicles as drug carriers, 483, 484
similarity to liposomes, 483
uses of, 484

Oil field chemicals, 305, 306
Optical properties
of colored pigment dispersions, 115–117, 151–154
of colorless (white) pigment dispersions, 76–90
Organic continuous polymer "emulsions", 306–330
as adhesives in textile printing, 328, 329

[Organic continuous polymer "emulsions"]
as adhesives for wallcoverings, 329, 330
as binders in textile printing, 320–322
as polyelectrolyte flocculants, 308–319
as thickeners for textile printing pastes, 322–329
Organic pigment dispersions, 144, 151–156, 164–172
aqueous pastes, 163
binder resin in, 165–167
chip form, 163
color strength of, 151–153, 156, 158, 162–164, 167–169, 173, 174
critical concentration of, 171
electron microscopy of, 154, 155, 173
flocculation control in, 164–168
flocculation effects on, 164, 171
flushed, 147, 169, 191–193, 200, 221
gloss of, 164
hyperdispersants for, 167, 169
level of dispersion in, 144, 152
in man-made fibers, 153, 210–213, 220, 221, 227, 237–242
masterbatch form for plastics, 164
mechanical properties of, 144, 164
opacity of, 153, 154
optical microscopy of, 172
in paints, 147, 153–154, 164, 165, 167, 171
particle shape in, 152, 171–173
particle size in, 151–154, 169, 172, 173
particle size analysis of, 172–174
particle size distribution in, 153
pigment/medium interface in, 165–168, 171
in plastics, 163, 164, 210–215, 220, 221, 227, 233–237
plastic viscosity of, 170, 171
in polar media, 163
preparations for plastics, 164
in printing inks, 153, 156, 164–171, 178, 179, 195, 200
[Organic pigment dispersions]
rheology of, 164, 168–172
scattering of light by, 152, 154
steric stabilization of, 165
strength of flocculate structure in, 168, 169, 171
synergist for, 167, 168
transparency of, 153
Organic pigments, 15, 143–176
abietyl resin in, 157, 161, 162, 169, 170
aggregate structure of crystals of, 158–162, 169, 170
aggregation of crystals of, 144, 156, 158–164, 169, 170
anisotropy of, 150, 151
azo metal salt, 147
azomethine, 147
chemical constitution of, 145–147
classical, 145–147, 210
color of, 144–147, 151–153
compaction of, 163, 215, 233
crystal growth of, 156–158, 160, 161
crystal lattice structure of, 148–151, 157, 158
crystal shape of, 150, 151, 153, 154, 158, 160–162, 171–173
crystal size of, 154–158, 161, 162, 169
deformability of crystals of, 149, 159, 163
dichroism of, 151, 153
disazo, 145, 147
dispersibility of, 158–163
effectiveness of, 144
electron microscopy of, 148, 149, 154–156
fastness of, 145–147, 151
high performance, 147, 211
high resolution electron microscopy of, 148
immersion calorimetry of, 161
intermolecular hydrogen bonding in, 149–151
mixed coupling of azo, 156
monoazo, 145, 147, 154, 171
nomenclature for, 145

[Organic pigments]
physical form of, 144, 147
physical properties of, 151–172
polycyclic, 147, 154
polymorphism of, 145, 147, 151, 154, 157
specific surface area of (by BET method), 159–162
surface polarity of, 151, 161
wettability of, 159
X-ray diffraction analysis of, 148–150, 154–157, 162
O/w (*see* Aqueous continuous)
Oxidic mixed phase pigments, 129–132

Packaging inks (*see* Liquid inks)
Paint films
absorption of light by, 44, 45
Benard cell formation in, 49, 50
brush marks in, 51, 57
coin marring in, 52
coloristic properties of, 14, 22, 44–50
color measurement of, 44, 45
critical packing volume concentration in, 49, 53–55
drying of, 17, 58
durability of, 11, 12, 14, 19–21, 52, 57, 99–103
elasticity of, 11
film build in, 15
flexibility of, 19, 20
floating in, 49, 50
flocculation of pigment and effects on, 46–52, 80–86, 89, 90, 99–102
flooding in, 48
flow in, 58
gloss of, 12, 52, 90, 100, 101, 164
laminate structure of, 11, 12
mattness of, 15
opacity of, 14, 47, 153, 154
particle packing in, 53–55
particle size of pigment and effects on, 45, 46
refractive index and, 14, 45–47
sagging of, 50, 51, 57
sandability of, 12–15
[Paint films]
scattering of light by, 45–47, 52
scattering of light by voids in, 55
thickness of, 12, 100–102
tinting strength of, 43, 44
Paints, 1–67
application of, 56–64
components of, 12–21, 44, 45
continuous phase of, 16–17
disperse phase of, 14–16
dispersion equipment for, 22–39
attritor, 27–30
ball mill, 22–27
bead mill, 30–33
charge-to-voids ratio in, 24, 25, 27, 33
colloid mill, 37
grinding media requirements of, 24, 25–28, 30, 33
high speed disperser, 34–36
high speed impingement mill, 37
pug mixer, 36, 37
roller mill, 23, 37
ultrasonic, 37, 39
economic aspects of, 5–8
film-forming resins for (*see* polymers used in)
formulation of
colloidal shock and, 22, 41–42
Daniel flow point and, 40, 41
millbase, 39–43
millbase let-down, 41–43
mixing in, 44–45
pigment-to-binder ratio in, 40
pigment wetting in, 40
future trends in, 59, 63, 64
health and safety aspects of, 59, 63
history of, 3–5
manufacture of, 21–45
pigment dispersion into, 21–39
polymers used in, 16–21
acrylic, 5, 19, 20, 59
alkyd, 5, 17–19, 45, 52
amino, 5, 20, 45
cellulosic, 5, 21, 59
epoxide, 20

[Paints]
glass transition temperature of, 20
melamine-formaldehyde, 5, 20
nonaqueous dispersions of, 16, 20, 57
oil length of alkyd, 17–19
phenolic, 5, 20, 21
polyester, 19
polyurethane, 5, 21, 45, 46
thermosetting, 5, 19, 20, 45
urea-formaldehyde, 5, 20
vinyl, 19
types of, 8–9
Paints as liquid colloidal dispersions
colloidal stability of, 16, 41–43, 50–52
color drift of, 47–48
critical packing volume concentration in, 55
depletion stabilization of, 41
false body in, 51
fineness of grind of, 43–44
flocculation of pigment in
assessment of, 52–53, 81–84
control of, 164, 165
effects of, 50–52, 57, 164
rate of, 50, 51
particle size assessment in, 43
particle size distribution in, 29
particle morphology and, 171
polymer adsorption in, 41–43, 45, 53
resuspension of sediment in, 50
rheology of, 15, 55–58
settling on storage of, 57
steric stabilization of, 16, 41–43
thixotropy of, 45, 51, 52, 57
viscosity of, 15, 50–51, 55–58
Paint systems, 11, 12
Paper coating mixes, 250–264
additives (miscellaneous) for, 264
adhesives/binders for, 257–260
application of, 247–250
by air-knife coating, 250
by blade metering, 247, 250
calendering of, 247, 266, 268
to cardboard, 250
by cast coating, 256, 258
dewatering of, 267
[Paper coating mixes]
by "off-machine coating", 250
by "on-machine coating", 247, 248, 250, 256
dispersants for, 262
insolubilizers for dried coatings of, 263, 264
latex adhesives for, 259, 260
pigment content of, 247
pigments for, 251–257
rheology of, 260–263
rheology modifiers for, 262, 263
solution type adhesives for, 258, 259
water retention agents for dried coatings of, 262, 263
Paper coating mixes, preparation of, 265, 266
formulations for, 265
pigment dispersion into, 265, 266
Paper coatings, 245–269
coat weight of, 247–250
electron microscopy of, 247
nature of, 245–246
properties of
absorbency, 246, 253, 255
adhesion to paper, 257–260
abrasivity, 254
brightness, 254, 255
gloss, 247, 266–268
holdout, 267
light scattering by voids, 254, 256, 257
mattness, 256
opacity, 254, 255
pick strength, 257
porosity, 254, 257, 266
smoothness, 266–268
stiffness, 246, 259
water resistance (lithographic printing), 257, 260, 263, 264
water retention, 259, 262, 263
wet strength, 257
wettability (lithographic printing), 259, 266
structure of, 266–268
types of, 246–250

Paper (uncoated), structure of, 245, 246, 267
Paste inks (*see also* Newsink), 179, 191–195
  bead milling of, 194, 202
  dispersibility of organic pigments in, 159–162
  formulation of, 191
  heat-set web offset, 191, 257
  large particles types in, 195
  lithographic printing of, 257
  manufacture of, 191, 193–194
  optical properties of, 169, 195
  pigment form for, 191–193
  printing performance of, 195
  properties related to pigment dispersion in, 195
  quality of pigment dispersion in, 193–194
  rheology of, 169–171
  sheet-fed, 191
  three-roll milling of, 193–195, 202
Perylene pigments, 147, 148, 154
Pharmaceutical dosage forms, dispersions as, 457–497
Pharmaceuticals, microencapsulation of, 299, 303, 304
Photographic emulsions
  alignment of T-grains in coated layers of, 410
  analytical techniques for, 407–409
  blending of, 427
  chemical sensitization of, 377, 393, 400–405
  chemistry of, 376–378
  coalescence ripening of, 393
  colloidal stabilization of, 377, 380
  composite emulsions (*see* core-shell, and epitaxial)
  core-shell types of, 394, 400–402
  crystal growth in, 392–394, 414–420, 422, 424, 425, 431–433
  dye sensitization of (*see* spectral sensitization of)
  gelatin in, 374, 376, 377, 379, 380, 382, 393, 398, 399
[Photographic emulsions]
  isoionic point in, 376, 377, 399
  monosize types of, 375, 392–436
  Ostwald ripening in, 393, 398, 414, 424
  pAg, definition of, 376
  preparation of, 392–394, 398, 399, 414, 427–432
  recrystallization in, 393
  reduction sensitization of, 377, 402, 403
  seeding of, 392, 414–420, 422, 423, 425
  sensitivity to light of, 375, 377, 390, 401
  sensitization of, 375, 379, 393, 400
  sensitizing dyes for, 378, 383, 388, 405–406
  silver halide content of, 377
  spectral sensitization of, 378, 383–385, 405, 406, 410, 429, 430
  technology of, 394–402
Photographic image (developed)
  acutance of, 382, 394, 410
  color saturation of dye image, 394
  contrast of, 375, 383, 394, 400, 429
  covering power of, 385, 386–388, 410
  dye image, 379, 382, 394
  granularity of, 379, 382, 432, 433
  morphology of silver image, 386
  sharpness of (*see* acutance)
  structure of holographic image, 388–390
Photographic material processing after exposure to light
  bleaching of silver image in, 379, 385
  chromogenic development of, 379
  chromogenic development, inhibition of, 432
  developability of, 379, 402, 410, 428, 432
  development of holographic film, 388–390
  development of monochrome, 378, 385–388

[Photographic material processing after exposure to light]
development of monochrome reversal, 385
fixing of, 378, 385
fogging of, 377, 403
holographic film processing, 388–390
mechanism of development, 386–388
monochrome development, 378, 385–388
silver dye-bleach processing, 394
tone reproduction, 386
Photographic particles (silver halide microcrystals)
adsorbed additives, effects on habit of, 397, 398, 399
aspect ratio of, 427
atomic force microscopy of, 376, 406, 409, 425
composite (*see* core-shell, and epitaxial)
core-shell, 392, 394, 400–402, 419, 423–432
defects in, 375, 390, 399, 400
DELTA type of, 432, 433
dislocations in, 427, 431
double-structure (*see* core-shell)
electron microscopy of, 376, 398, 408, 411, 418, 419, 433
electron paramagnetic resonance (EPR) of, 400
energy dispersive X-ray analysis of, 419, 425, 426
epitaxial, 397, 399, 414–421
habit of, 375, 393, 394, 397–400, 403, 404, 414–420, 422–424
halide boundaries (*see* Dislocations)
halide composition of, 375, 393, 397, 399, 414–420, 425, 426, 428–432
halide compositional distribution in, 399, 400–402, 414–420, 425–429, 431–433
heterojunctions in, 424, 432
hollow, 422, 423
interstitial silver ions in, 390, 400, 411
[Photographic particles (silver halide microcrystals)]
J-aggregation of sensitizing dyes on, 378, 405, 406, 409
photoelectron spectroscopy (XPS) of, 407
polymorphism of, 399
rhodium doping of, 400
scanning Auger microscopy of, 425
secondary ion mass spectroscopy (SIMS) of, 376, 399, 409, 425, 426
size of, 374–376, 382, 388, 391, 392, 397, 410, 419, 433
size distribution of, 375, 382, 391–393, 397, 414, 419, 424, 431
SMART type of, 406
solid-state properties of, 390–392
solvation effects on crystal habit of, 397
surface structure of, 398, 399
T-grains (*see* twinned tabular crystals)
twinned tabular crystals, 375, 382, 394, 409–427, 431, 432
X-ray diffraction (XRD) of, 408, 414–415, 425
Photographic properties of silver halide systems, 378, 379
latent image in, 378, 379, 388, 390–392, 400–403, 411, 424
photoelectron and, 390
photoelectron trapping in, 395, 396, 400–402, 405, 420
photohole and, 390
photohole trapping in, 403, 405
quantum efficiency of, 374, 378, 391, 392, 399
reciprocity of, 403–405, 429
sensitivity distribution (*see* Photographic image, contrast of)
sensitivity to light (*see* speed)
speed of, 375, 377, 391, 394, 401, 402, 406, 425, 430, 438
speed/granularity ratio of, 391, 402, 410, 411, 427

Photographic systems, 374, 379–385
aging of, 424, 425
black and white (*see* monochrome)
color, 379
color coupler dispersion for, 379
color film, 380, 381, 409, 428, 432, 433
future trends in, 433–436
holographic film, 388–390
layer structure of, 375, 380, 433
monochrome film, 380, 432, 433
monochrome paper, 380–385
paper base for, 378
"Polaroid" (one-step) type film, 382
polymeric support for film, 378, 394
reversal type of, 403, 428
silver dye-bleach film, 394, 406, 421, 422
technology of, 376–383
X-ray film, 382, 410, 432
Photon correlation spectroscopy (PCS) (*see also* Dynamic light scattering), 172, 173, 507–510, 520
Pigment Blue 15 (*see* Copper phthalocyanine pigments)
Pigment Red 57:1, 149, 153, 154, 155, 161, 169, 171
Pigment Yellow 12, 163, 173
Pigment Yellow 13-type pigments, 153, 154–157, 162, 169–171, 173
annealing of crystal structure in, 157
mixed coupling of, 156
sulfonated disazo derivatives of, 157
Plastic (polymeric) pigments, 256, 257
Plastics
dispersions for use in, 206–243
abrasiveness of, 210, 212, 213
calendering effect on, 223, 231
carrier resins/polymers for, 213, 214
color concentrate as, 206
compatibility with polymer, 225
custom colors as, 207, 232, 234, 236, 239, 240
dispersion, degree of, 215, 221–223
[Plastics]
for engineering thermoplastics, 238
heat resistance of, 210, 211, 224, 234
hindered amine light stabilizers in, 236
incorporation into polymers, 223, 224, 228–231
liquid carriers for, 213
liquid-form, 206, 213, 221, 231, 235, 240
masterbatches as, 164, 206, 212, 213, 234
mechanical properties of polymer, effect on, 222, 227, 234, 235–237
melt-flow index of, 223
meterability of, 206, 238
monoconcentrates as, 206, 211, 232–234, 237, 238
optical microscopy of, 222
particle size of, 222
physiological and ecological aspects of, 227, 237
pigment preparations as, 164, 206
for polyolefin, 234–237
for polyvinyl chloride, 231–234
processing aids for, 214
requirements of, 221–227, 231–238
requirements of colorants for, 211–213
rheological properties of, 214, 223
solid types of, 206, 215–221, 232
for styrenic polymers, 237
test methods for, 221–227
wetting of pigment in, 214, 227
incorporation of pigment dispersions into, 223, 228–231
liquid dispersions, 229–231
solid dispersions, 228–229
manufacture of dispersions for, 214–221
from flushed pigments, 220, 221
liquid-form dispersions, 221
mixing ingredients of, 215–217, 220

[Plastics]
thermoplastic polymers for, 207
colorants for, 210, 211
engineering thermoplastics, 208, 224, 238
glass transition temperature of, 211, 237
polyester, 233
polymer soluble dyestuffs for, 211, 237
polyolefin, 206, 207, 210, 213, 224, 234–237
polyurethane, 206
polyvinyl chloride, 206, 207, 211, 213, 223, 231–234
rubber, 211
styrenic polymers, 206, 207, 211, 213, 224, 237, 238
Plastics, pigmented,
bloom resistance of, 211, 212, 235, 238
color strength of, 210, 211, 215, 221, 233, 239
durability of,
effect of titanium dioxide pigment on, 95–97
electrical resistivity of, 232, 236
fogging by, 232
lightfastness of, 212, 225, 234, 236–238
mechanical properties of, 227, 234–237
migration resistance of color in, 212, 237, 238
opacity of, 210, 211, 222, 236
transparency of, 211, 222
weather resistance of, 212, 226, 235, 238
Polyelectrolyte flocculants, 308–319
mineral processing and, 314, 315
mode of action of, 308, 309
paper production and, 317, 318
physical form of, 308, 309
sedimentation and, 312, 313
[Polyelectrolyte flocculants]
sewage treatment and, 312, 313, 315–317
solids dewatering and, 313
viscosity of, 309–311
water pollution control and, 315–317
Polymer dispersions, 271–334
Polymer "emulsions" (*see* Aqueous continuous and Organic continuous)
Polymerization processes, 272–274
Printing inks, 177–204
abrasivity of, 187, 190, 197
flexographic printing, 179, 180, 198
gravure printing of, 179, 180, 187, 190
letterpress printing of, 191
lithographic printing of, 191, 195–197, 257
millbase optimization for, 202–203
substrates for, 179
varnish component of, 179
Publication gravure inks, 179, 190–191
control of flocculation in, 165–167
dispersibility of organic pigments in, 162

Quinacridone pigments, 147–149, 151, 156–158

Radiation-curable inks (*see* UV-curable inks)
Refinish paints, 8, 9, 21
Rheology modifiers (*see* Aqueous continuous)

Satin white pigments, 256
Saturation magnetization, 141
Screen printing inks, 179, 199
Secondary ion mass spectroscopy (SIMS), 376, 399, 409, 425, 426
Sewage treatment additives, 312, 313, 317
Silver halide dispersions (*see* Photographic emulsions)

Silver halide grains (*see* Photographic particles)
Solid/liquid separation,
 polyelectrolyte dispersions and, 313-319
 sedimentation and thickening, 312-313
 sludge dewatering, 313
Solution polymerization, 273
Sparkplug manufacture, 344
Suspension polymerization, 273
Synthetic fibers (*see* Man-made fibers)

Textile printing additives, 319-329
Textured wall finishes, 290
Thermoplastic polymers (*see* Plastics)
Thickeners, synthetic, dispersions of
 oil-based, 322-328
 water-based, 355
Titanium dioxide dispersions, 75-103
 absorption of light by, 76, 77
 accelerated weathering of, 93, 102, 103
 air inclusion in, 86, 87
 chalking of, 95
 color of, 45, 88
 critical pigment volume concentration of, 88
 durability of, 91-103
  film thickness and, 100-102
  flocculation and, 99-102
  pigment grade and, 96
  state of pigment dispersion and, 97-99
 embrittlement of, 94, 97
 flocculation in, 80, 81, 88-90, 99-102
  assessment of, 81-82
  effects of, 80, 81, 84, 88-90, 99-102
 flocculation gradient of, 81-84, 86, 99
 gloss of, 90, 95, 99, 100
 level of dispersion in, effects of, 80
 opacity of, 69, 76-88
 particle crowding effect in, 79
 particle population effect in, 79, 80
 particle size of, 76, 78
 photocatalytic degredation of, 91-95
[Titanium dioxide dispersions]
 photochemical protection of, 91-93, 97
 pigment volume concentration of, 78, 79
 reflectivity of, 77
 rheology of, 103
 scattering of light by (*see also* opacity of), 76-81, 86-88
 surface roughness of, 90
 undertone of, 88
 UV absorption by, 91-94, 102, 103
 UV transmittance by, 91
 weathering of, 95-103
 whiteness of, 69, 88
Titanium dioxide pigments, 14, 45-47, 51, 52, 69-104, 135, 141, 154, 178, 179, 185, 186, 254, 255
 abrasiveness of, 103
 alumina coating of, 75
 anatase form of, 70, 72, 75, 76, 88, 91
 beneficiated ilmenite for, 70
 brookite form of, 75
 calcination, 72
 chloride process for, 73, 74
 crystal size of, 88
 crystallographic properties of, 70
 easily dispersible grades of, 80
 effectiveness of, 70
 ilmenite for, 70
 manufacture of, 70-75
 mineral rutile for, 70
 nucleation of, 72, 74
 organic surface treatment, 75
 for paper coatings, 253
 polymorphism of, 75
 purity of, 70, 88
 raw materials for, 70
 refractive index of, 76, 77
 rutile form of, 70, 72, 75, 76, 88, 91
 silica coating of, 75
 sulfate process for, 70-72
 surface coating of, 74-75
 titanium slag for, 70
 titanium tetrachloride for, 72-74

[Titanium dioxide pigments]
transparent grades of, 78
X-ray diffraction analysis of, 75

Ultramarine pigments, 139
UV-curable inks, 179, 199–201
effect of pigment
on curing of, 200
on shelf life of, 201
manufacture of, 201, 202

Varnish 9, 20

Wallcovering adhesives, 329, 330
Water-based (*see* Aqueous)
Water pollution control additives, 315–317
Water purification additives, 312, 313
W/o (*see* Organic continuous)